AN ENVIRONMENTAL HISTORY OF SCOTLAND

Where Men No More
May Reap or Sow

AN ENVIRONMENTAL HISTORY OF SCOTLAND

Where Men No More May Reap or Sow

The 'Little Ice Age': Scotland 1400–1850

Richard D. Oram

John Donald

Title page image:
Muirburn smoke over the Strathbogie Hills, Aberdeenshire. Parliamentary legislation from the 1420s onwards was designed to curb the 'ill' of uncontrolled burning of upland grass and heather.

Opposite:
Scots Pine, Rohallion, Perth and Kinross, the self-seeded descendant of pines introduced here in the 1700s.

Page vi:
Doocot interior, Dirleton Castle, East Lothian. Nesting pairs were housed in every box that lined the interiors of such doocots. They provided their owners with a source of fresh meat, feathers and manure (for fertiliser and as a source of saltpetre).

First published in Great Britain in 2024 by
John Donald, an imprint of Birlinn Ltd

West Newington House
10 Newington Road
Edinburgh, EH9 1QS

www.birlinn.co.uk

ISBN: 978 0 85976 717 0

Copyright © Richard D. Oram 2024

The right of Richard D. Oram to be identified as the author of this work has been asserted by him in accordance with the Copyright, Designs and Patents Act, 1988

All rights reserved. No part of this publication may be reproduced, stored, or transmitted in any form, or by any means, electronic, mechanical or photocopying, recording or otherwise, without the express written permission of the publisher.

The publishers gratefully acknowledge the support of
the Scotland Inheritance Fund
towards the publication of this book

British Library Cataloguing-in-Publication Data

A catalogue record for this book is available on request from the British Library

Typeset by Mark Blackadder

Printed and bound in Britain by Bell and Bain Ltd, Glasgow

Having weathered the highs and lows with me as I fought to get this book written, it is to my wife Emma that I must dedicate this book. Without her, it would never have come together as it has done. It has been written with her unwavering support, love and encouragement, and she has walked with me through the alternating joys of new discovery and anger at our past (and continuing) ineptitude in how we have used, abused, exploited or cherished the environment of this land we call home.

Contents

Acknowledgements ... ix

Abbreviations ... x

Introduction ... 1

1 A New Normal? Subsisting at the Margins in the Early 'Little Ice Age' ... 11

2 Turning up the Heat: Fuel, Wood and Food Supply in the Fifteenth Century ... 43

3 An Age of Shocks and Transitions c.1500–1700 ... 79

4 Beasts and Birds, Trees and Peat in the Sixteenth and Seventeenth Centuries ... 110

5 Dawning Reason and Improving Practice: Alternative Visions 1700–1790 ... 155

6 Changeable Times: Climate and Weather in the Eighteenth Century ... 171

7 Improvement's First and Greatest Child: Woodland and Plantation 1700–1790 ... 192

8 Fuel: Improvement Fashion versus Practicality and Access 1700–1790 ... 233

9 Achieving Improvement amid a 'Sea of Waste':   251
   Agricultural Change to 1790

10 The Sting in the Tail: Rain, Snow, Frost and Drought   286
   at the End of the 'Little Ice Age' 1790–1850

11 Improvement Applied: Agricultural Transformation 1790–1850   310

12 Fuelling Scarcity, Gathering Abundance and Netting Loss: 338
   Peat, Coal, Kelp and Fish

13 Trees: For Beauty, Effect or Profit   366

Conclusion   392

Bibliography   397

Index   411

Acknowledgements

The debts of gratitude owed to friends, colleagues and a host of professionals recognised in the volume 'Scotland AD 400–1400' in this series cannot be repeated in full here, but it is important to restate my undying thanks for all of the advice, help, information, support and, above all, friendship that you have given me in the twenty-one years since I was first appointed at the University of Stirling. In this volume in particular, however, there can be seen the 'fingerprints' of my former student, friend and colleague, Dr Alasdair Ross, whose death in 2017 is still felt keenly by everyone who worked with him and whose acute observations and honest criticisms have been sorely missed in the writing of this book.

Special thanks go to my colleagues Catherine Mills, Michael Penman, Emma Macleod and Alison Cathcart, for their patient reading and feedback on texts that were still half-baked, for long, rambling conversations as I worked through my ideas, and for the good-natured put-downs that some of my wilder musings deserved. A special mention has to be reserved for my long-suffering best man and friend, Grant Harrison, who has accompanied me on many expeditions into the mountains and now knows more than he ever wanted to know about changing land-use, climate-related impacts, rewilding debates and the politics of the environment.

Abbreviations

List of Technical Abbreviations

AMOC
Atlantic Meridional Overturning Circulation: this is the major system of oceanic currents that carry warm water from the tropics northwards into the North Atlantic. It works like a conveyor belt that is powered by differences in water density that arise from changes in temperature and salinity. When the water flows north it gradually cools but also sees some evaporation, which causes the salt content of the remaining water to increase. The low temperatures and higher salt content cause the water to become denser. As the density increases, the dense water sinks deep into the ocean – several thousand metres beneath the surface – and begins to spread southwards. There, it begins to be drawn towards the surface, warming as it rises, and so starts a new circulation.

LIA
The 'little ice age': not a true ice age in global terms, this was a prolonged episode of regional cooling that was most pronounced in the northern hemisphere and in the regions fringing the North Atlantic especially. There is varying opinion on its inception, ranging from later fourteenth to later fifteenth centuries, but consensus on its end point in the mid nineteenth century. It was not an episode of continuous depressed temperatures or adverse weather, but saw a series of phases of greater or lesser intensity, such as the extreme years at the end of the seventeenth century or in the early 1800s.

MCA
Medieval Climate Anomaly: also known as the Medieval Warm Period, this was an era of generally warmer conditions in the North Atlantic region spanning the period from the mid tenth to mid thirteenth centuries. It was not a uniform event, in that peak temperatures occurred at different points in different regions, nor was it an unbroken era of benign conditions, seeing periods of stability interspersed with relatively brief episodes of extreme cold or storminess, as in the mid twelfth century.

NAO
North Atlantic Oscillation: this is the name applied to the changing relative relationships of the area of atmospheric high pressure centred on the Azores islands (the so-called Azores high) and low pressure over Iceland (the so-called Icelandic low). The interaction between these pressure systems exerts a strong influence on winter weather and climate patterns in Europe and North America, as it causes changes in the intensity and location of the North Atlantic jet stream – the flow of very fast winds high in the atmosphere that affect the movement of regions of low

pressure. The Meteorological Office des-cribes the alternating states of the NAO as:

Positive NAO phase
The positive NAO phase represents a stronger than usual difference in pressure between the two regions.

Winds from the west dominate, bringing with them warm air, while the position of the jet stream enables stronger and more frequent storms to travel across the Atlantic.

These support mild, stormy and wet winter conditions in northern Europe and the eastern USA. Conversely, northern Canada, Greenland and southern Europe are prone to cold and dry winter conditions.

Negative NAO phase
The negative NAO phase represents the reverse, with a weaker than usual difference in pressure.

Winds from the east and north-east are more frequent, bringing with them cold air, while the adjusted position of the jet stream leads to weaker and less frequent storms.

As a result, Europe and eastern US are more likely to experience cold, calm and dry winters. In contrast, northern Canada and Greenland will tend to be mild and wet.[1]

Abbreviated titles

ER	*The Exchequer Rolls of Scotland*
HMC	Historic Manuscripts Commission
NLS	National Library of Scotland
NRS	National Records of Scotland
RMS	*Registrum Magni Sigilli Regnum Scotorum*
RPS	Records of the Parliaments of Scotland

---

1 https://www.metoffice.gov.uk/weather/learn-about/weather/atmosphere/north-atlantic-oscillation.

Coppiced beech tree, White Loch of Myrton, Monreith, Dumfries and Galloway. This veteran of the later eighteenth-century planting by the Maxwells of Monreith has been coppiced early in its life and left to grow out into this dramatic, multi-trunk form.

Introduction

At the time of completing the writing of this volume (July 2023), it is not yet a quarter of a century since Brian Fagan popularised the scientific concept that the period from the fifteenth to mid nineteenth centuries had been what we now term a 'little ice age' (LIA).[1] He was by no means the first to use the description, François E. Matthes having first employed this characterisation of the era in his 1939 report on glaciers in the Sierra Nevada of California.[2] In the sixty years from Matthes to Fagan it had been used by geologists and climatologists to describe what was at first seen as a phase of marked deterioration in global weather patterns, subsequently reassessed as more of a North Atlantic/Northern European phenomenon, that followed the milder and stabler conditions now labelled the Medieval Climate Anomaly (MCA), explored in the volume 'Scotland AD 400–1400'. Abstruse scientific debate surrounding both what exactly constituted an 'ice age', little or not, and what was its precise chronological span, meant that over the following six decades discussion of the phenomenon had barely penetrated academic consciousnesses beyond the subject specialists. To read most historians' accounts of Britain and Europe written during the sixty years after Matthes coined the term, it would seem that for them the world existed largely in some unchanging environmental state where the only developments were driven by politics, religion and economics, and where subsistence crises were viewed purely in terms of human successes and failures. As a consequence of this academic *omertà*, it failed to register with the broader public, who – beyond an engaged few – although awakened to the concept of anthropogenic environmental change since the 1962 publication of Rachel Carson's *Silent Spring*,[3] still remained generally unreceptive to discussion of climate change, historic or contemporary, until the later 1990s. When considered against that background, we can appreciate how few books aimed at a broad readership have been so transformative of public understanding in so short a time as Fagan's *The Little Ice Age: How Climate Made History 1300–1850*, which became a best-seller and went into multiple

---

1 B. Fagan, *The Little Ice Age: How Climate Made History 1300–1850*, paperback edition (New York, 2002).
2 F.E. Matthes, 'Report of Committee on Glaciers, April 1939', *Transactions, American Geophysical Union*, 20:4 (1939), p. 518. In North America, this phase was later relabelled 'Neoglaciation'.
3 R. Carson, *Silent Spring* (Boston, 1962). The book and its influence on late twentieth-century environmentalism is discussed in the volume 'Scotland 1850 to COP 26'.

reprints and editions within a few years of its publication. We might quibble now with the finer points of his detailed argument and argue for more refinement using the greater body of data that is available to us today, but there is no question that the core argument of his work remains solid. Furthermore, we also need to recognise that Fagan's publication triggered a new wave of environmental history research and writing, as researchers re-examined the data from their own geographical regions. Since the early 2000s, therefore, there has been a surge of research seeking evidence for the possible impacts of the LIA on social, economic and political developments, and for its influence on cultural memory that preserves a record of the lived experience of an era of profound climatic and associated environmental upheaval. Despite a wealth of available data, however, apart from a handful of important studies of key episodes within this period,[4] Scotland has not hitherto participated widely in that flowering of environmental history research into the impacts of the 'little ice age'.

If any one descriptor for the general conditions experienced across this era can be applied, it would in all probability be upheaval. Upheaval is one of the most common themes in Scottish histories of the sixteenth to nineteenth centuries, from the religious conflicts of the Reformation and Covenant to the Jacobite rebellions and post-Union social, political and economic revolutions. Conventional histories of the wider British Isles and Europe take a similarly socio-economic, political and religious turn, but there has been a growing focus in those contexts on the contribution of climatic and wider environmental change to the upheavals experienced there, spinning off from and reacting to Fagan's overarching narrative of 'how climate made history'. Indeed, climate-stimulated social upheaval is now a dominant theme of many environmental histories of the later medieval and early modern periods in northern Europe, and the thinking behind them has influenced much of what follows in this volume. That concept of environmental triggers, however, is far from new. From the time that Dutch historian Johan Huizinga articulated his theory of an age of decline and decadence in later medieval culture and society, written in the context of the collapse of empires and the old European social and political order at the end of the First World War, there has been an established historiography of European socio-cultural breakdown and reconfiguration to match that articulated in English for the end of the Roman Empire in the works of eighteenth-century essayist Edward Gibbon.[5] Huizinga produced a vision of socio-economic disintegration and the failure of political structures to which the label 'medieval' is applied, and their replacement by what have come to be regarded as 'renaissance' institutions and thinking. He attributed a key cause of this process to the human catastrophes of the fourteenth century, from the mortalities brought first by famine and then by the recurring waves of epidemic disease, and worsened by the protracted warfare that scarred the century. The exploration of the changes that resulted from these environmentally driven phenomena, however, by both Huizinga and a host of scholars

---

4 The stand-out study is K.J. Cullen, *Famine in Scotland: The 'Ill Years' of the 1690s*, Scottish Historical Review Monograph Series 16 (Edinburgh, 2010).
5 The tradition found its initial full articulation in Huizinga's 1919 work *Herfsttij der Middeleeuwen*, which was translated into English in 1924 as *The Waning of the Middle Ages* (and various subsequent translations and editions as *The Decline . . .* or *The Autumn . . .* The influences of Huizinga's argument are evident in publications late into the twentieth century, long after his core theories had been subjected to criticism and revision. See, for example, G. Holmes, *Europe: Hierarchy and Revolt 1320–1450* (London, 1975) for the continuation of a tradition of decline and reconstitution across that period, prefiguring the emergence of the full-blown renaissance in the second half of the fifteenth century.

since 1919, remained profoundly anthropocentric and, even more strikingly, institutional in its focus. In essence, those explorations have been shaped by that general silence among academic historians of the late and post-medieval periods concerning the nature and effects of historic climate and environmental change; instead they continued to offer a picture of all change that mattered in this era as arising solely from human choices and decisions. It is as if climate, weather and environmental conditions played no part in framing those decisions by communities who were wholly dependent on their effects for their subsistence.

In broadly British historiography of this period, this presentation is part of what is termed a 'Whiggish' narrative, where social, economic, political and cultural change is viewed as on an unending upward trajectory of improvement driven by innovation, experimentation, science and reason. At its root lies the deep conviction that this was the era in which humanity finally achieved mastery of the environment and ceased to be constrained by its limitations. This confidence is epitomised in the concept of superior human endeavour applying the fruits of new knowledge and triumphing over the formless chaos of nature in the twin revolutions of the eighteenth and nineteenth centuries, Agricultural and Industrial.[6] Brian Fagan changed that traditional narrative, switching the focus from the nature of the transformations that occurred to the underlying factors that drove them. He first focused on the outcomes and then forced the gaze away from the new institutions that arose on the wreckage of the old to contemplate instead the impacts that climate change and disease had on communities, their daily lives and routines and, critically, on how they chose to respond to the challenges. Throughout his work, there is a conscious recognition of the risk of determinism that lies in that perspective, of seeing all change as a consequence of environmental factors, which shifts the weight of the work instead towards human agency in the responses: how exactly did European society respond to the successive body blows dealt to it by climate change and pathogens? The result was a study that looks to those responses as laying the foundations of the truly revolutionary social, cultural and economic changes that occurred between the fifteenth and nineteenth centuries, upon which our increasingly fragile contemporary society is built. In many senses, it is thus a Huizinga for the twenty-first century. In many others, however, as explored for the latter part of the era discussed in this volume in a second and more locally influential study – Fredrik Albritton Jonsson's *Enlightenment's Frontier*[7] – it is a warning for how those same forces acted in the past to impose limits on what can be achieved through human agency, despite the confidence placed in the ability of science and reason to triumph over wild nature. Those same forces, moreover, are still in play, and they can and will force modern global society to make hard choices as we confront the realities of rapidly unfolding climate change and ecological crisis.

But do the impacts of famine, disease and war of the fourteenth century mark the start of a new climate era as distinct from being the death knells of the older European social and cultural order? Fagan started his study confidently, as did Johan Huizinga, with the horrors of the mortalities of the fourteenth century. Fagan, however, traced the beginning of the new era to a point still further back in time, to the climatic deterioration that had become evident in the second half of the thirteenth century and entrenched by the beginning of the fourteenth. As

---

6 This tradition became the new orthodoxy with the 1959 publication of Lord Briggs' now classic study of the processes and achievements of Improvement: A. Briggs, *The Age of Improvement 1783–1867* (London, 1959).
7 F. Albritton Jonsson, *Enlightenment's Frontier: The Scottish Highlands and the Origins of Environmentalism* (Newhaven, 2013).

we saw in the closing chapters of the volume 'Scotland AD 400–1400', that had been an era characterised by instability, with wide fluctuations from decade to decade between temperatures and rainfall levels, which directly affected human welfare in terms of what we would call 'food security' and 'energy security' as global climate systems underwent a major realignment that affected atmospheric and oceanic circulation. Despite Fagan's introduction of it as a prelude to his 'little ice age', its transitional nature has meant that in most broad explorations of the environmental histories of the northern hemisphere it has not been clearly associated either with the preceding MCA or the following LIA, falling into a kind of limbo between two slightly better-defined climate eras.[8] In reality, it does not fully belong to either, constituting instead a time of transition from one state to the next, in which conditions fluctuated between extremes before making the final switch towards the generally poorer weather state of the LIA. For Scotland, in these volumes, I have chosen to explore the fourteenth century as the coda to the MCA, during which the transition to the 'new normal' of the fifteenth century onwards was worked through, rather than presenting it simply as a preface to those new conditions. It is thus with that period that the volume 'Scotland AD 400–1400' ends, drawing a line under the century-and-a-half period of climatic instability that had followed the decline of the MCA and marking the start of what proved to be its final switch into the 450-year-long episode of northern Europe's 'little ice age'.

Four and a half centuries is a longer span than would be expected in a conventional history of Scotland. The writing of Scottish history is generally still framed by a structure of periods that was first proposed in the later eighteenth century and, with only some tinkering around where subdivisions might fall, has remained in place to present. Reign-based and primarily political narratives still dominate the subject for the fifteenth, sixteenth and, to a lesser extent, seventeenth centuries, with even the *New Edinburgh History of Scotland* volumes for the later medieval and early post-medieval periods – both those published and those still being written – dividing around the deaths of kings or the ending of minorities. The impact of this politically driven method is seen most clearly in how the seventeenth, eighteenth and earlier nineteenth centuries are treated, where the dates 1603, 1690, 1707, 1745 and 1832 appear regularly as era-defining chronological parameters. The first marks the transfer of Scotland's monarchy to England and the emergence of a new style of government for Scotland, and the last is the date of the Reform Act (1832), which triggered the progressive movement over the next ninety years towards a full parliamentary democracy. For the daily lives of most people in Scotland, neither date has much – if any – meaning. They are iconic, yes, for what they mark in terms of socio-political and economic development, the structures of government and political culture, but their impact on the daily human experience of subsistence farmers and fishermen, on their interaction with land and water – and the resources in and on both – took decades, if not generations, to become manifest. This volume is focused instead on the evidence for the transformations that occurred across the longer term, as both Scotland's human population and the wider environment with which it interacted and upon which it was dependent responded to the turbulent climatic conditions of the northern hemisphere's 'little ice age'.

As we will explore in the following chapters, there are many problems with the 'little ice age' label and the connotations it brings. At the forefront is this: what, exactly, do we mean by a 'little ice age'? It conjures up mental images of a Narnia-like time of unending winter, with perpetual snow and ice

---

[8] This is the period to which Bruce Campbell has applied the label 'The Great Transition'. See B.M.S. Campbell, *The Great Transition: Climate, Disease and Society in the Late-Medieval World* (Cambridge, 2016).

through which human populations toiled in the face of endless hardship, poised on the knife-edge of a subsistence existence. Yet, the steady adjustment of the label from the capital letters of Little Ice Age to the lower-case letters and quotation marks of 'little ice age' suggests some discomfort with the term and its connotations, at least among the academic community. Was it, in fact, an ice age and, if so, why is it deemed 'little' – and little relative to what? Is that adjective a reflection of its short duration, or is it that the conditions experienced across its span were relatively minor in comparison to what we know of the extremes that occurred during the last glacial era of the Younger Dryas, a period of some twelve centuries that ended around 11,700 years ago? Or does it reflect a geographically more restricted context than the global ice ages of the Quaternary era? The answer is yes on all counts. In the words of the UN's Intergovernmental Panel on Climate Change in its 2001 Third Assessment Report, when looked at 'hemispherically, the "Little Ice Age" can only be considered as a modest cooling of the northern hemisphere during this period of less than 1°C relative to late twentieth century levels'.[9] An annual mean cooling of 1°C might seem 'modest' but it had significant repercussions for the European agricultural regimes of the time, increasing the likelihood of harvest failures, or at best substantially reduced yield levels, and shortening growing seasons, especially in areas that were already on the margins for agricultural exploitation. It did not mean, however, catastrophic year-on-year failure except during the most extreme episodes within the overall era. So, while there was significant readvance of Alpine glaciers during the coldest phases of this 'little ice age' and the possible development of incipient glaciers in some north-facing corries of the Scottish mountains, at no time was there permanent, perennial and widespread ice and snow cover in the British Isles and northern European plain, and the seas around Britain remained ice-free, unlike during the Younger Dryas. That playing down of the severity of conditions during the 'little ice age', however, does not change the fact that for those living through and describing such times, the cooling and its visible and experienced effects certainly seemed far from 'modest'.[10]

For Scotland, as explored in Chapters 1 and 3 of this volume, the contemporary 'voice' of the people articulating how those declining temperatures and the increasing weather volatility affected their existence is heard only intermittently, and usually through the medium of those towards the upper levels of society. They, of course, did not perceive change in terms of annual mean temperature variation but experienced it as 'good', 'fair' or 'bad' conditions, extremes of wet, drought, cold or heat, and the relative abundance or shortage in provisions that the harvest brought. There were certainly attempts by churchmen and royal councillors to rationalise the failures, for example as indications of Divine displeasure or as shortages caused by the illegal hoarding of foodstuffs by unscrupulous speculators. We remain, however, largely ignorant of how farming and fishing communities responded on a year-on-year basis to the growing void between their storehouse of traditional environmental knowledge that framed their interactions with land and sea and the increasingly inadequate reality delivered by their efforts. Despite the neo-romantic visions of some more recent historians, that some innate environmental awareness possessed by people in Scotland and in the Highlands in particular made them more

---

9   Digitally archived at https://web.archive.org/web/20060529044319/http://www.grida.no/climate/ipcc_tar/wg1/070.htm.
10 For a broad discussion of the trends since the end of the last episode of true glaciation, see G. Whittington and K.J. Edwards, 'Climate change', in K.J. Edwards and I.B.M. Ralston (eds), *Scotland After the Ice Age: Environment, Archaeology and History, 8000 BC – AD 1000*, paperback edition (Edinburgh, 2003), pp. 11–22.

attuned to and resilient in the face of such climate-driven change,[11] the 'little ice age' exposed the fragility of their existence and unsustainability of the systems of exploitation that they had been practising since the earlier Middle Ages. Because of the absence either of continuous chronicle, journal or diary reporting or of anything approaching a nationwide distribution of sources of data that can provide evidence of the growing impact of that void in knowledge and how people around Scotland responded to it, investigation of the first three centuries of this 'little ice age' must draw on a wide diversity of information types. Past failure to do so – and reliance on later secondary observations of extremely variable and usually doubtful historical accuracy – has left a gap that has been filled mainly by 'factoids' – those seemingly irremovable pieces of information that have been reported and repeated so often that they become accepted as fact – that recur in the popular and pseudo-scientific discourse.

But that gap is not unbridgeable. Although they are discontinuous, there are contemporary narrative records that do provide a wealth of information on both general weather events and more localised incidents, and on the broad consequences of those for human populations. They also illustrate, however, that just as in earlier episodes of cooler or warmer conditions, the experience of weather impacts might vary widely around Scotland. Latitude, altitude, aspect and geology could and did deliver very different outcomes, as accounts from observers on opposite sides of the Firth of Forth reveal for the middle decades of the seventeenth century.[12] From these, paired with the proxy climate records that featured so prominently in the climate and weather reconstructions of the volume 'Scotland AD 400–1400', we can assemble an occasionally detailed but more commonly broad-brush narrative of the main trends and their effects in different parts of the country. That broad-brush narrative serves as the contextual canvas against which are projected discussions of various issues: agricultural practice and the expansion or contraction of arable and pasture, the ebb and flow of sea-fishing, the condition of Scotland's woodlands, the quest for energy security and its impacts on the land, and changing balances of wild and domesticated animals. Innovation, experimentation, intervention and, above all, sustained responsive change together form the dominant sub-themes, but neither the responses nor the changes they delivered were uniformly positive or broadly beneficial. Indeed, in terms of their longer-term consequences, many have been profoundly negative and their effects have made an enduring impact on Scotland's environmental health.

In the exploration of those impacts in the following chapters we are confronted repeatedly with the extent to which we live not only in cultural landscapes but in a cultural environment. By that I mean that the environment of Scotland has been subjected to profound anthropogenic change and modification over millennia, to the extent that little remains that can be described as 'natural'. Despite regularly repeated references to Scotland's 'last wildernesses' or to surviving expanses of 'natural forest' and 'primeval' woodland, the hard reality is that almost no part of this land has been untouched by human activity over the last eight millennia. It is useful to reflect on the fact that long before 1400 there was little that was truly natural about any of Scotland's then remaining woodland. Indeed, despite the lingering legacy of Steven and Carlisle's pioneering but

---

11 Exemplified by J. Hunter, *On the Other Side of Sorrow: Nature and People in the Scottish Highlands* (Edinburgh, 1995), which was hugely influential at the time of its publication but whose reliance on the ahistorical musings of Frank Fraser Darling and his flawed reading of the historical ecology of the Highlands has undermined the value of its powerful argument for a more engaged dialogue between largely external conservationists and Highland communities.
12 See Chapter 2, pp. 88–91.

now seriously outdated *The Native Pinewoods of Scotland*, while 'to stand in [these woods] is to feel the past', the past they were feeling was no older than the later eighteenth century at best.[13] Upland areas that seem to us remote, unpeopled and untouched, largely as a consequence of the traumatic restructuring of rural society that started in Lowland districts in the seventeenth century and continued in the process we label the Clearances in the Highlands and Islands in the later eighteenth and nineteenth centuries, had already been settled and exploited since the fourth millennium bc. The earliest farmers of the Neolithic age, as touched on briefly in the introduction to the volume 'Scotland AD 400–1400', had made inroads on the land, clearing wood and scrub and bringing about changes to soil structure and chemistry, as well as to the plant species it supported, and these legacies remain with us six thousand years later. All orders of biota had already seen extinctions and introductions long before the fifteenth century. These have had unforeseen ramifications in more recent decades as human management regimes have changed. Moreover, some of the inshore coastal fishing grounds had already been plundered to exhaustion. In the era explored in this volume, the reach and intrusion of such human activity and its direct and indirect consequences attained a climax that was sustained through the remainder of the nineteenth and earlier twentieth centuries, as traced in the volume 'Scotland 1850 to COP 26'. Here, however, we navigate through some of the main processes that led to that climax. We also question some of the traditional assumptions and challenge the spurious factoid-based narratives that continue to circulate at present.

It is important that we do not lose sight of that cultural contribution to the 'making' of Scotland's still astonishingly diverse environment in current initiatives for the 'rewilding' of mainly upland areas.[14] Within the following chapters we will see how much of what we value in terms of landscape character and, indeed, how much of what we are trying to preserve, nurture and expand, is the fruit of many thousands of years of human–environment interaction. We must not forget, for example, that the definition of what constitutes 'ancient woodland' in Scotland is that it has existed continuously since before only 1750, based on its recording in the maps produced by General Roy.[15] That recording, however, does not mean that what Roy's surveyors saw was a remnant of 'wildwood' or even partly the product of natural reseeding and regeneration, but simply that there were trees there at the time of their visit. Its definition as 'ancient woodland' means merely that there is a continuous record of trees in that place over the subsequent 275 years. That span of time is only the average lifespan of a beech – all of which in Scotland were introductions of the last quarter of

---

13 H.M. Steven and A. Carlisle, *The Native Pinewoods of Scotland* (Edinburgh, 1959), p. v.
14 Rewilding has now been provided with a proposed definition by the Scottish Government, one that attempts to underscore the need to recognise and protect the delicate nature of past and current interaction between humans and wider environment and the active human agency within the rewilding process: see Scottish Government, *Defining Rewilding for Scotland's Public Sector: Research Findings* (published 4 July 2023) https://www.gov.scot/publications/defining-rewilding-scotlands-public-sector/. The definition offered in the report is: 'Rewilding means enabling nature's recovery, whilst reflecting and respecting Scotland's society and heritage, to achieve more resilient and autonomous ecosystems. Rewilding is part of a set of terms and approaches to landscape and nature management; it differs from other approaches in seeking to enable natural processes which eventually require relatively little management by humans. As with all landscape management, rewilding should be achieved by processes that engage and ideally benefit local communities, in line with Scotland's Land Rights and Responsibilities Statement, to support a Just Transition.'
15 General Roy Military Survey of Scotland 1747–1755 https://maps.nls.uk/geo/explore/#zoom=13&lat=56.28581&lon=-3.92479&layers=3&b=1.

the seventeenth century and later – and less than half the potential lifespan of an oak. For our most iconic 'national' tree, the Scots pine, a maximum age of around 500 years is possible, but most of the currently growing Scots pines are less than 300 years old; many pinewoods classed as 'ancient' are younger still and form part of planting operations undertaken by some of the great landed estates in the last decades of the eighteenth and first half of the nineteenth century. Naturally reseeded trees of this age, it must also be remembered, are largely new growth in areas of intense management, encouraged by the woodland exploiters in what we would now regard as sustainable forestry practice.

Contrary to the still oft-cited polemic of Frank Fraser Darling and the fantasies of the Sobieski-Stuart fraudsters, which are still repeated in the output of some engaged and environmentally conscious activists on social media and in the lazy journalism of press and broadcasters, Highland Scotland was not stripped of much of its surviving woodland from the later seventeenth century by fuel-hungry charcoal burners and ironmasters. Nor was it all then subject to still greater devastation from the late eighteenth century by overstocking of former afforested areas with a horde of 'woolly maggots', as George Monbiot has memorably stigmatised the Blackface and Cheviot sheep introductions. There have been claims of upland districts stripped of trees by overgrazing where woodland regeneration would be possible if that grazing was removed. In fact, many of these had been treeless due to complex interplays of climate change, grazing by non-domesticated animals, and human agency long before the 'little ice age', and many have never been exploited as sheep-range. Pressure from deer, moreover, became a serious issue only from the second half of the nineteenth century, when the owners of sporting estates began consciously to encourage an increase in the deer population to add value to their shooting-rents. The cold and the high precipitation characteristic of the most severe episodes of the LIA, on the other hand, need to be considered as factors in the accelerating decline of managed native woodland and montane scrub in the later Middle Ages and down to the early 1800s, because these conditions affected both rates of successful germination of new trees and the volumes of viable seed produced by the existing mature ones. Across northern and central European regions also, there was a marked drop in the tree-line altitude – up to 200m in parts of the Alps – especially during periods when snow cover in some of the more mountainous areas became perennial and glaciers began to readvance. In Scotland, where the winds and general storminess of our oceanic climate regime already meant adverse growing conditions for trees in exposed western mainland, Hebridean and northern island areas, and over the highest parts of the mainland mountains, the even poorer conditions of the LIA worked to depress natural woodland regeneration further. In those areas where a combination of climate change, growth of blanket peat coverage and grazing pressures from wild and domesticated animals had long before pushed altitudinal tree-lines below their post-glacial optimal levels, the LIA is likely to have contributed to a further reduction and made some of the surviving high-altitude woodland and montane scrub more vulnerable.[16] Where woodland expanded significantly during this era, it was most often in the sub-highland and mainly eastern fringes of the mainland mountains, and in almost every case in those districts

16 In the thirty years since Richard Tipping published 'The form and fate of Scotland's woodlands', *Proceedings of the Society of Antiquaries of Scotland*, 124 (1994), pp. 1–54, understanding of the rise and fall of altitudinal tree-lines and the factors driving that process has evolved considerably, but his general observations – and caveats – remain valid. Altitudinal records for several species have recently been shattered, see S.H. Watts, 'High mountain trees: altitudinal records recently broken for eleven different tree species in Britain', *British and Irish Botany*, 5:2 (2023), pp. 167–179.

it was the result of estate planting rather than natural reseeding.

Where the greatest expanses of semi-natural woodland persisted into the early post-medieval period, chiefly in the central and south-western Highlands, coppice in oakwoods and reseeding in pine forest ensured that these organisms regenerated and in some cases thrived until they ceased to be valued as a resource in the later nineteenth and twentieth centuries. None of the surviving areas of such regenerated woodland is truly 'natural', but their longevity as a landscape element and the richness of their associated ecosystems have led to their veneration as almost natural, as opposed to self-seeded beech woods or the wind-spread creep of more recently introduced species of spruce and fir. This at best 'semi-natural' character does not devalue them (perhaps if they were labelled 'semi-anthropogenic' our perception of them would be more negative?), and the complex ecologies they shelter constitute some of our most vibrant and diverse plant, invertebrate, bird and small mammal communities. But surely their cultural character and the practices that sustained them merit better recognition alongside their environmental significance? Yet, in many of the current programmes that look to 'rewild' Scotland, that cultural dimension has been figuratively airbrushed and literally bulldozed out of the narrative, an act of conscious eradication and obliteration of memory that the Scottish Government's new definition seeks to prevent in future.[17] And it is much needed, for the fact that not all past anthropogenic impacts on the environment have been negative and that there is much that can and should be celebrated barely registers in current media discourse, regardless of whether it is pro- or anti-rewilding. I can but hope that the contents of this volume might help to stimulate a more balanced conversation of the future form of our landscape and the diversity of environments it embraces.

Even in a book of this size, however, not everything can be covered and even themes and topics that are included can most often be treated only superficially. As in the volume 'Scotland AD 400–1400', an early decision was taken to concentrate on a series of four recurrent themes. As mentioned earlier, climate change and the shifting weather patterns it delivered provides the context for exploration of a core suite of topics. Woodland, as the previous paragraphs perhaps more than hinted at, constitutes one strand of that suite. Arable and pastoral practices and the impact of technological developments and experimentation with plant and animal species forms another. The quest for fuel security and the physical and ecological impacts of extractive processes provide another major theme. The fourth strand is fishing, both sea and freshwater, and the multiple ecological niches affected by the health or fragility of aquatic resources. Woven through them, however, are interleaved explorations of other topics that rise to the forefront at particular points across the span of the 'little ice age', from the resurgence of wolf packs as 'enemies' of human populations during the episodes of most extreme weather in the fifteenth and sixteenth centuries, to the irruption of *Phytophthora infestans* (potato late blight fungus) in the 1840s. Some of such components are intended to stimulate further enquiry and to highlight where work has been targeted and where it has not, hopefully to encourage wider and better integrated interdisci-

---

17 For a measured critique of this trend, see H. Deary, 'Restoring wildness to the Scottish Highlands: a landscape of legacies', in M. Hourdequin and D.G. Havlick (eds), *Restoring Layered Landscapes: History, Ecology, and Culture* (Oxford, 2016), pp. 95–111. See also short comment pieces: P. Barrett and H. O'Regan, 'Environmental archaeology in the wild: making space for the past in the new conservation movement', *The Archaeologist*, 119 (2023), pp. 3–5. N. Whitehouse, E. Jenkins and P. Barrett, 'Rewilding, the historic environment and moving beyond "wilderness"', *Archaeology Scotland*, 47 (Summer 2023), pp. 14–17.

plinary research in future. Others, like the brief discussion of bees and beehives, are intended to highlight different perspectives on some of our contemporary concerns and to flag how both the globalisation of commodity production and trade – in this case of sugar from British-, Dutch- and French-controlled, slave-worked plantations in the Caribbean – and the revolutionary discovery of the process for making paraffin wax combined to initiate a dramatic decline in the keeping of honey-bees. Insecticides, pathogens and fungus infestations have added a crisis dimension to bee decline, but for honey-bees we might pause to reflect how past populations may have been maintained at unnaturally high levels by human demand for their honey and wax, and the consequences for those populations of the introduction of cheaper and more easily obtained alternatives. In that one small example, we catch a glimpse of the mesh of interdependencies and unforeseen ramifications that arise from even seemingly inconsequential actions that constitute our endlessly interconnected environment. Here too is a signal to heed, as we contemplate the temptation for 'improvement', however we define it, from genetic modification to rewilding. All have potential consequences that we cannot foresee.

CHAPTER I

A New Normal? Subsisting at the Margins in the Early 'Little Ice Age'

―――

Esope myne authour makis mentioun,
Of twa myis and thay wer sisteris deir,
Of quham the eldest duelt in ane borous toun:
The vther wynnit vponland weill neir:
Richt soliter, quhyle vnder busk, and breir:
Quhilis in the corne, in vther mennis skaith,
As owtlawis dois and leuit on hir waith.

This rurall mous in to the wynter tyde,
Had hunger, cauld, and tholit grit distres,
The vther mous that in the burgh can byde,
Was gild brother and made ane fre burges.
Toll-fre als but custum mair or les,
And fredome had to ga quhair euer scho list,
Amang the cheis and meill in ark and kist.[1]

When Walter Bower, the early fifteenth-century chronicler-abbot of Inchcolm Abbey in the Firth of Forth, looked back on the reign of King Robert III (r.1390–1406) from the famine-, disease- and weather-ravaged 1440s, he described Scotland as having then had 'an abundance of provisions'.[2] As we explored at the end of the volume 'Scotland AD 400–1400', the later fourteenth century in Scotland was far from having been a golden age of plenty. Bower's rose-tinted view of the decades either side of 1400, it is likely, arose not from a reality of significant positive difference but from unfavourable comparison of his childhood experiences and memories of what were already poorer conditions than had prevailed in the middle decades of the century with the kingdom's more recent experience. He would not have known that the plagues, hunger and worsening weather that he was witnessing were to be regular occurrences

---

1 Extract from Robert Henryson, 'The taill of the vponlandis mous, and the burges mous', in G.G. Smith (ed.), *The Poems of Robert Henryson*, vol. 2 (Edinburgh, 1906), p. 15.
2 Walter Bower, *Scotichronicon*, vol. 8, ed. D.E.R. Watt (Aberdeen, 1987), p. 63.

for the next four centuries. Nor would he have understood that his perception of a past time of plenty just a generation earlier simply saw less bad conditions as a lost Golden Age. Just like those Scots who had endured the war-torn and disease-ridden half century down to the 1350s and looked back longingly on the seemingly bountiful harvests and peaceful rule of King Alexander III from the grimness of the Wars of Independence and the rains, harvest failures and famines of the Wolf Solar Minimum,[3] he saw only better times.

What Bower was living through were the opening decades of a profound realignment of global climate systems as the planet entered what some climate scientists have labelled a 'little ice age'.[4] It is a label that conjures up visions of extremes – snow and ice, storminess, 'frost fairs' on frozen rivers, blighted harvests and mass starvation – but the lived reality of this era is still hotly debated. We recognise now that it was not a time of unrelenting severity lived in the shadow of harvest failures and subsistence crises but one of short episodes of extreme conditions, with what euphemistically might be termed 'bad outcomes' for many of the human population, set against a wider trend of poorer yields and higher livestock mortalities worsened by generally cooling climate conditions. The headline events of crop failures and famines, outbreaks of disease or months of unseasonably poor weather that Bower and other chroniclers narrate were what people endured and remembered, despite the fact that they were intermittent and rarely nationwide in their impacts. What was less obvious to them – especially from their insulated vantage points in well-heated and well-supplied monasteries – was the long, slow decline in annual mean temperatures, the almost imperceptibly shortening growing season, or how some tried and tested traditional agricultural methods were becoming less reliable at delivering good results. Crops were not failing everywhere or annually, but their yields were becoming noticeably poorer in both volume and quality. Cold springs delayed germination and stunted growth, while indifferent summers and cold, wet autumns led to harvest times falling later in the year, and some grain crops were gathered while still green. When such circumstances continued year-on-year, those working the land would soon have learned to take opportunities when they came, to pivot away from reliance on crops that required better environmental conditions towards ones that were better suited to the poorer weather, and to adopt new strategies of managing flocks and herds where fodder resources had become scarce. It was in this period that there was an accelerating transition towards investment in oats, away from wheat and barley, as a response to the changed conditions. From their socially elevated positions – Bower was an Augustinian abbot and a royal councillor – our chroniclers witnessed the political ramifications of the symptoms of climate change, understanding social disturbance as a result of 'over-mighty' nobles vying for territorial dominance, or seeing food shortages as the product of greedy merchants hoarding supplies. They did not see such issues as stemming from the same cause: a long-term failure of traditional agricultural practices to deliver the resources needed to sustain the established economies and cultures of lordship in both Highland and Lowland

---

3 The Wolf Minimum and its consequences are discussed in the volume 'Scotland AD 400–1400', Chapter 9.
4 The concept of a major planetary climate episode that profoundly influenced the course of human political development was popularised in the early 2000s by Brian Fagan's eponymous study *The Little Ice Age*. Although now normally expressed within inverted commas and uncapitalised to emphasise its conceptual and adjectival nature, distancing discussion from the more polemical aspects of Fagan's thesis, his view of the half millennium of upheaval that he saw as consequent upon the global 'cold snap' remains highly influential in both popular and academic discourse on historic and current climate change.

Chronicling the early years of the 'little ice age'

In his chamber beside the tower of his abbey on its windswept island in the Firth of Forth, the chronicler Walter Bower, abbot of the Augustinian monastery on Inchcolm (b.1385, d.1449), wrote the voluminous Latin account of Scottish history that we know as *Scotichronicon*. Although he was concerned primarily with political and ecclesiastical events, and with lengthy moralising for the benefit of his readers, he also included some of the most detailed descriptions of weather events to survive from medieval Scotland. For the events of most of the historic period he covered in his work, he was reliant on older sources that he copied from at length, several of which describe extreme events and other weather phenomena in places far from Scotland. Some of his earlier sources, however, had been composed by well-informed individuals in senior ecclesiastical or royal household office in Scotland, and details of the worsening conditions at the end of the MCA were drawn largely from such high-quality earlier sources. Some of this writing must be used with caution, for we must also remember that the purposes of the chroniclers in recording such events were usually moral and religious, with the events being cast as incidences of Divine retribution for the sins of humanity generally or of an individual people whose lands and resources suffered most. We can have greater confidence in Bower's reports of extreme weather events, epidemics, livestock deaths and harvest failures in the first half of the fifteenth century – the critical period of transition into the 'little ice age' – which were written from personal observation and the experience of family, friends and fellow clerics. In the topics he covered we can see his personal, direct access to reliable information, reaching back into the last decades of the fourteenth century: for example, in his extended account of the 1358 flood in Haddington in East Lothian, the burgh where was born in 1385. The anecdotes embedded in that section of *Scotichronicon* were no doubt informed by his local knowledge. The early fifteenth-century material, such as the report of the structural damage and deaths arising from the impact of the 1410 storm on the cathedral-priory at St Andrews, would have reached him through his Augustinian and wider ecclesiastical networks. Although written in a summary, stylised manner, his descriptions contain acute observation of the conditions faced in the extreme winters of the 1410s and 1430s, and his reference to the 'fearfully cold' days of winter 1434/5 probably reflects a personal experience of the weather in his exposed island monastery. His accounts range from the higher than normal mortality among wild animals to the inability to grind grain for human consumption because the mill-races were frozen. For a few brief years, Bower's words shed light on the lived experience of ordinary Scots as they struggled to subsist in the steadily worsening climatic conditions.

Inchcolm Abbey from the south.

Scotland. As climate systems and environmental conditions reconfigured, so too did the fabric of Scotland's political structures, social hierarchies and economic regimes. But, like the transitioning climate regime, while most people saw and felt the short pulses of intense and extreme weather which punctuated the era, the real changes that delivered long-term decline crept forward slowly, almost imperceptibly, but relentlessly.

Human responses within Scotland to generally deteriorating conditions, for which evidence survives, were reactive or opportunistic. Violence, both physical and applied through the coercive weight of law and its enforcement, was one obvious manifestation of social disturbance. Competition for productive land, both cultivable and pasture – those most precious of natural resources – was a key driver of that violence. And the competition extended to other resources, from woodland and peat-yielding muirs to game animals and fish. That some people were driven to physical violence in their efforts to maintain traditional lifestyles provides the eye-catching headlines for modern narratives. Looking beyond them, however, we see a steadily growing volume of litigation concerned with encroachment on rights or actual wrongful dispossession. That material emphasises how many more cases were settled through arbitration or judgement in tribunals rather than through resort to blows. We see parliamentary legislation developed to address what were perceived to be sources of stress, from tree-planting policy and encouragement to sow more food crops, to safeguards for poor tenants in times of crisis. It was from these acts that flowed the litigation offering remedy for forceful occupation of property and its cultivation or grazing by trespassers, imposed quartering of one lord's men on another's dependants, unauthorised cutting of peat or wood, or intrusion on hunting or fishing rights. What we see through the lens of court actions is a society experiencing great stress, dislocation and disturbance, in which resource conflict seems, perhaps misleadingly, to be ubiquitous. Those same records and surviving financial accounts, however, also show that there were some who could capitalise on their access to in-demand resources and staple commodities to prosper accordingly. And yet, underlying all of this, we see that the majority continued to subsist in bad times and channel their energies towards bettering their lot when opportunity and conditions allowed.

Climate and weather: a slide into a 'little ice age'

How – if at all – did the people of early fifteenth-century Scotland understand the beginning of this new climate era? Extreme weather events are usually the most headline-grabbing of historical records for climate change, but for Scotland in the 1400s weather records from any historical sources – chronicles, charters, royal administrative or financial records, estate rentals, and so on – are sparse. Nevertheless, as in the earlier periods explored in the volume 'Scotland AD 400–1400', material from geographically widespread locations in mainland Britain and Ireland can throw light on general trends from which we can build models for how different parts of Scotland fared at those times. Collectively, the kinds of event described in such sources, coupled with proxy climate data, indicate that a major systemic shift was underway across the North Atlantic by around 1400. Climatologists suggest this shift led, among other things, to a reversal of trends in North Atlantic weather towards a degree of storminess unseen since the ninth century, and to an abrupt, marked deterioration in weather conditions.[5] What that meant in

---

5   L.D. Meeker and P.A. Mayewski, 'A 1,400-year high resolution record of atmospheric circulation over the North Atlantic and Asia', *The Holocene*, 12 (2002), pp. 27–66; P.A. Mayewski et al., 'Holocene climate variability', *Quaternary Research*, 62 (2004), pp. 243–255.

reality was a greater frequency of storm events and an increase in their violence, with the exposed Atlantic coasts, Western and Northern Isles affected worst, but with the North Sea coasts battered by intensifying northerly and easterly wind storms and tempests. With this rising storminess came higher levels of precipitation generally, but with the storm events leading to greater flooding and waterlogging of land, or producing deep snow cover that prevented cultivation, blanketed winter grazing and smothered livestock. Some of the impacts would have taken many years to make themselves manifest. For example, the lowering of altitudinal tree-lines through poorer seeding, germination and growing conditions, and threat of wind-cast or limb-loss in storm events, were as likely to have been outcomes of the colder and, critically, wetter conditions in the early decades of the LIA as they are suggested to have been in prehistory.[6] While farmers might be acutely aware of poorer crop growth in the annual cycles of sowing, growth and harvest, the slower growth or regeneration of woodland might not have been so immediately recognisable. All such trends could be endured, but when it became clear that they had become the 'new normal' in the weather, Scotland's people did not remain passive victims but began to adapt their agricultural practices to the new conditions.

Can any clear point of switch in climate conditions be traced in the narrative records? Our richest and geographically closest documentary records are the interrelated western Irish accounts, like the material copied from medieval manuscripts into the seventeenth-century 'Mac Carthaigh's Book' or the more contemporary Connacht and Loch Cé annals. All these sources record deteriorating weather down the Atlantic-facing west of the British Isles from around 1405–7.[7] The first significant event was recorded for 1405 in accounts of very rainy and tempestuous weather in winter, which brought widespread property damage across western districts. The year 1407 was reported as blighted by similarly poor conditions and, consequently, high weather-related losses among the livestock that was a mainstay of the economies of the Scottish and Irish regions worst hit by the weather. The Atlantic seaboard experience, however, was not representative of elsewhere in mainland Britain. We lack any equivalent of the post-1660 weather and temperature reconstructions for central England, but abundant contemporary English records of the early 1400s, especially from south-eastern districts, offer qualitative narratives that report wide regional variation. The St Alban's based monk, Thomas Walsingham, for example, wrote of a summer of heat, drought and pestilence in Hertfordshire in 1407, while western England enjoyed significantly cooler but markedly wetter conditions. Walsingham's fevered summer was followed first by a bitterly cold and long-lasting winter with snow on the ground for four months, and then a late spring in 1408.[8] His account describes what was for the southern British Isles the only truly extreme winter there of a more protracted episode that extended across Europe from France to Russia for some years after c.1405. That long winter of extreme cold and precipitation brought the conditions immortalised

---

6 Tipping, 'Form and fate', pp. 14–15 for outline discussion of 'pluvial' episodes and their impact in early prehistory on woodland extent.

7 Annals of Connacht http://www.ucc.ie/celt/published/T100011/index.html AC1407.17; Annals of Loch Cé http://www.ucc.ie/celt/published/T100010A/index.html LC1407.15; Mac Carthaigh's Book http://www.ucc.ie/celt/published/T100015/index.html MCB1405.27.

8 D. Preest (ed.), *The Chronica Maiora of Thomas Walsingham (1376–1422)* (Woodbridge, 2005), pp. 349 (pestilential summer of 1407), 359 (bitter winter). For weather impacts in 1407/8, see K. Pribyl, *Farming, Famine and Plague: The Impact of Climate in Later Medieval England* (Cham, 2017), p. 182 and data tables in Appendix, pp. 250–251.

in 1412 in the snowy February scene painted by the Limbourg brothers for the *Très Riches Heures du Duc de Berry*.[9] On the basis of this limited record, then, it could be argued that there was a brief 'buffering' from the worst impacts of the deteriorating conditions in southern and eastern districts facing the North Sea, but that episode was transitory and gave way to winters of equal severity across the whole of these islands.

Those chronicle accounts of an initially marked southeast–northwest split in weather conditions across the British Isles can be modelled against more recent, instrumentally recorded conditions, supported by historic data from climate proxies. Such models suggest that there was high winter NAO in the early 1400s, with a stronger westerly airflow that delivered intense cyclonic rainfall. Such a phenomenon is strongly evidenced in multiple proxy data types from across a region spread from Greenland to the Siberian Ural Mountains, which has been interpreted as a northern hemisphere symptom of a global climate transition postulated as underway by *c*.1400. That transition arose from a large-scale reorganisation of the motors that drive atmospheric circulation around the northern hemisphere. The transition is evident in the climate proxies from several Highland Scottish localities spread from the sea-loch sediments of Loch Sunart on the mainland west coast, to the speleothems – stalagmites and stalactites – of Uamh an Tartair in the Traligill cave system in Sutherland, and the pine tree-ring sequences of the northern Cairngorms.

It must be emphasised that the conditions revealed in these proxies are the local experience of weather trends and are not hard and fast evidence for more general effects across the country. We can, however, look at how similar trends in more recent sections of the climate proxy record correlate with the instrumental records from elsewhere in Scotland, and suggest how the impacts felt at the sources of the proxy data would likely have been manifest in Lowland or more easterly districts. Considered in that way, two of our main proxy records point to the likely character of Scotland's winter-time experiences in the fifteenth century. For the first half of the century, the Traligill speleothem record suggests that conditions likely to have delivered severe winter weather were prevalent down the western side of the country.[10] Comparing the types of weather revealed in proxy and instrumentally recorded evidence from the north-west Highlands and eastern districts for the later eighteenth and early nineteenth centuries with this Traligill proxy data for the 1400s, it is likely that what was occurring was the tracking east of successive waves of moisture-laden fronts from the Atlantic and their meeting intensely cold easterly airflow from the Siberian winter high-pressure cell; the result in the 1810s to 1840s was heavy snowfall over northern and probably north-eastern Scotland,[11] and we might reasonably expect similar conditions to have occurred in what seems to have been almost identical climatic circumstances four centuries earlier. Tree-ring data from the northern Cairngorms certainly suggests that a sharp drop in annual summer temperatures commenced around 1375 and reached its lowest level before 1410, but it then recovered somewhat between 1410 and 1420.[12] Even collectively and comparatively, it is still slender and relatively blunt-edged evidence, but it does support a view that Highland and Hebridean districts,

---

9   J. Longnon and R. Cazelles, *The Très Riches Heures of Jean, Duke of Berry* (New York, 2008).
10 C.J. Proctor et al., 'A thousand year speleothem proxy record of North Atlantic climate from Scotland', *Climate Dynamics*, 16 (2000), pp. 815–820 at 818 Fig. 4.
11 A.G. Dawson, *So Foul and Fair a Day: A History of Scotland's Climate and Weather* (Edinburgh, 2009), p. 105.
12 M. Rydval et al., 'Reconstructing 800 years of summer temperatures in Scotland from tree rings', *Climate Dynamics*, 49, issue 2951 (2017), pp. 2951–2974 at 2959 Fig. 4.

St Andrews Cathedral Priory, looking east along the north cloister alley to the south transept. In January 1410 the south gable of the transept collapsed in a storm, the debris falling onto the dormitory, parlour and chapter-house, crushing and killing the sub-prior of the monastery.

if not all of Scotland, faced similar weather trends to the wider northern Atlantic and European region.

Those broader regional trends continued after 1410 in a brief episode of warmer but variable conditions. Higher summer montane temperatures in the Cairngorms were one aspect of these unstable weather patterns through the 1410s. That was accompanied, however, by what seems to have been a steadily intensifying sequence of isolated storm events that signalled a general increase in overall storminess. Although Bower reported northerly winds devastating eastern Scotland in mid January 1410 in a tempest of (to him) unprecedented severity that wrought devastation on woodland, houses and churches – including the collapse of the south transept gable of St Andrews Cathedral – and caused the wreck of numerous ships, his account presents it as a memorable one-off event.[13] It is not until 1413 that a second great storm, this time westerly, was reported in the Annals of Ulster as driving ships towards the Scottish coast and sinking many with great loss of life.[14] Again, it sits as an isolated incident

13 Bower, *Scotichronicon*, vol. 8, p. 75.
14 Annals of Ulster https://celt.ucc.ie/published/T100001A/ 1379–1541, U1413.5.

The Udal, North Uist, seen from the summit of Clettraval. Increasing erosion of the machair on the west (Atlantic-facing) side of the peninsula progressively inundated the settlement and its surrounding fields from the fifteenth century.

in the documentary record. This time, however, the event can be seen through climate proxies as one peak standing out amid the generally upward trajectory of North Atlantic storminess. It was an outstanding episode amid the progressively strengthening winds and storms responsible for the spectacular sand-blow events which at this time overwhelmed areas of machair cultivation in the Western Isles from Cnip in Lewis to Borve on Barra's west coast. Here, the 1410s saw the beginning of a decline that ended with the abandonment of long-established and high-status settlements, including the ancient estate centre at the Udal at the northern tip of North Uist.[15]

15 S. Dawson, A.G. Dawson and A.T. Jordan, 'North Atlantic climate change and Late Holocene windstorm activity in the Outer Hebrides, Scotland', in D. Griffiths and P. Ashmore (eds), 'Aeolian archaeology: the archaeology of sand landscapes in Scotland', *Scottish Archaeological Internet Reports*, 48 (2004), p. 32. https://doi.org/10.9750/issn.2056-7421.2011.48. For North Atlantic storminess, see above, note 5. The continuing impact on the machair in the sixteenth and seventeenth centuries is discussed in Chapter 3.

High upland temperatures were symptomatic of a strongly positive summer NAO in 1419 which, after the precipitation of earlier years, brought drought to Scotland and Ireland. As the summer droughts of the early 2020s in the north-west of the British Isles have shown, significantly reduced precipitation in this region can lead quickly to problems for human and animal populations, despite the perception of general wetness associated with those mountainous districts. In the mountains, as our recent experience confirms, dry conditions were problematical for both peasant communities and large estates whose flocks and herds were driven there for summer pasturing. Many watercourses throughout the Highlands are spate streams, strongly seasonal in character, whose flows rise and fall rapidly and dry up quickly, reducing watering options for livestock. Such drier conditions, however, were perfect for an abundant Lowland harvest.[16] The climax came in 1419, however, with the highest summer temperatures in the Cairngorms since the 1370s; after that, it cooled steeply and steadily into the middle of the century.[17] During this new phase, snow is likely to have lingered long on the highest mountains and perhaps never wholly

---

16 Bower, *Scotichronicon*, vol. 8, p. 113; Annals of Connacht, AC1419.15; Annals of Ulster, U1419.20.
17 Rydval et al., 'Reconstructing 800 years of summer temperatures in Scotland', p. 2959 Fig. 4.

The Northern Corries of the Cairngorms, seen from Tom Pitlac near Boat of Garten. The colder, snowier conditions of the fifteenth century saw cornices and snowfields last later into the year, keeping local temperatures lower and suppressing summer pasture grass growth.

vanished in summer from north-facing corries. Winds from the mountains would have remained bitterly cold and the sustained low air temperatures would have curtailed vegetation growth, especially on seasonal upland pastures, and also had a negative impact on germination of tree and scrub seeds in montane districts. It seems also that the 1420s saw a switch in winter conditions in the north-west Highlands, with the long sequence of high winter NAO ending abruptly, to be replaced by an episode of low NAO that was the most pronounced since the extended lows of the eleventh century.[18] According to Bower, although the following year (1420) was temperate, starting with a cold, dry 'northern' winter and rainy spring, its harvest 'failed to a great extent'.[19]

A wet autumn in 1421, which saw violent downpours recorded in normally drier Lothian,[20] brought a second poor harvest and consequent grain shortages throughout the region. Nothing in this record suggests that there had been any repeat of 1407/8's icy extremes; winter, however, was coming.

Plummeting temperatures first seen in the severe winter of 1422/3 became a regular experience for around three decades across most of northern Europe from the Urals to Scandinavia.[21] This entrenched cold added to the stresses of persistently wetter summers and autumns on the continent by recurrent high summer NAO.[22] There is, however, no corresponding report of such winter severity in the north or west of the British Isles at this date.

18 Proctor et al., 'A thousand year speleothem proxy record', p. 818 Fig. 4.
19 Bower, *Scotichronicon*, vol. 8, pp. 113, 117. For the epidemic of 1420, see below, p. 33.
20 Bower, *Scotichronicon*, vol. 8, p. 125.
21 H. Lamb, *Climate, History and the Modern World,* 2nd edition (London, 1995), pp. 202–206.
22 Fagan, *Little Ice Age*, p. 83.

Certainly, Irish sources present no noteworthy deterioration in conditions until 1425, when the Ulster annalist observed that poor weather began late that year, extended through winter until May 1426, brought widespread cattle deaths, delayed the new season's ploughing for some weeks, and caused unwontedly high human mortality.[23] For Bower from his Scottish east-coast perspective, however, there was nothing unusual in the weather to report beyond a spell of stormy conditions in late autumn 1425.[24] There are, however, hints in non-narrative sources that all was not well, especially in upland districts north of the Tay, where it is possible that deteriorating winter conditions and extended periods of cold, snowy weather contributed to the re-emergence of competition for resources that had been little heard of since the thirteenth century.

It is unlikely to be coincidence that there was a resurgence in the consciousness of many Scots of a threat that had greatly receded since the early thirteenth century, the wolf. While the hunting of wolves that we saw in the volume 'Scotland AD 400–1400' ordained in thirteenth-century deeds had brought their near eradication in parts of Southern Upland Scotland, the evidence of the later fourteenth-century English Gough Map attests to their abundance in the north-western Highlands.[25] As the largest apex predator apart from man surviving into the later medieval period in northern mainland Britain, it is likely that they had continued to compete with humans for prey within the deer-rich uplands. During harsher times, however, such as were experienced in the winters of the early fifteenth century, when Bower records high mortalities among wild animals, they probably turned to domestic livestock for food. It is in this context that we should see a 1428 act of parliament instructing lords of baronies to organise hunts each spring to kill wolf whelps – in effect culling the packs – and to hold wolf hunts four times annually or as often as wolves were seen on their land.[26] The reiteration of the legislation in 1458 following a second run of extreme winters is not unexpected but on this occasion it was aimed explicitly at the destruction of wolves, and the period from 25 April to 1 August was defined as 'the time of the whelps', during which three hunts at least were to be organised.[27] And there was clearly an expectation on the part of lords that such hunts would happen and their tenants would equip themselves appropriately for them. Lease conditions from the 1480s into the 1550s on the Glenisla properties of Coupar Angus Abbey, where the monks kept their horse-stud and largest sheep flocks and herds of cattle, required tenants to keep 'a leash of good hounds' for hunting wolves and foxes.[28] Gaelic tradition also records local landowners, or in the case of Margaret Lyon, Lady Lovat, their wives, organising hunts to exterminate wolf packs on their lands. Lady Lovat is said to have overseen their eradication from the area between Loch Ness and the Aird (the district running west from Inverness to Beauly) in the third quarter of the fifteenth century.[29] Cumulatively, such material points to a zone of collision between humans and wolf packs all around the interface between the central

23 Annals of Ulster, AU1425.5.
24 Bower, *Scotichronicon*, vol. 8, p. 255.
25 For the Gough map, see http://www.goughmap.org/map/ where a mountain-top in north-west Sutherland (at extreme left of the sheet) is crowned by a wolf figure and the text 'hic habundant lupi' (here wolves abound).
26 RPS 1428/3/6 [accessed 20 January 2022].
27 RPS 1458/3/36 [accessed 20 January 2022].
28 C. Rogers (ed.), *Rental Book of the Cistercian Abbey of Cupar-Angus*, 2 volumes (London, 1879), vol. 1, p. 236; vol. 2, pp. 107, 141, 142, 143, 144, 251.
29 A.E.M. Wiseman, '"A noxious pack": historical, literary and folklore traditions of the wolf (*Canis lupus*) in the Scottish Highlands', *Scottish Gaelic Studies*, 25 (2009), pp. 95–142 at 115–116.

Glenisla looking north from near Tulchan Lodge. This glen-head section of the valley was the location of Coupar Angus Abbey's sheep-ranges and horse-stud, but it was also most exposed to wolf predation from the mountain plateau into which it cuts.

Highland massif and the neighbouring lowland zones.

Collectively, most written and climate-proxy records evidence a major cooling event commencing by the early 1430s at the latest and lasting for decades, with temperatures not again reaching fourteenth-century highs until around 1490–1500. This start of this decline is noted by Bower, who observed that the eastern Lowlands' first extreme winter came in 1431/2, although northern and western parts of Scotland might already have been experiencing poor winters for much of the previous decade. For him, that winter was 'particularly hard, with ice and gales' and brought high cattle, sheep and wild animal mortalities.[30] It began a run of what various narrative sources present as especially poor years, with Irish annals reporting famine throughout Ireland by summer 1433,[31] and deteriorating conditions through 1434 culminated in a worse winter for 1434/5. Both Bower and these Irish sources agree on the severity of the times, with the former reporting 'fearfully cold weather' lasting three and a half months, during

---

30 Bower, *Scotichronicon*, vol. 8, p. 267.
31 Annals of Loch Cé, LC1433.3; Annals of Ulster, U1433.6. The severity of this period is reported across Europe. For a summary, see Fagan, *Little Ice Age*, pp. 83–84.

Deer skeleton, Glen Kinglass, May 2011. The high mortality rate among deer in the winter of 2010–11, when many starved as a result of the deep snow cover preventing them from reaching the foggage below, possibly replicated the conditions reported by Walter Bower in 1431/2.

which time mill lades froze and prevented the milling of grain.[32] For the Ulster annalist this was 'the year of the great frost', during which extreme cold persisted for around seven weeks, the great loughs of Ireland could be crossed safely on foot and high numbers of wildfowl perished.[33] Unsurprisingly, Bower reported dearth in 1435 and that shortages of food were especially acute in south-eastern districts,[34] although that regional information is likely to be a bias arising from his location and information

---

32 Bower, *Scotichronicon*, vol. 8, p. 293. This claim has been viewed sceptically in the past, with most scholars thinking of the large stream lades of the eighteenth and nineteenth centuries, but excavation at, for example, Gogar in Lothian, has indicated that the mill-streams of the medieval period were relatively shallow and narrow: J. Morrison, R. Oram and A. Ross, 'Gogar: archaeological and historical evidence for a lost medieval parish near Edinburgh', *Proceedings of the Society of Antiquaries of Scotland*, 139 (2009), pp. 229–56.

33 Annals of Ulster, U1434.6. Similar accounts are offered in Annals of the Four Masters http://www.ucc.ie/celt/published/T100005C/index.html M1435.3, Annals of Loch Cé, LC1434.2. The Annals of Connacht 1434.8 report that the frost lasted from St Andrew's Day (30 November) to the Feast of St Berach (15 February), during which time the ground was too hard to plough or prepare for the spring sowing.

34 Bower, *Scotichronicon*, vol. 8, p. 299.

sources. It is, nevertheless, evident from these geographically dispersed accounts that the winters were becoming colder, stormier and longer, with growing seasons in spring and summer curtailed and harvests diminished. This would also have been one of the worst times for natural reseeding of woodland and the production of viable seed on mature trees, although that impact perhaps did not become visible until years if not decades after the coldest and wettest episode, which might have contributed in part to the first parliamentary legislation to stimulate tree-planting happening only in 1458 (see Chapter 2). Unsurprisingly, the extended episode of significantly poorer annual conditions through the 1430s introduced unwonted degrees of stress into human social relationships as the agricultural basis of the economy contracted. We should not forget the contribution of these early decades of the LIA to the tensions within the kingdom that ended with the assassination of King James I at Perth in the bitterly cold February of 1437.

It is unfortunate that both Bower and the Irish annals end or cease to record weather information at more or less this point. We lack any equivalent sources of similar quality for the remainder of the fifteenth century in Scotland, with only a handful of relatively brief and mainly politically focused accounts existing. The nearest there is to a continuation with detail similar to Bower's, moreover, has a geographically different locus – west-central and south-west Scotland – and is less concerned with the socio-political affairs of the court and eastern Lowlands that dominated Bower's work. Unfortunately, it only extends our detailed climate or weather data down to the end of the 1430s and more superficially into the late 1450s. This narrative source is the anonymously written 'Auchinleck Chronicle',[35] whose author was often as interested in reporting weather events and epidemics as he was in recording the political machinations of the time. Picking up where Bower breaks off, its author confirms that the poor years continued to the end of the 1430s. Its chronicler, indeed, reported what he termed 'the land Ill' which, if we can rely on his account of dearth in 1439 that brought widespread death and suffering, was a very poor growing season that had seen the failure of the 1438 harvest.[36] That brief continuation apart, however, he provides otherwise only general notes on weather events and for the remainder of the century we become reliant on climate proxies and the hints and implications in non-narrative record sources.

Turning again to the proxy record, we can see evidence to reinforce the negative message presented in the chronicle accounts and the wolf legislation. Data from Traligill indicates that low winter NAO persisted, delivering a more northerly airflow and colder conditions over the northern mainland and islands. The northern Cairngorm tree-ring data suggests too that cooler summer conditions prevailed. Once more, these two datasets suggest adverse winter and summer conditions through the 1440s and into the 1450s, although the summertime mountain temperatures were recovering after the mid-1430s lows.[37] There is a lag in the record of similar cooling in the bottom water temperature data from Loch Sunart, with the most rapid shift towards colder temperatures commencing c.1445 and extending into the 1490s.[38] That lag suggests that any

---

35 For the Auchinleck Chronicle see C.A. McGladdery, *James II*, 2nd edition (Edinburgh, 2015), Appendix: The Auchinleck Chronicle, pp. 261–276.
36 Ibid., p. 262.
37 Proctor et al., 'A thousand year speleothem proxy record', p. 818 Fig. 4; Rydval et al., 'Reconstructing 800 years of summer temperatures in Scotland', p. 2959 Fig. 4.
38 A.G. Cage and W.E.N. Austin, 'Marine climate variability during the last millennium: the Loch Sunart record, Scotland, UK', *Quaternary Science Reviews*, 29 (2010), pp. 1633–1647 at 1640.

atmospheric warming had not delivered significant improvement in the wider weather regime of high rainfall that contributed to the lower salinity of the sea-loch.[39]

Can the consequences of this climate episode be traced elsewhere in the documentary record? Parliamentary records of the period confirm that poor harvests and consequent episodic shortages of victuals recurred in the 1450s, although details of specific incidences do not survive and it is likely that dearth rather than full-blown famine was more common. The legislative evidence for continuing stresses commence with a January 1450 act prohibiting the hoarding of foodstuffs, which reflected a belief that such behaviour was causing shortages rather than being a panicked reaction to it.[40] Further legislation against hoarding in 1452 reveals that supply difficulties had persisted despite parliament's efforts to legislate out of the crisis.[41] Poor late autumn weather, which probably affected the harvest of late-sown grain crops, is implicit in our one chronicle account of weather events, the Auchinleck Chronicle's report of heavy rain in the western Southern Uplands in November 1453. The chronicler narrates how unprecedented rainfall brought extensive flooding in the valley-bottom agricultural districts of lower Clydesdale, but he does not mention explicitly the destruction of the grain crops.[42] It is probably not entirely coincidental, however, that the 1454 parliament saw legislation enacted to encourage the importing of foodstuffs.[43] This evidence and further indications of extensive social dislocation and socio-economic distress from the later 1440s and 1450s suggest that the weather impacts on food production were not only severe but also protracted.

We can perhaps see the episode of solar (in)activity known as the Spörer Minimum as worsening this phase of poor conditions through the mid and later fifteenth century. The physics of how exactly the reduced sunspot activity of such a minimum affected the key global climate drivers like the NAO or the El Niño/La Niña oscillating system in the Pacific is still debated. There seems, however, to be some interaction with atmospheric oscillations such as these, potentially intensifying global climate trends that were already in motion. It is generally argued that this solar minimum episode spanned around a century between about 1460 and 1550,[44] although others have argued for a start as early as the severe winters around 1400 and entering its coldest phase between 1430 and 1455.[45] All, however, agree that the second half of the century saw some of the poorest and coldest northern hemisphere weather conditions of the medieval period. For Scotland, the worst episode occurred in a succession of harvest failures in the 1480s, with legislation in 1482 and 1486 revealing that urgent action was taken to relieve the threat of famine.[46] In its timing, this legislation after three decades of silence on such issues hints at a recent and profound emergency. Volcanic forcing is postulated as adding to the intensity of switches in climate dynamics in this period and might explain why such

---

39 Cage and Austin, 'Marine climate variability', pp. 1643–1644.
40 RPS 1450/1/22 [accessed 19 January 2022].
41 RPS 1452/6 [accessed 19 January 2022].
42 McGladdery, *James II*, p. 270.
43 RPS 1453/3 [accessed 19 January 2022].
44 J.A. Eddy, 'The Maunder Minimum', *Science*, 18:192 (June 1976), pp. 1189–1202 at 1196, 1199.
45 Y. Jiang and Z. Xu, 'On the Spörer Minimum', *Astrophysics and Space Science*, 118 pts 1–2 (January 1986), pp. 159–162; H. Miyahara, K. Masuda, Y. Muraki, H. Kitagawa and T. Nakamura, 'Variation of solar cyclicity during the Spörer Minimum', *Journal of Geophysical Research*, 111:A10 (2006), A03103. M.G. Ogurtsov, 'The Spörer Minimum was deep', *Advances in Space Research*, 64 (2019), pp. 1112–1116.
46 RPS A1482/3/8, 1486/3/2.

Noltland Bay and Links, Westray, Orkney. The low-lying sandy areas of Westray and adjoining Sanday experienced severe and worsening episodes of erosion, sand-blow and sea-flood as the fifteenth century progressed.

measures were taken at this time.[47] A specific crisis event within a longer-term episode of climatic deterioration was possibly linked with a large eruption of Mount St Helens in Washington State on the Pacific seaboard of the USA, dated to 1479, with a secondary and smaller eruption in 1482.[48] The first of these events was marked by a pyroclastic eruption which threw up several million tonnes of aerosol particulates into the high atmosphere, potentially reducing solar radiation reaching the Earth's surface and having similar effects to the 1257 Samalas eruption discussed in the volume 'Scotland AD 400–1400'. Under normal circumstances the impact might have been minimal but, in the already cooler conditions of the Spörer Minimum, it was perhaps sufficient at northern latitudes to make the difference between marginal success and failure for the grain harvest while the dust lingered in the atmosphere. In Orkney's low-lying and sandy northern islands, this eruption may have been the catalyst in the potent mix of weather conditions and agricultural practice that from the 1480s onwards saw extensive erosion

---

47 K.R. Briffa, P.D. Jones, F.H. Schweingruber and T.J. Osborn, 'Influences of volcanic eruptions on northern hemisphere summer temperatures over the past 600 years', *Nature* 393 (1998), pp. 450–455.

48 U.S. Geological Survey Fact Sheet 2005–3045 (2005), U.S. Geological Survey and the U.S. Forest Service – Our Volcanic Public Lands: Pre-1980 Eruptive History of Mount St. Helens, Washington
https://pubs.usgs.gov/fs/2005/3045/#:~:text=Kalama%20Eruptive%20Period%20(A.D.%201479,form%20during%20the%20Kalama%20Period. [Accessed 19 January 2022].

Ceardach Mor, Baleshare, North Uist. Major storms in the 2010s accelerated erosion of Baleshare's Atlantic-facing western shore, exposing the remains of settlement dating from the Neolithic to the early Middle Ages. Each 'step' in the eroded dune face marks an era of ancient settlement.

of areas of sandy-soiled arable; sea-flood inundation and sand-blow events led to wry observations in rental documents of land 'blawin to Birrowne' (blown to Bergen).[49]

Wind and weather erosion and deposition

Orkney's experience of erosion is a well-documented instance of the negative interplay of human and wider environmental factors in these early stages of the 'little ice age'. There and in the Western Isles, the increasing North Atlantic storminess mentioned above contributed to substantial losses of land that had been prized for centuries for settlement and agriculture. Some was overwhelmed by sand-blow, but more was washed away by storm and tidal action over frighteningly short periods. In the 1490s, persistent cells of winter and summer high pressure returned the weather to a sequence of cold, dry winters and dry summers, where the already damaged Orkney fields were exposed to the perfect conditions to accelerate the erosion of once productive land. The machair of the Uists and Benbecula in the Western Isles was similarly affected, with rapid erosion commencing there around 1350; settlements like Bornais and Cille Pheadair were abandoned in the later fourteenth or early fifteenth centuries in

---

49 W.P.L. Thomson, *History of Orkney* (Edinburgh, 1987), p. 130; A. Peterkin (ed.), *Rentals of the Ancient Earldom and Bishoprick of Orkney* (Edinburgh, 1820), pp. 84, 85, 87–90, 93–99.

favour of more physically secure locations on the 'rock and loch' landscapes of the Uist blacklands.[50] Numerous researchers have written at length about Heisgeir, an island now four miles west of the present North Uist coast but which was reputedly connected by a low-lying sandy spit to North Uist and Benbecula.[51] This land-bridge was destroyed in only a few years in the first half of the century, leaving just the rocky outcrop of Heisgeir itself. Down the Outer Hebridean machair strip, erosion of old land surfaces has continued, with associated exposure of human burials and settlement debris emerging from below sea-eroded dunes on, for example, the west side of Baleshare, Eoligarry on Barra and Sheader on Sandray; although they are mainly of prehistoric and earlier medieval date, they suggest that these tales of major losses of the settled area to the sea are grounded in folk memory.[52] It is possible that some losses involved 'barrier islands', low-lying sandy bars that buffer the impact of Atlantic breakers, whose sands are swept away and redeposited in a rolling sequence over centuries if not millennia. For the inhabitants of the threatened 'mainland islands' (North Uist, Barra and Sandray), the disappearance of such familiar seascape elements – possibly fixtures for generations before being washed away – would have been traumatic.

Major sand-blow and erosion events also occurred on the east coast in the early 1400s. Local tradition records the overwhelming of the village and parish church of Forvie on the north side of the Ythan estuary in Aberdeenshire during a nine-day storm in 1413. Although that date or event cannot be corroborated from the historical record, excavation in the 1950s produced evidence to support an end to occupation around 1400.[53] During the excavation, areas of rig-and-furrow cultivation were identified adjoining the village and it seems that farming on the sandy links, which had been productive during the stable era of the MCA, had seen accelerating erosion through the fourteenth century. Worse, though, came probably from stripping of stabilising turf layers and gorse root mats from the adjacent dunes for fuel, soil-deepening materials and building needs; those practices destabilised the dunes and caused the drifting sand that led to the abandonment of the village. Similar erosion and deposition occurred at Rattray in Buchan and, in the sixteenth and seventeenth centuries, at Culbin in Moray and Eldbotle in East Lothian, discussed in Chapter 3 below. In all cases, from Orkney to the Uists or the North Sea coast, the common denominator was not just continuance of exploitative practices that had developed in better times, but their intensification. Cultivation may have expanded rather than contracted in the later fourteenth and fifteenth centuries as peasant farmers sought to compensate for shortfalls in yields that the deteriorating weather had brought. The consequences of that decision were disastrous.

Erosion was not only a coastal phenomenon. Alongside the potential of landslip or riverine encroachment such as affected Dumbarton regularly through the later medieval and early post-medieval

---

50   D.D. Gilbertson, J.-L. Schwenninger, R.A. Kemp and E.J. Rhodes, 'Sand-drift and soil formation along an exposed Atlantic coastline: 14,000 years of diverse geomorphological climatic and human impacts', *Journal of Archaeological Science*, 26 (1999), pp. 439–469 at 439; N. Sharples (ed.), *A Norse Farmstead in the Outer Hebrides: Excavations at Mound 3, Bornais, South Uist* (Oxford, 2005), p. 195.

51   For example, Dawson, *So Foul and Fair a Day*, pp. 107–108.

52   Gilbertson et al., 'Sand-drift and soil formation', pp. 442–443.

53   W. Kirk, 'Prehistoric sites at the sands of Forvie, Aberdeenshire: a preliminary examination', *Aberdeen University Review*, 35 (1953), pp. 150–71; Kirk, 'Sands of Forvie', *Discovery and Excavation in Scotland, 1957* (Dundee, 1957), p. 4; Kirk, 'Sands of Forvie', *Discovery and Excavation in Scotland, 1958* (Dundee, 1958), p. 2; Kirk, 'Forvie', *Discovery and Excavation in Scotland, 1960* (Dundee, 1960), p. 2.

Forvie Kirk, Sands of Forvie, Aberdeenshire. The simple parish church of the coastal settlement was overwhelmed by sand and abandoned with its surrounding settlement and fields after an extreme sand-blow episode in the early 1400s.

periods,[54] erosion manifested itself most visibly in the silt deposits in the lower courses of the major rivers. The increased rainfall that affected the west and north of the country during high winter NAO episodes is likely to have brought frequent spates on rivers such as the Findhorn, Spey and, especially, the Tay, whose tributary headwaters rose on the east-facing downslope of the mountains of western Breadalbane, almost within sight of the Atlantic coast. The Tay's catchment was a locus of intense human activity, from woodland clearance for timber or to create new arable land in the upper reaches of the Tummel, Tilt and Garry, to intensified agricultural activity – mainly for arable cultivation – on the lower reaches of the Isla, Earn and Tay itself. The sandbanks that are still a feature of the firth below the Tay's confluence with the Earn were not a new phenomenon in the fifteenth century, but an increase in sediment transportation in the river enlarged them and created new obstacles for sea-going ships trying to reach Perth.[55] In such circumstances, it is no coincidence that the fifteenth century saw Dundee and

---

54   For Dumbarton's long struggle to contain the waters of the Leven, see E.P. Dennison, 'Burghs and burgesses: a time of consolidation?', in R.D. Oram (ed.), *The Reign of Alexander II 1214–49* (Leiden, 2005), pp. 253–284 at 279–281.

55   R.D. Oram, 'Estuarine environments and resource exploitation in eastern Scotland c.1125–c.1400: a comparative study of the Forth and Tay estuaries', in E. Thoen, G.J. Borger, A.M.J. de Kraker, T. Soens, D. Tys, L. Vervaet, H.J.T. Weerts (eds), *Landscapes or Seascapes? The History of the Coastal Environment in the North Sea Area Reconsidered*. Comparative Rural History Network, volume 13 (Turnhout, 2013), pp. 353–377 at 370–371.

Mugdrum Island and sandbanks, upper Tay estuary. Although Mugdrum is a long-established fixture of the river, first recorded in the 1200s, the shifting sandbanks downstream of it were an unpredictable hazard to shipping heading for Perth.

its silt-free port outstrip its upstream rival as the point of landfall or loading for trade that once passed through Perth.

More generally, as research in the Bowmont Valley in the Borders has suggested, the climatic stresses from the 1400s onwards brought erosion, but such events were less prominent than the headline incidents. These stresses triggered soil erosion at higher altitudes in this north Cheviot valley and led to changes in the behaviour of local watercourses with more spate events, as well as the carriage and deposition of more silt, and rapid shifts in the channels,[56] but fields and grazing land suffered little obvious loss of topsoil. Analysis of blanket peat on summit ridges around the valley has shown that there was continuous inwashing of mineral sediment starting in the early fourteenth century and continuing through the fifteenth and down to present, consequent on higher rainfall and movement of water through and over the peat. The peat blanket itself was subject to gully erosion on the higher ground, again starting in the fourteenth century. Despite the higher precipitation of this era, that break-up of the peat cover brought increased drainage of the blanket peat and slowed its growth. Gully formation in the peat promoted faster run-off, in turn causing deeper gully erosion and loss of soil on the steeper valley slopes, where there was little or no cultivation, but gentler slopes and the lower ground in the valley seem to have escaped any significant erosion. There, no noteworthy loss of soil was identified in the cultivated ground, which has been interpreted as evidence for the success of soil conservation measures that had developed in the thirteenth and

---

56 R. Tipping, *Bowmont: An Environmental History of the Bowmont Valley and the Northern Cheviot Hills, 10,000 BC – AD 2000* (Edinburgh, 2010), pp. 197–198.

fourteenth centuries.[57] These findings need to be tested in other contexts but the Bowmont example is strongly suggestive of the resilience of agricultural practices that had developed in the eastern Borders hill-country through and since the more benign era of the MCA and their success in stabilising and preserving soil in more adverse conditions. As the sand-blow events recorded in areas of sandy soil attest, however, continuity of long-established agricultural practices adopted in better times could also deliver adverse outcomes when local weather conditions changed. Traditional environmental knowledge, it could be argued, was a two-edged sword that produced different experiences in agricultural communities who had inherited different cultural practices and who lived in different geological and climate zones.

Epidemic disease: 'an exaltation of virulence'

Throughout the early stages of the 'little ice age', Scotland's population remained substantially smaller than it had been before the later fourteenth-century plague epidemics. Climate, however, was not the primary agent that placed a brake on recovery in population levels. Although the numbering of successive waves of plague that had characterised its reporting in Scotland and all across western Europe since the 'Great Mortality' ended soon after 1400, disease continued to claim the lives of thousands with every recurrence.[58] Fresh outbreaks were still greeted with horror and dread but, with the end to numbering, it was as if the disease was coming to be accepted as an irregular visitor who had to be accommodated if it could not be avoided. It did not mean that plague disappeared from the consciousness of the chroniclers and later diarists, who still narrated the course and impact of its recurrences through to its last Scottish visitation in 1645–8.[59] Our main sources of written evidence for later plague visitations and responses to them, however, shift to the formal records of parliament and organs of local government. In those, we see for the first time the implementation of measures to prevent or contain the spread of the disease, including what might now be labelled as public or environmental health procedures strikingly similar to many of the steps implemented by governments through the 2020/21 phase of the Covid-19 pandemic.

A major step towards effective plague management came in the parliament of October 1456, when laws empowering holders of local jurisdictions to implement containment and prevention measures were enacted.[60] While the principal hopes were still placed in a Divine release from epidemic, the legislation set out practical steps to protect the wider community. These steps included quarantining or expulsion of those suspected of being infected from their homes to temporary encampments on what became known by the sixteenth century as 'foul muirs', control on movement between burghs, and

---

57 Ibid., p. 198.
58 R.D. Oram, 'Disease, death and the hereafter', in E.J. Cowan and L. Henderson (eds), *Handbook of Scottish Medieval History* (Edinburgh, 2011), pp. 196–225 at 209.
59 For discussion, see R.D. Oram, '"It cannot be decernit quha are clean and quha are foulle." Responses to epidemic disease in sixteenth- and seventeenth-century Scotland', *Renaissance and Reformation*, 30:4 (2007), pp. 13–40.
60 RPS 1456/7 [accessed 18 January 2022]. The full enactment does not survive in the parliamentary record but the surviving statement of proposed refinements to the acts by the clerical estate members of that parliament and the regular rehearsal of the measures in burgh council records through the later fifteenth and sixteenth centuries enables their general tenor to be reconstructed. J.F.D. Shrewsbury, *A History of Bubonic Plague in the British Isles* (Cambridge, 1971), seems to have been unaware of this legislation, the first of its kind implemented anywhere outside mainland Europe.

Ardchattan Priory, Argyll and Bute, section of a fifteenth-century grave slab with carving of a partly decomposed corpse with a toad gnawing its bowels. Macabre monuments like this were one development of the disease-ravaged era.

cleansing of possibly infected portable material by boiling, immersion in running or salt water, or exposure to frost conditions, and for properties the fumigation or, in extreme cases, burning of houses. The efforts of the burgh council in Edinburgh in the 1490s to enforce these measures and take additional steps to protect townsfolk are the best-known example of the implementation of the 1456 act, but even in smaller burghs from Peebles – where traffic into and from the burgh was regulated in 1468 – to Irvine and Elgin, councils regularly took swift and ruthless action to protect their community.[61]

Plague, however, was not the only epidemic disease to afflict Scotland through the fifteenth century. For the first time since before 1349, chroniclers found space in their accounts to report on new

---

61 J.D. Marwick (ed.), *Extracts from the Records of the Burgh of Edinburgh*, volume 1, *AD 1403–1528* (Edinburgh, 1869), pp. 72, 74–78 etc. For discussion of the Edinburgh legislation, see Oram, 'Responses to epidemic disease'; Oram, 'Disease, death and the hereafter', pp. 211–213; Shrewsbury, *History of Bubonic Plague*, p. 151. For Peebles, see W. Chambers (ed.), *Charters and Documents Relating to the Burgh of Peebles, with Extracts from the Records of the Burgh. AD 1165–1710* (Edinburgh, 1872), p. 157.

and equally dreadful diseases that were ravaging the human population. In 1402, for example, Walter Bower recorded a major outbreak of some form of dysentery in eastern Scotland, with both urban and rural populations from East Lothian to Angus affected. He described how victims died 'in profluvio ventris', which translates as 'with outflowing of the bowels' (i.e. diarrhoea).[62] The epidemic possibly provided convenient cover for the murder of the heir to the throne, David, duke of Rothesay, who was said to have perished of dysentery while a prisoner in his uncle's castle at Falkland in Fife.[63] The impact of this disease was severe enough to interrupt the normal operations of government, with Exchequer business disrupted when custumars – the collectors of the customs on exported goods – from burghs such as Stirling failed to appear at the audit.[64] It spared no one, regardless of rank, and in Dundee two of the burgh's bailies, the senior figures among the burgh's merchant elite, were reported by a surviving bailie to have died from it.[65]

It is possible that this 1402 epidemic was a first occurrence of another soon-to-be frequent virulent epidemic disease, whose next recorded appearance was in 1439. In that year, the high death-rate of a famine was worsened by what the author of the Auchinleck Chronicle labelled 'the wame ill' (*wame* (Scots) = stomach), a deadly gastro-intestinal disorder.[66] According to his account, this new sickness killed as many people as any previous pestilence, with a second wave of the illness spreading from England later in the year that caused the death within twenty-four hours of any who contracted it. From the Auchinleck chronicler's description of the epidemic and its high mortality, modern medical historians have suggested that the pathogen at work was *Vibrio cholerae* – cholera, sweeping through the human population in 'an exaltation of virulence'.[67] Although records of subsequent episodes are lacking for the remainder of the fifteenth century, the resurgence of cholera in later centuries suggests that it might have become endemic.

Cholera was just one possible development amid the new climate order of the fifteenth century, but another was almost certainly an upswing in what appears to have been respiratory infections. As our recent experience of Covid-19 and its variants has demonstrated, the most virulent of such pathogens are spread easily through close and regular human contact. In 1420, Scotland was swept by such a disease, possibly the same as an unspecified plague in Norfolk that year, for trade between the Scots and eastern English ports such as King's Lynn and Great Yarmouth continued whenever peace prevailed between the kingdoms. This new illness was even more frightening because of how it differed from what had come before and it defied the measures that folk had begun to adopt to keep themselves safe. Bower records in *Scotichronicon* that the new illness caused many deaths among all social classes throughout the kingdom and that the common folk called it the 'qwhew' (pronounced *whew*), a name that seems to be onomatopoeic, that is it mimics the sound made by people suffering from it. We do not know exactly what this disease was but, interestingly, Bower associated it with prevailing weather conditions, which he noted had been 'very dry and northern' (i.e. cold with northerly winds) over winter followed by a wet spring. These conditions, he advised, were known to lead to 'summer-fevers, eye-inflammations

---

62 Bower, *Scotichronicon*, vol. 8, p. 45.
63 Ibid., p. 39.
64 *ER*, vol. 3, p. 553.
65 Ibid., p. 579.
66 McGladdery, *James II*, p. 262.
67 Shrewsbury, *History of Bubonic Plague*, pp. 150–151.

and dysenteries'.⁶⁸ Bower's noting of the *whew* sound suggests some kind of pulmonary infection that caused breathing difficulties. Since there is no subsequent reference to it in any medieval source, it would seem that the population that caught and survived the disease at the time gained immunity to future outbreaks and its impact was therefore lessened.

Scotland seems to have avoided another new and extremely virulent disease that ravaged England from 1405 until at least 1408. Contemporary English chroniclers gave wild estimates of the levels of mortality, one claiming that 30,000 perished in London alone, but that suspiciously round number probably simply means 'a very great many'.⁶⁹ The persistence of this disease through the winter months – the epidemic in London started in mid October or thereabouts and continued into late winter – differs from the clearer seasonality associated with bubonic plague and suggests that this was something different. The autumn start in London and the Thames estuary region has led to its identification as cholera, typhoid fever, or even a form of 'malignant dysentery' such as was also said to be affecting Bordeaux and south-western France at this time, which would have become almost self-sustaining once water supplies were infected, and passed from person to person through poor hygiene in markets and in food preparation.⁷⁰ Given that the epidemic was widespread beyond estuarine south-east England, however, we should look at alternatives. One London-based chronicler recorded that 'men and bestys were gretely infectyd with pockys', which perhaps gives us a stronger indication of what was at work: smallpox or some variola variant.⁷¹ Certainly, it has been suggested that 'smallpox apparently began to slowly gather momentum' in Europe in the fifteenth century, reaching Iceland *c.*1430 and Sweden in the middle of the century, and killing thousands in Paris in 1438.⁷² It is not known when this smallpox wave first reached Scotland but perhaps we should see it in 'the swiftly-spreading pestilence' that Walter Bower recorded in Edinburgh in February 1431 and Haddington in 1432 rather than that disease outbreak being some form of epidemic influenza, as has been proposed.⁷³ Given the continuing preoccupation with plague, however, there is little descriptive reporting of these more exotic-sounding fifteenth-century epidemic diseases. We remain ignorant, for example, of the identity of those conditions labelled locally in Selkirk as 'the boch' (perhaps some form of vomiting virus if 'boch' is equivalent to the later Scots 'boak/boke' = to vomit) and in Perth as 'the mallochis pest' that swept through those burghs in the early 1500s, other than that they were not bubonic plague.⁷⁴

We are on certain ground with the agent in one of the final fifteenth-century epidemics, *Treponema pallidum*, better known as syphilis. While it is now

---

68   Bower, *Scotichronicon*, vol. 8, p. 117.
69   Shrewsbury, *History of Bubonic Plague*, p. 141.
70   Ibid., pp. 141–143.
71   J. Gairdner (ed.), *Historical Collections of a Citizen of London in the Fifteenth Century (William Gregory's Chronicle)* (London, 1876), p. 87.
72   D.R. Hopkins, *The Greatest Killer: Smallpox in History*, revised edition (Chicago, 2002), pp. 28–29.
73   Bower, *Scotichronicon*, vol. 8, pp. 265, 277. Shrewsbury, *History of Bubonic Plague*, p. 150. Shrewsbury was citing the earlier work on plague by Creighton and accepted his proposal uncritically, suggesting that it was a revisitation of the disease called 'le qwhew', which spread through Scotland in 1420. In October 1431, it was agreed that the Exchequer auditors would gather at Perth the following February 'if the pestilence is not there' and at St Andrews if it was, which gives some sense of its progressive spread through the heart of the kingdom: RPS 1431/10/2 [accessed 20 April 2020].
74   J. Imrie, T.I. Rae and W.D. Ritchie (eds), *The Burgh Court Book of Selkirk 1503–45*, volume 1 (Edinburgh, 1960), p. 48; J. Eagles (ed.), *The Chronicle of Perth* (Llanerch, 1996), p. 21.

Inchkeith from the west. With separation and confinement seen as among the most effective methods of containing the spread of epidemic disease, the Forth islands' proximity to the most populous urban centres of eastern Scotland made them ideal for isolating victims of plague and syphilis.

fairly well understood from the archaeological and historical record that what is likely to have been a non-venereal variety of this bacterium was present in Europe, including Scotland, for many centuries before the pan-European epidemic of the 1490s, the venereal form of the disease struck a continent whose population had no previous exposure to it.[75] Associated traditionally with Christopher Columbus's return from the New World, with his sailors bringing back the bacterium which they then introduced to the prostitutes of the port cities of the Mediterranean, by 1497 in Aberdeen it was being labelled as 'the infirmity cummout of Franche and strang partis', emphasising its movement along trade and communication networks.[76] It is unlikely to have been the first sexually transmitted disease to make its presence felt in Scotland but it was the first to see systematic efforts by local authorities to stamp it out, signalling its exceptional virulence and the fear it was generating. Aberdeen's encounter with syphilis in 1497 saw burgh council enactments clamping down on prostitution, while Edinburgh used legislation similar to anti-plague measures in an attempt to have all infected townsfolk come forward for transportation to and confinement on Inchkeith in the Firth of Forth.[77] None of these efforts was successful and by the early 1500s incidences of what was by then known as *grandgore* were recorded throughout the kingdom and in all social classes.[78] Even with the current resurgence of syphilis globally, the discovery of antibiotic treatments for *Treponema pallidum* in the mid twentieth century has removed most of the fear that once surrounded this disease. But at the close of the Middle Ages, its impact in burghs that were attracting increasing numbers of economic migrants was potentially of epidemic proportions, killing or incapacitating an untold number of Scotland's younger, sexually active and economically productive people.[79] Its role in slowing Scotland's post-bubonic plague pandemic population recovery and economic growth in the sixteenth century remains unresearched.

75 Oram, 'Disease, death and the hereafter', pp. 213–214.
76 J. Stuart (ed.), *Extracts from the Council Register of the Burgh of Aberdeen, 1398–1570* (Aberdeen, 1844), p. 425.
77 Marwick, *Edinburgh Burgh Records*, vol. 1, pp. 71–72.
78 J.D. Comrie, *History of Scottish Medicine*, vol. 1 (London, 1932), p. 201.
79 For a detailed analysis of its urban impact and levels of morbidity and mortality among especially the socially and economically marginal, see J. Fabricius, *Syphilis in Shakespeare's England* (London, 1994).

Famine and feast: the hungry fifteenth century

Movement of people and materials which aided the spread of disease also produces records that illustrate wider issues affecting Scotland's environment and people. Data relating to Scottish trade and commerce provides much of the evidence that corroborates the hints of hardship, dearth and famine contained within the narrative accounts and parliamentary record. Evidence for the prices of grain and bread from 1398 to 1414 in Aberdeen, for example, demonstrates the cycle of price fluctuations related to harvest yields, with within-year spikes as high as an increase of 100 per cent in 1410 and average price rises lying in a band from 11 to 20 per cent; such movements continued across the century.[80] Although the prices of many processed foods were regulated by burgh council 'assize', that of grain was not; prices were instead negotiated publicly between buyers and sellers.[81] As a port trading with grain-producing regions in the southern Baltic, and one whose prices are recognised as generally cheap in comparison to other major burghs, Aberdeen could expect to see some buffering against the extreme price fluctuations caused by local failures of the harvest. But within-year price fluctuations of 50 per cent and more in Aberdeen show that this protective effect was lost and indicate more general supply difficulties in those parts of northern Europe with which the burgh traded. Parliamentary legislation from as early as 1401 to encourage the movement of victuals for sale 'where[ever] seems to be most expedient in the kingdom' speaks of efforts to alleviate local shortages.[82] It is interesting to note that Aberdeen in 1401 saw a 50 per cent increase in wheat prices, so perhaps here we glimpse the situation at which this legislation was aimed.

Among James I's parliamentary acts from 1425 was one aimed ostensibly at stopping profiteering through artificial inflation of the prices of foodstuffs in places where the king's court was based. Lying as it does amid other acts attempting to both regulate and stimulate international trade, however, this act probably had fundamental issues of supply availability at its root.[83] The repeated efforts to encourage the import of food staples, including cereals, coupled with acts from 1426 requiring landowners to sow specified minimum quantities of wheat, peas and beans or face heavy fines, suggests a general perception among the kingdom's rulers of persistent critical shortages of basic foodstuffs that could be remedied by an expansion of cultivation.[84] Here, then, we might be glimpsing one of the standard responses to shortfalls in yields, the breaking in of new ground simply to grow more of the crop of which there was a perceived deficiency, or the more intensive cropping of land that was in the past sown only intermittently.

Prices for wheat, barley, oats and malt can be used as a proxy for harvest successes or failures but are a rather blunt instrument given the patchy regional survival of data. The generally upward trends through the later fourteenth and first quarter of the fifteenth centuries, however, even when allowing for currency debasement, help to provide some context for the James I legislation encouraging sowing. What data we have from the 1430s, however, suggests that extreme weather events that accompanied the then-current climate transformation may have slowly

---

80 E. Gemmill and N. Mayhew, *Changing Values in Medieval Scotland: A Study of Prices, Money, Weights and Measures* (Cambridge, 1995), p. 150 Table 10.
81 Ibid., pp. 57–65.
82 RPS 1401/2/12 [accessed 20 January 2022].
83 RPS 1425/3/10 [accessed 20 January 2022].
84 RPS 1426/39 [accessed 20 January 2022]. It is striking that this act was repeated in 1458 (RPS 1458/3/29) shortly after the long period of supply crisis in the early to mid 1450s.

wiped out any gains brought about by the 1426 act. Indeed, the high prices – up to 108 pennies[85] per boll for wheat in Aberdeen in 1438 against an average below 50 pennies per boll through the 1400–24 period – indicate a serious short-term subsistence crisis and confirm the evidence of Bower and the Auchinleck Chronicle for the harvest failures and famines of the late 1430s.[86] But it needs to be stressed that although they brought death and great suffering for many, these episodes were of short duration; quick price reversion to lower, more stable levels helps to underscore that. The price levels of 1438 were not again reached – and surpassed – until the 1450s, another decade for which the historical record indicates acute shortages and widespread social disorder.[87]

Rather than continuous and unrelenting crisis, therefore, we see a gradually rising trend of socio-economic distress traceable from the 1420s. One symptom was an increase in able-bodied beggars, targeted by legislation in 1425,[88] which is suggestive of people failing to subsist lawfully at society's margins. Many were peasant farmers who became dislocated economic migrants and sought better lives in the burghs.[89] Insecurity of tenure worsened in times of subsistence crisis, when small tenants struggled to pay rents and feed their families, but landlords were also offloading landed property to relieve debt burdens and generate cash. Such conditions underlie legislation in 1450 that tried to give greater security to 'the poor people that labour the ground' and to protect their tenancies where there was a change of lord.[90] This was a response to people being driven from the land and into beggary, adding to the landless drifters whose existence so alarmed the kingdom's rulers. Indeed, its pairing with a law against 'sorners, overliers, masterful beggars' which empowered local officials to take draconian action against such men reveals a perception that these groups formed a real and growing threat to social order.[91] Sorning, the forceful imposition of yourself on another for your food and shelter, although linked to support mechanisms used by Highland chiefs to maintain military retinues, was a form of predatory behaviour that increased sharply at this time. In some instances, it was a planned offloading of the burden of maintaining militarised retinues in times of protracted dearth. Such behaviour is alluded to in a 1458 act against men who 'occupy spiritual and temporal lords' lands masterfully'.[92] Further action against vagrancy occurred in the aftermath of the poor harvests and

---

85 In the pre-decimal system there were 12 pennies in a shilling and 240 pennies in a pound.
86 Gemmill and Mayhew, *Changing Values*, pp. 153–153 Table 11. Barley prices are less plentiful but show a spike in the 1430s (p. 166 Table 14). Malt prices show a less pronounced spike in the 1430s but rose sharply in places like Dundee in the 1450s (p. 184 Table 17). Oat prices, which survive from a number of locations around Scotland in all periods, reveal some strong regional variations, suggesting better harvests in the north-east, for example, in the later 1430s but with eastern Scottish prices generally significantly higher than western in the 1450s (pp. 192–193 Table 20).
87 Gemmill and Mayhew, *Changing Values*, pp. 153–153 Table 11; p. 166 Table 14; p. 184 Table 17; pp. 192–193 Table 20.
88 RPS 1425/3/22 [accessed 20 January 2022].
89 That this was perceived as a problem focused on the burghs is confirmed by reinforcing legislation in 1428 designed to ensure compliance on the part of local officers with the requirements of the 1425 act: RPS 1428/3/5.
90 RPS 1450/1/16 and 1450/1/17 [accessed 20 January 2022].
91 RPS 1450/1/20 [accessed 20 January 2022].
92 RPS 1458/3/26 [accessed 20 January 2022]. R. Nicholson, *Scotland: The Later Middle Ages*, paperback edition (Edinburgh: 1978), p. 383. This issue is discussed in R. Oram and P. Adderley, 'Lordship and environmental change in central Highland Scotland c.1300 to c.1450', *Journal of the North Atlantic*, 1 (2008), pp. 74–84 and R.D. Oram and P. Adderley, 'Lordship, land and environmental change in West Highland and Hebridean Scotland c.1300–c.1450', in S. Cavaciocchi (ed.), *Economic and Biological interactions in Pre-Industrial Europe from the 13th to the 18th Centuries* (Florence, 2010), pp. 257–268.

political turbulence of the 1452–5 period, when economic necessity and the socio-political fallout from the fall of the Black Douglas family displaced yet more vulnerable people.[93] That similar legislation was repeated in 1478[94] amid successive poor harvests confirms that much of this vagrancy arose from those at society's economic margins slipping over the edge.

An insight into the social tensions arising from poor yields and landlord demands on tenants to continue to pay rents in kind is provided by an account of a court action at Perth in February 1457.[95] The complainant was the abbot of Scone, who alleged that cain[96] payments in wheat due to the abbey from the lands of Inchmartine in the Carse of Gowrie had 'not been paid of sufficient wheat, but of tares, impurities and other useless seeds' and he showed the court a sample of what his two principal tenants there had delivered to him. The contamination referred to may have been through field or corn marigold (*Glebionis segetum*), an invasive species which seems to have become a widespread agricultural pest in the later Middle Ages and which tenants were instructed to eliminate from their fields through the second half of the fifteenth century. From 1468, tenants on the Coupar Angus Abbey estate had the obligation inserted into their leases to keep the land free from 'guld', as the corn marigold was named in Scots.[97] A surviving section of a late medieval law code contains two clauses relating to guld, the first warning that tenants who allowed the plant to flourish on their land should be punished like someone who had led an enemy host into the kingdom, and the second setting a fine of one butchered sheep for every guld plant.[98] From this scale of fine, clearly guld contamination was regarded as a serious problem and it is these laws that were probably referred to in some Coupar Angus leases from 1478 and 1503 as 'the guld law'.[99] Men whose land was found to be 'foul' with the weed were fined heavily – 10 to 20 shillings – and then required to clear their land three times annually thereafter (24 June, 1 August and at the start of the harvest).[100] Whether or not the delivery of contaminated seed to Scone was simply sharp practice on the part of two local lords, who sought to capitalise on potential profits from selling more of their better-quality wheat on the market during a time of supply-driven inflation, or an act of desperation on the part of their tenants, who had insufficient foodstuffs on which to subsist if they paid their rents with the best grain, this case points to deficiencies in the quality of the grain yields in one of the kingdom's normally fertile districts. Between declining yields, poor-quality or contaminated seeds and severe penalties for permitting invasive weed species to proliferate on agricultural land, if these problems were replicated in more marginal areas, the displacement of tenants and drift into vagrancy hinted at in the parliamentary legislation become understandable.

Cumulatively this evidence points to sustained difficulties with supply of dietary staples and cycles of harvest failure, dearth and price inflation that for many Scots who lived at the socio-economic margins

---

93 RPS 1458/3/27 [accessed 20 January 2022].
94 RPS 1478/6/88 [accessed 20 January 2022].
95 RPS A1457/1 [accessed 20 January 2022].
96 Cain is an annual render in kind due to a lord from his peasant tenants.
97 Rogers, *Cupar-Angus Rental*, vol. 1, p. 143 (see index in vol. 2 under 'guld' for multiple examples).
98 T. Thomson (ed.), *The Acts of the Parliaments of Scotland*, volume 1, A.D. 1124–1424 (Edinburgh, 1844), Appendix V, pp. 750–751.
99 Rogers, *Cupar-Angus Rental*, vol. 1, pp. 213, 256.
100 Ibid., p. 228.

Field or corn marigold (*Glebionis segetum*) is nowadays encouraged in field-edge set-aside strips, but in the fifteenth century was viewed as an infesting pest that devalued grain fields. (Image ID: B3B9J0 WILDLIFE GmbH / Alamy Stock Photo)

fuelled a drift into vagrancy, violence and social disorder. A final dimension of the violence generated through these sustained stresses was the 'masterful occupation' of another's land mentioned above. Although the survival of the records of the courts of the Lords Auditors from the early 1470s to late 1480s skews perceptions of a sudden upswing in actions concerning arable encroachments on grazing land and commons, such behaviour was already a perceived social ill by 1458 and, by the act that empowered sheriffs and bailies locally to move against perpetrators, would have been heard previously in more local tribunals whose records are now lost. The legal ill they were addressing was clearly not itself new, but the spread of complaints – from the Borders, Lothian, Lanarkshire, Fife and Aberdeenshire – points to the extent of the stress within Scotland's agricultural zone. The number of such cases through the 1470s and 1480s supports the view that these decades saw continuing deficiencies in harvest yields in many parts of Scotland and multiple instances of 'self-help' or sheer opportunism to address that situation. As the fifteenth century drew to a close, the rural poor in Scotland were faced with choices: licit or illicit expansion of their agriculture to grow the food needed to subsist and pay rent; abandonment of their tenancies and a slide into vagrancy, which was then punished severely; or straightforward starvation.[101]

Expansion of the area under cultivation seems to

---

101 Nicholson, *The Later Middle Ages*, p. 572.

The northern edge of the Cheviot Hills, looking south from Cessford. The upper slopes and parts of the plateau are today used mainly for rough grazing, sport shooting and recreational walking, but the fifteenth century saw attempts to increase occasional cultivation of small arable patches to boost falling yields on lower-altitude fields.

have been a common response to falling yields. In some districts this expansion took the form of new assarts,[102] a process possibly reflected in the numerous and rising volume of complaints of unlicensed breaking of land – usually on pasture and in upland districts – and its ploughing and sowing by men from neighbouring communities. Practices that had evolved in some regions in previous centuries might have made assarting relatively straightforward, but entailed a loss – perhaps sometimes temporary – of land used for other purposes. In the Cheviot Hills, for example, the local employment on the upland grazing of a form of the widespread medieval practice of enclosing sheep and cattle at night in folds on the outfield for 'tathing', concentrating their dung and urine on these patches of grazing, probably created small blocks of fertilised ground that could be cultivated for one or two seasons before being left to revert to grass.[103] Such practices might lie behind the patches of high-altitude cereal cultivation of twelfth-century and later date identified in the palaeoenvironmental research project in the

---

102 An assart is a new intake of land for cultivation, cleared from woodland or pasture.

103 For folding, see R. Shiel, 'Science and practice: the ecology of manure in historical retrospect', in R. Jones (ed.), *Manure Matters: Historical, Archaeological and Ethnographic Perspectives* (Farnham, 2012), pp. 13–24 at 21; F. Watson and P. Dixon, *A History of Scotland's Landscapes* (Edinburgh, 2018), p. 87–88.

Bowmont Valley south of Yetholm, and evident also in the isolated blocks of cultivation noted by archaeological field survey around Southdean, 20km to the south-west of Bowmont.[104] Significantly, such cultivation techniques continued to be practised through the fifteenth and sixteenth centuries and, in some areas, into the nineteenth century.[105] It was perhaps patches of cultivation like these that the early fifteenth-century acts limiting the season for muirburn were partly intended to protect. What the Bowmont Valley evidence does not reveal, however, is whether the upland cultivation was perennial or represented short and discontinuous episodes of exploitation on ground temporarily enriched through tathing, opportunistic in the same way as its better-known application on outfields. What it also does not tell us is how productive these small areas of upland cultivation were and whether the yield levels justified the energy devoted to their production, especially when weather conditions were adverse. If the continued investment in their cultivation – and the apparent pressure to extend its scale

---

104 Tipping, *Bowmont*, pp. 192–194; P. Dixon, *Southdean, Borders: An Archaeological Survey* (Edinburgh, 1994), pp. 10–17.
105 Tipping, *Bowmont*, pp. 198–199. For comparative models of late medieval upland and marginal agriculture, see C. Shepherd, *The Late Medieval Landscape of North-East Scotland: Renaissance, Reformation and Revolution* (Oxford, 2021), chapter 2.

later in the fifteenth and sixteenth centuries – is in any way a measure of this, the answer is clearly yes. Oats and bere[106] will grow in poor weather conditions and at altitude, although the shorter growing season limits grain development and ripening and raises questions of yield volume and quality. The gradual proliferation of corn-drying kilns in late medieval rural settlements points perhaps to the harvesting of green or rain-saturated crops, and increasing references to poor grain quality or contamination with weed-seeds, as the abbot of Scone complained, probably illustrate the general experience of those living on the agricultural margins.

• • •

Although the evidence for deteriorating climatic conditions after 1400 explored in this chapter is fragmentary and often indirect, we can see that right across Scotland the fifteenth century marked the beginning of a new and progressively worsening era. It might not have been an age of endless winter, as the 'little ice age' label still tends to project into the popular imagination, but it was one marked by a protracted slide into colder and wetter conditions, in which obtaining satisfactory harvests became increasingly difficult and where old methods of production were clearly failing to meet the needs of even the plague-suppressed population numbers of late medieval Scotland. While chronicle accounts are sparse, parliamentary legislation, litigation and price data present a litany of failure and anxiety. Those least resilient in the face of successive crop failures and livestock losses faced a choice of vagrancy or starvation, and some chose to take what they needed from the better-off rather than face the consequences of a drift to the furthest social margins. In such nutrient-poor times, when the prices of grain staples were inflated beyond the reach of the poorest consumers, it is unsurprising that epidemic disease was rampant and continued to inflict a terrible toll on a nutritionally deprived and potentially immuno-compromised population. As we shall explore further in the next chapter, however, these were only the first of a series of environmental challenges faced by Scotland's people at the start of the 'little ice age'.

---

106 Bere is a six-row variety of barley (*Hordeum vulgare* subsp. *hexastichum*), still grown in the Northern Isles and some parts of the Western Isles.

CHAPTER 2

Turning up the Heat: Fuel, Wood and Food Supply in the Fifteenth Century

———

> Concerning the planting of woods and hedges and sowing of broom, the lords think it profitable that the king charge all his freeholders, both spiritual and temporal, that when making Whitsundays they statute and ordain that all their tenants plant woods and trees and make hedges and sow broom after the quantity of their mailings in places convenient, under such pain and unlaw as the baron and lord shall modify.[1]

Modern awareness of the multi-layered complexity of the impacts of climate change on all aspects of the environment upon which we depend has increased our understanding of the daily challenges and opportunities faced by the people of later medieval Scotland. While the onset of the 'little ice age' can be readily recognised in the reports of adverse weather conditions we explored in the previous chapter, and its impact can be read in the records of legal actions and legislation designed to curb abuses that were believed to be worsening the common weal, it is important to recognise that these were far from being the only indicators of increasing climate-related environmental stress. Colder and wetter conditions changed how the land could be used and the seas could be exploited, limiting the growing range of certain crops, shortening the growth season of grass, affecting the ability to secure adequate supplies of fuel, or reducing the number of days that could be spent safely in fishing grounds. People were perhaps slow to respond to these changing conditions, placing their faith in the tried and tested methods of their traditional environmental knowledge and praying – often literally – for a return to the more familiar discomforts of former times. But we must also recognise that people had the ability to make choices and to respond to new circumstances, for example in the types of crops they grew or how they managed their livestock, abandoning some practices and types of produce that their forefathers had exploited successfully in different conditions and adopting others that were better suited to the colder, wetter and greyer days of the early 'little ice age'. It was not all negative, however, for oppor-

---

[1] RPS 1458/3/28.

tunities also emerged, most notably in the new presence of shoals of migratory fish around the Scottish coast as changes in ocean temperatures led to movement of their feeding and breeding grounds. Most visibly, though, it is in the record of human responses to the unstable conditions of the fifteenth century that we see the greatest evidence for changes in practice and a move to give greater resilience. Legislation can be a blunt instrument and often fails to deliver the desired outcomes, but parliamentary acts from the 1420s onwards that aimed to stimulate the planting of woodland, for example, indicate that our ancestors also understood that change was a game played out over long periods. It is with those long-term shifts made in response to the climate transition of the 1400s and the resultant environmental changes that this chapter is concerned, with the focus falling on some of the resources most necessary for the maintenance of social cohesion amid the stresses of political, economic and environmental upheaval.

Fuel

After the provision of their dietary staples and the expenditure of labour on cultivation, the greatest outlay of human effort by our ancestors was focused on securing fuel for heat, light and cooking. Against a backdrop of colder conditions – occasionally intensely cold over extended periods – it would be unsurprising to find that consumption of firewood, peat and coals across all social classes increased. Domestic consumption, however, was only part of the equation, for environmental givens of earlier ages – such as the expectation that crops would be harvested fully ripe and dry – could no longer be taken for granted. As mentioned in Chapter 1, securing the necessary volumes of basic foodstuffs such as cereals to subsist across the following year required maximisation of the grain yield, regardless of its quality or condition. Cereal types that were more dependent on higher temperatures and moderate rainfall, like wheat, seem to have been planted less frequently and continued as important crops principally in regions such as Lothian, where better conditions were more likely, but elsewhere the trend was towards greater dependence on the cold- and wet-tolerant varieties of oats. If unripe and wet, all grains required to be dried to prevent them from rotting, becoming fungus-infested, or germinating, and to render them capable of being milled. Such things as corn-drying kilns, therefore, and the need for a greater expenditure of thermal energy in the colder, damper conditions of the fifteenth century, perhaps added to pressures on sources of adequate fuel that had been affecting some parts of the country since the later 1100s.[2] If they did, however, there is little clear evidence for any response to such pressure until the closing decades of the fifteenth century, when references to conflict especially over peat supplies begin to occur with greater frequency. It was, however, the sixteenth century before there is any sense of a widespread problem in securing adequate supplies of domestic fuel.

As with the language surrounding harvest failures and famine, it is perhaps going too far to refer in terms of a crisis to what was clearly an evolving situation that spanned centuries. Nevertheless, the provision of one of the fundamentals of daily living in the northern parts of Europe was an issue that affected – largely negatively – the everyday lives of the vast majority of the Scottish population at the most basic of levels. We need, however, to distinguish between two separate problems. First, there was a long history of perceived shortage of fuel in certain areas, mainly in proximity to long-established population centres with high levels of consumption. The issue here is ease of access to fuel rather than shortage. Second, both in such areas and in areas where fuel supplies remained relatively abundant, poor weather made

---

2   See the volume 'Scotland AD 400–1400' Chapters 8 and 10 for evidence of twelfth- to fourteenth-century supply issues.

Corn-drying kiln, Newton, Glenalmond, Perth and Kinross. This probably late seventeenth-century example indicates the effort, in both new construction and provision of appropriate fuel, that was needed to ensure that all grain that was harvested in the poor climatic conditions of the 'little ice age' could be made suitable for consumption by human populations.

the securing of adequate quantities of certain fuel types much more difficult. Just as we need to temper our views of harvest failures and dearth to recognise that food supply for many Scots was precarious in most years, so we should recognise that the securing of ample thermal energy sources to cook, heat homes or process materials was a perennial struggle that was marginally better or worse in any given year. It was not necessarily the case that there was no fuel to be had – in comparison to many parts of Europe, much of Scotland was energy rich[3] – but that the bulk of the supplies were located increasingly at a distance from the main places of their consumption, or that they were in a condition that made them difficult to prepare for use. Added to this was an inhibitor to the development of a commercial trade in fuel between supply-rich and demand-heavy areas, in that most rural peasant and lower-status urban households still met their annual fuel needs through the expenditure of their personal labour. In that sense, fuel was 'free', but the labour cost of winning it was high and increasingly so when weather deteriorated. For urban elites and the upper ranks of society generally, however, there was an increasing shift towards

3 Oram, 'Arrested development?'.

Cardoness Castle, Dumfries and Galloway. Although now lacking its great lintel stone the cavernous void of the chamber fireplace of the castle's principal room still conveys a sense of the conspicuous consumption of fuel – in this case probably wood billets or faggots – burned by the nobility in a display of their power and wealth.

satisfying their fuel needs through market supply, which in time stimulated the introduction of, and growing demand for, alternative fuel types, coal especially.[4]

As the well-known January and February scenes from the early fifteenth-century *Très Riches Heures* of the Duc de Berry illustrate, fires and the warmth they generated were important to rich and poor alike, for survival through winter extremes and for the essential processes of domestic, craft and industrial activity. For magnates like that French royal duke, or the king and greater nobles in Scotland, securing the fuel that was burned in the fires of their palatial residences was not a direct or personal concern; their stewards and their clerks were responsible for obtaining adequate supplies. For most peasant households, from the seemingly comfortably-off depicted in the *Très Riches Heures* to those poised on the subsistence knife-edge, however, ensuring that there was a sufficient store of fuel laid down to last through a full annual cycle was one of the biggest labour commitments. As the February *Très Riches Heures* scene

---

4 Oram, 'Environmental change, resource conflicts and social change', pp. 213–214, 215, 219, 220–223.

presents, when stocks ran low in the course of a period of extreme climatic deterioration and weather instability, finding any additional fuel, let alone readily combustible material, was a real challenge. In a country where peat was the primary fuel type, wet weather added to the difficulties in ensuring that there was adequate provision of such combustible material literally to keep the home fires burning, let alone continue craft and industrial processes upon which livelihoods depended.

From our climate records, the first change that would have had an adverse impact on most households was that it was colder and remained so for longer in most years across the century. There were some short periods of yet more intensely cold winters and delayed springs, followed by cooler and wetter summers, which increased and extended the need to secure adequate fuel supplies. Such conditions intensified through the 1470s and 1480s, which correlates with the increase in expressed concerns over fuel provision in the historical record. What such conditions translated to in the lived experience of the majority of Scots is not noted in the records of the time, beyond the eye-catching comments of Walter Bower in his descriptions of the extreme winters of the 1430s.[5] From his well-supplied, warm chamber, the abbot may have been wholly unaware of the reality of his abbey's tenants' experience of living through the extended winters and springs of the bitterly cold northerly winds that he reported. It is only in the more abundant records from later centuries, discussed in the following chapters, that we are provided with the graphic illustrations of the problems of peat dependency in an era of wetter summers and colder winters. Quite simply, wet peats will not burn and saturated peat will lose its cohesion and disintegrate, making its transportation and storage even more difficult. Poor cutting, drying and storing practices added to the difficulties, again as

'February' from the *Très Riches Heures du Duc de Berry* (Limbourg Brothers, public domain, via Wikimedia Commons), showing the peasants' struggle for warmth and the means to provide it during the extreme winters of continental Europe at the start of the 'little ice age'.
(https://upload.wikimedia.org/wikipedia/commons/0/02/Les_Tr%C3%A8s_Riches_Heures_du_duc_de_Berry_f%C3%A9vrier.jpg).

illustrated in the record from later centuries of the repeated efforts to regulate such things and prevent the physical degradation of the mosses from which it was cut.[6] It was not that peat was unobtainable, the problem was more one of ensuring that you had

---

5 Bower, *Scotichronicon*, vol. 8, pp. 117, 267, 293.
6 Oram, 'Abondance inépuisable?'.

cut, dried and stored sufficient quantities for your likely annual needs.

As discussed in the volume 'Scotland AD 400–1400', the records of earlier centuries have already shown that individuals and communities in lowland areas especially had been adopting greater mixes of different types of fuel to ensure that adequate amounts of thermal energy sources were available across the year.[7] At Edinburgh, ready access to local supplies of wood and brush was already restricted by the twelfth century, which meant that the general population had become heavily dependent on peat by the 1200s. As the closest accessible mosses to the burgh had become depleted in the fourteenth century, those who could afford it had begun to burn coal.[8] Similarly, Perth's access to fuel of all types had become problematical before the end of the thirteenth century, and the absence of coal sources in the burgh's immediate hinterland meant that its use remained restricted to the wealthiest members of the community into the nineteenth century.[9] Large coastal towns such as Dundee and Aberdeen, to which coal could be carried in relative bulk from the Lothian or Northumberland coalfields already by the thirteenth century, had perhaps started an earlier transition to a greater mix of fuel types from multiple sources.[10] That security, however, was not available to smaller and more distant communities, where transport costs and the absence of a significant number of consumers who could afford the higher price fuel kept demand limited. While great lords, including the king, higher nobility and senior clerics, could afford to have coal transported to their residences, the bulk of the population even in locations close to the mines remained heavily dependent on peat.

Dependence on peat does not mean that it was used to the exclusion of all other fuels. Mixes of fuel types employed in some lowland contexts are recorded in property grants. On the Erskine lands in Clackmannanshire in 1441, for example, a grant of a small tenancy near Alloa included windfall wood from the Forest of Clackmannan, plus peat and turf in the local moss.[11] These three different fuel types had widely varying calorific values and thermal efficiency and may have been used in different domestic contexts for different purposes. Turf, for example, the poorest thermal energy source, was perhaps used for keeping domestic hearths burning overnight or when more intense heat and flame was not wanted; wood was used for fire starting and where quick heating at a higher calorific output was needed; peat provided the main thermal output for cooking and heating in most households. Charcoal was also available but was reserved for technical usage – for example, metalworking – where the highest temperatures were required. Archaeological evidence from Perth suggests just such selection of fuel types for different purposes in an urban context, between peat, turf, smallwood and charcoal.[12] Even in the extensive zone of blanket peat coverage through the Southern Uplands, Highlands, Hebrides and Northern Isles, mixes of types were employed, including dried seaweed and cattle or sheep dung, although the bulk of ash deposits are from peat and turf, confirming their predominant use. The later fourteenth-century

---

7 Oram, 'Environmental change, resource conflicts and social change'.
8 Ibid., pp. 213–214.
9 Ibid., pp. 214–216, 218. For a more general discussion of later medieval experience of general shortages, see R.D. Oram, 'Perth in the Middle Ages: an environmental history', in D. Strachan (ed.), *Perth 800* (Perth, 2011), pp. 19–28.
10 Oram, 'Environmental change, resource conflicts and social change', pp. 216, 222–223.
11 NRS GD124/1/527.
12 G.W.I. Hodgson, C. Smith et al., *Perth High Street Archaeological Excavation 1975–1977*, Fascicule 4: *Living and Working in a Scottish Medieval Burgh. Environmental Remains and Miscellaneous Finds* (Perth, 2011), p. 84.

house at Bornais in South Uist illustrates this kind of diversity, with peat dominating in hearth deposits, but with wood, dung, cereal-processing waste (chaff, husks etc.) and other dried plant remains identified.[13] The wood included probably a small amount of material sourced from the limited patches of woodland that still existed in the islands by that date, but the majority was burnt fragments from wooden utensils or structural timbers that had reached the end of their recyclable potential, or dried heather that was probably cut and stored chiefly for fuel use. Even with such diversity of potentially combustible materials, however, the worst years of the fifteenth century perhaps pushed many households to the uttermost margins of their capacity to secure the volumes of fuel necessary to subsist with any degree of comfort.

It is likely that the ability to obtain fuel varied widely on a regional basis and locally within regions. Using the eighteenth-century rule-of-thumb of an average of 200 'loads' of peat being needed annually to fuel a single domestic hearth and with each 'load' approximating 1 cubic metre, we can gain some sense of the level of inroad that three centuries of cutting and consumption would have had on the mosses around most urban communities. Larger household operations, for example monastic communities, received substantially larger supplies, as in the forty cartloads of cut and dried peat to be supplied to the Dominicans of Perth annually,[14] equivalent to around 360 cubic metres of peat, as a supplement to their market purchases of fuel. It is likely that the thousands of cubic metres of peat cut annually from the mossland closest to most urban centres had seen these convenient sources depleted long before 1400, adding significantly to the time expended on travelling to and transporting materials from the peateries. Deteriorating weather in the fourteenth century as much as shortages locally may have stimulated the transition from peat to coal in the salt manufactories around the Forth estuary.[15] It is likely that the shift in emphasis from the mainly peat-burning sleeching sites on the carses of the upper tidal zone to the coal-burning sea-water boiling operations of the outer Firth accelerated through this period, aided possibly by a reduction in the supply of Biscay salt caused by the poor conditions for solar evaporation on the Atlantic coasts of Brittany and Gascony. Increased estate coal output to feed the saltpans seems to have produced a surplus that some merchants saw as an opportunity to supply the fuel-hungry burghs. Thus, mines around Tranent and on the south Fife coast were shipping some of their product to Edinburgh before the end of the fourteenth century.[16] It was perhaps this response to new opportunities that saw the beginning of the switch in priorities for coal supply from saltpans to urban consumers, with the burgh markets receiving the higher-quality large coals and the salters receiving the smaller, less marketable pieces of coal that came to be referred to as 'panwood'.

Edinburgh's richer burgesses were among the more fortunate, however, for communities located in the interior of the country, far from the coalfields, faced an uphill struggle in identifying and securing access to fuel sources as their traditional supplies became depleted. Some indication of the stresses on fuel supply in eastern Borders districts, for example, is provided by the steady increase in records of access to new sources being sought by consumers and of disputes between property-holders over boundaries that ran across areas of valuable peat. An illustration

---

13 Sharples, *A Norse Farmstead in the Outer Hebrides*, p. 179.
14 *Regesta Regum Scottorum*, volume 5, *The Acts of Robert I 1306–1329*, ed. A.A.M. Duncan (Edinburgh, 1988), pp. 173, 227.
15 R.D. Oram, 'The salt industry in medieval Scotland', *Studies in Medieval and Renaissance History*, 3rd series, 9 (2012), pp. 209–232.
16 Oram, 'Environmental change, resource conflicts and social change', pp. 220–222.

Knock of Alves, Moray, from the south. The low ground covered by harvested fields south of the wooded hillside, located 5km west of Elgin, had become the closest source of good-quality fuel available to the burgesses before 1500. All peat had been stripped to the underlying gravels within a further two hundred years.

of the search for new supplies is provided by a grant made by Robert Ker, abbot of Kelso, to his kinsman Walter Ker of Caverton on 1 October 1478. This award recognised Walter's general service to the abbey but it also singled out his gift to the monastery of rights to take fuel from the moor of Caverton some 6km south of Kelso, naming the fuel resources from there as 'le turf, pete, hathir, cole and brume', in return for which the abbot gave him the offices of justiciar and bailie of Kelso's baronial jurisdictions.[17] From the abbey's records, it appears that it had been exploiting the sources of fuel that had been granted to it in the second and third quarters of the twelfth century, principally on the moor of Ednam,[18] but those reserves had become severely depleted by the end of the thirteenth century, when the monks were obliged to reiterate their rights to, control over, and protection of, their wider fuel supply.[19] The value of the new peatery is revealed in its rapid employment as a source of supply to their key dependants as well as for their own needs. For example, access to the fuel from the Caverton source was provided in 1488 to support a chaplainry that Walter Ker had endowed in 1475 within the abbey church at Kelso, with a subsequent extension of the income assigned to the chaplain making up for a shortfall in his original

---

17 Historic Manuscripts Commission (HMC), *14th Report*, Appendix, vol. 3, *The Manuscripts of the Duke of Roxburghe* (London, 1894), no. 35.
18 For the Ednam award, see G.W.S. Barrow (ed.), *The Charters of David I: The Written Acts of David I King of Scots, 1124–53, and of His Son Henry, Earl of Northumberland, 1139–52* (Woodbridge, 1999), no. 183.
19 A.A.M. Duncan, *Scotland: The Making of the Kingdom*. The Edinburgh History of Scotland, volume 1 (Edinburgh, 1975; paperback edition 1978), pp. 342–345 for careful recording of peat-supplying and carriage dues required of tenants, carriage of coal from Berwick, etc.

provision and allocating him an allowance of fuel from the recently acquired mosses.[20] Close to fuel-hungry Perth, a second illustration is provided by records from July 1483 of a conflict between the abbot of Scone and a local laird, Thomas Wardrope of Gothens, over possession of a peat moss at Arnbathie in the Sidlaw Hills north-east of the burgh.[21] Here, the basis of the dispute was ostensibly a clash between Scone and Wardrope over the extent of rival ownership in a moss that lay on the boundary between their respective properties, but an underlying issue seems to have been the latter's interest in its commercial value for supplying peat to nearby and fuel-poor Perth. Whatever the driver for the conflict, the root of the dispute lay in competition for an increasingly scarce commodity within relatively close proximity to a source of established and growing demand.

Even where fuels other than coal remained at least notionally abundant throughout the medieval period, the prevailing weather conditions added to the effort needed to ensure that adequate supplies were obtained. Removing the water content of peat and dung is an essential stage in the process of preparing them for the fire, and even heather and wood require to be dried to enable them to be burned easily. Clearly, conditions were never so extreme that no fuel could be dried at all and the absence of evidence either for wider conflict over supplies, or of significant numbers of cold-related deaths, should caution us against exaggerating the depth and reach of any supply crisis. Instead, we should see the provision of thermal energy sources for household uses as an increasing problem – indeed, as a growing struggle – in a broad spectrum of difficulties that added to the grinding burden on most peasant families in their daily effort to subsist in the first extreme episode of the 'little ice age'.

Grassland, grazing and livestock

While we nowadays most often see peat as an element of upland environments, even in this era of diminishing supplies few communities in the late Middle Ages would have been obliged to seek their fuel in the hills. The principal mossland areas exploited for fuel by urban and lowland rural populations were raised bogs in low-lying districts, few of which have survived relatively unscathed into the modern era and many of which were drained or stripped in the eighteenth and nineteenth centuries or, more recently still, planted with trees. Rather than for its possible fuel content, for most of Scotland's people in the later Middle Ages, the upland zone was perceived as the location of the main areas of pasture, usually exploited seasonally. While grass will continue to grow in all but the coldest of conditions – nothing beneath a minimum 6°C daytime threshold – it is far from clear what general impacts the deteriorating conditions of the fifteenth century had on the wider range of upland vegetation across Scotland. Localised studies, such as Bowmont, have identified continuity of trends that had commenced in the fourteenth century, mainly arising from increased soil acidification; they do not suggest any rapid or profound changes in plant ecologies or human exploitation regimes. Acid soils, however, encouraged the spread of less productive vegetation, manifest as a decline in grass cover and further expansion of expanses of *Calluna* heath that had begun to develop by the thirteenth century.[22] There was also evidence

---

20 HMC, *14th Report*, Appendix, vol. 3, *Roxburghe*, no. 20.
21 NRS GD68/1/8. For the position of the moss on the boundary between the lands of the regality of Scone and the Gothens property, see NRS GD190/3/52.
22 Tipping, *Bowmont*, pp. 195–196.

Sidlaw uplands looking west from Pole Hill towards Arnbathie and Oliverburn, Perth and Kinross. Nowadays mainly agricultural land, the lower ground beyond the lochan was a disputed peatery in the fifteenth century, fought over by competing landlords.

for increased waterlogging of soils and associated 'bog burst' and gulley erosion at higher altitudes, but this again was not linked to significant changes in the local vegetation or in modes of human exploitation.[23] Indeed, the Bowmont project saw no clear evidence for any widespread ecological changes or for impacts on the human population and its use of the grazing land before the later seventeenth and eighteenth centuries. What has not been established through such projects, however, is the effect of the general cooling of the fifteenth century and later on the rate and volume of growth, especially during extreme episodes when a northerly and easterly airflow prevailed. In such conditions, when spring started later and autumn came early, the growing season for grass in traditional shieling grounds in Highland corries or on the sheep and cattle ranges of the Southern Uplands was shortened. The same suite of vegetation remained but perhaps in less lush abundance than in the preceding warmer era. With daytime temperatures rising above that critical grass-growth threshold of 6°C for fewer days annually, over time herders perhaps changed the date of the traditional move onto and from summer pastures. In addition, they possibly revised down their assessment of the carrying capacity of pasture, or risked the degradation of their summering grounds and the poorer nutrition and health of their livestock. It is unclear, however, if the increased variability in weather across the century established long-term trends that would have been immediately visible to farmers or encouraged novel responses from them.

23 Ibid., p. 197.

Most, it seems, tried to persist with the tried and tested practices of their inherited environmental knowledge.

In south and south-eastern districts, rather than innovation there was apparently continuation and perhaps even intensification of traditional grassland management methods. Fire continued to be employed as a mechanism to stimulate new growth on upland grazing, as it had been perhaps for centuries before 1400, but the climatic shift around that date possibly encouraged its wider use when drier periods allowed.[24] The short-term benefits of burning were understood, but, as discussed in the volume 'Scotland AD 400–1400', the long-term negative impacts in respect of waterlogging and acidification were not. Extensive use of muirburn may have accelerated a slow decline in grass quality. Concerns over its impact seem to have been voiced by major landholders through general council and parliamentary meetings by the end of the fourteenth century, presumably reflecting direct experience of loss arising from this practice or the complaints brought before them by their tenants. Practical considerations directed at better management and regulation of the practice, rather than outright banning of something that was still regarded as generally beneficial in terms of stimulating growth, probably drove the introduction of parliamentary legislation to limit the use of muirburn to defined seasons. The fire risk to the areas of upland cereal cultivation that are known from the south-eastern parts of the kingdom seems to have been a significant concern. A first attempt to regulate

24 Ibid., pp. 172, 194–197.

Muirburn on the Liddesdale moors, Dumfries and Galloway. Seasonal burning of the uplands is a still-controversial practice, now employed mainly to stimulate new growth of heather on grouse moors.

burning of upland muirs and heath was made in the parliament of February 1401, when an act became law limiting the period of muirburn to the month of March only and imposing a 40-shilling fine on transgressors.[25] Uncontrolled muirburn and the damage it could cause was regarded as a serious issue. This can be seen in the additional measures included in that act to ensure that the lord of the affected land took action – the fines collected were awarded to them as an inducement – but with higher authorities empowered to step in if the competent local court failed to act. In May 1424 and with the legislation repeated in March 1458, parliament again prohibited the starting of fires for muirburn from the end of March until after harvest time, with tightened and strengthened penalties for those convicted of starting fires illegally.[26] There is an implicit understanding in these reiterations that the fires were most likely set by people of low socio-economic status who lacked the means to pay the fines imposed. That they were to be imprisoned instead for forty days points to impoverished farmers at the lower end of the peasant

25 RPS 1401/2/15 [accessed 21 January 2022].
26 RPS 1424/22, 1458/3/39 [accessed 21 January 2022]. These acts were repeated as late as 1685, usually associated with other 'ills' such as the burning of mosses and the illicit installation of salmon cruives and nets.

Controlling and regulating muirburn

Periodic burning of grassland, heath and moor was intended to remove old tussocks, dead vegetation and scrubby *Calluna*, and to stimulate the growth of fresh fodder for grazing animals. As the many recent examples of wildfire devastation across Scotland illustrate, however, there was the risk of fire's uncontrollable spread and the destruction of crops, woodland, wildlife and homes that lay in its path, especially after periods of prolonged drought and during episodes of strong winds that fanned the flames. Efforts to control and regulate muirburn punctuate the parliamentary record, setting down basic conditions allowing the practice and providing the legal framework for the competent local courts to take action against tenants who breached those conditions or caused loss to others. The 1458 act formed the basis of all control action down to 1746, when private local jurisdictions were swept away.

> Item, where it was decreed in a parliament of our sovereign lord that last died [James I] on 26 May 1424 that no man make muirburn after the month of March until all the corns are shorn, under the pain of 40 s. to be raised to the lord of the land of the burner, and if he has nothing [with which] to pay, he is to be punished with 40 days in prison, and if the lord of the land does not raise the pain, nor punish the trespasser as said, the justice clerks by indictments shall cause such trespassers to be corrected before the justice and punished as is written before, in this parliament the estates approve this statute and ordain it to be kept in time to come (RPS 1458/3/39).

In this short enactment we can see summarised the 'closed' season for muirburn, lasting from the end of March to the variable date of the end of harvest time annually. This was essentially the period when desirable crops were growing and when lowland grass areas had resumed growth after winter dormancy. Although it is not explicit in the act, the aim was to protect the resources upon which all subsisted, from the land's lord down to the humblest tenant, with the size of the fine and the term of imprisonment pitched at a level that would have given even the wealthiest peasants cause to pause.

spectrum, possibly lacking rights to a share in the better-quality common grazing, who were willing to take the risk in an effort to stimulate grass growth on the moors and uplands to feed their small flocks and herds.

Even without the stimulus of muirburn, grass will continue to grow in most circumstances, provided the minimum growth temperature threshold is reached. Cattle and sheep, however, do not necessarily thrive in marginal grass-growth conditions, and availability of adequate fodder has long been recognised as having had a major influence over age compositions of flocks and herds and the development in the Western Isles of practices such as autumn calf-slaughter.[27] The slaughter of calves is generally understood as a means of extending and maximising milk production in societies where dairying dominates food production, as occurred in many parts of upland and Hebridean Scotland. It was a response to limited availability of outside grazing opportunities for overwintered livestock or the shortage of fodder for beasts kept indoors, but it was also

---

27 F. McCormick, 'Calf slaughter as a response to marginality', in C. Mills and G. Coles (eds), *Life on the Edge: Human Settlement and Marginality* (Oxford: Oxbow Books, 2006), pp. 49–51.

influenced by the availability of summer grazing opportunities. Livestock were grazed on summer pastures for two principal reasons other than simply keeping them away from growing crops: for gaining body-mass for meat or for supporting milk production for dairying. Both are affected by poor weather, especially where wind-chill is worsened by wet conditions, when the animals use a greater proportion of their daily calorific intake on maintaining body heat.[28] The consequent energy deficit – where the animals expend more energy than the calories obtained through their feed – has an impact on fertility and gestation, with an increased loss of foetuses in cold and undernourished cows. Animals that have been exposed to poor weather through the summer months are likely to enter winter in poor condition, which increased over-winter mortality levels or encouraged radical autumn slaughtering solutions to permit the limited available resources to be devoted to the best stock. The cooler and wetter summers, delayed springs and shortened autumns of the 'little ice age' represented a veritable perfect storm for the cattle economy of Atlantic-facing Scotland, where the average annual temperature drop might not have been sufficient to prevent grass growth, but, worsened by wind-driven rain, was more than sufficient to affect animal welfare. So, while there was no recognisable shift in the plant ecologies of the upland pastures and Hebridean grazing land, or in the efforts of the human population to exploit them, the general condition of the animals grazed there probably declined.

Were the episodes of cooling sufficiently long for any decline in biomass production to have become appreciable and entrenched anywhere in Scotland? The Bowmont Valley example suggests that it was not until the nadir of the Maunder Solar Minimum in the last quarter of the seventeenth century that any significant change in management practices occurred there. Elsewhere, we lack comparable data to identify general trends. It is, however, unlikely that grazing in areas further north and west was unaffected by increasing Atlantic storminess, cold and wet. In lower-lying areas of the Hebrides, the buffering effect of the ocean delivers a generally mild winter climate in terms of temperatures; grass growth in North Uist extends over ten months in modern conditions. Foggage (feeding on the dead grasses and decaying straw left on the land over winter) on the cultivated zone of the Uist machair, where there was grass growth occurring after the cereal harvest, reduced the need to find alternative sources of winter fodder, and the milder conditions in the islands meant that animals could in theory be left outside for most of the year to maximise the benefit of this grazing resource.[29] But greater storminess and increased rainfall in summer and winter created the conditions that delivered a greater risk of energy deficit in the livestock, reducing the gain from the availability of the winter fodder. In the north and north-west Highlands and the exposed uplands of north-eastern Scotland, this same deterioration would have been replicated, perhaps giving us some insight on one important underlying factor that generated the economic pressure that fed into the political instability of these regions through the fifteenth century.[30]

Although there is a generally poor impression of

---

28 Ibid., p. 50.
29 D. Serjeantson, *Farming and Fishing in the Outer Hebrides AD 600 to 1700: The Udal, North Uist*, Southampton Monographs in Archaeology, New Series 2 (Chandlers Ford, 2013), p. 30.
30 For discussion, see Oram and Adderley, 'Lordship and environmental change in central Highland Scotland'; Oram and Adderley, 'Lordship, land and environmental change in West Highland and Hebridean Scotland'; R.D. Oram, 'Between a rock and a hard place: climate, weather and the rise of the Lordship of the Isles', in R.D. Oram (ed.), *The Lordship of the Isles* (Leiden, 2014), pp. 40–61.

Rain over the North Uist hills. The higher rainfall, colder conditions and stronger winds coming in from the Atlantic brought major challenges for livestock farmers in the Western Isles, leading to the adoption of new techniques to manage their herds' health and viability.

meadow quality and hay production in the pre-Improvement era, throughout the Middle Ages grass was cultivated for harvest as hay to augment foggage. Hay cultivation was practised mainly in lowland areas and zones where there was a cultural tradition of hay production, like in the heavily Scandinavianised areas of Caithness and the Northern Isles. Dealers in hay and fodder were present in most Lowland towns by at least the early fifteenth century, when the fire risk of their stores in the still largely timber- and-thatch urban built environment was recognised by act of parliament.[31] Edinburgh's market for hay, straw, grass and horse 'meat' (oats) was fixed in 1477 to a location at the west end of the Cowgate that is still called the Grassmarket.[32] The main customers of this fodder market were commercial stable operators who in 1508 were commanded by Edinburgh's council to keep adequate supplies of 'guid stufe and fresche' for sale at 4 pennies per stone weight.[33] The crown, magnates and great religious corporations, however, maintained meadows for their own fodder needs or leased them profitably to tenants. As in earlier centuries, these meadows were in natural low-lying wet areas or seasonal wetlands – as in the bishop

---

31 RPS 1427/3/8 [accessed 27 January 2022].
32 Marwick, *Edinburgh Burgh Records*, vol. 1, p. 34.
33 Ibid., p. 117.

Howmore machair, South Uist, looking across the machair and blackland to the mountains. Foggage, like the stubble in the field to left of centre, provided essential feed for overwintering cattle.

of Dunkeld's meadows at Dalguise on the Tay flats north of Dunkeld, in the wetlands around Cardney, in the marshy ground at the Loch of Clunie and at Kinnoull on the river at Perth[34] – rather than being deliberately flooded water-meadows of the southern English and European variety. These seasonal meadows, although subject to the vagaries of the floodwaters, were nevertheless productive and clearly valued by their owners. In the 1430s to 1450s, for example, royal financial accounts show payments for cutting and winning 41 acres of hay-producing meadow on the wet carse of the Forth just west of Stirling; other royal meadows are recorded at Row near Doune, Restalrig, Logie, Frew, Falkland, Edinburgh, Dundonald and Auchtermuchty.[35] All provided hay primarily for stables at the nearest royal residences rather than for the king's flocks and herds more generally. Leases into the mid sixteenth century on the Coupar Angus properties in Glenisla, however, required tenants to provide 'meit to our scheip in hay and foddir' to supplement foggage for the monks' flocks.[36] This stipulation was a late refinement in the

---

34 R.K. Hannay (ed.), *Rentale Dunkeldense. Being Accounts of the Bishopric (A.D. 1505–1517) with Mylns 'Lives of the Bishops' (A.D. 1483–1517)* (Edinburgh, 1915), pp. 82, 91, 99, 107, 109, 122, 124, 132, 164, 168, 173, 188, 190, 204–205, 211, 217, 276–277, 284.
35 *ER*, vol. 4, pp. 592–593; vol. 5, pp. 219, 309, 346, 385, 397, 473, 475, 479, 480, 535, 590, 595, 616, 635, 677, 689.
36 Rogers, *Cupar-Angus Rental*, vol. 2, p. 261.

abbey's leases, possibly made in response to difficulties in securing adequate foggage or fodder in the glen's higher-altitude pastures over winter during the worst years.

An indication of shortages of fodder is perhaps to be seen in reports of thefts of gathered amounts of animal feed, but it is unclear if this increase in references to fodder thefts was a consequence of a general shortage of supplies or an artefact arising from the existence of a set of records from a new court system established in the reign of James III that heard such cases. The courts of the Lords Auditors in the 1470s to 1490s dealt with multiple instances of theft or destruction of 'threaves' of fodder (1 threave = 2 stooks = 24 sheaves), but the circumstances of these losses are not always explicit.[37] Regardless of why this phenomenon occurred, what is clear is that wherever possible owners kept meadows within their direct possession, were at pains to ensure that any alienation was for time-limited periods or carried residual rights for them and their heirs, and pursued vigorously any encroachment on or theft of their produce. Thus, for example, in 1455 the laird of Swinton in Berwickshire gave his local vicar a life-rent of a meadow in return for daily prayers for him and his family, but with the property explicitly to be returned to the Swintons when the vicar died or demitted either the vicarage or the

---

37 See for example, the 960 threaves of fodder spoiled from Lord Drummond: RPS 1490/2/83 [accessed 27 January 2022] or the 100 'fodders of hay' destroyed in Calder: RPS 1479/3/125.

The Dean Water at Grange of Aberbothrie, Perth and Kinross. Post-medieval reclamation and improved drainage have largely obliterated the seasonal wetlands that flanked the river as it flowed through land attached to the main granges of the Cistercian abbey at Coupar Angus. It was from here that the monks received the bulk of their hay needs.

meadow specifically.[38] In 1410 the Scotts of Balwearie in Fife, who also held property in south-eastern Perthshire, granted two acres in their meadow at Flawcraig in the still boggy northern edge of the Carse of Gowrie to the Dominicans of Perth, requiring the friars to secure their permission before subletting it to tenants.[39] In both cases, the granter retained reversionary rights in a way that was rarely done with other forms of real property, underscoring the probable scarcity of meadow and the high value of the hay it produced in an era when grazing quality was diminishing as the climate deteriorated.

Woodland, trees and timber

While meadow and hay were highly prized resources, trees and timber held even greater value in those parts of the country where woodland had been scarce for centuries already. As explored in the volume 'Scotland AD 400–1400', the extent of woodland in any part of Scotland across the whole of the medieval period is a subject that invites much speculation based on very little substantial evidence. The frustrations of our medieval records plenty of references to woodland and to timber but no reliable

---

38 NRS GD12/38.
39 NRS GD79/1/15. The Scotts' permission was secured in 1460 when the Dominicans gave Flawcraig meadow to the neighbouring laird of Fingask: GD79/1/24.

quantitative data that indicates its extent – were a contributory factor behind the fixing of the start-date of the most recent major environmental-historical analysis of Scotland's native woodland at 1500.[40] As outlined in the overview of the extent and character of Scotland's native woodlands before 1500 in that study, the perception of timber shortages throughout the eastern Lowlands that is implicit in records of disputed ownership of woodland, measures to preserve specific stands of trees, or to encourage the development of a long-distance trade in wood from the early thirteenth century onwards grew in the late fourteenth century and generated a formal parliamentary response in the 1420s. This legislative response occurred before the impacts of the deteriorating climatic regime of the fifteenth century could have made their presence felt as limiters on woodland regeneration. That is strongly suggestive of a longer-term decline that had at least as many anthropogenic as environmental causes and, especially in the Lowlands, in the areas of most intensive agricultural expansion and rising population, was almost entirely a result of centuries of human action. Certainly, there was a deeply ingrained belief among Lowland chroniclers, poets and legislators – perhaps reflective of a wider popular tradition – that the Highlands were abundantly provided with woodland, at least in comparison to their own districts.[41] The laws are also an indication of where the need was felt most keenly; it was a continuing issue of supply of wood and theft or destruction of a valuable commodity that was legislated for, not any explicit recognition that a remedy might be the planting of more trees. In the March 1425 parliament of King James I, two acts were passed which addressed aspects of wood thefts.

The first instructed the justice clerk, when he received indictments, to make specific inquiries about men who stole green wood, or peeled the bark from trees, so destroying the woods. Anyone who was convicted of such a crime was to pay a fine of 40 shillings to the king for the unlaw, and also make reparation to the owner of the trees.[42] This was a significant level of fine and was clearly intended to be punitive and preventative. It was paired with a second act concerning the discovery of stolen wood in the land of another lord and provision for its return and for the trial of the trespasser in the court of the injured party, i.e. ensuring that he rather than the lord of the trespasser received the fine for the trespass,[43] which indicates belief in a newly significant and increasing problem of thefts from growing woodland. Both laws treated trees and woodland more generally as elements of real property for which compensation was paid for damage inflicted on them, their illegal felling or theft of their wood. The intention of the 1425 act was to empower lords to protect their woodland resources or secure appropriate compensation; law was here a reactive remedy for a perceived problem, not a proactive solution for the underlying shortages that were drivers of illicit encroachment on areas of woodland.

It was another generation before the question of planting to reduce dependence on foreign imports and confront the problem of illegal felling was addressed directly, in a parliamentary act of 6 March 1458. That act was the first parliamentary legislation anywhere in Britain aimed explicitly at encouragement of tree-planting, and also planting of hedges and sowing of broom. Its date might not be purely providential, for it would have been in the 1450s that

---

40 T.C. Smout, A.R. MacDonald and F. Watson, *A History of the Native Woodlands of Scotland, 1500–1920* (Edinburgh, 2005): see chapter 2 for an overview of the palaeoenvironmental, archaeological and historical explorations of the pre-1500 period.
41 Smout et al., *History of the Native Woodlands*, pp. 39–41.
42 RPS 1425/3/11 [accessed 17 January 2022].
43 RPS 1425/3/12 [accessed 17 January 2022].

the suppression of tree growth – both natural seeding and germination of new trees and regeneration of coppice – that commenced with the onset of exceptionally cold and wet conditions in the 1430s would have become fully evident (see Chapter 1, pp. 14–27). Under its provisions, new leases were to be granted only on the condition that the tenant planted a prescribed number of trees and in-demand scrub species.[44] Although its subject-matter was clear, like the earlier legislation it was studiously vague in setting out how this new policy would be implemented locally. It is far from prescriptive, stating simply that the lords of parliament had deemed it profitable that the king charge all his freeholders, both spiritual and temporal, that when making new tenancy agreements they were to require their tenants to 'plant woods and trees and make hedges and sow broom after the quantity of their mailings in places convenient'. Essentially, there was no clarity around required extents of planting or its location, only a requirement that it was to be in some way proportionate to the size or value of the property-holding and located in places within it that were 'convenient'. Convenient for what is not defined, but we should probably assume that it was on land that was not otherwise given over to another productive use and was best located for the benefit of the occupants of the property. The only reference to enforcement was that the landholder could apply 'such pain and unlaw as [they] shall modify' to tenants who failed to adhere to this requirement. The fruits of those sanctions were presumably the incentive on the landholders' part to apply the act rigorously and on the tenants' part to comply with that demand.

In common with most Scots laws where the subject-matter was property, the 1458 act empowered landholders to take particular action but it did not obligate them to do so. Neither king nor parliament wished to create circumstances that could be viewed as interference in heritable private property and jurisdictional rights or the freedoms of the landholder to deal with those properties and rights as they saw fit. It is characteristic of legislation that persists into the present. As with much legislation in a kingdom where legal franchises were widespread, this was effectively an enabling act that the holders of regality, barony and free forest jurisdictions could apply as rigorously as they chose. With the rights to exercise the jurisdictional powers contained within the law delegated to men across the kingdom, the crown was here providing a carrot to landholders to address a problem that was widely recognised, but the only stick was for the landlords' use, being wielded against tenants who failed to comply with the terms of their leases. There is evidence for such action, as in the termination of a lease in the baron court of Keith in 1488, when the tenant was found to have grubbed up a considerable area of woodland in the barony, ploughed it and prepared it for cultivation, and then sowed a grain crop.[45] This forfeiture has been presented as an action against unauthorised assarting[46] during a period when extension of cultivation was one response to declining yields. The critical factor in the decision, however, was the wanton destruction of the woodland without the permission of the superior lord of the property. The act, therefore, was only as effective as the landholders who chose to apply it, tempered always by the judgement

---

44 RPS 1458/3/28 [accessed 17 January 2022].
45 G. Donaldson (ed.), *Protocol Book of James Young, 1485–1550* (Edinburgh, 1952), p. 29 no. 135. The case is discussed in J.M. Gilbert, *Hunting and Hunting Reserves in Medieval Scotland* (Edinburgh, 1979), p. 234, where Gilbert linked the forfeiture of the tenant to an enactment recorded in the early fifteenth-century Harleian MS of Scots laws (British Museum, MS Harleian 4700 (Old Scots Laws), f. 282r), which gave the power of forfeiture to lords against tenants who destroyed woods by burning or selling it and who built on the lord's lands without authorisation.
46 Smout et al., *History of the Native Woodlands*, p. 39.

of the tenants as to whether the penalties likely to be levied against them would be costlier than the burden of making the plantations. At Keith, the scale of destruction was apparently sufficient to trigger the intervention of the superior and was deemed sufficiently egregious to merit termination of the tenancy. Elsewhere, tenants may have calculated on a fine at a lower monetary value than their anticipated profit from the extended area of cultivation. Such considered trade-offs must have been manifold across the kingdom, but they have left few visible traces in the historical record.

What effect might the 1458 act have had on landlords' policies towards planting on tenanted property? Although it is not referred to specifically, there is some correlation between the passing of the legislation and the appearance from the 1460s in lease terms of specific requirements on tenants to plant trees. In what look very much like later 'improving' leases, dated June 1468, the monks of Coupar Angus stipulated that their tenant at Aberbothrie, Cotyards and Coupargrange in Strathmore were to plant ash trees and osiers or willows.[47] It was only in 1472, however, that there is explicit reference to such a requirement for tree-planting on tenanted lands as being an ordination of the statutes of the abbot's court.[48] Frustratingly, the surviving record does not reveal if the abbot was using the powers delegated by the parliamentary legislation or had introduced a similar act in the abbey's barony court on his own initiative. By 1472/3, Coupar Angus leases required tenants to provide the trees with 'hanyng and defens',[49] perhaps indicating that contrary to the royal injunction in the 1458 act, the first plantations had been unenclosed and suffered through grazing or illicit wood-cutting by third parties. The monks placed particular value on the woodland at Campsie on the Tay 11km southwest of their abbey, yet here a 1474 lease permitted their tenant to graze an unlimited number of his cattle in the west part of the wood 'if only without damage to the wood'.[50] By a new lease in 1483, however, the abbey's tenants were instructed to keep Campsie woods clear of their cattle,[51] which might indicate that their earlier injunction had been fruitless. It emerges, however, that the woodland of Campsie was already enclosed by a wall, with tenant access to the walled area being regulated by the lease terms; in 1483 the tenants were required to build walls round half of the wooded area.[52] These 'walls' may not have been entirely of stone, for 2km from the abbey precinct at Keithick the monks' 1472 lease required the planting of hawthorn by their tenant to form 'hainings' (enclosures by hedges, ditches or walls), around the tree-plantations, following the 1458 act's terms.[53] North of Rattray at Middle Drimmie, the tenant was to put the land to 'al polici efter her pouar that is to say in biggyn and plantatioun of treys, eschis, ozaris, and sauch, with hanyngis and defens of thirs', while at Cotyards in 1473, in addition to the trees, the tenants were to plant broom parks – a second stipulation of the 1458 act – and orchards 'with hanyng and defensuris of the said plantacionis'.[54] Although there is no explicit evidence for their haining until 1541, that the broom plantations were described as 'parks' indicates that they were protected by dykes or hedges like other woodland plantation

---

47 Rogers, *Cupar-Angus Rental*, vol. 1, pp. 141–142, 150.
48 Ibid., p. 163.
49 Ibid., pp. 164–165, 166–167.
50 Ibid., p. 222.
51 Ibid., p. 237.
52 Ibid., pp. 227, 237, 242.
53 Ibid., pp. 162–163.
54 Ibid., pp. 167, 171.

Whitefield, Kirkmichael, Perth and Kinross. This ancient ash tree, one of six adjacent to the ruined sixteenth-century house of Whitefield, is possibly over four centuries old and reflects one laird's response to parliamentary acts requiring tree-planting.

Broom (*Cytisus scoparius*). Viewed largely as an invasive 'pest' since the eighteenth century, the brilliant yellow flowers and dark green stems of the Scottish broom were once a more common feature of the agricultural landscape. Grown for its fodder and fuel value, broom plantations and hedges were a significant new feature of the rural landscapes of fifteenth-century Scotland.

and tenants were instructed to keep their cattle out of the enclosed areas.[55] Sixteenth-century arrangements record that broom was planted as a crop on ground that had been used for two years for grain production and then seven years for broom, pointing to probable enclosure of parts of the outfield for these purposes. Although it was used at times as fodder or bedding in byres and stables, the main use of cut and dried broom was as fire-kindling or fuel,

one 1558 lease describing it specifically as 'for the hearths and ovens of the abbey'.[56]

Enclosures to protect woods were hardly innovative. One 1425 Melrose Abbey document describes a 'bra and dyke' (bank and ditch) with a 'yet closand and opynand for the caring of tymmer' (opening and closing gate for protecting (or carriage) of timber) around a wood belonging to the canons of Dryburgh.[57] Earlier charter material indicates that

---

55 Ibid., vol. 1, p. 246; vol. 2, pp. 16, 169.
56 Ibid., vol. 2, p. 169.
57 C.S. Romanes (ed.), *Selections from the Records of the Regality of Melrose*, volume 3 (Edinburgh, 1917), p. 224.

the enclosure had already existed before 1300, but it was clearly a well-maintained and still-functioning structure two centuries later. Such early enclosures, however, were the work of proactive landlords who had recognised the need to protect valuable components of demesne resources that would otherwise have to be secured on the open market; the fifteenth-century legislation shifted the obligation to protect plantations onto the tenants, who were made responsible also for planting the trees. The key word in the leases is 'policy', which in this context means 'improvement'. For Coupar Angus, planting of trees and broom and provision of hawthorns for fencing or hedging were intended to increase the quality and value of its properties, and to stimulate the wider economy through production of in-demand materials. The fragility of that improvement in a wider economic environment, where growing timber of any size was a coveted resource, however, might be reflected by the sudden inclusion from 1472 of protective requirements in lease formulae. For the 1458 act to have had a measurable impact, other landlords needed to be as proactive and responsive as the monks of Coupar Angus.

Although it was in the sixteenth century that the lamentations of the kingdom's rulers concerning timber shortages reached a peak of shrill anxiety,[58] the fifteenth-century legislation nevertheless exposes deep concern over a critical decline of domestic woodland. Stripping aside the rhetoric and polemic in the language of the acts and the leases, the legislation was a response to a long-term but progressive decline in the availability of mainly large timber-producing trees as opposed to coppice. Here, we might be seeing an unforeseen outcome of centuries of coppice management that had seen most woodland in Lowland districts at least reduced to blocks of smallwood coppice with isolated, mature standards that produced the seed for regeneration. As those standards aged, died or were felled, with most of the remaining trees having growth of less than twenty-five to fifty years, the reproductive or regenerative capacity of the woodland became increasingly precarious, worsened by the poor climatic conditions in which few tree-seeds set and where seedlings and saplings were especially vulnerable to weather extremes. Mature trees were also likely to have experienced increased damage, including major limb-loss, and widespread woodland loss is likely to have occurred during the more frequent and intense storms, as reported by Walter Bower. Summer storms, when the trees were in full leaf and waterlogged soils loosened the root systems, were likely to have had a greater impact than winter events. The greatest casualties are also likely to have been among those of seed-producing age, further diminishing the natural regenerative capacity of woodland.

With reduced availability of accessible large, timber-yielding trees, the importation of building timber that was already happening in east-coast ports in the thirteenth century had continued to expand in scale through the fourteenth and reached a new peak in the fifteenth century. The timber demands of James I's new palace at Linlithgow in the early 1430s and repairs at royal castles were such that wood imports in several east-coast burghs were taken for the king's work, like the fir boards obtained at Perth in 1434/5 or the Prussian planks bought at Leith for Stirling Castle.[59] But neither the developing legislation, the stringent lease terms, nor the increased reliance on imported timber means that there were no large-growing timbers available anywhere in Scotland; as with peat, the underlying issue was one of location and accessibility. The king and magnates could source timber in portions of their estates remote from where they needed them for use, and they had the means to bring them to where they were necessary. Thus, in 1435 James I sent his barge

---

58 See below, Chapter 3, pp. 125–126, 129.
59 *ER*, vol. 4, pp. 613, 626.

'to northern parts' of his kingdom to transport timber south for royal use, probably mainly at Linlithgow and the king's new Carthusian priory at Perth.[60] It is likely that this wood was felled in the glens leading into the Moray Firth, from where timber used in the royal castle at Inverness in 1412–16 was earlier obtained.[61] Following the forfeiture of the Black Douglas family in 1455 and the annexation of their Moray and Easter Ross estates to the crown, the forest of Darnaway and the oak reserve there that had provided the timber for Darnaway castle's great hall roof in the 1380s[62] gave James II and his successors a royal timber source close to the northern burghs. As we will see becomes clearer in later centuries, extensive woodland dominated by large oaks and pines remained in parts of the kingdom away from the main urban centres in the eastern and central lowlands; the problem was one of accessibility rather than availability.

The preponderance of masonry in Scottish later medieval building traditions has tended to obscure the continuing high level of demand for timber. Even the largely stone high-status and domestic buildings constructed in this era required significant volumes of wood for floors, roofing and panelling, as well as for the jettied galleries on houses such as those illustrated in Bower's mid-fifteenth-century depiction of Stirling.[63] As surviving remains like the possibly earlier fifteenth-century roof of Alloa Tower show,[64] oak – both domestically procured and imported – continued to be the most favoured timber type used for all of these purposes. At Alloa, most of the trees used were around forty to fifty years old and probably were felled in the nearby royal forest of Clackmannan, where re-expansion of the woodland area, or at least no further contraction, had followed the release of assarting pressures as a result of the fourteenth-century population crash; the tree ages would support regeneration from the 1350s/1360s.[65] The use of the softer sapwood as well as the hard heartwood of the timbers, however, suggests that suitable trees were not so plentiful as to allow carpenters to discard the less desirable portions of the trees and that the wood might have been used while still green rather than being seasoned before use. This same use of green wood was evident in the roof timbers of a residence in Brechin, where native oak felled in spring 1470 and used immediately in the construction of a house had been re-used for the rafters and collars of its early post-medieval replacement.[66] The fact that native oak was being used in Brechin rather than imported timbers that were readily available through nearby Montrose might indicate that woodland in the Angus glens, which were key timber sources for the shire into the later seventeenth century, were already exploited for local needs in the fifteenth.[67]

---

60 Ibid., p. 626.
61 Ibid., pp. 163, 255.
62 Discussed in the volume 'Scotland AD 400–1400'.
63 A. Crone and F. Watson, 'Sufficiency to scarcity, 500–1600', in T.C. Smout (ed.), *People and Woods in Scotland* (Edinburgh, 2003), pp. 60–81 at 71; Corpus Christi College, Cambridge MS 171, f. 265, reproduced as the frontispiece in Bower, *Scotichronicon*, vol. 6.
64 T. Ruddock, 'Repair of two important early Scottish roof structures', *Proceedings of the Institute of Civil Engineers*, 110 (1995), pp. 296–307 at 297.
65 Crone and Watson, 'Sufficiency to scarcity', p. 81.
66 A. Crone, N. Grieve, K. Moore and D.R. Perry, 'Investigations into an early timber-frame roof in Brechin, Angus', *Tayside and Fife Archaeological Journal*, 10 (2004), pp. 152–165.
67 See below, Chapter 3, p. 138. For woodland in Glen Esk at the end of the sixteenth century, including what seems to be an enclosed plantation on the south side of the glen beside Invermark Castle, and in upper Glen Clova, see Timothy Pont, [North Esk; South Esk] (front) – [Pont 30] https://maps.nls.uk/view/00002327.

Timothy Pont's map of the upper part of the valley of the River North Esk in Angus: NLS, Timothy Pont, [North Esk; South Esk] (front) Pont 30, https://maps.nls.uk/view/00002327. (Courtesy of the National Library of Scotland)

In areas where there was already very limited availability of native growing timber or access to nearby sources, such as the south-eastern Borders, parts of Lothian and Fife, there was continuing investment in the formation of parks. Ostensibly for the enclosure of deer for sport, most were probably as important as reserves of growing timber, wood pasture and fodder. One such was the royal park at Falkland in Fife, which perhaps had earlier medieval origins but which was redeveloped by King James II from 1451 onwards.[68] It is interesting to note here that in 1458/9, in the immediate aftermath of his new legislation, the king ordered the construction of an enclosure around 'the ward and woods' of Falkland, while his son James III instructed repairs to banks and ditches around meadows and woods

---

68   J.M. Gilbert, 'Falkland Park to 1603', *Tayside and Fife Archaeological Journal*, 19/20 (2014), pp. 78–102.

Oak tree, Cadzow Oaks, Chatelherault Park, Hamilton. The oaks at Cadzow were possibly planted in response to the shortages of such timber locally by the first half of the fifteenth century. The trees stand on land previously given over to cultivation and settlement.

there.[69] The coincidence of the redevelopment of the woodland reserve at Falkland and the recorded need for timber to be brought in for repairs and new building work at its castle could point to the enclosures being for the future local supply of such large trees.[70] Sixteenth-century references to haggs[71] and the recorded delivery of fuelwood to the castle in the 1470s, however, suggests instead that what woodland existed there at that date was coppice-wood and may have included fast-growing species like alder and hazel as well as in-demand oak. The king was not alone in new woodland planting, as the parks of the Douglases at Dalkeith and Hamiltons at Cadzow illustrate. At the latter, trees dated dendrochronologically to at least 1444 were planted on an area of former rig-and-furrow cultivation, illustrating

69 ER, vol. 6, p. 556; vol. 8, p. 296; Gilbert, 'Falkland Park', pp. 86–87.
70 Gilbert, 'Falkland Park', p. 87.
71 Haggs are enclosed blocks or compartments within an area of woodland, which would be felled or coppiced individually in a sequence and left to regenerate, usually over a cycle of between fifteen and twenty-five years.

Oak tree, Dalkeith Park, Lothian. Rising from a massive coppice stool, the individual trunks of this old-growth tree are each over 550 years old, which suggests that the stool from which they have sprouted is several centuries older.

graphically the emparkment of former zones of arable in the emptier post-plague landscape.[72] At Dalkeith, massive coppice stools with later and now fully mature growth indicate that the oaks there were managed from a much earlier date, but were still being worked through the fifteenth century and beyond. Based on this record of attempts at regulation, conservation and protection, traditional representations of the later Middle Ages as seeing a free-for-all in woodland clearance are wide of the mark. Instead, the impact of climatic deterioration, adding to the effects of centuries of steadily mounting human pressures, seems to have provided a stimulus for the first attempts to introduce a planting, conservation and protection policy for trees. Nevertheless, it is also clear that by the start of the 'little ice age', the extent of woodland cover in Scotland outside such reserves had continued to contract and in most lowland districts was much less than at present.

72 M. Dougall and J. Dickson, 'Old managed oaks in the Glasgow area', in T.C. Smout (ed.), *Scottish Woodland History* (Edinburgh, 1997), pp. 76–85 at 77–81.

Fish

Shifting our focus to the waters, we can see that in common with its terrestrial resources, Scotland's freshwater and sea-fisheries experienced a change in modes of human exploitation in the fifteenth century, as the 'little ice age' began to affect the migratory behaviour of fish. In the case of the seas, intensification of operations was largely at the hands of fishermen from north-western mainland Europe. This resurgence followed an apparent decline, noticeable especially in the archaeological record for the Western Isles fisheries from sites such as Bornais in South Uist in the decades either side of 1400. That decline is not immediately explicable, but possibly arose from the inadequacies of local boat-building traditions in coping with the deepening North Atlantic swell and the increasingly unpredictable weather at this time.[73] The loss of a single fishing-boat and its crew was potentially catastrophic for the small farming and fishing communities of the Outer Hebrides, and it is likely that fewer days were risked in the deeper offshore fishing grounds than during the calmer conditions of the MCA. The Bornais excavation data, moreover, suggests that most fish being caught by fishers there in the later fourteenth and earlier fifteenth century were apparently being consumed locally rather than traded.[74] This local consumption points towards fishing as part of a subsistence regime, which might suggest that any international demand for Hebridean processed herring and cod that had apparently driven an expansion of fishing in earlier centuries had waned; the export of processed fish identified in thirteenth-century archaeological contexts was over. We do not know if this was the result of changes in ocean temperatures and fish migration patterns, or if it was a consequence of mid-fourteenth-century and later population collapse in mainland Europe reducing demand for fish.

In the seas, however, some 'little ice age' shifts brought longer-term gains. For example, oceanic temperature changes towards the end of the fourteenth century meant that rapid cooling of the North Atlantic waters around Iceland led to a movement south in the main breeding grounds for cold-sensitive gadid species, especially cod. This increased their numbers around the north mainland coast in Caithness and the waters off the Northern Isles. In 1487, custom was paid in Arbroath, Crail, North Berwick and Pittenweem on the export of 2,200 'mulones' (suggested to be a name for cod or, far less likely, for basking sharks), perhaps indicating the southern range of these fish in commercially exploitable numbers and of the size sought by merchants.[75] In comparison to the exports from the northern cod fishery of this time, however, this east-coast trade was insignificant and possibly represented an opportunistic add-on to the better-established herring fishery. Excavation evidence from east-coast burghs, especially Perth, has revealed evidence for consumption of cod by local populations, suggesting that much of the North Sea catch was perhaps consumed locally. Most of these bones, however, were recovered during excavations in the later 1970s and 1980s, before the development of isotopic analysis that could identify the sea regions in which the fish lived. Given the domination of the trade in cod by fish caught in more northerly Atlantic waters and the findings of modern isotopic analysis of bone from excavations in England and northern Europe, it is

---

73 R.D. Oram and P. Adderley, 'Re Innse Gall: A Norse colony in the Irish Sea and Hebrides?', in S. Imsen (ed.), *The Norwegian Domination and the Norse World c.1100–c.1400* (Trondheim, 2010), pp. 125–148 at 128–129, 134. Sharples, *A Norse Farmstead in the Outer Hebrides*, pp. 157–158, 176, 179, 192–194.

74 Sharples, *A Norse Farmstead in the Outer Hebrides*, pp. 157–158, 176, 179, 192–194.

75 *ER*, vol. 9, pp. 536, 538, 539. The Firth of Forth is today effectively fished out for deep-water species like cod: T.C. Smout and M. Stewart, *The Firth of Forth: An Environmental History* (Edinburgh, 2012), pp. 7, 121, 270.

likely that some of the bones recovered were from imported Icelandic and Norwegian 'stockfish' rather than fresh-caught in the coastal fisheries.[76]

At the same time as the revival in the cod fishery and for other mainly demersal species, migratory pelagic fish species, especially herring, were being caught at volume. Initially, as in previous centuries, it was the Forth ports that dominated this operation, as revealed in customs collected mainly at Crail down to the 1430s. There is, however, a break in our records for the 1440s to 1480s, after which time they show that it was the western Scottish herring fishery that yielded the highest numbers.[77] On the basis of the surviving data, we cannot identify any specific reason for this switch in dominance, whether environmental or political. These abundant stocks led in the fifteenth century to renewal of the international commercial trade in fish from this region, once demand had resumed after the mid-fourteenth-century epidemics.[78] This time it was foreign boats – mainly Dutch, Breton and Basque – larger, better built and able to weather the rougher sea-conditions of the time, that fished in Hebridean waters. Local chiefs, however, profited through imposition of 'tolls' and the supply of processing materials including salt and barrels. The king and his councillors were aware of this siphoning off of the potential profits from the sea-fisheries and as early as 1471 had passed acts to encourage lords and burghs to build fishing vessels and procure nets 'for the common good of the realm and the great increase of riches to be brought within the realm from other countries'.[79] Like so much of the early legislation, however, there is no indication that the first act saw any appreciable growth in the Scottish fishing-fleet. What changed matters was understanding of the volume of potential revenue being lost to foreigners, which led to a more coercive reissue of the act, with targeting of 'strong idle men' to crew the vessels.[80] Again, there is no evidence that the act was effective. Instead, the even larger Dutch herring 'busses' that were introduced by the sixteenth century reduced dependence on Scottish middlemen for their supplies. Vessels of that scale, accompanied by their own support ships, could carry the salt and barrels for the fish. Until then, though, regional nobles enjoyed new wealth to lavish on material displays of their status from the profits of fish. New building work at older castles is one manifestation of this wealth, such as the remodelling in the years either side of 1500 of the Mhic Lachlainn chieftain's principal residence and lordship centre at Castle Lachlan in Argyll.[81] It is tempting to see this investment as a result of participation in the exploitation of the Loch Fyne herring fishery, from which the Mhic Lachlainns' Campbell neighbours were profiting in partnership with lowland Scottish and foreign merchants by the late 1400s.

Bonds between Edinburgh merchants and Breton fish-traders in the 1480s to supply salt, barrels and, on occasion, fish, delivered to them in 'the lochis at

---

76 A.K.G. Jones, 'The fish bone', in D. Perry et al. (eds), *Perth High Street Archaeological Excavation 1975–1977*, fascicule 4: *Living and Working in a Medieval Scottish Burgh. Environmental Remains and Miscellaneous Finds* (Perth, 2011), pp. 53–67.

77 Gemmill and Mayhew, *Changing Values*, pp. 318 and 322 Table 27. The authors highlight the return to an apparent dominance of the eastern fishery in the sixteenth century but do not seem aware that many of the fish customed in eastern ports at that period had been caught in the west, processed aboard the *crears* and shipped back to the Forth: Smout and Stewart, *Firth of Forth*, p. 28.

78 M. Rorke, 'The Scottish herring trade, 1470–1600', *Scottish Historical Review*, 84 (2005), pp. 149–65 at 150–153, 155.

79 RPS 1471/5/10 [accessed 21 January 2022].

80 RPS A1493/5/21 [accessed 21 January 2022].

81 *The Royal Commission on the Ancient and Historical Monuments of Scotland. Argyll: An Inventory of the Monuments. Volume 7: Mid-Argyll and Cowal: Medieval and Later Monuments* (Edinburgh, 1992), no. 118.

the west sey', signal the revival of the herring fishery in the sea-lochs at the head of the Firth of Clyde.[82] There is evidence for this trade as already established by the third quarter of the thirteenth century, then involving south-west English merchants,[83] before its collapse in the fourteenth and revival in the fifteenth centuries. The Campbells of Argyll moved fast to participate in – if not control – the reviving fishery and c.1450 built a new castle at Inveraray on Loch Fyne to oversee and secure a cut of the operation.[84] By 1474, Colin Campbell, 1st earl of Argyll, had secured royal permission for a chartered burgh adjacent to the castle to give his family still greater control over the fishing profits, while permission for a second Campbell burgh at Kilmun was granted before the end of the century to enable similar exploitation of the fisheries in Loch Long and the Holy Loch. In these initiatives, Earl Colin was ostensibly a royal agent collecting the king's share of profits from a previously unregulated and, consequently, uncustomed trade. In reality, the Campbells of Argyll were among the chief financial beneficiaries of the major expansion in operations and consequent growth in customs revenue from the fishery.

It is evident from royal financial records from the last quarter of the fifteenth century that the crown was exercising considerable control over the west-coast fisheries.[85] Royal collectors of custom payments were visiting multiple locations around the western sea-lochs and islands, underscoring the scale of the fishery and the involvement in it of clan chiefs.[86] At its peak in 1487, the royal custumar at Dumbarton accounted for duty on 397 lasts of herring, equalling 4,760,000 processed fish, which is probably only a fraction of the numbers caught in Clyde and Hebridean waters.[87] The gradual decline of the value of the west-coast fishery in crown revenues after 1487 is not an indication of any contraction in its scale or the depletion of the shoals but was a symptom of the ability of the well-organised Dutch fleets in

Castle Lachlan, Argyll and Bute. The principal residence and lordship centre of the Mhic Lachlainn chiefs saw significant development in the later fifteenth century, with the introduction of expensive new fireplaces and windows in its lodgings probably paid for from the profits of the fish trade.

---

82 For example, NRS GD430/43, GD430/185.
83 S.I. Boardman, *The Campbells 1250–1513* (Edinburgh, 2006), pp. 295–300; S.I. Boardman, '"Pillars of the community": Clan Campbell and architectural patronage in the fifteenth century', in R.D. Oram and G.P. Stell (eds), *Lordship and Architecture in Medieval and Renaissance Scotland.* (Edinburgh, 2005), pp. 123–60 at 146.
84 Boardman, 'Pillars of the community', pp. 135–136.
85 Boardman, *The Campbells*, p. 297.
86 Boardman, 'Pillars of the community', pp. 146–8 and notes.
87 *ER*, vol. 9, pp. 542–543. One last of herring = 12,000 fish. By way of contrast, the herring fishery in the Forth the same year accounted for only 4 last and 4 barrels = c.50,000 fish: *ER*, vol. 9, p. 540.

particular to operate without input from local agents or suppliers, rendering it impossible to police – or tax – their activities. The later fifteenth century, therefore, had seen a brief efflorescence of Scottish interest in the west-coast fishery and the profits of the harvest of the seas before foreign fishermen gained a near monopoly and drew the profits to their ports.[88] It was another three centuries before the indigenous fishing operation began to be developed on a scale that enabled Scottish fishermen to compete with these foreign specialists.

As in earlier centuries, rather than the herring and wider sea-fishery, it was the salmon fishery and the international trade in processed salmon, especially to England and Flanders, that was more consistently the most important to Scotland.[89] Lords, both temporal and ecclesiastical, and burghs were jealous protectors of their salmon fisheries and probably collaborated to ensure that the protective legislation that had been issued down to 1400 was augmented and enforced through the fifteenth century, starting in 1401 with the imposition of punitive 100 shilling fines for catching salmon in the close season.[90] Indeed, there is more legislation concerning salmon than for woodland protection or increase of cereal production combined, which is a sure signal of the economic significance of this resource. That legislation ranged from enforcement of the close season, regulation of the placing of fish-traps, cruives and yairs,[91] to protection of immature fish to ensure that spawning stocks were maintained. With prices per barrel averaging 34 shillings down to 1437 and 48 shillings in the 1490s, it is understandable that there was frequent litigation between landowners with conflicting fishing rights and owners of upriver and downriver fish-traps, cruives, yairs and nets. Indeed, salmon fisheries were often the one component in a property portfolio that landowners would go to extraordinary lengths to preserve when the rest was being alienated. This is exemplified by the abbot of Balmerino's energetic defence of his monastery's Tay fisheries in the tidal estuary between Dundee and Newburgh.[92] Inland fisheries were of major importance, with seasonal nets and traps placed on the Tweed, the Forth, the Tay and its tributaries, the Angus Esks, Dee, Don, Ythan, Deveron, Spey, Findhorn, Ness, Farrar, Conon, Carron, Shin, Clyde, Leven, Ayr, Doon and all the rivers feeding into the Solway Firth. Foreign observers regularly commented on the volumes of salmon consumed by people of all social ranks within Scotland in the fifteenth century, but much of their catch seems always to have been destined for international trade. Monasteries such as Balmerino, however, were not simply conservators of their fishing rights but active encroachers on others', as the well-researched case of the dispute between Cambuskenneth Abbey and the adjacent burgh of Stirling attests.[93] Here, the

---

88 For the involvement of east-coast fishermen in the western herring fishery in the seventeenth century, see below, p. 281.
89 Gemmill and Mayhew, *Changing Values*, pp. 303, 307–309 and Table 55. In 1498, the Spanish envoy Don Pedro de Ayala commented on the great quantities of salmon being exported: P. Hume Brown (ed.), *Early Travellers in Scotland* (Edinburgh, 1891), p. 44.
90 RPS 1401/2/11 [accessed 21 January 2022].
91 A cruive is a dam or weir for trapping fish on a river. A yair is a tidal version, where fish are trapped in a pool formed behind the barrier as the tide falls.
92 R.D. Oram, 'A fit and ample endowment? The Balmerino estate 1228–1606', *Comentarii Cisterciensis* (2008), t. 59 fasc. 1–2, pp. 61–80.
93 R.C. Hoffmann and A. Ross, 'This belongs to us! Competition between the Royal Burgh of Stirling and the Augustinian Abbey of Cambuskenneth over salmon fishing rights on the River Forth', in E. Bhreathnach, M. Krasnodębska D'Aughton and K. Smith (eds), *Monastic Europe AD1100–1700: Communities, Landscape and Settlement* (Turnhout, 2020), pp. 451–476.

River Tay from Kinnoull Hill, Perth and Kinross. This tidal stretch of the river, extending from just below Perth to the start of the main estuary east of Newburgh in Fife, was divided into multiple salmon fisheries held by the lay landowners and monastic interests on either side of the Tay, including the abbeys of Balmerino and Lindores.

abbey not only illegally occupied portions of the river whose fisheries belonged to the burgh, while taking regular court action to exclude the burgesses from interference in waters that the abbey had controlled since the mid twelfth century, but also actively encouraged others to encroach on the burgh's rights.

Behind such actions lay money, and salmon was one of the biggest earners of foreign currency in late medieval Scotland. Salmon was already being exported in volume through Aberdeen in the first half of the fifteenth century, with an average of 465 'Hamburg' barrels going out annually in the hands of foreign merchants in the decade after 1429.[94] The proportion of exports that was being handled by foreign merchants declined after 1439, and by 1445 native merchants had taken over much of the formerly foreign-handled trade and were exporting 780 barrels from the burgh's total of around 1,200 that year, when Aberdeen accounted at exchequer for around 100 lasts (1 last = 12 Hamburg barrels);[95]

---

94 M. Rorke, 'Scottish Overseas Trade, 1275/86–1597', unpublished PhD thesis, University of Edinburgh (2001), p. 194.
95 Rorke, 'Scottish Overseas Trade', p. 195.

the 'Hamburg' barrel in Scotland was defined by law as having a volume of 14 gallons.[96] This expression by volume does not allow us to calculate the size of the catch but it was probably in excess of 200,000 adult salmon at that date.[97] Salmon was also exported through Dundee, Dysart, Perth and other small ports, but with their trade totalling only two-thirds the annual average of Aberdeen, that burgh was clearly recognised by merchants as the principal trading burgh for salmon. By the start of the last quarter of the century, Scotland's customed exports were on average at least 2,100 barrels but, with the custom on salmon exports from ports north of Aberdeen having been leased and so unrecorded in the exchequer accounts, this is likely to represent only around two-thirds of total exports.[98] None of these figures, of course, reflects the volume of salmon being consumed in Scotland and, while we must always take with a pinch of salt oft-quoted but unverifiable tales of fish eaten to repletion and laws having to be enacted to prevent lords and employers feeding salmon to their servants more than twice or three times a week, we can be sure that it constituted a major element of ordinary diet. All told, the salmon fishery perhaps constituted the most heavily exploited aquatic environmental resource in fifteenth-century Scotland.

• • •

Although the image of unrelenting awfulness presented in popular descriptions of the 'little ice age' is not supported for Scotland by either the historical record or the climate proxy data, what we can see is a progressive downward slide in annual weather trends interspersed with extended periods of extreme conditions. A shift towards wetter, cooler – and in winter snowier and frostier – norms, however, certainly made life harder still for those who already existed on the social and economic margins. This was true especially for those living in island or coastal areas, where increased storminess and episodes of drought caused erosion of vulnerable, sandy-soiled land, or in upland districts where each snowy winter brought a resurgence of wolf predation on their flocks and herds. The grain harvests upon which most of Scotland's population depended perhaps failed entirely only on a small number of occasions across the century but, when they did, that triggered periods of dearth and possibly of famine, with high mortality rates and episodes of social violence, disturbance and population displacement. All of these factors fed into undercurrents of discontent among Scotland's noble community, whose levers of political, social and economic power were dependent upon the flow of rents from peasant tenants or the profits from the grain produce, wool, hides and fish of their estates. It is not simply environmental determinism to see the flashpoints in crown–magnate relations in the reigns of James I, James II and James III as occurring against a backdrop of climate-driven environmental pressures. The geographical spread of tensions, from the rising then disintegrating lordship of Clann Domhnuill in the Hebrides and west and northern Highlands, the destruction of the Black Douglases in the Southern Uplands and Moray Firth lowlands, to the multiple local disputes between lesser noblemen from Buchan to Galloway, reveals the ubiquity of those pressures.

There is no doubt that the marginal 'knife-edge' between agricultural success or failure narrowed critically during the periods of greatest weather instability in the fifteenth century. But it was the longer-term shifts that stoked a perception of fragility in a rural

---

96 First enacted in 1478 to ensure standardisation of volumes sold abroad: RPS 1478/6/87. This standardisation was confirmed in 1489: RPS 1489/1/5 [accessed 21 January 2022].
97 *ER*, vol. 9, pp. 539, 541, 544, 545.
98 Rorke, 'Scottish Overseas Trade', pp. 195–196.

economy that often staggered along at little more than subsistence levels, rather than those spectacular one-off events recorded in the surviving narratives. The large – and growing – body of evidence for rearguard management of the dwindling numbers of trees and for legislative successes in encouraging new plantation overturns long-held popular belief in a late medieval free-for-all in destructive clearance of Scotland's already depleted woodland cover. So too, more recent research into grassland questions equally long-held views of a crisis in the pastures. But that slightly more positive picture does not mean that everything was rosy, for woodland cover had diminished and supplies even of coppice-wood from managed reserves were in some areas precarious, while grazing resources were everywhere coming under sustained pressure. Vegetation, however, including in more temperature-sensitive upland zones, is resilient and shows little evidence for changes in species composition in the face of changing weather conditions until long after the fifteenth century. This fact underscores how the headline events described by chroniclers such as Walter Bower need to be balanced by recognition that even the most exposed of upland shieling grounds would have recovered quickly from short-term, one-off impacts. Although the grass kept growing, though, it is unclear if it continued to grow as much and for as long each year as annual mean temperatures declined. But farmers were not passive in the face of crisis, and abandonment of their holdings in favour of a life of vagrancy was probably a choice made in only the most extreme need. Instead, perhaps responding to parliamentary acts encouraging higher levels of sowing of food crops, we see evidence for efforts to extend and intensify cultivation to make up shortfalls in grain yields by growing more on short-term intakes from outfield or upland grazing. But such efforts were not uncontroversial, as complaints of illicit assarts on other people's lands and encroachment on common grazing areas makes clear.

But this 'better than we thought' picture also needs to be viewed carefully. There were certainly winners, especially those who managed to secure controlling stakes in the herring and salmon fisheries in the second half of the century. For them, fifteenth-century climate change brought dividends. Those who depended on grain crops, however, faced choices between increased efforts at cultivation, perhaps just to stand still, or chancy reliance on harvests of unpredictable quality and declining volume, supplemented by short-term assarts on low-grade soils. Livestock numbers also seem to have remained buoyant but there are question marks against the health and productivity of flocks and herds. The energy deficit between calories gained through grazing and those expended on maintaining body heat during the colder and wetter summer seasons affected everything from volumes of milk for the dairy products that were central to food cultures throughout the Highlands and Islands, to meat quality and quantity for noble and burgess consumers; from the fertility of breeding stock, to the carrying of foetuses to full term; and, of course, even the production of the dung used to fertilise cultivated ground. Farmers faced choices in how to manage these changes, perhaps opting to increase grazing numbers in the same way as they managed declining grain yields through expanding the cultivated area. In other, less arable regions they chose a regime of calf-slaughter to keep stocking numbers at levels that the available fodder could sustain. In all such choices, we see at work the 'new normal' of the deepening 'little ice age'.

Linlithgow Palace, West Lothian, viewed from the west across Linlithgow Loch. The palace and its wider land- and waterscape setting is perhaps the ultimate example of the Stewart kings' command of the environmental resources of their realm, from the eel-fishery in the loch or hay and fodder yield of its park, to the Lanarkshire lead of its roofs and the Highland oak for its beams and joists.

CHAPTER 3

An Age of Shocks and Transitions
c.1500–1700

Upon the utmost corners of the warld,
and on the borders of this massive round,
quhaire fate and fortoune hither hes me harld,
I doe deplore my greiffis upon this ground;
and seeing roring seis from roks rebound
by ebbs and streames of contrair routing tyds,
and phebus chariot in there wawes ly dround,
quha equallye now night and day devyds,
I call to mynde the storms my thoughts abyds,
which euer wax and never dois decress,
for nights of dole days Ioys ay euer hyds,
and in there vayle doith al my weill suppress:
    so this I see, quhaire euer I remove,
    I chainge bot sees, but cannot chainge my love.[1]

A new century brought no positive change in the climate trends that had become entrenched in the late 1400s. The descent into the first of the prolonged troughs of significantly colder and wetter weather of the 'little ice age', which had commenced with the onset of the Spörer Minimum in the late fifteenth century, accelerated and deepened in the 1500s. It was not unrelentingly bad, for there were lengthy episodes of more benign annual conditions across the next two hundred years to break the otherwise consistent grimness of this era. We can see retrospectively, however, that these were extended periods of still poor if not adverse conditions rather than times of relative positivity. When times were less poor, our sources speak of productivity rather than abundance, but as with our earlier caution about not reading famine into every mention of harvest failure, this bounty was relative and the benefits unequally spread; hunger, deprivation and disease had been and remained the normal fare for many of Scotland's people who subsisted on marginal agricultural productivity. Weather conditions unfavourable to

---

1 William Fowler (1560–1612), 'In Orkney', in E. Dunlop and A. Kamm (eds), *The Scottish Collection of Verse to 1800* (Glasgow, 1985), p. 35.

crop growth, animal husbandry and fishing afflicted the terrestrial, coastal and marine environments of Scotland in a seemingly endless cycle of precariousness. In addition to successive climatic hammer-blows, the human experience of these frightful times was worsened by the continuation of some land- and resource-management practices that had arisen in better days long past. Further pressure came from an increasing trend towards production for monetised markets rather than for local consumption or exchanges and renders in kind. Together, these factors exacted a terrible price from human populations who depended upon the land's resources for sustenance and shelter. They also inflicted a heavy toll on wild and domesticated animals and on woodland battered by unseasonal storms, and further scarred the face of the land itself. For many looking back from the 1700s across these two centuries, harshness outweighed comfort and perhaps seemed endless in its impacts. From the perspective of the eighteenth century, the sixteenth century appeared to have begun with crop failure and food shortages and the seventeenth had ended with a decade of the worst weather in living memory, with the country locked in the grip of a devastating and very real famine. For the six generations living through these times, any suggestion that this was just a further episode of climate transition would have been incomprehensible. To them, the hunger, want and death that characterised this shocking age of transitions were ever-present and very stark realities.

It is difficult to reconcile this dark image with the traditional historical depiction of this period as witnessing the flourishing of Scottish renaissance culture amid the political and religious turmoil of the Reformation, civil wars and revolutions. By any measure, it was an era of outstanding achievement: from the 'Aureate Age' of the reign of James IV, with the poetry of William Dunbar, music of Robert Carver, and the architectural achievements of Linlithgow Palace or Stirling's great hall; the glamour and glitter of James V's court, enlivened by the poetry and drama of Sir David Lindsay; Mary's brief but brilliant Francophile household; James VI's cultured court of humanist scholars and 'Castalian Bands'; and into the post-1603 era of absentee rulers, lightened by the poetry of Montgomery, the learning of Alexander Seton and Robert Leighton, the architecture of John Mylne and Sir William Bruce, or the remarkable Aberdeen and Edinburgh medical networks of the later seventeenth century. Indeed, in its emergent practices of 'policy' and 'politeness', meaning the improvement or betterment of estate infrastructures, introduction of new approaches to land and resource management and planting regimes, all seen as wealth-generating stimuli, it was a time of transition that presaged the new order of the eighteenth century. But the shocks and dearth of the present far outweighed the promise of better times to come for Forth-side farmers in the 1500s seeing hard-won land reclaimed by spate flood and storm surge; for communities in the Western Isles and Orkney witnessing their livelihoods literally blown away by Atlantic gales; for Perthshire shepherds losing flocks to bitter frosts and drifting snow, or for Morayshire householders unable to find sufficient dry fuel to warm and feed their families.

From Spörer to Maunder: solar minima, climate and weather impacts

Why was this two-century period so marked in the frequency and severity of climate-driven impacts? The main causes can be seen in how the sixteenth and seventeenth centuries were bracketed by two episodes of very low solar activity, the Spörer Minimum (*c.*1460–*c.*1550), which we have already encountered, and the Maunder Minimum (*c.*1645–*c.*1718). A direct correlation between such minima and the extremely cold conditions experienced across the northern hemisphere which coincided with them has not yet been established firmly but is strongly suspected. It is possible that their co-occurrence with periods of low NAO and weakened AMOC, in an interplay of reduced solar radiation with the Earth's atmospheric and oceanic circulation, combined to deepen already cooler, wetter and more unsettled weather conditions into extremes of cold and precipitation.[2] The solar minima themselves, however, were not unbroken episodes of poor conditions, as is illustrated by less bad years from the 1660s to 1680s during the slide to the depths of the Maunder in the 1690s. Nor was the century between the two a time of surplus or surfeit bathed in glorious weather, as the cyclonic storms of the late sixteenth century and famines of the 1600s, 1620s and 1630s likewise show. The inter-minimum weather conditions may have been less unfavourable than what came before or after, but the epidemics of the era, culminating in the high mortalities of the last Scottish plague outbreak in 1645–8, just as the Maunder Minimum commenced, took any gloss off the human lived experience of these years.

Progressively improving records across the sixteenth and seventeenth centuries mean that weather becomes the most readily visible environmental factor affecting Scotland. Even the suites of non-narrative financial accounts from the earlier 1500s can be used to confirm that trends established during the first decades of the Spörer Minimum continued in the new century. As an illustration, exchequer and burgh record data suggest that a wet summer in 1501 heralded the first of several harvest failures that contributed to price inflation for the grain that formed the dietary staples of the bulk of Scotland's people.[3] No exact accounts exist for the scale of the resulting hardship or the levels of hunger-related deaths, but human skeletal remains from probably lower-class adults who had been children around this time reveal widespread markers of nutritional deficiency that had affected their immature teeth and bones.[4] References to 'the sterility of the land', the consequent deferral of payments due from affected districts, and high market prices for grain indicate that crop and harvest failures recurred in 1505.[5] That there is little record of such failures in the 1510s is not a sure indication that conditions had improved, for prices remained high across the decade. Less bad conditions possibly led to a slight improvement in yields, but other factors intervened to contribute to widespread social and economic disturbance; plague had returned at the end of the 1490s to spread in waves around the kingdom for the next twenty years, and its impact and containment kept busy the record-keepers at local level.[6]

---

2 Eddy, 'The Maunder Minimum'; H. Miyahara et al., 'Gradual onset of the Maunder Minimum revealed by high-precision carbon-14 analyses', *Scientific Reports*, 11:5482 (2021).

3 Gemmill and Mayhew, *Changing Values*, Table 11 (p. 157), Table 14 (pp. 170–171), Table 16 (p. 186) for wheat, barley and malt prices. Oats and oatmeal prices show greater volatility in the 1500s, with generally lower prices in the north and west reflecting differential regional harvest success or failure. See Table 20 (pp. 195–196) and Table 22 (pp. 208–210).

4 See, for example, the burials discussed in J. Franklin, C. Troy, K. Britton, D. Wilson and J.A. Lawson, *Past Lives of Leith: Archaeological Work for Edinburgh Trams* (Edinburgh, 2019), pp. 98, 105–106, 128–129, 166–167, where enamel hypoplasia of the teeth, cribra orbitalia and porotic hyperostosis occurred at levels above the national average.

5 The failure of the crops in Rannoch in 1505 required a respite of payment of the dues owed to the bishop of Dunkeld from the clergy of the district: Hannay, *Rentale Dunkeldense*, p. 13.

6 Oram, 'Responses to epidemic disease'. The detailed records of Edinburgh's response to this wave of epidemic are well known, but less well known are the accounts of its spread throughout the country, from Irvine and Glasgow in the west, Strath Tay in central sub-Highland Scotland, Aberdeen in the north-east, or Selkirk in the Southern Uplands. See J. Bain and C. Rogers (eds), *Liber Protocollorum M. Cuthberti Simonis AD 1499–1513 and Rental Book of the Diocese of Glasgow*, volume 1 (London, 1875), pp. 6–7, 68–69; A. Myln, *Vitae Dunkeldensis Ecclesiae Episcoporum* (Edinburgh, 1831), pp. 40, 43; Imrie, Rae and Ritchie, *The Burgh Court Book of Selkirk*, vol. 1, pp. 48, 54.

As the epidemics waned in the late 1510s, the condition of harvests once more preoccupied the minds of officials. The impact of climatic deterioration can be seen in successive poor harvests across 1523–5; the particularly wet summer of 1525 was recorded in Perth as 'the deir symmer' on account of price inflation produced by the grain shortages that followed back-to-back crop failures in the preceding years.[7] Shortages re-emerged in the mid 1530s, but legislation enacted to address the perception of mounting crisis assumes that human action – hoarding or illicit export – was the primary cause of supply issues rather than yield levels. The spike in grain prices that commenced in the late 1520s climaxed in 1535,[8] the year that parliament introduced legislation intended to relieve inflationary pressures. King James V's 1535 act prohibiting the sale of all varieties of foodstuff to English merchants spoke of how 'a great part of the inland [was] robbed of their goods and the same had and sold in England by thieves and traitors, whereby all manner of stuff has grown to a great price and dearth, and now our sovereign lord putting order of justice, rest and tranquillity amongst his lieges and that plenty of goods may grow amongst them' introduced a foreign trade ban for staple foodstuffs.[9] Further related legislation in 1541 indicates that poor weather, harvest failures and delayed grass growth had continued to affect both the arable heartland of the kingdom and its upland grazing, but the ban on dealing in fish could suggest that sea conditions were also poor.[10] Further enactments in 1552 concerning prices and the supply of goods reveal that the shortages continued into the next decade.[11]

Was this period afflicted by more widespread harvest failures than previous eras or were other factors at play to exacerbate the consequences of individual events? There is little doubt that the general climatic trend towards cooler, wetter and stormier conditions increased the precariousness of life for most of Scotland's people. Summer and autumn rainfall affecting harvests during the worst years of the Spörer and Maunder minima probably stimulated increased use of the corn-drying kilns, whose presence had become established in the later thirteenth and earlier fourteenth centuries; they became more numerous in the fifteenth century and proliferated ubiquitously across this period. Documented references to kilns in rural contexts, as opposed to excavated remains, grow in number from the later sixteenth century.[12] While some kilns were associated with malting for brewing and baking, it is probable that the majority were essentially for drying grain that had been harvested when not yet fully ripe or in wet weather, to prevent further loss through rot or fungal infestation and enable it to be milled. The widespread use of such kilns is both a proxy for the prevailing poor weather and an indicator of positive action taken to optimise the volume of the yield suitable for human consumption. This was a positive response to adverse conditions, but other human decisions added to social vulnerability.

Specialisation, especially towards livestock, was encouraged by two factors: an increasing alignment

---

7   Eagles, *Chronicle of Perth*, p. 21. For prices of grain and grain products, see Gemmill and Mayhew, *Changing Values*, Table 11 (p. 158), Table 14 (pp. 173–4), Table 17 (p. 187), Table 20 (pp. 197–198) and Table 22 (p. 212).
8   Gemmill and Mayhew, *Changing Values*, Table 11 (p. 159), Table 12 (p. 175), Table 20 (p. 199) and Table 22 (p. 213). Again, the prices show patterns of greater availability in some areas and increased demand in others, especially the main urban markets like Aberdeen and Edinburgh.
9   RPS 1535/34 [accessed 29 November 2021].
10 RPS 1540/12/70 [accessed 29 November 2021].
11 RPS A1552/2/1 [accessed 29 November 2021].
12 RPS 1587/7144 [accessed 7 January 2022].

Craignavar, Glen Almond, Perth and Kinross. The grassy mound is the partly collapsed remains of the corn-drying kiln and barn of the adjoining settlement that is first recorded in 1701 and had been abandoned before the time of the 1841 census. The largest of five such kilns in the upper section of the glen – and possibly the largest in Scotland – it is testimony to the volume of grain being produced in this upland location during the 'little ice age'.

of Lowland Scotland's rural economy to satisfy market demand for specific commodities, and a switch by landlords towards cash rents as opposed to payment in agricultural produce. This trend accelerated through the sixteenth century as Lowland organisational practices were applied by the crown and nobility to their Highland estates, or Highland landowners began extractive exploitation of their estates' resources to supply Lowland and overseas markets. This development marked an economic shift where the use-value of the land's resources to local consumers was supplanted to an increasing extent by their exchange-value in distant markets.[13] The impact of this shift from what we would in modern terms label 'sustainable' management to essentially unsustainable exploitation has been exam-

13 R.A. Dodgshon, *From Chiefs to Landlords: Social and Economic Change in the Western Highlands and Islands, c.1493–1820* (Edinburgh, 1998), chapter 5.

ined in detail for Scotland's woodland, but is less well understood for upland grazing or the grazing/arable balance in parts of Highland and Hebridean Scotland where the new commercialism is most evident.[14] What is least well understood is the interplay of these economic trends and the physical impacts of the weather extremes of the 1500s and 1600s on cereal yields and general biomass production, livestock health and fertility. Profits from wool and cattle hides were such that more areas of former or potential arable land were put to pasture, increasing dependence on grain purchased from producers elsewhere, principally in the southern Baltic region. When such external sources of staple foodstuffs were also negatively affected – as happened in the 1690s – the fragility of the new networks of market-oriented dependency was revealed with devastating effect.

It is no surprise that protracted periods of supply crisis affecting staple foodstuffs witnessed escalations in social and political discord in the kingdom, much as similar episodes in the fifteenth century had done. In the first half of the sixteenth century, crown rents and ecclesiastical teinds were gathered with rigour. The resulting distress among tenants and lesser landowners, who saw unsustainably high portions of their meagre yields siphoned into others' hands, contributed to the gathering volume of hostility directed at the secular and ecclesiastical establishment. Although normally attributed to political factors, principally James V's introduction of regular taxation of the clergy in the 1530s, the scramble to alienate land at feuferme[15] for a fixed rent and the hunger of the tenantry to acquire that land points to the uncertainty of incomes for the great landowners. It also suggests a mixture of desperation and opportunism on the part of peasant farmers, or profiteering among aristocratic and merchant-burgess speculators.[16] It would be overly simplistic and deterministic to claim that cycles of crop failure and famine were the primary motors that drove both noble and monastic willingness to divest from direct exploitation of their properties and feufermers' readiness to take on those risks, albeit with the potential for greater profit for themselves, but they were important contributors. Amid the religious fervour of the 1550s, pressure for a further flow of landed assets from the hands of the Church helped to fuel support for reform of the Church. Social, economic and political stresses linked closely to the environmental impact of climate change, it can be argued, contributed directly to the most profound upheaval of the sixteenth century in Scotland: the political and religious revolutions of the Protestant Reformation and its protracted aftermath.

Observing weather across and between the minima

Behind these social, economic and political stresses there lies a wealth of evidence that does not survive from earlier periods. Our understanding of the human lived experience of the repeated shocks to Scotland's environmental metabolism delivered during this central phase of the 'little ice age' is enhanced especially by the 'voice' given to individuals and communities by the written record of the conditions and the consequent suffering they endured. Locally detailed documentary records rather than the one-off headline events of earlier chronicle and

---

14 T.C. Smout and F. Watson, 'Exploiting semi-natural woods, 1600–1800', in T.C. Smout (ed.), *Scottish Woodland History* (Edinburgh, 1997), pp. 86–100. For assumptions about grazing, see the critique in section 1 of A. Ross, 'Scottish environmental history and the (mis)use of soums', *Agricultural History Review*, 54 (2006), pp. 213–228.

15 The grant of land to a tenant on a heritable basis, in return for an annual money payment and payment of a larger sum on the succession of an heir.

16 J. Wormald, *Court, Kirk and Community: Scotland 1470–1625* (London, 1981); reprinted (Edinburgh, 1991), pp. 52–54.

annal sources – even the observant and well-informed Walter Bower – start to survive in increasing numbers from the mid sixteenth to later seventeenth centuries. Such sources provide coverage spanning the country from Easter Ross to Lanarkshire in the period c.1550 to c.1685, bridging the decades between the Spörer and Maunder minima and the most extreme period of the Maunder. They detail weather impacts, crop growth and harvest conditions, and the effects of those on the population in general, set into the wider social and political contexts of their regions and the nation generally.[17] Many parts of the country are less fortunate in written record survival, but climate proxy data from the Cairngorms, north-west Highlands and Argyll coast enable broad comparisons to be made with the better-documented areas.[18]

Durations of extreme events and local observation of their direct impact on human and animal populations are key elements in these sources that were generally lacking in the accounts of earlier ages. Not only do these later sixteenth- and seventeenth-century sources provide a level of detail that enables scalable comparison, but they also provide data to which we can relate from our modern experience, especially of extreme weather events. Earlier sixteenth-century records are frustratingly sparse in comparison, meaning that the start of this period is visible in less high-resolution detail than its end. Nevertheless, the climate proxy data of the speleothem record from the Traligill basin in northwest Sutherland goes a long way towards filling the gap. That data signals that higher than average winter precipitation due to high NAO conditions prevailed into the middle of the century; those deliver strong westerly airflow and cyclonic rainfall.[19] This proxy record correlates well with the first of the detailed written reports of widely varying annual weather trends, those given in the central Perthshire *Chronicle of Fortingall*. Its accounts start with winter 1554/5 and end in 1577, narrating a veritable see-saw between dearth and abundance. Its first report is of widespread heavy snowfall and cold conditions persisting into the spring, which tallies also with tree-ring data from the Cairngorms that is suggestive of a slide into cold, unstable conditions across the summer growth period.[20] Coming close on the heels of the Spörer Minimum, this record emphasises that colder temperatures continued despite the end of that solar episode. Severe winters and cold springs, which affected both autumn and spring sowing, fruit blossoming and grass growth resumption in the spring, in fact lasted into the 1560s.[21] Such conditions fed price inflation and possibly artificially worsened the food shortages that underlie the reissue in 1563 of earlier enactments targeted at addressing dearth

---

17   See, for example: Anonymous (ed.), *The Black Book of Taymouth* (Edinburgh, 1855); *Chronicles of the Frasers: The Wardlaw MS.*, ed. W. Mackay (Edinburgh, 1905); Eagles, *Chronicle of Perth*; *The Diary of Alexander Brodie of Brodie, MDCLII–MDCLXXX., and of His Son, James Brodie of Brodie, MDCLXXX–MDCLXXXV: Consisting of Extracts from the Existing Manuscripts, and a Republication of the Volume Printed at Edinburgh in the Year 1740* (Aberdeen, 1873); A.G. Reid (ed.), *The Diary of Alexander Hay of Craignethan 1659–1660* (Edinburgh, 1901); L. Lamont (ed.), *The Diary of Mr John Lamont of Newton 1649–1671* (Edinburgh, 1830); John Nicoll, *A diary of Public Transactions and other Occurrences, Chiefly in Scotland from January 1650 to June 1667*, ed. D. Laing (Edinburgh, 1836).

18   Rydval et al., 'Reconstructing 800 years of summer temperatures'; R. Wilson, N.J. Loader, M. Rydval, H. Patton, A. Frith, C.M. Mills, A. Crone, C. Edwards, L. Larsson and B.E. Gunnarson, 'Reconstructing Holocene climate from tree rings: the potential for a long chronology from the Scottish Highlands', *The Holocene*, 22:1 (2011), pp. 3–11; Proctor et al., 'A thousand year speleothem proxy record'; Cage and Austin, 'Marine climate variability'.

19   Proctor et al., 'A thousand year speleothem proxy record', p. 818 and Fig. 4.

20   Anon., *Black Book of Taymouth*, pp. 124–125; Rydval et al., 'Reconstructing 800 years of summer temperatures', p. 2960.

21   Anon., *Black Book of Taymouth*, p. 129.

Ben Lawers and the Glen Lyon hills from Creag Uchdag. The snowy conditions of January 2011 replicated the extreme weather of the winters recorded in the *Chronicle of Fortingall* from the 1550s to 1570s.

of victuals.[22] Summer 1567 brought a reversal of the previous decade's weather, with a parched spring and summer delivering an 'evyl haryst' (evil harvest) with low grain yield[23] and an inflationary spike in grain prices until good harvests in 1569 and 1570 brought a brief respite.[24] Winter 1570/1, however, reverted to the late 1550s' pattern, causing widespread livestock deaths. Harsh late-winter frosts then left the ground too hard to plough until the beginning of April.[25] A run of poor conditions resumed with 1574's wet and windy summer and poor harvest, leading into a winter of extremes that lasted until late March 1575. It seemed at first that 1576 would be a better year but rains from June to September spoiled the promising harvest; foul winter and early spring weather delayed the sowing of oats until the fourth week of March 1577; and the wet weather continued all summer.[26]

The Fortingall chronicler's gloomy narrative continues in the *Chronicle of Perth*, which not only details weather events particular to the burgh but also narrates the impact in the lower Tay basin of weather events in its highland catchment. We see this in its record of winter 1578/9, where above-average precipitation in the southern Highlands brought spate conditions to the Tay. This event caused severe flood damage downriver and the partial collapse of the bridge at Perth; a second in January 1582 destroyed more of the bridge.[27] These spates, however, were minor in comparison with flooding in

22 RPS A1563/6/6 [accessed 29 November 2021].
23 Anon., *Black Book of Taymouth*, p. 135. This reduced rainfall in the 1560s and reversion to high precipitation in the 1570s is evident in the Traligill data: Proctor et al., 'A thousand year speleothem proxy record', Fig. 4.
24 Anon., *Black Book of Taymouth*, p. 136.
25 Ibid., pp. 137–138.
26 Ibid., pp. 141–142.
27 Eagles, *Chronicle of Perth*, p. 23.

Inner Firth of Tay from Culteuchar Hill, Perth and Kinross. The extreme winters of 1607/8, 1614/15 and 1623/4 saw this lower tidal reach of the river freeze from side to side as far east as Erroll, where the estuary is 1.5km wide, and the ice was reportedly thick enough to support the weight of a man.

January 1615, which saw inundation of all low-lying areas around the burgh.[28] Although it brought no similar flood, winter 1607/8 was one of the harshest on record, better known as the first recorded instance of the Thames freezing over at London and a 'frost fair' being held on the ice.[29] Around Perth, severe frosts extended from early December 1607 to late March 1608 and it was recorded that the Tay – a river whose discharge rate is three times that of the Thames – could be crossed on the ice as far downriver on its tidal estuarine reach as Erroll in the Carse of Gowrie, where it is 1.5km wide.[30] A second, more intense episode of extreme cold followed the January 1615 flood, with the Tay again freezing over at Perth.[31] This cold period led into a long episode of heavy snowfall, from March into May, a sting in the tail from what was possibly the most extreme winter for decades throughout the British Isles.[32] Autumn 1621's heavy rainfall again resulted in major flooding in Perth and ruined the local grain harvest.[33] Winter

---

28 Ibid., p. 40.
29 T. Dekker, *The Great Frost. Cold Doings in London, Except it be at the Lotterie. With Newes Out of the Country. A Familiar Talke betwene a Country-man and a Citizen Touching this Terrible Frost and the Great Lotterie, and the Effects of Them. The Description of the Thames Frozen Over* (London, 1608).
30 Eagles, *Chronicle of Perth*, p. 36.
31 Ibid., pp. 40–41.
32 The winter of 1614/15 in England was recorded as that of the 'Great Snow', with flooding following the snow-melt and then a drought that led to widespread crop failure. See T. Dekker, *The Cold Yeare 1614: A Deepe Snow: In Which Men and Cattell have Perished . . . Or of Strange Accidents in this Great Snow* (London, 1615); L. Veale, G. Endfield and J. Bowen, 'The "Great Snow" of Winter 1614/1615 in England', *Weather*, 73:1 (2018), pp. 3–9. The Traligill data is indicative of sharply rising precipitation through the 1610s, reaching a climax in the middle years of the century: Proctor et al., 'A thousand year speleothem proxy record', Fig. 4.
33 Eagles, *Chronicle of Perth*, pp. 44–45.

1623/4 brought a further visitation of extreme cold, with the Tay freezing over for nearly four weeks in January/February.[34] These events seem to have been the result of previously unparalleled episodes of protracted and intense cold. The 'little ice age' had entered its mature state.

That last extreme winter represented the nadir of that particular episode, for the following years saw a welcome but short-lived period of warmer, more benign conditions that raised false hopes of a return to marginally better conditions. Although the 1630s continued to bring some of the coolest summers of the seventeenth century in north-eastern Scotland,[35] milder winter weather prevailed in east-central Scotland for almost a decade. The respite lasted only until January 1635, when a three-week snowfall led to a reported 2 metre covering around Perth and the freezing of the Tay for four weeks. The severity of that episode was compounded by widespread inability to move food supplies from where they were stored to where they were needed, or to mill grain due to the freezing of mill lades, just as had been experienced in the worst years of the fifteenth century.[36] Technological and infrastructural capability was unable to provide the levels of resilience necessary to overcome the challenges of the weather, but in the eyes of most people the fault lay in human failures that were attributable to satanic agency or Divine displeasure. Against such a backdrop of adverse environmental and economic conditions, sharpened by the psychology of the age, the political and religious tensions of the 1630s become more explicable.

Social upheaval arising from similarly adverse conditions can also be seen in the earlier seventeenth-century material gathered in the Fraser chronicle. Although written in the 1670s, it drew heavily on earlier accounts and provides important evidence for extreme events in the northern Highlands. Importantly, it reveals that crop failure in the 1600s was not solely a consequence of cold-delayed springs and wet autumns but could arise from milder conditions that favoured plant-attacking fungi and parasites. Its account for 1602, for example, reports dearth and famine throughout the central and northern Highland region following the failure of the barley, oats and pease crops, the staples upon which most Highlanders subsisted; the fruiting crops were blighted by fungal infestation that thrived in warm but damp conditions.[37] Only the abundance of dairy products, the chronicler claimed, presumably buoyed by the ample grass growth delivered by the same mild, damp weather, enabled the region to escape a more catastrophic level of famine deaths. The severity of this event, however, was sufficient to stimulate social destabilisation, the same source reporting that the dearth triggered an episode of escalating plundering raids and bloodshed between clans. Chiefs, he claimed, were driven by necessity to make up shortfalls in the food supplies of their own stores or from their clan's resources, which supported their culture of conspicuous consumption and exchange, returning to past practices of predation on neighbours whose stores were no more plentiful.[38] As in the fifteenth century, the marginality of survival and the fragility of lordly power in parts of the north and west Highlands was here exposed, as was the recourse to predatory behaviour with its consequent escalating spiral of social and political discord.

The early years of the Maunder Minimum are spanned by the diaries of the Edinburgh-based John Nicoll and John Lamont of Newton in south-east

---

34 Ibid., p. 46.
35 Rydval et al., 'Reconstructing 800 years of summer temperatures', p. 2960.
36 Eagles, *Chronicle of Perth*, p. 51.
37 *Chronicles of the Frasers*, p. 236.
38 Ibid., p. 239.

The East Lothian plain from the slopes of Arthur's Seat above Duddingston, Edinburgh. The diary accounts of John Nicoll record the impact of protracted droughts in the early and mid 1650s on water supply and crop yields across the district around Edinburgh and throughout East Lothian, a region noted normally for its agricultural productivity.

Fife. These provide informative details of weather events on both sides of the outer Firth of Forth – and further afield in Scotland – and their impact on crops, wild vegetation, buildings and shipping. Some information is incidental but reveals interesting insights on weather events, such as Nicoll's bloody rain ('thair rayned bluid') in May 1650, which points to a southerly airflow carrying Saharan dust at high altitude.[39] More, however, contain detailed accounts of weather-related impacts on growing crops or local dearth caused by the passage of armies and the price inflation that followed.[40] The synchrony of the two diaries enables the regionality of effects to be observed, with Fife and Lothian experiencing similar weather but with different local outcomes. Lamont noted, for example, that summer 1652 saw an early harvest in Fife and Tayside and was the hottest and driest in living memory, which stunted cereal crops, parched the grass and ended pea-flowering in mid July rather than late September.[41] Nicoll likewise observed the early start to the harvest on account of the hot and dry summer weather, but noted a bountiful rather than a failing harvest.[42] Both Nicoll and Lamont, however, agreed on the unseasonably warm autumn and early winter, which saw trees bud and birds begin to build nests in November. Although some modern scholars have suggested that rain run-off from the hard-baked fields led to winter floods,[43]

---

39 Nicoll, *Diary*, p. 16.
40 Ibid., pp. 31–32 for the impact of the armies on Lothian east of Edinburgh in August 1650, compounded by rain and windstorms before the harvest. Ibid., p. 75 for price inflation.
41 Lamont, *Diary of Mr John Lamont*, pp. 53–54.
42 Nicoll, *Diary*, pp. 98, 100–101.
43 See, for example, Dawson, *So Foul and Fair a Day*, p. 113.

both diarists noted how drought into early 1653 brought water shortages in and around Edinburgh, with public supplies failing.[44] The hot summer and mild winter conditions were repeated in 1653 and 1654, with the grain harvests being higher than average, bringing a drop in prices and securing a good supply to the market through into 1655, but water supplies in Lothian again ran short through drought.[45] The south-west, Nicoll noted by way of contrast, experienced higher than average rainfall across these years.[46] Climate proxy data from the Cairngorms region confirms that its summers in the 1650s were warmer than in the 1630s but not matching the warmth and drought experienced further south and east, while the Traligill speleothem record evidences high NAO and westerly airflow conditions, which delivered the apparently wetter conditions prevailing in the west.[47] The north-west Sutherland data, however, shows a falling trend through the second half of the century into low winter NAO conditions that brought more northerly and easterly winds, heralding colder and poorer conditions for the eastern side of the country.

This shift is recorded in both Lamont's and Nicoll's diaries. Both men noted how the exceptionally hot and dry years ended in 1655. For Lamont, they ended with heavy rains that brought destructive floods in August; Nicoll saw the 'foull and filthie' February, which left waterlogged fields in Lothian that were unfit for ploughing or sowing, as marking the turn.[48] The deterioration continued, exemplified by 1658, when February to June saw northerly and easterly winds bring frost, snow and intense wind-chill. December 1658 saw heavy snow and rapid thaws triggering spates and flooding in Fife's rivers, accompanied by storm surges in the Firth of Forth with inundation of coastal areas from Kirkcaldy east to Wemyss that destroyed saltpans and houses.[49] January, February and June through to September 1659 were equally stormy, with heavy rains and high, south-westerly winds. Particularly severe summer episodes caused shipwrecks, heavy flooding around the Esk and Water of Leith, destruction of watermills and growing timber, and extensive damage to the grain crop in Lothian.[50] In March 1661 these woes were compounded by severe winds and rains which triggered more flooding, coastal inundation and, more damaging yet, the washing away of the recently tilled soil and the new-sown seed-corn with it.[51] The diary records of shipwrecks, fishing vessels capsized in harbour, industrial saltpans devastated, homes destroyed and fields stripped bare provide graphic testimony of the human misery that followed such extreme events.

Although John Lamont recorded the 1660s as an increasingly stormy decade, with a great tempest in October 1669 marking a climax of destruction around the Firth of Forth and northwards into the uplands of the Tay and Earn basins, he made no observations about harvests and yields.[52] John Nicoll, however, while recording the same winter storms and autumnal tempests as Lamont, spoke of highly

---

44 Lamont, *Diary of Mr John Lamont*, p. 61; Nicoll, *Diary*, pp. 102–103, 105.
45 Lamont, *Diary of Mr John Lamont*, p. 63; Nicoll, *Diary*, pp. 120, 122, 130, 131, 134, 138.
46 Nicoll, *Diary*, p. 138.
47 Rydval et al, 'Reconstructing 800 years of summer temperatures', p. 2959, Fig. 5; Proctor et al., 'A thousand year speleothem proxy record', Fig. 4.
48 Lamont, *Diary of Mr John Lamont*, pp. 114–115; Nicoll, *Diary*, p. 149.
49 Lamont, *Diary of Mr John Lamont*, p. 138; Nicoll, *Diary*, p. 212.
50 Lamont, *Diary of Mr John Lamont*, pp. 148–149; Nicol, *Diary*, pp. 247, 249–250, 264.
51 Lamont, *Diary of Mr John Lamont*, p. 167.
52 Ibid., pp. 193, 202, 224, 225, 234, 246, 251, 253–254, 265–266, 266–267. For discussion of the 1669 event, see Dawson, *So Foul and Fair a Day*, p. 120.

productive harvests and mild winter conditions. In 1662, Nicoll described conditions as 'wondrous blissed' and 1664 was noted by him as having a hot summer, a productive early harvest, and an abundance of fruit.[53] It is only in January 1665 that he noted any significant change towards severe winter conditions, but a warm summer followed to deliver another good harvest, leading into another notably mild winter.[54] This record does not sit well with the temperature reconstructions modelled from the Cairngorm tree-ring data, which suggest that the 1660s in that location had seen sustained low summer temperatures across the decade; this part of the sequence, however, has one of the lowest levels of confidence for accuracy.[55] Alexander Brodie of Brodie's diary, from its Moray perspective, confirms that the same patterns of weather were experienced north of the Cairngorms as in Fife and Lothian,[56] which underlines the discrepancy within the modelled temperature trends from the tree-ring evidence.

There is no similar question mark against the Cairngorm tree-ring data for the 1690s, which indicates that the region's coldest summer temperatures of the century were experienced in that decade.[57] At Traligill, the same decade saw a reversal of the winter precipitation trends, with a return to a high winter NAO signalling the onset of wetter and windier conditions.[58] This shift marked the start of the decade that saw what are commonly termed 'the Seven Ill Years', which brought extreme weather, harvest failures and widespread famine both to Scotland and to much of northern Europe.[59] It is unfortunate that none of our narrative records that provided so much information for weather events earlier in the century extend beyond the 1680s. Indeed, such is the paucity of contemporary record for these times that as recently as 1991 the climatic contribution to the disasters of the period could be dismissed in less than a single sentence. The impact of the period was written off as 'probably no worse than the much more regular subsistence crises' of the late sixteenth century, and the fact that it managed to kill off only 5 per cent of the population nationally was presented as nothing of significance in the grander scheme of things.[60] We have instead been dependent on less comprehensive references, which afford little insight into the events and processes that brought the transition into crisis or into the unfolding catastrophe that affected all of Scotland, and which generally look back from the vantage point of the post-crisis years and attempt to rationalise the events they had witnessed. Our diary sources are clear, however, that conditions had deteriorated through the later 1670s and 1680s, with their reports of increasingly stormy summer and winter weather along the Atlantic seaboard leading to sea-floods, erosion and, critically, harvest failures and the loss of access to fisheries.[61] Across northern Europe, including Scotland, we can also see that progressively worsening winters became extreme in the early 1690s. The infamous events of the February 1692 Glencoe Massacre, for example,

---

53 Nicoll, *Diary*, pp. 362–363, 386–387, 406, 428.
54 Ibid., pp. 429, 430, 444, 448.
55 Rydval et al., 'Reconstructing 800 years of summer temperatures', pp. 2959 (Fig. 5), 2960. For discussion of the correlation or divergence of the NCAIRN data from other datasets in Great Britain and Europe, see pp. 2960–2970.
56 *The Diary of Alexander Brodie*, p. 308.
57 Rydval et al., 'Reconstructing 800 years of summer temperatures', pp. 2959 (Fig. 5), 2960, 2970.
58 Proctor et al., 'A thousand year speleothem proxy record', Fig. 4.
59 The most detailed analysis of this era is K.J. Cullen, *Famine in Scotland: The 'Ill Years' of the 1690s* (Edinburgh, 2010).
60 M. Lynch, *Scotland: A New History* (London, 1991), p. 309.
61 These events are summarised in Dawson, *So Foul and Fair a Day*, p. 121.

Scots pine above Glen Derry, Southern Cairngorms. Tree-ring data suggests that temperatures plummeted in the mountains in the 1690s, with evidence of suppressed summer growth pointing to late-lingering snow cover, frosts and colder air conditions. Snowfields on the Cairngorm plateau would have kept spring and summer temperatures low in the neighbouring lowland districts.

occurred against a backdrop of winter snowstorms and intense cold.[62]

This slide to the lowest temperatures and worst harvest failures of the seventeenth century was possibly worsened by the eruption of Mount Hekla in Iceland, which started on 13 January 1693 and lasted for over seven months.[63] This eruption expelled a vast volume of tephra, which was carried at high altitude as far south and east as Norway and Scotland; the aerosols reduced already falling levels of solar radiation and worsened the cold. Hekla did not cause the cold episode, but it accelerated the downward trend and helped in the rapid expansion of sea-ice around Iceland. The ice contributed to the entrenchment of a low NAO and the extension of the Iceland low pressure cell much further to the south as the Azores high weakened. The result was a run of increasingly severe winters and poor, wet and cold summers and autumns. Those conditions curtailed hay and grass growth for livestock fodder, made crop failures more probable, and caused the effective collapse of both the Shetland cod fishery and the west-coast herring fishery. A final visitation from the eruption was a wave of sulphurous fogs in summer 1694 that blasted the growing crops and grasslands, the effects of which lasted into the early 1700s.[64]

The harvest failure of 1694 marked the start of the worst phase of this dire period. By 1696, reports within letters between estate owners and their local officers detail the mounting dearth in regions around Scotland spread from Lothian to Ayrshire, which contained what were widely regarded as the nation's most productive farmland.[65] Attempts by Hugh Montgomerie of Eglinton to secure a remission on his contract as sub-tacksman for excise in south-western districts in the late 1690s detail the impact of the failure of the grain crop nationally on the malt and brewing operations from which he was meant to draw excise duty.[66] Similar appeals from his counterpart, George Mackenzie, in 1697 in respect of the north Highlands, illustrate the ubiquity of the crisis.[67] Shetland, which already had experienced harvest failures in the 1680s, also suffered successive failures in the 1690s, which led to supplications from the islands' landowners in 1696 for assistance from the administration in Edinburgh.[68] Setting aside the polemic within contemporary accounts, such as those of Sir Robert Sibbald and Andrew Fletcher of Saltoun, observations in private correspondence confirm those authorities' accounts of population displacement as the starving poor became roving refugees in search of food; many who did not die of hunger became prey to epidemic disease, falling and perishing at roadsides and then lying unburied.[69] The shortage of victual experienced in Scotland was, however, ubiquitous, and attempts to import grain to at least alleviate some of the worst effects were hindered by the widespread harvest failures elsewhere in the

---

62 P. Hopkins, *Glencoe and the End of the Highland Wars* (Edinburgh, 1998).

63 The most comprehensive discussion of Hekla and the history of its eruptions is by the Icelandic vulcanologist Sigurður Thorarinsson, *Hekla: A Notorious Volcano,* translated by Jóhann Hannesson and Pétur Karlsson (Reykjavík, 1970); R. D'Arrigo, P. Klinger, T. Newfield, M. Rydval and R. Wilson, 'Complexity in crisis: the volcanic cold pulse of the 1690s and the consequences of Scotland's failure to cope', *Journal of Volcanology and Geothermal Research,* 389 (2020), 106746.

64 A very detailed account of this event was published in 1835 by the Cromarty-based geologist, Hugh Miller. See H. Miller, *Scenes and Legends of the North of Scotland or the Traditional History of Cromarty* (London, 1835), p. 259.

65 NRS GD406/1/4128, GD406/1/4279, GD406/1/10855.

66 NRS GD3/10/4/1.

67 NRS GD26/7/439.

68 NRS E41/24.

69 NRS GD406/1/9080; Sir R. Sibbald, *Provision for the Poor in Time of Dearth,* 1st edition (Edinburgh, 1699); Second Discourse in Andrew Fletcher, *Two Discourses Concerning the Affairs of Scotland; written in 1698* (Edinburgh, 1698).

The north-west coast of Westray in the Orkney Islands, looking east across the sands at Grobust towards Rackwick, with the island of Papa Westray beyond. In rentals from the early 1500s, these districts were recorded as almost valueless on account of wind erosion and inundation by the Atlantic.

British Isles and across northern and western Europe. Competition drove up prices of already scarce commodities, placing grain beyond the reach of a cash-poor country. Ironically, in the midst of this subsistence crisis, national 'fasts and humiliations' were ordained by parliamentary proclamation, continuing into the early years of the new century.[70] Where human agency was failing, trust was instead placed in God to relieve the nation from the grip of the worst disaster within living memory.

Wind and water erosion

With the weather extremes described above came accelerating deterioration of land conditions, much related to the agricultural practices of the time, and major episodes of erosion and loss to the sea. Thirteenth-century records, discussed in the volume 'Scotland AD 400–1400', show that storm surge and sea-flood were not new experiences for the populations of exposed coastal districts. After 1500, however, we have evidence for increasing rates of occurrence across this period, often the consequence of specific extreme weather events. Much of our evidence relates to districts most exposed to Atlantic storms in the Northern and Western Isles, but significant events affecting districts in the south and east of the country indicate that the buffering effects of the western mountains and islands did not entirely mitigate the threat: northerly and easterly winds in winter and south-westerlies in summer and autumn also battered the eastern Lowlands. Among the worst affected areas was Orkney, where the Atlantic storms that wrought the wind erosion of arable land that was first recorded in the 1480s continued unabated into the 1500s.[71] The rentals for the island of Westray for 1500–8, coinciding with the episode of poor

---

70 RPS 1698/7/78 and 1703/5/127 [accessed 17 December 2021]; NRS GD150/3381.
71 R.D. Oram, 'From "Golden Age" to Depression: land, lordship and environmental change in the medieval Earldom of Orkney', in H.C. Gulløv (ed.), *Northern Worlds* (Copenhagen, 2014), pp. 203–214 at 205–206.

weather noted earlier, for example, record extensive areas of arable on the sandy soils at Rackwick as 'blawin' (blown away by the wind), while a large part of the arable land and links grazing on the smaller island of Papa Westray was noted as 'our blawin with wattir and sand' (over-blown with water and sand) as a result of dune deflation and wind-driven waves, and farms on Sanday were likewise 'blawin' with many lying unoccupied due to their unproductive condition.[72]

Similar trends can be seen in the Western Isles, where increasingly frequent and intense storms from the first quarter of the fifteenth century marked the start of a long process with peaks of arable loss in the early sixteenth and seventeenth centuries.[73] As discussed earlier, the sandy barrier islands and machair down the exposed Atlantic-facing west coast of the Uists have experienced cycles of breakdown and redeposition across the last 14,000 years. Considerable and accelerating erosion occurred from the mid fifteenth century onwards, however, especially after the early 1500s and again in the solar minima of the seventeenth century.[74] Analysis of sediments from Loch Lang on the eastern side of South Uist has revealed deposition of erosion materials there from around 1510 onwards, attributed to increased grazing nearby around that date.[75] This expansion of grazing into an area that had seen little use throughout the post-Iron Age period points to pressure on traditional summer grazing grounds further west. It is unclear whether erosion proceeded slowly and progressively or in short bursts of rapid and significant loss through the following decades. By the early 1540s, North Uist had lost at least two merk-

---

72 Peterkin, *Orkney Rentals*, pp. 84, 85, 87, 88, 90, 93–101.
73 Dawson et al., 'North Atlantic climate change'.
74 Gilbertson et al., 'Sand-drift and soil formation'.
75 K.D. Bennett, J.A. Fossit, M.J. Sharp and V.R. Switsur, 'Holocene vegetational and environmental history at Loch Lang, South Uist, Western Isles', *New Phytologist*, 114 (1990), pp. 281–298.

Hecla, Ben Corodale and Beinn Mhor, South Uist, looking from the machair near Rubh' Aird-mhicheil across the blackland towards the upland pasture where summer grazing intensified in the sixteenth century as increasing storminess and sand-blow made the lower pastures less viable.

lands of the sixty at which the island was then valued. This land, described as 'lying destroyed through the flowing of the sea', lay in the highly productive machair on the west side of the island. In 1542, the crown recognised the impact of this loss and reduced the rental value.[76] Further major episodes struck Uist in the first quarter of the 1600s, where 'the oldest men report this Isle to be much empayred and destroyed be the sands ovirblowing and bureing habitable lands, and the sea hath followed and made the loss irreparable'.[77] At the Udal in North Uist, it was likely the appalling weather of the 1690s, culminating in a great gale in 1697, that brought occupation at this formerly high-status settlement finally to an end beneath an overburden of 6 metres of sand-blow.[78]

Westerly gales also brought the progressive abandonment of coastal agricultural land on the mainland. Three examples from across the seventeenth century illustrate the impact. The earliest is from the eastern end of Aberlady Bay, East Lothian, where archaeological evidence indicates that the dune systems were destabilised in the later Middle Ages through agricultural encroachment, the stripping of turf and bent[79] for domestic needs, and extensive rabbit burrowing.[80] A 1627 report on the parish of

---

76 ER, vol. 17, p. 557. E. Beveridge, *North Uist: Its Archaeology and Topography*, facsimile edition (Edinburgh, 2018), p. vi.
77 'Notes and Observations of dyvers parts of the Hielands and Isles of Scotland', in A. Mitchell (ed.), *Geographical Collections Relating to Scotland Made by Walter Macfarlane*, volume 2 (Edinburgh, 1907), pp. 509–613 at p. 530.
78 I.A. Crawford, *The West Highlands and Islands: A View of 50 Centuries. The Udal (North Uist) Evidence* (Cambridge, 1986).
79 Bent is the coarse grass found in coastal districts. The bents are open areas of usually sandy ground covered with such grass.
80 J. Morrison, R. Oram and F. Oliver, 'Ancient Eldbottle unearthed: archaeological and historical evidence for a long-lost early medieval East Lothian village', *Transactions of the East Lothian Antiquarian and Field Naturalists' Society*, 27 (2008), pp. 21–45; E. Hindmarch, R. Oram, G. Haggarty, D. Hall and J. Robertson, 'Eldbotle: the archaeology and environmental history of a medieval rural settlement in East Lothian', *Proceedings of the Society of Antiquaries of Scotland*, 142 (2012), pp. 245–300.

AN AGE OF SHOCKS AND TRANSITIONS C.1500–1700

East Lothian coast from North Berwick Law to Fidra. The mainland coast opposite Fidra, from Longskelly Point to Gullane Point, was subject to repeated inundation in the early seventeenth century by sand from the destabilised dune systems edging Gullane Bay and Eldbotle.

Dirleton noted that the minister's glebe in adjoining Gullane parish was 'so overblown with sand as the largest aiker therof hes not bein manured these five yeiris bygone' and that the estimate of value given to it by the assessors was unlikely to stand for long as 'the samyne is licklie to be overblown more with sand'.[81] The same report listed seven husbandlands in Eldbotle, a small settlement north-west of Dirleton adjacent to the coastal dune system, but reported that three of them and six further acres of arable were 'two part . . . waist being overblown with sand'.[82] Before the end of the century, Eldbotle had been abandoned. The second case, from March 1661, involved a single sand-blow event inland from the links along Largo Bay and near Elie in Fife, accompanied by a major sea-flood, which buried new-ploughed and sown fields.[83] The final case, dating from the 1690s, occurred at Culbin on the Moray coast west of Findhorn Bay. Despite long-held traditions of a single catastrophic event overwhelming a prosperous arable estate, Culbin's demise came as the culmination of a protracted process spanning decades, if not centuries, that gradually destroyed the economic viability of this small property. Again, the progressive encroachment of sand onto the arable there, that had been a problem for many years, accelerated in the face of the prevailing strong westerlies. Sand-blow probably worsened the local impact of the weather-related harvest failures of the early 1690s, finally to defeat the efforts of Culbin's occupants to eke a livelihood from the land.[84] At both Eldbotle and Culbin, the common denominator was a disastrous interplay of human and environmental factors, where human exploitation of the turf and

---

81 Anonymous (ed.), *Reports on the State of Certain Parishes in Scotland, 1627* (Edinburgh, 1835), pp. 110, 115.
82 Ibid., p. 116.
83 Lamont, *Diary of Mr John Lamont*, p. 167.
84 S. Ross, 'The Culbin Sands: a mystery unravelled', in W.D.H. Sellar (ed.), *Moray: Province and People* (Edinburgh, 1993), pp. 187–204; S. Ross, *The Culbin Sands: Fact and Fiction* (Aberdeen, 1992). For the laird of Culbin's petition to parliament for financial relief, see RPS 1695/5/225 [accessed 9 December 2021].

Forth carse at Airth looking west towards Alloa and the Ochil Hills, with sea-dyke and banks at left centre. The land fringing the inner estuary of the Forth on both sides of the river suffered repeated sea-flood in the later 1630s.

marram of the dunes damaged the root systems that stabilised the sand and created weaknesses in the mat that were exposed to devastating effect in the stormier conditions of the late 1600s.

As these experiences illustrate, windblown sand was not the only threat to coastal zones. Inundation of low-lying districts by storm surges was increasing across this same period. Driven by the stronger winds, the seas were reclaiming areas of drained former wetland, eroding sandy zones like the Outer Hebridean machair or areas of salt-grass grazing, and re-salting estuarine flats that had been taken into cultivation in previous centuries. On the mainland, the most visible inroads occurred in the carses of the Forth in the east, around the Laich of Moray in the north-east, and the rivers that fed into the Solway Firth in the south-west. On the inner Forth estuary between Grangemouth and Airth, carseland that had been drained and settled earlier in the Middle Ages stood abandoned by the 1530s 'since a great part of the lands . . . are submerged daily by inundation of the sea and the rivers known as Carron and Forth'.[85] Here, the risk was heightened by the increased threat of spate floods from those rivers during the decades of increased rainfall and snow-melt in the first quarter of the century, especially when such conditions coincided with high tides and storm surges like those recorded as causing similar floods as long ago as the 1260s. On the opposite side of the Forth estuary at Ferryton in Clackmannanshire, severe weather resulted in late 1636 in the breaching of the dykes that protected low-lying land on the carse, leading to their inundation.[86] This event occurred during a run of especially poor years that affected the whole country, characterised by the winter weather extremes reported in the *Chronicle of Perth*. The following year, William Kerr, 3rd earl of Lothian, observed in a letter to his father that 'there hath beane sutch inundations, floodes and wyndes, as no man living remmembers the lyke'.[87]

---

85 *RMS*, vol. 3, no. 3016.
86 NRS GD124/17/203.
87 NRS GD40/2/14/10. Anonymous (ed.), *Correspondence of Sir Robert Kerr, First Earl of Ancram, and his son William, Third Earl of Lothian*, volume 1, *1616–1649* (Edinburgh, 1875), p. 98.

Borrowmoss, Wigtown, Dumfries and Galloway, looking north-east over the carseland towards Cairnsmore of Fleet. What is now used mainly as pasture was, in the later seventeenth century, contested land exploited for its fuel resources and arable potential.

Expanding or contracting arable

Losses of coastal arable and grazing land in the later seventeenth century increased competition for remaining resources and fuelled litigation where competition spilled over into conflict, or where one party's intentions for the land in dispute clashed with the nature of its use by the other. The greater numbers of surviving court records from this period as opposed to earlier times might skew the picture through the high volume of cases involving such clashes, but this could also be a symptom of the mounting pressure on populations caused by the deteriorating weather. In some cases, the link with extreme weather events was explicit. For example, in 1687 at Borrowmoss on the west side of the Cree estuary north of Wigtown, the earl of Galloway's tenants were accused of removing dykes around their arable land and extending cultivation into areas from which the burgesses of Wigtown secured turf and peat for building, roofing, fuel and garden soil deepening, digging up in the process the highway northwards from the burgh. Wigtown's council took strenuous action to remove the transgressors, but the earl claimed in their defence that this encroachment had been a response to losses to the sea of the land upon which his tenants had subsisted.[88] If true, his claim points to significant marine transgression on the west side of Wigtown Bay during the stormiest periods of the Maunder Minimum.

At Borrowmoss, the response to loss of arable land through marine encroachment on one side was to expand cultivation on the other to make up the loss and maintain the grain yield that the earl's tenants needed both to pay rents and to subsist. Expansion of the cultivated zone, however, was not a new response to falling yields, and across the two centuries after 1500 it seems to have been a common recourse of lairds and their agricultural tenants in the face of dearth. Transgression onto other people's land to break new ground into cultivation, or unauthorised and illegal intrusion into another's property, were sources of disputes recorded in court rolls from the early 1500s. The peaks of such litigation appear to

---

88 Extract from Council Minute Book of Wigtown and transcription of letter in response from the earl of Wigtown, printed in G. Fraser, *Lowland Lore; or the Wigtownshire of long ago* (Wigtown, 1880), p. 20.

Dronner's Dyke and the failed draining of Montrose Basin

Severe weather brought an abrupt end to one of the most ambitious land reclamation projects in later seventeenth-century Scotland, the drainage of around half of the area of the tidal basin that forms the inner estuary of the River South Esk. For its last 1.5km before its discharge into the North Sea, the river's channel is squeezed between the north-eastern end of the ridge of hills that separates the valley of the Lunan to the south from that of the Esk and the 2km-long spit of land occupied by the town of Montrose. Inland to the west of that constriction, the estuary forms a broad expanse some 2.5km north to south and 3km east to west: the Montrose Basin. At full tide, the whole of this area is submerged and in winter forms one of the most important wildfowl habitats in eastern Scotland. At low water, however, the basin empties to leave a plain of sand and mud with only the river flowing through its deeply scoured channel close to its southern edge, and the channel of a tributary stream flowing in from the north-west. The keen-eyed might spot another feature at low tide, running east–west for over 1km close to the centre of the basin. This is the remains of what is known as the Drainer's or Dronner's Dyke.

The basin has probably been exploited for its fish and wildfowl resources for millennia, and for the reeds that fringe its margins. By at least the later twelfth century, salt was also being sleeched from the sandy foreshore at its north-western edge near Dun. This was a long-exploited source of diverse riches from which many who lived around its edges benefited, controlled by local lairds but drawn upon by their tenants. That changed sometime around 1670, when the Montrose burgess Robert Raitt devised a scheme to drain the northern part of the basin to create over 800ha of new agricultural land. Rights to that area of the basin were secured from the Erskines of Dun and a co-partnery business was formed to raise capital for the project, with a man identified as a Dutch engineer (his surname suggests

that he might actually have been Swedish) employed to create what was essentially a polder. The aim was to erect a continuous barrier dyke of rammed earth or clay on a stone foundation, fronted by a wooden bulwark retained by driven timber piles. Sluices allowed for the draining of the streams feeding in from the north but prevented the rising tide from re-filling the reclaimed area. When John Slezer visited Montrose in 1678 and drew the sketch for his engraving of the burgh from the south-west, he showed the eastern landfall of the new dyke close to the Forthill at the south end of the town. Construction appears to have been completed before the time of Slezer's visit and the business partners were looking to see a return on their investment as the land behind the dyke was flushed of salt and converted progressively into arable land. Shortly afterwards, however, possibly in the severe storms of winter 1681/2 (Lamb, *Historic Storms*, p. 50), the dyke was breached and the water reclaimed the whole of the basin (A. Jervise, *Memorials of Angus and the Mearns* (Edinburgh, 1861), p. 61). Their finances exhausted and facing challenges from various parties who claimed that the work had affected their fishing rights and other interests in the river, the reclaimers abandoned their effort and the ruined dyke was rapidly dismantled by materials-hungry salvagers and the strong currents of the South Esk. The weather had defeated human effort and ingenuity at its first testing.

Above. Montrose Basin, Angus, looking north from Rossie Mills. The remains of Dronner's Dyke can be seen as a darker line running left to right just above the centre of the image. The darker appearance comes from the seaweed growth that clings to the stone foundation of the dyke.

Below. John Slezer's 'Prospect of the Town of Montrose', drawn in 1678 and published in 1693 in his *Theatrum Scotiae*. The piled dyke can be seen as a broad structure of planks retained by upright posts running to the left of the scene from the shore. (National Library of Scotland)

coincide with periods of heightened environmental stress. These were not *prima facie* cases of disputed occupancy rights, although some were contested on that basis, but cases of straightforward intrusion. The Acts of the Lords of Council from 1501 to 1503, for example, include cases involving unauthorised and illegal occupation of land in agriculturally marginal zones at the upland–lowland interface, such as Obney (Perth and Kinross), Invernochty and Bellabeg in Strathdon (Aberdeenshire), or Glencairn (Dumfries and Galloway), but also in more productive Lowland areas, like Ardownie in south-east Angus.[89] Business recorded in the sheriff court book of Fife for the 1515–22 period includes actions where lords sought the removal of neighbouring landowners' tenants from new arable intakes made on their land, or the award to them of the corn 'won' on that ground.[90] During the worst years of the 1530s and early 1540s, Scottish peasant farmers in eastern Roxburghshire were crossing the recognised Anglo-Scottish borderline in the Cheviot Hills to break ground into cultivation on the English side, in some districts reportedly clearing several hundred acres for new cornland.[91] While some commentators might seek a political explanation for this development in the context of Anglo-Scottish border warfare, this is too simplistic a reading of both the historical and the palaeoenvironmental records. The encroachment onto English land recorded by Henry VIII's agents seems to have coincided with an expansion of agriculture in upland districts around the Bowmont Valley on the Scottish side of the border, visible in the palynological record. This combined evidence is strongly suggestive of mounting pressure to maximise grain yields by increasing the extent of cultivation during a time of intense environmental pressure.[92]

A further phase of expansion of arable acreage occurred amid the poor years at the end of the sixteenth century. Legislation from 1600 relates to apparently widespread encroachment onto the commonties shared between different estates and assigned by landlords for common use by their tenants' peasant communities, large parts of which had seen recent ploughing and enclosure for private use.[93] This was an age-old issue, surfacing recurrently in records from the thirteenth century onwards and representing a conflict between the reservation of land for common use – principally for grazing and for obtaining muirland resources like broom and gorse – and individuals seeking either to maximise personal profit during times of high demand for cereals or, simply, to ensure their family's survival during times of dearth. Such attempts to restrict arable expansion or enclosure were reversed in 1661 through legislation intended to boost the Scottish economy in the wake of the financially crippling Civil War era, when Charles II's government secured an act 'for planting and enclosing ground'.[94] This was

---

89   A.B. Calderwood (ed.), *Acts of the Lords of Council*, volume 3, *1501–1503* (Edinburgh, 1993), pp. 19, 59, 247, 269–270.
90   W.C. Dickinson (ed.), *The Sheriff Court Book of Fife 1515–1522* (Edinburgh: Scottish History Society, 1928), pp. 19, 33, 70–71, 78–79.
91   Report by Sir Robert Bowes and Sir Ralph Ellerker, 2 December 1542, in J. Hodgson, *A History of Northumberland*, part 3 volume 2 (Newcastle-upon-Tyne, 1828), pp. 171–248 at 175, 176, 177–178.
92   R. Tipping, 'Cereal cultivation on the Anglo-Scottish Border during the "Little Ice Age"', in C.M. Mills and G. Coles (eds), *Life on the Edge: Human Settlement and Marginality*. Oxbow Monograph 100 (Oxford, 1998), pp. 1–11; R. Tipping, 'Palaeoecology and political history: evaluating driving forces in historic landscape change in southern Scotland', in I.D. Whyte and A.J.L. Winchester (eds), *Society, Landscape and Environment in Upland Britain*. Society for Landscape Studies supplementary series 2 (2004), pp. 11–20.
93   RPS 1600/11/24 [accessed 20 December 2021]. Commonties were areas where multiple landowners held shared rights of access for grazing, fuel and other organic materials, with their rights usually exercised through their tenants.
94   RPS 1661/1/348 [accessed 20 December 2021].

not aimed solely at boosting cereal cultivation, for the act encouraged the creation of cattle parks, the planting of lint and hemp (the latter for canvas manufacture), and the planting of more in-demand tree species, such as oak, elm, ash and plane, as well as more traditional timber types, like willow.[95] The main purposes, however, were to increase the kingdom's food security and production of primary commodities that were either imported at significant cost or whose regional availability was severely restricted, as well as boosting the volume of produce for international trade. The result was further inroads into areas of marginal land, especially coastal and upland zones, many of which had experienced only extensive rather than intensive exploitation following the population crash of the later medieval period.

Against the general picture of seeming helplessness in the face of deteriorating conditions, with only top-down interventions by the crown offering some respite, some landowners were proactive in seeking solutions that would bring general betterment and an increase in productivity (and profits). These efforts must be seen alongside the continuing evolution of the infield/outfield systems that had emerged in the later Middle Ages; that basic model was further refined through the seventeenth and into the eighteenth centuries in response to regional variations in soil, climate and market opportunity, or as new agricultural equipment and fertilising techniques were adopted.[96] Alongside the wider use of peas, whose nitrogen-fixing benefits were not understood but whose soil-enriching qualities were widely recognised, one of the key developments of the first half of the seventeenth century was the widespread adoption of liming in Lothian from the 1620s onwards. This process altered the soil pH, increased fertility and enabled more areas of poorer-quality land to be brought into regular cultivation;[97] it was to be a further three centuries until the ecological consequences of adjustments to soil chemistry were widely recognised. More immediately, the trend towards liming led to an expansion of arable at the expense of pasture, especially for production of in-demand wheat for the burgeoning urban market in Edinburgh, the bulk of whose populace was no longer involved directly in the cultivation of their own dietary staples. Around the major population centres, the environmental transformation seen traditionally as a consequence of 'Improvement' agriculture was already far advanced before the end of the seventeenth century.

Elsewhere, improvement and recovery took more physical forms. The countess of Mar's efforts to recover arable and coastal grazing on the Forth carse at Ferryton, mentioned above, illustrates one response: placing the onus for reclamation on a tenant. Others, however, responded differently to the recurrent flooding of valuable arable and pasture. For example, landholders whose properties adjoined the Inchaffray Pow in Strathearn, whose canalisation in the Middle Ages had created expanses of rich farmland around Inchaffray Abbey, entered a common bond in 1641 to contribute to the annual scouring of the watercourse.[98] Their aim was to mitigate, if not entirely prevent the regular flooding that arose from the combination of decades of neglect that had led to the clogging of the pow's channel by silt and reeds and increased rainfall. It was their tenants who provided the labour, but they agreed collectively to maintain this initiative for the general improvement of agriculture locally. In Aberdeenshire in 1685, it was the major landowners again who collaborated to secure an act of parliament designed to increase the planting of peas and beans, since they were recog-

---

95   I.D. Whyte, *Scotland Before the Industrial Revolution: An Economic and Social History c.1050–c.1750* (Harlow, 1995), p. 144.
96   R.A. Dodgshon, *Land and Society in Early Scotland* (Oxford, 1981), pp. 184–195.
97   Whyte, *Scotland Before the Industrial Revolution*, pp. 138–140.
98   RPS 1641/8/430 [accessed 9 December 2021].

nised to 'contribute to the improving and fattening of the ground', and prohibiting the local practice of 'delving, tilling and casting up' of turf to make a form of manure, a practice that was widely recognised as destructive to grassland, 'swarded ground' and meadow.[99]

As we have already seen, attempted soil-enrichment occurred across good times and bad, but particular efforts were made during periods of declining yields to enhance productivity. Around Scotland's urban communities and in the kailyards associated with rural housing, centuries of effort to 'good' the land are evidenced by extensive areas of anthrosols, where the soil has been deepened and fertilised by the admixture of a range of household midden and latrine waste, street-sweepings, byre waste, animal slaughter, fish processing and industrial by-products, turf, old roofing thatch and the like.[100] Outcomes were not always positive, for much of the erosion and sand-blow that occurred in Orkney and around the mainland coastline was worsened by traditions of turf-stripping from sandy links to provide organic materials that could be dug into cultivated ground as soil-enrichers.[101] At Marwick in mainland Orkney, it is estimated that nearly 200,000m³ of grassy turf and topsoil (c.280,000 tonnes) was cut by hand between the twelfth and nineteenth centuries and transported 5km from common land in the upland district to be added as enrichment to the cultivated ground of the nearby farms, mixed with animal manure and sea-ware.[102] This process devastated the upland, leaving an ecologically impoverished zone of poor-quality grassland over the nutrient-poor subsoils left exposed by removal of surface layers.

Although some landowners secured legislation through parliament before the end of the seventeenth century to prevent the devastation of grassland areas through the blanket stripping of turf, the practice was still noted by travellers in the third quarter of the eighteenth century. In Moray, for example, where overzealous stripping of the turf layers from the coastal dunes at Culbin had had cataclysmic consequences, in 1758 the English traveller Sir William Burrell noted that the land was largely arable with scarcely any pasture to be seen. Here, he observed what he identified as 'large commons' that could have provided extensive pasture for sheep flocks, but everywhere these had been scalped to provide turves for roofing. By this practice, he commented drily, 'these wastelands are entirely bereaved of the green sward'.[103]

In Orkney, analysis of soil content and structure in the cultivated zone reveals that its cultivation intensified in the early post-medieval period, especially on garden plots and areas in private rather than common exploitation. Along with the technological innovations in plough, harrow and harness types adopted in the islands at this time, this is indicative

---

99 RPS 1685/4/84 [accessed 20 December 2021].
100 R.D. Oram, 'Waste management and peri-urban agriculture in the early modern Scottish burgh', *Agricultural History Review*, 59 part 1 (2011), pp. 1–17; D.A. Davidson, 'Soils as cultural resources', in G. Fellows-Jensen (ed.), *Denmark and Scotland: The Cultural and Environmental Resources of Small Nations* (Edinburgh and Copenhagen, 2001), pp. 171–180; D.A. Davidson, G. Dercon, M. Stewart and F. Watson, 'The legacy of past urban waste disposal on local soils', *Journal of Archaeological Science*, 33 (2006), pp. 778–783; Hindmarch et al., 'Eldbotle', illus. 3 and pp. 281–282.
101 I.A. Simpson, 'Relict properties of anthropogenic deep top soils as indicators of infield management in Marwick, West Mainland, Orkney', *Journal of Archaeological Science*, 24 (1997), pp. 365–380. See also Oram, 'Waste management and peri-urban agriculture', p. 16; Oram, 'From "Golden Age" to Depression', pp. 207–211. For extreme examples of the impacts, see D.A. Davidson and I.A. Simpson, 'Soils and landscape history: case studies from the Northern Isles of Scotland', in T.C. Smout and S. Foster (eds), *The History of Soils and Field Systems* (Aberdeen, 1994), pp. 66–74.
102 Davidson and Simpson, 'Soils and landscape history', p. 69 and following.
103 J.G. Dunbar (ed.), *Sir William Burrell's Northern Tour, 1758* (Edinburgh, 1997), p. 62.

Lower Strathearn, Perth and Kinross, looking south-east from Milquhanzie Hill over the valley of the Inchaffray or Madderty Pow.

of the Orcadian response to climatic deterioration and environmental degradation. It was a response replicated in urban and rural contexts across mainland Scotland, where research into manuring practices on backland plots in burghs and peri-urban fields has revealed that turf stripped from nearby grassy areas provided the bulk of anthrosol-forming materials.[104] Soil profiles from locations as widespread as Wigtown, Lauder, Pittenweem, St Andrews and Nairn, in particular the last two, bear striking similarities to those from Marwick and show that the greatest efforts in these places were also focused on private gardens and fields rather than on areas of common cultivation. They are also testimony to the vast quantities of turf and topsoil removed through processes that effectively flayed the landscape of the burghs' hinterlands in one of the biggest processes of anthropogenic environmental transformation before the industrial era.

Returning to Borrowmoss in Wigtownshire, we encounter inroads made by members of one community on grazing, and land used for sourcing fuel by others, as a recognised 'ill' for which legal remedy was sought. In that specific case, however,

---

104 For detailed discussion, see Oram, 'Waste management and peri-urban agriculture'.

Deerness, east Mainland, Orkney. The turf-stripped expanse on the cliff-top north of Skaill is a legacy of the disastrous interplay between anthropogenic and natural processes, where the top layers were removed to deepen and enrich cultivated ground nearer the settlements, leaving the poorer subsoils to be eroded by wind and weather.

we also see an issue that recurs through most recorded cases of damage to arable and pasture in the sixteenth and seventeenth centuries, the continuation of exploitative practices that had evolved in times of more benign weather. In most cases, it was the stripping of the turf layer to provide soil-deepening materials but which rendered the land useless for grazing or exposed the underlying soil or peat to erosion. This was one of the ills that the 1685 act against tilling and delving was intended to prevent. It was, however, the disastrous interplay of human cultural practice and climate change at Culbin in Moray in the depths of the 'little ice age' that stimulated a second piece of legislation to halt avoidable damage in the most erosion-vulnerable locations. There, it was removal of dune-stabilising scrub, mainly for fuel needs, winter fodder and fencing materials, that was the primary focus of the legislation, but it should be seen alongside the provision of the 1685 act prohibiting the stripping of turf to make soil-enriching mulch. The response to the loss of the Culbin farmland was the 1695 'Act for preservation of meadows, lands and pasturages, lying adjacent to sand-hills'.[105] The law had responded, but local practice seemed little inclined to recognise the new legislation and there is scant evidence for action against those who broke its provisions.

Turf-stripping was not ubiquitous in coastal

---

105   RPS 1695/5/189 [accessed 9 December 2021].

Baleshare Beach, North Uist, Western Isles. The deep banks of sea-ware that build up along the ocean-facing coast of the islands were for centuries used to enrich and stabilise the calcareous shell sands of the machair.

districts, nor was it the sole source of organic material for agricultural fertiliser. In some coastal locations, especially where there were easily accessible and gently shelving beaches, the soil profiles reveal substantial inputs of shell sand and sea-ware. The latter, the deposits of fresh or partly rotted, wave-dislodged seaweed fronds washed up on beaches, has not been identified in analysis of areas of Neolithic to later Iron Age anthrosol formation, but its use commenced in the Middle Ages and it remained traditional soil-enrichment material down to the twentieth century in districts as widely dispersed as East Lothian and Fife, Shetland, the Western Isles, and Ayrshire.[106] Contrary to notions that sea-ware fertilising was mainly a sixteenth- and seventeenth-century development, which is an artefact of the skewing of dates of material detailing its use,[107] there is clear evidence for its employment in the fifteenth century. Possibly the earliest recorded reference to sea-ware dates from 1479/80 and relates to the lands of the archdiocese of St Andrews around the Firth of Forth.[108] That record was a confirmation of rights

106 A. Fenton, 'Seaweed as fertiliser', in J.R. Coull, A. Fenton and K. Veitch (eds), *A Compendium of Scottish Ethnology*, volume 4: *Boats, Fishing and the Sea* (Edinburgh, 2008), pp. 135–150.
107 Ibid.
108 RPS 1479/10/12 [accessed 22 December 2021]. For general discussion, see Oram, 'From "Golden Age" to Depression', pp. 209–210.

enjoyed from at least the 1440–65 period, a date which correlates well with the possibility that its greater availability was a consequence of the stormier conditions of the Spörer Minimum, which had turned sea-ware into an economically important asset whose ownership needed to be established. Indeed, the greater frequency with which it was included in charters from the 1540s onwards points to its emergence as an in-demand commodity of significant worth. Recognised as particularly effective for increasing the fertility and water-retention of sandy soils, beyond even good-quality farmyard manures, due to its particularly high potassium content and richness of organic matter, its lack of lime and phosphates required admixture of other materials to avoid the negative consequences of chemical imbalance in the soil. Nevertheless, the remarkable 'bounce' given to yields from sea-ware-enriched ground ensured that its application was regular and at volume. In most regions, its use expanded from the late 1400s, when North Atlantic swell deepened and extreme weather began to batter the North Sea and Atlantic coasts, delivering larger volumes of dislodged seaweed. Importantly, if its use as soil-enrichment relates to its greater abundance due to the increased storminess of the late fifteenth and early sixteenth centuries, it also illustrates the rapidity with which coastal agricultural communities embraced new opportunities amid profoundly unsettled times. Its greater use signals human adaptation to the changing environmental conditions and circumstances of the 'little ice age', with the inherent conservatism of traditional environmental knowledge being counterbalanced by resilient opportunism.

· · ·

An air of general crisis and anxiety pervades our record of the sixteenth and seventeenth centuries. Opening and closing with extended periods of extreme weather and human misery through famine and epidemic, and punctuated by repeated legislative efforts to mitigate shortages of foodstuffs, fuel and building materials, it is an era that is easy to characterise as the nadir in the last two millennia of Scotland's environmental history. *Prima facie*, there is little to challenge that view; it seems to be confirmed by the contemporary voice of those who witnessed and experienced the repeated shocks of climate-related crop failures and consequent famines, declining biomass production to sustain flocks and herds, and extremes of cold, heat and wet, without adequate fuel supplies to warm homes, cook food or process materials. Likewise, the language of desperation that permeates parliamentary acts and council decrees that aimed to conserve or augment resources perceived as under pressure is so persistent as to convince of the reality and ubiquity of crisis. Indeed, until recently, the seemingly unimpeachable authority of those contemporary voices helped to form some of the most powerful images of the environmental and ecological havoc wrought by humans on the face of the land; of woodland cover reduced to a post-glacial minimum through uncontrolled exploitation by voracious peasants, avaricious land-owners and uncaring commercial agents; of ignorant, culturally conservative farmers locked in dependence on traditional methods that further exhausted nutrient-depleted lands or stripped one area of its resources to augment those of another; of blind continuation of practices that were more suited to the milder conditions of former times; of competitor species hunted to extinction; and of peat mosses cut to oblivion through the insatiable demand for fuel.

But there is another side to the coin, albeit one that changed environmental pressures rather than removed them. Yes, woodland was still in retreat and the extremes of the solar minima of this era added to the human pressures, shrinking the altitudinal range of in-demand tree species and preventing regeneration in areas exposed to extremes of wind and weather. But repeated legislation and the spread of market-oriented economic thinking awakened landowners to the value of planned and protected plantations, laying the groundwork for a re-expansion of woodland in the eighteenth and nineteenth centuries. Selective planting was already encouraged in the fifteenth century, and by the late seventeenth century, as we shall see in the next chapter, it was timber value that drove choices and started trends that changed the character of Scotland's woodlands. Farming, too, is now understood as less conservative and more sophisticated than presented in traditional models framed by the prejudices of eighteenth-century Improvers. Experimentation and specialisation were already transforming the shape of the agricultural landscape, and even the very chemistry of the soil in some regions was changing as farmers used nitrogen-fixing legumes or added lime to fields. One of the most common themes to emerge from the weather record and the evidence for its impact on the manner in which Scotland's rural populations exploited the land is of the unrelenting toil to subsist in generally grim conditions; but there is a second, counterpoint theme that highlights widespread resilience and experimentation. As the material reviewed in this chapter has highlighted, even in the middle of some of the worst conditions of the period, communities were seeking to expand their agricultural operations and find new means to enrich the soil of their land. Expansion for some was principally a reaction to the shortfall in yields that the deteriorating climatic regime brought about, but for others it was a response to the rising market demand that supply shortages stimulated; desperation for some was entrepreneurial opportunity for others. Enclosure, too, was becoming more systemised and substantial in parts of the country, changing how the land within was used but creating new ecological islands from which wider transformation spread. Much of this change was driven by reaction to the long-term outcomes of past exploitation regimes. Some were purely economic and political choices, but others were pragmatic responses to system stresses and clearly changed environmental conditions. The shocks of the depths of the 'little ice age' helped to stimulate such responses. The innovation they encouraged marked the beginning of the great transition that grew in the following century.

CHAPTER 4

Beasts and Birds, Trees and Peat in the Sixteenth and Seventeenth Centuries

———

The dew as diamonds did hing
Upon the tender twistis ying,
Our-twinkling all the trees;
And ay where flouris did flourish fair,
There suddenly I saw repair
Ane swarm of sounding bees.
Some sweetly has the honey socht,
Whill they were cloggit sore;
Some willingly the wax has wrocht,
To keep it up in store.
So heaping with keeping,
Into their hives they hide it,
Precisely and wisely
For winter they provide it.[1]

Moving on from our discussion of the evidence for weather extremes and their impacts on Scotland's human population and their agricultural activities across the sixteenth and seventeenth centuries, the focus in this chapter shifts to wildlife and the natural resources of woodland and wetland. Even the most superficial reading of the written record can afford clear insights on the consequences for all of these things of the same extreme weather that blighted crops and devasted domestic herds and flocks. We see the quest for human food security spilling over into increased competition for sources of wild foods, along with increased control and management of them. These sources were principally the deer herds and wildfowl that were valued for their elite hunting opportunities, but also hunted illicitly by those who saw them as competition for their crops as well as an opportunity to supplement meagre diets. Competition for game meat, however, was not solely from human communities, as the most prolonged and coldest episodes of the sixteenth century also witnessed a sharp rise in action against Scotland's last

---

1 From Alexander Montgomerie, 'The Cherry and the Slae', in Alexander Mountgomery, *The Cherry and the Slae, with other Poems* (Glasgow, 1751), p. 4.

remaining packs of its largest wild predator, the wolf. At the opposite end of the scale, it is no coincidence that this period also saw an apparent decline in bee numbers and a consequent proliferation of thefts of hives, as the prized beeswax and honey produce fell into short supply. It is likely, moreover, that the dips in bee numbers also affected pollination of the fruit trees that were important features of gardens of even relatively humble houses. The dramatic fluctuations in weather extremes showed no favours to any living organism, human, domesticates or wild animals.

As already touched on in Chapter 3, the changing climate had effects on all forms of vegetation. In our sources, we can perhaps see evidence for such impacts in efforts to address rising pressure on woodland resources in the most populous districts of the kingdom. While the contraction of the area under trees in the Lowlands, Southern Uplands and round the eastern edge of the Highlands had continued throughout the later medieval period, the sixteenth century brought a perfect storm of increased demand for timber, bark and firewood. That rising demand coincided with intensifying grazing pressures and a matching era of colder and wetter conditions that were unfavourable to the regeneration of trees through natural seeding. It is in this context that we should understand the repeated legislation that first encouraged and then required woodland planting. It is also in this response that we should perhaps recognise the first moves towards one of the earliest and most important manifestations of 'improvement' that were being made in the last quarter of the seventeenth century: the creation of the first plantations of both native and non-native species on a scale larger than anything encouraged or required by the fifteenth- or sixteenth-century parliamentary legislation. It is no accident, either, that it is in that period of the severest weather impacts in the late 1600s that the first detailed reports were drawn up of the extents of existing woodland in all parts of the kingdom. The reports survive in the extensive collection of regional descriptions that were prepared for those seeking to exploit more systematically the available resources located in parts of Scotland distant from the main centres of consumption.[2]

The chapter concludes with exploration of the pressures on the supply of the main source of household thermal energy across most of Scotland and the islands, peat. It is ironic that peat might have been the one necessary commodity that grew well in the colder and wetter conditions of the 'little ice age', but that wetness served to intensify the shortages of available and usable fuel at a time when the colder weather triggered an increase in demand for longer periods each year. Wet weather and poor cutting techniques added to supply problems and also contributed to deterioration of the condition of the mosses during the worst episodes in the seventeenth century. Not surprisingly, against this backdrop of deteriorating weather and diminishing availability of adequate supplies of peat, there was a marked increase in competition for exclusive control of fuel resources, along with an escalation of violent confrontations where such efforts were challenged by others who claimed rights in this essential commodity. For the poorest and most marginalised in society, the grimness of their everyday existence became grimmer still as the changing climate delivered its new prescription of wetter, colder and stormier weather.

Reserving game

Nothing in our extensive historical record for this period indicates that any radical revision was made

---

2 Many of these accounts are published in Mitchell, *Macfarlane Geographical Collections*. For the woodland recorded by Timothy Pont at the end of the sixteenth century, see T.C. Smout, 'Woodland in the maps of Pont', in I.C. Cunningham (ed.), *The Nation Survey'd: Timothy Pont's Maps of Scotland* (Edinburgh, 2006), pp. 77–92.

in the laws surrounding the hunting of select 'game' animals and their reservation as 'sport' for a privileged few that had first been introduced in the twelfth century (see the volume 'Scotland AD 400–1400'). Yes, there had been successive changes in the territories designated as 'forest' through the medieval period, with some areas being 'disafforested' and ceasing to function as hunting land and others being brought into the legal status of 'forest', but most areas that continued as protected hunting land into the seventeenth century were already established as such by the early 1400s. This apparent stability, however, does not mean that the crown officers and the foresters and parkers of the select nobles to whom the right to have a forest or game reservation had been extended were oblivious to the changing environment on their property and its impact on the wild animals within it. Responses to such impacts might be physical, for example in efforts to improve woodland cover to provide shelter and better grazing opportunities for deer, but were as frequently legal, in terms of tighter protections, greater powers of exclusion, and increased penalties for poachers and other transgressors of private hunting rights. Reissued and confirmed parliamentary acts are a particularly important proxy for environmental conditions in that they provide indications of where and when pressure was recognised on particular species. The 1401 and 1458 acts to protect hares 'in time of snow', for example, were reaffirmed in 1535, 1552, 1567 and 1579,[3] dates that follow extended periods of exceptionally poor annual weather, but subsequent acts down to 1707 were designed mainly to control their hunting with guns, dogs and nets, poaching, and illicit close-season hunting. The sixteenth-century confirmations and their

Mountain hares (a) in summer colours in Glen Turret, Perth and Kinross, and (b) in winter colours in Glen Clunie, Aberdeenshire.

coincidence with sustained phases of poor summers and harsh winters, however, also point to some increase in incidences of illicit hare-hunting in winter months. This increase was perhaps a response to wider food shortages in rural districts but possibly also was due to a perception of hares as competitors for human winter forage foods and animal fodder.

Most of the sixteenth-century legislation concerning deer of all kinds likewise represented a continuation of medieval hunting laws and practice. Much of it concerned poaching within emparked or forest enclosures,[4] but from mid century the focus switched to the prevention of hunting with guns.[5]

---

3 RPS 1535/22, A1552/2/20, A1567/12/67, 1579/10/37 [accessed 31 December 2021].
4 For example, RPS 1504/3/30, 1535/19, A1555/6/37 [accessed 31 December 2021].
5 RPS A1551/5/3. This act was regularly restated down to the July 1587 parliament of James VI, where a strengthened act decried the ineffectiveness of the previous enactments and the failure of the competent magistrates to take action against perpetrators: RPS 1587/7/53. Ironically, James IV is recorded in 1508 as firing at the seabirds around the Isle of May with culverins from his ship: N. Macdougall, *James IV* (East Linton, 1997), p. 198.

According to 1551 enactments, the use of firearms had entirely destroyed the sport of the nobility, implying that the hunters were not members of the privileged elite to whom deer and wildfowl hunting had been reserved since the 1100s. Although effectively a reassertion of traditional rights reserved to the nobility, underlying the act there is concern over depletion of game which might point to over-hunting of species that were also under pressure from deteriorating climatic conditions. James VI's 1594 act for the better maintenance of royal parks and hunting land, which aimed to boost the numbers of deer and wildfowl for his personal sport, does allude to conditions whereby deer and people had come into collision for the same food resources.[6] It prohibited the slaughter of deer that had strayed during times of severe weather into barnyards and other private enclosures 'seeking their food'. In the hand-to-mouth existence endured by many peasant farmers in the later 1500s, it is unlikely that the king's increasingly shrill attempts to protect his favourite leisure activity had much impact on their survival instinct to kill their competitors.

Game birds and gannets

Wildfowl, too, had seen a raft of fifteenth-century legislation designed to protect their hunting value, but the first sixteenth-century legislation was their inclusion in a 1551 act to stop illicit hunting of all game with guns.[7] That act and two statutes of February 1552 that fixed the price of game-bird meat and curtailed the slaughter of lambs, partridge chicks and rabbits, spoke of 'the great and exorbitant dearth' of wildfowl throughout the kingdom, both migratory and non-migratory species.[8] The first 1552 act lists wild and domesticated fowl that were prized for their meat, the former including cranes, swans, barnacle, brent and greylag geese, plovers, all varieties of 'muirfowl', black and red grouse, all 'pout' (young partridge), curlews, wood-cock, skylarks, small songbirds, snipe and quail. Pheasant was added in a 1594 act, indicating that the new, exotic game species had by then been introduced in royal and aristocratic parks.[9] James VI remained sufficiently convinced of the devastation of game birds for hunting through illegal shooting and netting that parliament in 1621 was persuaded to tighten controls on the hunting of partridges, red and black grouse, heath-hens, ptarmigan, quails and wood-grouse (capercaillie) specifically.[10] These lists provide some insight on one reason for the perceived dearth of wild birds: the broad range of species being hunted; but fowling might have increased during the recent years of poor agricultural yields to supplement food supplies and – especially in the case of migratory geese – to reduce any threat to winter- and spring-sown grain crops. The coincidence of reissues of such legislation and periods of extreme weather-related stress on human food supplies is strongly suggestive of an upswing in illicit wildfowling at such times.[11] The second 1552

---

6   RPS 1594/4/31 [accessed 31 December 2021].
7   RPS A1551/5/3 [accessed 31 December 2021]. A 1599 act referred to increasing use of snares, nets and fowler dogs to circumvent earlier attempts to control wildfowl hunting, although some privileged individuals had explicit permission to net fowl, as in the case of the nets erected on the sands at St Andrews in the early 1600s for the archbishop's use: RPS 1599/5/1, 1612/10/69 [accessed 5 January 2022].
8   RPS A1552/2/2, A1552/2/15 [accessed 5 January 2022].
9   RPS 1594/4/31 [accessed 5 January 2022].
10 RPS 1621/6/42 [accessed 5 January 2022].
11 The run of legislation culminates with acts in 1698, 1705 and 1707 to reinforce earlier legislation and to reduce hunting through a prohibition on the sale of game birds in meat markets: RPS 1695/5/1, A1705/6/11, 1706/10/458 [accessed 5 January 2022].

Game-bird conservation in the sixteenth century

Two acts of the Scottish Parliament from the 1550s illustrate perceptions of a shortage of wildfowl (and other sources of meat), some notion of the reasons for that 'dearth', and measures introduced to prevent shortages in future. Both acts sought to impose a three-year moratorium on the widespread slaughter of suites of designated animals and birds, reserving the right to do so only to elite households or, when they were being hunted, only where they were hunted by lords who possessed the right to hunt with hawks. That the 1552 and 1555 acts ran consecutively in respect of wildfowl is strongly suggestive of some perceived crisis in the numbers of such birds during the 1550s, possibly as a consequence of over-hunting and poaching during a time of poor weather and associated shortages of more common food sources, but also possibly caused by some pathogen similar to avian flu.

> Item, forasmuch as the dearth of sheep, rabbits and wild meat daily increases, through the slaughter of the young lambs, young rabbits and young pouts of partridge or wild fowl, and to eschew such dearth in time coming, it is devised, statute and ordained by [James Hamilton, earl of Arran], my lord governor, with the advice of the three estates of parliament, that no manner of person or persons take upon hand to buy any lambs, to slay and bring to market to be sold, and that no lambs be slain by whatsoever persons except noble and great barons' houses for their meat for the space of three years, under the pain of confiscation of all such persons' goods and punishment of his person at my lord governor's will; and that no manner of person take upon hand to slay any young rabbits or young pouts except gentlemen and other nobles with hawks, or buy the same in market or otherwise during the said space under the pain foresaid (RPS A1552/2/15).
>
> Item, it is statute and ordained that no pouts, partridges, plovers, muirfowls, ducks, drakes, teal or golden-eye be slain until the feast of Michaelmas [29 September] yearly, under the pain of £10 to be taken and paid by the doer and breaker of this act to our sovereign lady [Queen Mary] and applied to her use, and that every earl, lord, baron, freeholder and other gentlemen, each within their own bounds, take the breakers of this act and hold them until the time that they find caution for payment of the said sum, and this act is to endure for the space of three years (RPS A1555/6/33).

That these acts were not repeated after 1558 perhaps indicates that there had been an appreciable recovery in the numbers of wild birds, or that the holders of hunting rights had become more effective in the policing of illicit hunting on their land. Their reinforcement over this period and the still more stringent hunting controls imposed by parliamentary acts through the seventeenth and early eighteenth centuries, however, helped to cement the perception of game birds as 'elite food' and remove them from the ordinary diets of most Scots.

act, and its various subsequent ratifications that exempted 'gentlemen' hunting with hawks or on their own property, point to the biggest inroads being at the hands of the non-noble populace, for whom wild meat was an essential replacement for their otherwise deficient or unaffordable staples.

It was not just for their meat that wild birds were exploited. Feathers, especially of geese and wild ducks like the eiders that abounded in the Hebrides, were highly prized. Bird fat was also a prized commodity, especially from the 'solan geese' or gannets that fed on the herring shoals at the mouth of the firths of Forth and Clyde, sought after as a medicinal treatment for rheumatic and arthritic pain.[12] Probably hunted for millennia for their feathers, meat and grease, such was the value of these

---

12 The most detailed recent environmental history discussion of the Bass gannets is in Smout and Stewart, *The Firth of Forth*, chapter 9.

Gannets on the Bass Rock, Lothian, photographed in May 2013. Known as 'solan geese', the young of the species were prized for their fat, which was obtained through boiling down the netted birds and was used as a base medium in many medical ointments and unguents.

birds that lords who controlled locations where they were found in number took measures to protect their 'property'. On Ailsa Craig at the outermost reaches of the Firth of Clyde, the Kennedy lords of Dunure and earls of Cassillis who had acquired the island from Crossraguel Abbey at the Reformation, built and maintained a tower there to regulate the exploitation of both the gannets and the fishery.[13] It was in the Forth estuary, however, that the most intensive exploitation of these birds occurred by the later Middle Ages. So important had this activity become that in 1491 Sir Robert Lauder, lord of the Bass, had sought papal permission to elevate the chapel in his castle on the Bass Rock into a parish church,[14] supposedly because of the difficulties in reaching the mainland parish church at Auldhame during stormy

[13] That Crossraguel maintained a chapel on Ailsa Craig in the fifteenth century points to both a population – perhaps seasonal – who needed spiritual care, but also teinds that could be collected: see NRS GD25/1/26. In 1629, John, earl of Cassillis, leased the fishery of the island to one Thomas Weir of Dunure for 200 merks for one year. Weir was to live on the island, perhaps in the tower, with six men for fishing: NRS GD25/8/215. The Kennedys were still receiving 'solan geese' for consumption in the mid eighteenth century: NRS GD25/9/18.

[14] *Calendar of Entries in the Papal Registers Relating to Great Britain and Ireland: Papal Letters*, xv, 1484–1492, ed. M.J. Haren (Dublin, 1978), no. 719.

weather, but in reality to give him exclusive rights to the teinds of gannet fat. His efforts were challenged by the nuns of North Berwick, who claimed rights to the annual teind of barrelled fat payable to Auldhame, and the dispute was resolved only in 1542.[15] The Lauders had their monopoly on the gannets confirmed by a commission in January 1584 and ratified again in 1592. Despite intensified predation, the birds were described as occurring on the Bass in greater numbers than almost anywhere else in the kingdom. These acts, in a move which shows an awareness of the impact of unmonitored inroads on a natural resource, also forbade all fishermen and sailors from the coasts of Angus, Fife and Lothian to catch the birds with nets, baited hooks and lines in the waters around the island. This, the act intoned, was because it was the mature – and relatively low-value – gannets that were being killed, reducing the breeding population and the production of the prized, fat-rich young.[16] Like much early 'sustainability' legislation, however, the act was designed to protect the profits of the principal supplier rather than to safeguard the species. Possession of the Bass came into royal hands in the seventeenth century and exploitation of the gannets became a profitable sideline for members of the crown garrison of the island's castle. In 1668, the Chancellor issued a lease of the right to collect 'all the young solan' on the Bass to Charles Maitland, the deputy governor of the castle, for a period of five years.[17] The continuance of this practice across these centuries perhaps provides some evidence for the effectiveness of the legislation to regulate the annual cull of the young birds. We have no idea of the numbers of birds slaughtered, but the toll was evidently sustainable given the continued value of the right into the later seventeenth century.

Honey and wax

Another species prized for its by-products, the honeybee, was also subject to legislation throughout this era. Although refined cane sugar had been imported to Scotland from the eastern Mediterranean for elite use since at least the early fourteenth century,[18] honey was the only natural sweetener available to most Scots until bulk processing of imported Caribbean sugar cane commenced in the Glasgow 'sugaries' in 1667. Scotland's 350-year addiction to refined sugar could be seen as contributing to the decline in bee populations, as the growing availability and relative cheapness of the cane and later beet sugar, coupled with its easier conversion into other products like rum or candy, reduced demand for honey. Production of honey was, of course, also very much subject to the vagaries of the weather in Scotland, whereas the imported sugar came from plantations that were unaffected by the more northerly North Atlantic and European climate phenomena of the LIA. On the one hand, cold and delayed springs affected the activity of hives, keeping the bees that survived the episodes of most intensive cold within the hive itself. On the other, these conditions also delayed the appearance of the wildflowers and blossoms on which the bees depended for their sustenance. A poor growing season was generally accompanied by low bee numbers and a consequent low volume of honey.

Beeswax was also in heavy demand for good-quality candles, especially those used in churches and the houses of the social elite, and for the seals with which documents were still authenticated through the sixteenth and seventeenth centuries. Beeswax candles were much preferred to the smoking and smelly tallow candles before the discovery of significantly cheaper and bulk-manufactured paraffin wax

---

15 NRS CH7/39; GD103/2/44; I.B. Cowan, *The Parishes of Medieval Scotland* (Edinburgh, 1967), p. 15.
16 RPS 1592/4/167 [accessed 5 January 2022].
17 NRS GD160/186 (28).
18 *ER*, vol. 1, pp. 96, 97, 119, 140, 144 etc.

Edzell Castle, Angus, the inner face of the eastern wall of the pleasance. The chequerboard-like small square openings were filled with flowers, but the large openings at a lower level between the chequer sections were recesses into which the woven straw or wicker 'skeps' (hives) for the bees were set.

in the 1830s. Most households of any status, therefore, are known to have kept multiple hives in the gardens adjacent to the house, evidenced by the 'boles' or mural recesses to contain skeps; some are still to be seen, for example in the early seventeenth-century walled pleasance at Edzell Castle. Although the pollination value of bees was not understood until its discovery by the German botanist Rudolf Jakob Camerarius at the end of the seventeenth century, they were normal introductions to gardens and orchards where they obtained the nectar for their honey from blossoms. Wild sources of honey and wax were probably already outnumbered by domestic hives in the earlier Middle Ages, and the domestic hives were classed as sufficiently important in economic and social terms to be regarded in law as items of private property. It is as such that they first appear in parliamentary legislation; James V's 1535 act to protect owners of doocots, yards, rabbit-warrens, parks and ponds specifically addressed the theft of bees or hives from private enclosures.[19] The 1535 act was renewed and refined regularly down to

19 RPS 1535/20 [accessed 5 January 2022].

1661, its five iterations down to that final year again coinciding, interestingly, with episodes of extreme weather that had perhaps inflicted high mortality on the flying insect population and driven people to steal viable sources of two essential commodities from luckier neighbours.

Purging that 'noxious pack'

Foremost among the animals that re-emerge in the records during this era, after a period of limited or no notice, is the wolf.[20] After the 1428 and 1458 acts that formalised wolf hunting and placed the onus for organisation of the hunts on local administrations (see above, pp. 21–22), wolves effectively disappear from crown or parliamentary records for almost a century. It may be an artefact of the greater survival of local documentation from the later medieval and early post-medieval periods, but there is an upsurge of references to wolves dating from the coldest episodes of the Spörer and Maunder minima. This renewal of references to wolves after the lull that followed the burst of mid-fifteenth-century sources discussed earlier mainly concerned predation on the sheep flocks of landowners in the eastern Perthshire and Angus glens. These references provide a glimpse of the impact of the extreme winters of the 1550s on wolves' wild prey, forcing packs into territory with more abundant hunting opportunities in the domesticated flocks and herds found there. This material relates mainly to lease conditions on properties in the south-east Highlands, such as the records for the Glenisla lands of Coupar Angus Abbey. Tenancy agreements from the 1480s onwards acknowledge the 1428 and 1458 acts in their requirement that the tenants should 'rise' and join their neighbours 'in defence' of the abbey's flocks from wolves as well as human predators. This requirement became an especial consideration after a decree of the abbot in October 1546.[21] A 1552 tack of lands in upper Glenisla adds further detail to the conditions imposed by the monastic landlords, revealing that their tenants were required to keep a 'leash of good hounds' for hunting foxes and wolves when needed.[22] Such documents are strongly suggestive that the hill-country of the Highlands' southern edge saw regular visitations from wolf packs during the worst winters of the early sixteenth century.

Although references are plentiful for north-east Perthshire and northern Angus, wolves were said to be more abundant in the northern and central Highlands than on their southern and eastern edges. Identified as locally hunted beasts in Caithness in a 1543 appeal to King Henry VIII of England, and again in a 1588 description of Strathnaver, they appear consistently as a threat to livestock and even to the human population.[23] This threat to human travellers was reported by the English chronicler, Ralph Holinshed, who claimed that the 'Spittals' constructed along Highland land routes were, among other functions, for the overnight protection of travellers against wolf attack.[24] No verifiable attacks on human victims are recorded, but individual acts of wolf predation on livestock in central Highland districts are noted in records. These include references to an ox killed by a pack near Inverness in 1570 and the two-year-old cow, four mares and a year-old horse slain in Breadal-

---

20 Wiseman, 'A noxious pack'.
21 Rogers, *Cupar-Angus Rental*, vol. 1, p. 236 (lease of the lands of Inverharity in Glenisla); NRS GD16/36/3. The act, which was directed against inroads of thieves and Highland raiders as much as wolves, was reaffirmed in October 1589 and arrangements were made for watch parties as late as July 1608.
22 Rogers, *Cupar-Angus Rental*, vol. 2, pp. 107, 141. See also p. 251 for similar requirement on the tenants of Nether Illrik.
23 Wiseman, 'A noxious pack', pp. 112–113.
24 Ibid., p. 115.

Assynt's mountainous interior, looking west from Ben More Assynt, Highland. Since they were first depicted as a refuge of wolves on the medieval Gough Map, the mountains of western Sutherland were reputed as the haunt of dangerous wolf packs into the seventeenth century.

bane in 1594.[25] The numbers of such occurrences, however, appear surprisingly low against popular traditions of widespread wolf-depredation.[26] What is more striking is how close to major population centres wolf packs were roaming during these periods of winter severity, highlighting further the impact of extreme weather on their normal prey. Severe conditions which reduced hunting of their naturally occurring wild prey were pushing wolves and humans into direct competition; for the wolves, the result was renewed persecution.

There was a further peak in action against wolves during the run of poor years at the start of the seventeenth century. A legal complaint from 1608, for example, referred to the firing of a hagbut while in pursuit of a wolf in Assynt.[27] In Moray, various men were charged with carrying firearms between 1617 and 1623, but secured acquittal on their argument that they were engaged in wolf hunts.[28] In Breadalbane, act 15 of the 1621 barony court required all tenants to make four 'croscattis' (short-shafted spears with cross-bars below the head) 'for slaying of the

---

25 W. Mackay and H.C. Boyd (eds), *Records of Inverness*, volume 1 (Aberdeen, 1911), p. 197; Anon., *Black Book of Taymouth*, pp. 289, 298.
26 The folk traditions discussed in Wiseman, 'A noxious pack', pp. 115–117 contain motifs common to most European regions with wolf populations.
27 *Register of the Privy Council of Scotland*, volume 8, *1607–1610*, ed. D. Masson (Edinburgh, 1887), p. 331.
28 *Register of the Privy Council of Scotland*, volume 23, *1622–1625*, ed. D. Masson (Edinburgh, 1896), p. 303.

Mam na Cloich Airde shieling ground, bealach between Glen Dessary and Finiskaig glen, Wester Ross, Highland. High-altitude shieling grounds such as this were under pressure from over-grazing during periods of reduced grass growth in the depths of the 'little ice age'.

wolff yeirly in tyme cuming'.[29] References to wolves, however, dwindle through the seventeenth century, suggesting that hunting for their skins and to reduce possible predation on livestock had made serious inroads on their number.[30] Even if the tradition recorded in the biography of Sir Ewen Dubh Cameron of Lochiel is incorrect, that Scotland's last wolf was shot by him in 1680 near Killiecrankie, there are multiple local traditions of the slaying of the last wolf in the 1680s or 1690s.[31] These folktales confirm that Scotland's apex wild predator was the most significant casualty of the Scottish people's struggle to survive through the worst years of the 'little ice age'.

29 Anon., *Black Book of Taymouth*, p. 356. For croscattis, see Wiseman, 'A noxious pack', p. 114.
30 Wolfskins were an export commodity listed in the 1661 act for bullion equivalence: RPS 1661/1/338 [accessed 20 December 2021]. The fact that the value was given in multiples of dickers (bundles of ten skins) suggests that the numbers of animals being killed annually was sufficiently high to merit inclusion in the bullion valuation 'alphabet'.
31 Wiseman, 'A noxious pack', pp. 123–131. As Wiseman points out, however, numbers may have fallen so low in this period as to push it to marginal existence, and extinction possibly occurred in the following century.

Pasture and livestock

With wolves, communities could take direct action to eradicate what they perceived as a threat to their wellbeing. There was nothing that they could do about annual temperature variations, which were the far greater threat to human populations, for everything that they grew, raised or caught and consumed was dependent on solar radiation and warmth. Even the most common of vegetation, grass, requires temperatures above a minimum threshold to grow, and drops below that threshold for any extended period could affect both the volumes of grass available and the overall quality of the grazing. Grass is resilient and will grow under most conditions. The exceptions are extreme drought and winter daytime temperatures under a 6°C threshold, when the cold and lack of sunlight slows or halts its growth. Low temperatures persisting into the spring and starting to occur earlier in the autumn could also significantly reduce the growing period. For communities with access to both summer and winter pasture, usually with an altitudinal difference as well as horizontal distance between the two, the ability of grass to grow in mild winter conditions could enable cattle and sheep to be grazed outdoors all year with little or no fodder supplement, or only relatively short periods in the byre during the worst seasonal weather. The extremes of the Spörer and Maunder minima disrupted those certainties by removing the seasonal predictability of grass growth that is a central strand of traditional environmental knowledge for all livestock-rearing cultures. The delayed springs of the early and mid 1500s, repeated frequently in the seventeenth century, affected transhumance regimes through slowing new grass growth in the high summer pastures, while waterlogging in summer and autumn increased damage to shieling grounds through trample pocking and destruction of the root mat, leading to erosion especially on hillsides. Even in areas of more horizontal transhumance like the Uists, where summer grazing from west-side communities on the eastern moorland areas intensified during the sixteenth century, loch sediments preserve a record of probable livestock-triggered erosion of land whose carrying capacity for grazing had been exceeded as the climate deteriorated.[32] Cooler summers, as reflected in the Cairngorm tree-ring record for the 1630s and 1690s,[33] probably also saw an earlier end to growth on upland pastures, but possibly not before hungry cattle, sheep and goats had over-grazed the vegetation. Together, such delayed starts and earlier ends to the growing season on summer pastures intensified pressure on lower-altitude areas of foggage.

Can we see evidence for responses to pressure on grazing land in the sixteenth and seventeenth centuries? It has been argued that the calculation of soums, the notional pasture extents required for the support of fixed livestock numbers at different seasons, was adjusted periodically to reflect the reduced or enhanced carrying capacity of the land and so influenced stocking densities or flock and herd sizes.[34] There is some sixteenth-century evidence for such adjustment in the Southern Uplands, where the souming of the common grazing in the barony of Carnwath was recalculated in 1542 on advice from 'sartane of the eldest & of best ondirstandis' (certain of the eldest men of best understanding), specifically to reflect what 'the cowmownd ma beir' (the common may bear),[35] in a striking display of confidence in the traditional environmental knowledge

---

32 Bennett et al., 'Holocene vegetational and environmental history at Loch Lang'.
33 Rydval et al., 'Reconstructing 800 years of summer temperatures in Scotland', p. 2960 and Table 2.
34 See discussion and references in Ross, 'Scottish environmental history and the (mis)use of soums'.
35 W.C. Dickinson (ed.), *The Court Book of the Barony of Carnwath, 1523–1542* (Edinburgh, 1937), p. 210; A.J.L. Winchester, *The Harvest of the Hills: Rural Life in Northern England and the Scottish Borders, 1400–1700* (Edinburgh, 2000), p. 83.

of community elders. Thereafter, those tenants with common grazing rights could only put the re-assessed number of animals 'as efferis to his mallyng' (as appertains to his rented holding) onto the common pasture. Such responses were perhaps replicated elsewhere but, despite a greater survival of record evidence post-1600, we lack long runs of annual or at least regular statements of souming levels that would show patterns of adjustment made over years and decades to make us confident that landowners and their tenants were sensitive and reactive to environmental change, or indeed other social, economic or political factors.[36]

Soums also have been used as evidence to support claims that a shift towards more market-oriented management of estates in the southern and central Highlands commenced in the mid seventeenth century.[37] While the evidence of the soums in this regard is more ambivalent than the proponents of that thesis have argued, there are grounds for seeing a more commercial approach to livestock management on, for example, the Breadalbane and Atholl estates, which benefited from proximity to lowland markets. Commercial opportunity lurked beneath Campbell and Murray advocacy of the establishment of a new royal forest in Glen Almond in the 1620s – Lord Glenorchy even offering to introduce deer to the glen for the king's sport – but both he and Sir Patrick Murray were interested chiefly in securing the keepership of the new forest and that office's associated control over rights of grazing there.[38] By the end of the century, Glenorchy's Breadalbane and Murray's Atholl heirs were running large sheep flocks on the pre-existing fermtouns within the short-lived royal forest. This use intensified in the eighteenth century as the multiple-tenancy touns were consolidated into single-tenant farms.

Campbell and Murray used status and connection to secure legal title to what they wanted but, as with the arable encroachments noted in Chapter 3, there is also evidence for communities seeking to augment grazing resources by simply putting their animals on other people's ground. Thus, amid the run of poor winters recorded in the southern Highlands in the 1550s, the abbot of Coupar Angus required his forester in Glen Brighty at the head of Glenisla to ensure that no men from Strathardle or Glen Shee brought their herds from their pastures over the watershed at the head of the glen and grazed their cattle on the abbey's summer pasture there.[39] Similar pressures for access to grazing underlie a legal dispute in this same period between three landowners on the border between Fife and Kinross-shire, where one party was accused of digging up the march markers

---

36 Ross, 'Scottish environmental history and the (mis)use of soums', p. 217.
37 Dodgshon, *Land and Society*, pp. 255–265; R. A. Dodgshon, 'Livestock production in the Scottish Highlands before and after the Clearances', *Rural History*, 9 (1998), pp. 19–42.
38 NRS GD112/39/32/14: Sir Patrick Murray to Lord Glenorchy, thanking him for his offer to stock the glen with deer and to make it a forest; GD112/39/33/27: Lord Glenorchy to Sir Patrick Murray '. . . albeit his majestie hes mony forests in Scotland Ingland and Irland trewly I beleiff he hes not ane forrest that he may chace ane dear downe ane watter saxtein mylis of lench upon hors bak and a hundredth pundis starling will mak the way that he may gallop in his cotch'. For the royal extension of the forest and negotiations with neighbouring lairds over their compensation for loss of commonty rights, see NRS GD124/10/243, GD124/10/250, GD124/10/255, GD124/10/260, GD124/10/292, GD124/10/301, GD124/10/321, GD124/10/324. See also, for example, NRS GD112/29/8 no. 18, for Glenorchy's grazing land around Auchnafree in the upper glen, GD112/39/19/9 for his cattle there, and GD112/39/33/11 for the marching of lands along the whole north side of the glen. Glenorchy and Murray interests there had clashed as recently as 1622: GD112/39/32/9. Murray was appointed sole keeper in March 1623 but promised to work to secure Glenorchy any second keepership: GD112/39/33/5; GD112/39/33/11.
39 Rogers, *Cupar-Angus Rental*, vol. 2, pp. 267–268.

to extend his pasture at the expense of the others.[40] Again, there is always a degree of opportunism to be expected in such cases, yet the proliferation of litigation over pasture during times of subsistence crisis does equally suggest expediency born of desperation.

Some livestock management practices established before the end of the fifteenth century persisted after 1500, especially trends towards specialisation in stock types on royal and larger monastic and aristocratic estate complexes. There was regular and large-scale flock and herd movement between components of those estates and the gearing of the estate economies towards bulk delivery of produce to the central household or to market. As in earlier centuries, the crown was a key player in this livestock economy, running its own large flocks and herds on its demesne properties spread from Galloway and Ettrick to Islay and Ardmeanach, or drawing rents on its pastures that were let to tenants.[41] Seasonal movement on a grand scale was intended to maximise the potential of summer and winter pasture to support large numbers of animals, which enabled their owners to deal at scale in the wool, hides and meat markets. Some of the most detailed evidence comes from the Coupar Angus rentals, which record how by the late 1540s the abbey's sheep flocks – usually around 500 head per flock – were being sent annually in late spring to Glenisla and Glen Brighty from the winter foggage grounds in central Strathmore, returning to the lowlands after harvest time.[42] The specialisation in sheep on the central Borders estates of the earls of Buccleuch by the 1630s also represented a continuation of practice from the former Melrose Abbey properties that the Scott family had acquired.[43] Like the Coupar Angus estates, the emphasis in particular components of the Scotts' estate in the Teviotdale and Ewesdale uplands was on wool rather than meat production, maintaining the traditional patterns established by the Cistercian monks in both cases. Elsewhere, however, there was a marked trend towards livestock management for meat and hides but also for dairying. In Galloway, it was for the meat trade that some important property-owners began to develop their grassland assets at the expense of arable.

Although the main expansion of Galloway's cattle-oriented regional economy has been seen as occurring post-1661, when Charles II's enclosure act facilitated the privatisation of grazing and the construction of large cattle 'parks', western Galloway was already functioning in the late 1500s as a source of live beef cattle.[44] This role developed further in the early 1600s, when several regional landowners were tied closely to the Ulster Plantation and began the movement – sometimes illicitly – of cattle from their Irish to their Scottish properties. This connection helped to further stimulate regional specialisation later in the century. A leader in this late seventeenth-century commercialisation of grazing, dairying and meat supply was Sir David Dunbar of Baldoon in the Machars, south of Wigtown, who established grassland parks in the properties adjoining his castle on which he pastured around 1,000 head, including calves. He ran a 200-strong dairy herd and fattened around 400 beef cattle and oxen for sale

---

40 NRS GD254/31.
41 For example, a herd of 96 live cattle was transported to the mainland from Islay in 1506: *ER*, vol. 12, p. 709.
42 Rogers, *Cupar-Angus Rental*, vol. 2, pp. 236, 261, 268.
43 For Buccleuch's sheep in Roxburghshire, see NRS GD224/393/6 nos 2, 3, 15, 16, GD224/394/6 nos 3–5, GD224/395/6 (chamberlain's accounts).
44 Whyte, *Scotland Before the Industrial Revolution*, pp. 141–142. For live beef cattle from Wigtownshire, see the dispute between the burghs of Dumfries and Wigtown brought before the privy council in December 1598: *Register of the Privy Council of Scotland*, volume 5, *1592–1599*, ed. D. Masson (Edinburgh, 1882), pp. 505–506.

Belted Galloway bull, Mochrum, Dumfries and Galloway. Although the Belted Galloway was only formalised as a distinct breed in 1921, it was descended from the animals that had made the region famous as a source of beef cattle by the sixteenth century.

either to drovers or direct in northern English markets.[45] He seems to have been exceptional in the scale and precociousness of his development; Andrew Symson in fact referred to his Baldoon park as 'the mother of all the rest'. Other families, including the Herons of Kirroughtrie and the Blairs of Rusco in the Stewartry, were also early specialists, developing their parks from initial grazing leases granted to men involved in the droving of cattle from Ireland.[46] Before the end of the century, from Ayrshire and Galloway to the Lothians and the Merse, or northwards through Renfrewshire,

---

45 A. Livingston, 'The Galloway Levellers: A Study of Their Origins, Events and Consequences of Their Actions', unpublished MPhil (Research) thesis, University of Glasgow (2009), pp. 20–21; Andrew Symson, 'A large description of Galloway and the parishes in it', in Mitchell, *Macfarlane Geographical Collections*, vol. 2, pp. 99, 107.

46 Livingston, 'The Galloway Levellers', pp. 25–30. For detailed explorations of the trade, see I.D. Whyte, *Agriculture and Society in Seventeenth Century Scotland* (Edinburgh, 1979); D. Woodward, 'A comparative study of the Irish and Scottish livestock trades in the seventeenth century', in L.M. Cullen and T.C. Smout (eds), *Comparative Aspects of Scottish and Irish Economic and Social History 1600–1900* (Edinburgh, 1976), pp. 147–164.

northern Lanarkshire, Dunbartonshire and along the southern edge of the Highlands to Angus, a landscape of grass parks had been developed to accommodate growing numbers of beef and dairy cattle and to service a gathering volume of animals being driven south from the Highlands. A key step towards the commercialised farming of the 'Improvement era' had been taken already long before 1700, but it was in the eighteenth and nineteenth centuries that the ecologically transformative impacts of this specialisation would become widespread and entrenched.

A final dimension of grazing and livestock to be considered here is the long-term shifts that were consequent on the major climatic cooling that occurred in Maunder Minimum. The impact of the extremes of cold and storminess on woodland (discussed below) are well understood, but the general resilience of grass has meant that it has rarely entered the discussion. Possible contraction of the viable grazing season in upland summer pastures has already been identified as one consequence. It is likely that in the 1690s this shortened season caused the permanent abandonment of some of the highest-altitude shieling grounds in the central Highlands, as the nutrient-rich grasses failed to recover adequately to sustain large herds. It is also likely that grazing at much lower altitudes in the Western Isles was also affected, with more weeks over winter and spring failing to reach the 6°C threshold for grass growth.[47] With few available sources of nutrients other than grass or seaweed for overwintering cattle even before the seventeenth century crises, the succession of poor years in the 1690s may have helped to further entrench the already well-established dairying practice of calf-slaughter.[48] At one end of the country, there was steady movement towards enclosure and the intensification of grazing opportunities in districts less affected by the dire weather of the 'Seven Ill Years'; at the other, exposed to the full force of nature's extremes, livestock farming was locked tightly in practices that had developed centuries if not millennia earlier.

Trees and woodland: from maximum depletion to first planting 'policy'

As for earlier centuries, the state of Scotland's woodland has been used as a key measure for the scale of environmental change, habitat loss and general ecological vitality around Scotland in the two centuries after 1500. Our record evidence from this era most often presents a far from positive picture, which has led to a long historiographical – and 'romantic' – tradition of the 1500–1700 period as marking the absolute nadir for Scotland's woodland.[49] The pressure on woodland can be inferred from this era's repeated legislation to protect trees and encourage new planting, and from litigation against illegal felling, barking and peeling of bark for tanning. Accounts of increasing imports from Scandinavia and the Baltic, or the securing of sources in remoter parts of Scotland's interior, further convey a sense of deepening anxiety in the kingdom's rulers at the continuing inroads on, and poor management of, the remaining woodland expanses within their realm. They also appear to underline the inability of both local and national jurisdictions to stem the tide of woodland loss, and the ineffectiveness of James

---

47 See discussion in Serjeantson, *Farming and Fishing*, pp. 30–31.
48 For calf-slaughter, see ibid., pp. 56–57; McCormick, 'Calf slaughter as a response to marginality'.
49 For a critique of this tradition, see Smout et al., *History of the Native Woodlands*, pp. 20–25, 45–47. The interplay of climate and tree growth in this era on mainland European woodland is explored in K.R. Briffa, P.D. Jones, R.B. Vogel, F.H. Schweingruber, M.G.L. Baillie, S.G. Shiyatov and E.A. Voganov, 'European tree rings and climate change in the sixteenth century', *Climate Change* 43 (1999), pp. 151–68.

I's 1428 and James II's 1458 legislation to protect woods and encourage planting. The lamenting tone of James IV's March 1504 parliamentary act, which describes how 'the woods of Scotland are utterly destroyed',[50] has conditioned historical perceptions of later legislation and has also become the launching point for many modern calls for reafforestation. Here, it seems, is proof positive in the historical record for the scale of anthropogenic impacts on the Scottish landscape at the beginning of the sixteenth century. The hyperbole of the act's language, however, is now recognised as being more of a rhetorical flourish designed to contextualise the proposed legislation, and the act itself is understood as being perhaps more representative of the king's frustration at the lack of availability of the large and expensive timbers he wanted for his ship- and palace-building projects than of the reality of the state of the nation's woods.[51] Lack of availability of the most in-demand trees closest to the areas of highest demand does not necessarily mean their total absence, nor does it mean that other species and forms of managed growing timber were also absent. Multiple records contemporary with this parliamentary lament for Scotland's lost forests reveal that apparently extensive woodland existed in many parts of James IV's kingdom; the issues were accessibility and suitability for his purposes.

A brief digression is necessary here to explore that question of location and suitability of the woodlands in what were regarded by the rulers, increasingly based as they were in the central Lowlands, as outlying districts of Scotland. As shown by James I's dispatch of his barge in 1435 to bring timber south from the northern districts of his kingdom (see above, pp. 66–67), there was knowledge within royal administrative circles at least of areas of potentially exploitable woodland in the Highlands. What was not well understood – except to local residents – until later in the sixteenth century, however, was the physical extent and diverse composition of such woodland. A major advance in knowledge of the scale and nature of the northern woods was brought about by the first systematic mapping project to be undertaken in Scotland, the work of the mathematician, topographer, cartographer and Kirk minister Timothy Pont, which he carried out between c.1583 and 1614.[52] It has been observed that there is a possible correlation between Pont's survey and the passing of parliamentary legislation in the Convention of January 1609 to control the development of new ironworking sites in Highland districts using charcoal from 'certain woods in the highlands, which woods by reason of the savageness of the inhabitants thereabouts were either unknown or at the least unprofitable and unused'.[53] Lowland authorities, it seems, had woken up to the economic potential of this resource and were stepping in to assert control over its future exploitation before private interests intervened opportunistically.[54] The interest and intervention of the crown's officers, however, were driven by a wish to regulate development and, of course, to profit from that regulation, not simply to conserve the trees. It is clear that they believed the wood was appropriate for a range of uses from structural timbers to charcoal production. It is the sequel to

---

50   RPS A1504/3/116 [accessed 7 January 2022].
51   Smout et al., *History of the Native Woodlands*, p. 45.
52   The most comprehensive discussion of the maps is in the collection of essays I.C. Cunningham (ed.), *The Nation Survey'd: Timothy Pont's Maps of Scotland* (East Linton, 2001). The surviving maps and texts are in the National Library of Scotland and can be accessed digitally through https://maps.nls.uk/pont/index.html.
53   M. Stewart, 'Using the woods, 1600–1850 (1): The community resource', in T.C. Smout (ed.), *People and Woods in Scotland* (Edinburgh, 2003), pp. 82–104 at 87; RPS A1609/1/10.
54   Smout et al., *History of the Native Woodlands*, p. 229.

this 1609 act rather than its context, though, that has most exercised the minds of antiquarians, historians and, more recently, restoration ecologists, for what followed has been held up as one of the most destructive episodes in the history of the native woodlands of the north-western Highlands specifically.

The coming of the first ironmasters

Since the nineteenth century, much of the discourse surrounding the fate of Highland woodland in the seventeenth and eighteenth centuries has pointed accusing fingers at the impact of Lowland Scottish and English ironmasters who established furnaces in areas with abundant woods from which to make their primary smelting fuel, charcoal.[55] At the forefront of this discourse was the record of the activities of Sir George Hay, a prominent Lowland speculator who had already been involved in exploitative development initiatives in the Western Isles. Having secured a royal commission and licence in 1610 to make iron and glass anywhere within Scotland, and in 1611 having bought a thirty-one year lease of woodland on the Mackenzie estates in Wester Ross, he has been presented in the past as the agent responsible for the devastation of the semi-natural woods that Pont had recorded in the district around Loch Maree.[56] Pont had described the loch as being

> compasd about with many fair and talle woods as any in al the west of Scotland, in sum parts with Hollyne, in sum places with fair and beautifull fyrrs of 60, 70, 80 feet of good and serviceable timmer for masts and raes, in other places ar great plentie of excellent great oaks, whair may be sawin out planks of 4 sumtyms 5 feet broad. All thir bownds is compasd and hemd in with many hils but theis beautifull to look on, thair skirts being al adorned with wood even to the brink of the loch for the most part.[57]

Nearby in the coastal district, Pont also described an abundance of 'very fair firr, hollyn oak, elme, ashe, birk and quaking asp, most high, even, thick and great'.[58] In his deal with Mackenzie, which also covered the woods around Loch Carron, Loch Alsh and Loch Duich, Hay secured the rights to almost all of these species, defined in the deal as all of the oak (one cutting only), pine (leaving sufficient trees for Mackenzie to build his galleys), ash, elm and aspen, half of the birch, hazel, holly and lesser trees. Despite this natural abundance, by 1624 it was being reported to Hay that the furnace had ceased operations and that the woods around Loch Maree were by then inadequate as a dependable source of fuel for work to resume.[59] How much of this was due to felling for charcoal manufacture or arose from Hay's wider timber-cutting operations is unclear and unknowable,[60] but it is evident that by the time the last efforts to establish ironworking in this district petered out in the 1660s, the woodland in the district had been transformed.

---

55 For discussion of the nineteenth- and twentieth-century historiography, see Chapters 8 and 9 below. The development of these furnaces was explored in detail in J.M. Lindsay, 'The iron industry in the Highlands: charcoal blast furnaces', *Scottish Historical Review*, 56 (1977), pp. 49–63.

56 The history of Hay's short-lived industrial operations at Loch Maree have been discussed in detail in Smout et al., *History of the Native Woodlands*, pp. 229–236.

57 NLS, Pont Maps of Scotland, ca. 1583–1614 – Pont texts, 'Lochew and Letyr-ew', pp. 119v–120r https://maps.nls.uk/pont/texts/transcripts/ponttext119v-120r.html.

58 Ibid.

59 Smout et al., *History of the Native Woodlands*, p. 232.

60 For Hay's timber business and its activities on the Mackenzie lands, see ibid., p. 199.

Oak-dominated broadleaf woodland, near Kinlochewe, east end of Loch Maree, Highland. The species diversity of this woodland is similar to the suite of oak, pine, ash, elm and aspen listed in Sir George Hay's 1611 wood contract.

But was the woodland that Pont had so graphically described 'wildwood' in any sense and does the traditional narrative of its rapid wasting and permanent destruction bear scrutiny? From the stipulation concerning the single felling of oak within Hay's thirty-one-year lease and the retention of pines suitable for Mackenzie's shipbuilding needs, it seems likely that oak at least was already being managed on a rolling coppice basis and pine was also conserved for lordly use. Given wood management practices elsewhere in the Highlands – and across Scotland – and recorded from earlier centuries, it is unlikely that the other named species were not coppiced or managed on a felling and regeneration basis by the local population long before 1600. New wood-supply agreements made in the 1620s are also strongly indicative of established coppice management on the Mackenzie estate.[61] Collectively, while not conclusive evidence, this record is strongly indicative of the Loch Maree and neighbouring woods as long subject to human exploitation, and thus 'semi-natural' at best, rather than some remnant of a primeval 'wildwood'. In one recent assessment of the impact of the charcoal manufacturing and timber operations on these woods, it has been suggested that the early seventeenth century saw the conversion of old growth

61 Ibid., p. 235.

woodland into secondary growth, but the loss of the older, established trees exposed the new growth to the hazards of uncontrolled grazing.[62] With no evidence of haining or enclosure of any kind, the woodland extent is likely to have reduced in range and possibly become ever more fragmented, but while the nature of the woods was certainly changed through these operations it was not wholly lost.[63] Indeed, substantial woodland with high-quality trees was identified around Loch Maree in the eighteenth century as ripe for extraction and there is still an important biodiverse woodland around the loch today.

Abundance and scarcity: the case for woodland protection

Returning now to our discussion of the legislation, we can see in light of the evidence presented at the end of the sixteenth century by Pont, and from the activities of speculators like Hay, that much of it was framed with a Lowland context in mind. The language of utter destruction present in James IV's lament for the woods of his kingdom, moreover, was a hyperbole directed at the condition of woodland in only a small part of the realm. Northern and western Highland lords might well have raised an eyebrow at the generalised picture of wasted forests presented in the legislation. It is even more likely that they and their officers would not have recognised in the coppice-managed woodland resources that existed on their lands the kind of destruction and failure of regeneration that have been read into this act's words.

There is, however, a negative to balance the positive. Pont's maps – and the early seventeenth-century maps of Robert Gordon of Straloch – indicate that while James IV's language was misleadingly inappropriate for large parts of the Highlands, it was perhaps very appropriate for Lowland districts. How widespread landlord responsiveness in those parts of the kingdom had been to James II's 1458 act concerning woodland, and, consequently, its effectiveness as an instrument of change, can be measured in the 1504 act's increase of the fines for cutting and burning 'living wood'. The relative ineffectiveness of that 1504 act as well is emphasised by the need for a further restatement and expansion of the 1458 provisions in James V's 1535 parliament.[64] From this repeated legislation, there was clearly both a perceived problem and a reality of scarcity on at least a regional basis. Setting aside the mantra of utter destruction, it must be acknowledged that woodland depletion, especially in the Southern Uplands and central/eastern lowland districts, had clearly continued despite the 1458 act. In fact, in some areas depletion had possibly been hastened by royal policy rather than slowed or prevented.[65] To counter the perceived decline, the 1535 legislation was more detailed than its precursors, in part reflecting the growing reach and intrusiveness of king and parliament into areas where the public and private spheres overlapped, but also through recognition of the 1458 act's vagueness. It provided a clear exposition of the 'policy to be had within the realm' in respect of planting of woods, making of hedges, orchards and yards, and sowing of broom. Going beyond the provisions of James II's act, it made the additional requirement that 'every man spiritual and temporal within this realm having a hundred pound land of new extent per year, and may spend so much, where there are no woods or

---

62 Ibid., pp. 235–236.
63 T.C. Smout, 'Highland land-use before 1800: misconceptions, evidence and realities', in T.C. Smout (ed.), *Scottish Woodland History* (Edinburgh, 1997), pp. 5–23 at 16–17.
64 RPS 1535/16 [accessed 29 November 2021].
65 Smout et al., *History of the Native Woodlands*, p. 39.

forests, he should plant woods and forests and makes hedges and enclosures for himself extending to three acres of land, and above or under as his heritage is more or less in places most suitable'. Landlords were thus encouraged to plant on their demesnes and to take direct action to do so rather than simply offloading responsibility onto their tenants, as the 1458 act had encouraged. James V's parliament had presumably recognised that the 1458 act's enabling powers had been relatively ineffective in securing the desired result. Under the 1535 act landlords retained coercive powers over their tenants, being empowered to 'cause every tenant of their lands that has the same in tack and assedation to plant upon their dwelling-site yearly for every merk land one tree'. Failure to comply, however, was judged against the lord, not the tenant, with a *pro rata* fine of £10 (with the norm being a £100 land) imposed for each failure. Again, unlike the 1458 act, the new legislation established monitoring processes, empowering the king to make an annual inquisition into compliance and ordering all men to begin planting from the next season.

Given James V's reputation for rapaciousness and extortion, this act could be dismissed as another example of his revenue-raising efforts. As with the 1458 act, Coupar Angus Abbey again provides evidence for the response to the new requirements, with the monks adopting its terms swiftly into their own lease-making arrangements. The general language of the agreements remained much as they had been since the 1460s, instructing tenants to plant ashes, osiers and willows, and protect them adequately with 'defensouris',[66] but from 1541 the clause relating to these plantations referred to both 'the actis and statutis of our courtis and actis of parlyament'.[67] The date correlation between the 1535 act and the change in lease language is suggestive and points to the new legislation as having greater effectiveness as a mechanism to require planting. The fact that the leases also explicitly referred to the fines that would be imposed for breaches indicates that it was the coercive weight of the 1535 act that was valued by landlords and which truly differentiated it from its predecessor. Its impact, however, was short-lived if the Coupar Angus rental agreements are indicative of more general practice; by the 1550s all reference to the 1535 act and its associated penalties had vanished from new leases. It was a further seventy years before another effort at directive legislation was attempted.

A new act in 1607 was designed specifically to safeguard particular forms of property and to improve the process for actions against people who damaged such property or allowed it to be damaged.[68] The preamble stated that 'considering how woods, parks and all sort of planting and fencing decay within this realm, and how dovecots are broken, bees stolen, men's proper lochs and ponds despoiled of fish, to the great hurt and prejudice of the country and decay of policy', the new act was designed to ratify and approve all previous parliamentary legislation that aimed at their conservation. Going beyond the 1535 act's enforcement provisions, however, it identified competent courts to deal with lawbreakers; they were to be called either before the Privy Council or the ordinary magistrate, at the option of the complainer. The penalty that could be imposed by these two levels of tribunal differed significantly and it was clearly intended that the former should deal with major disputes between landlords themselves, while the ordinary magistrates were to deal with minor contraventions at a more neighbourly level. Surviving court records suggest that the level of action raised under the terms of the 1607 act was generally low and could simply involve claims over single trees or generalised damage arising

---

66 Rogers, *Cupar-Angus Rental*, vol. 2, p. 7.
67 Ibid., pp. 14, 17, 19.
68 RPS 1607/3/17 [accessed 29 November 2021].

Lone willow, Aberbothrie, Perth and Kinross. This ancient tree stands by the River Dean on the lands of Coupar Angus Abbey's grange at Aberbothrie and is perhaps a descendant – or a remarkable survival – of willows planted by the monks' tenants.

The Hercules roundel from the 'Stirling Heads' Collection, Stirling Castle. The discs were cut from composite sheets of Polish-Lithuanian oak, whose fine-grained structure was favoured over native Scottish oak by wood carvers for the smoothness of its finished texture.

from uncontrolled grazing by the defendant's livestock. At Melrose in the Borders in 1608, for example, a Gattonside tenant was fined £6 10 shillings – described as the value of one ash tree – to recompense a man in nearby Bowden for his loss.[69] At Branxholm and Trinity Lands near Hawick, again in the Borders, however, the earl of Buccleuch in 1627 used the act to pursue three Hawick men who had felled one hundred of his trees. He claimed that 'notwithstanding the laws enacted against the cutting and destroying of green wood, policy and planting', his woods and parks which he had caused to be carefully preserved and protected within a haining, 'have been meddled with and depleted'.[70] This was a major incursion and presumably easy to evidence, but for the most part the levels of fine remained so low that the charges seemed more like enforced belated payment of the true value of the property taken or damaged than any kind of punitive impositions designed to deter future infractions.

Behind such acts was a conscious effort to increase the extent of growing timber in the south and east of the kingdom and reduce the dependence of those regions on imports of Baltic oak and Scandinavian pine. The much-cited description of the inroads made on both the woodland of Fife and supplies of imported timber to construct James IV's warship, the *Great Michael*, between its first conception in 1505 and completion in 1512, although probably overstated nevertheless reflects the fragility of the remaining broadleaf woods in east-central Scotland. According to the Fife chronicler Robert Lindsay of Pitscottie, whose account, produced more than half a century after the vessel's completion, preserves local popular memory rather than official record, 'this scheip was so greit statur and tuik so mekill timber that scho waistit all the wodis in Fyfe except Falkland wode, by all the tymmer that was gottin out of Noraway'.[71] Pitscottie's particular exception of Falkland is interesting, for the reservation of this crown woodland preserve perhaps points to the immaturity of its oaks following an earlier episode of felling rather than reflecting the king's protection of his own property. The crown's preserves of growing timber in the royal parks were evidently still depleted or inadequate for the building requirements of James V in the 1530s, when most of the major

---

69 Romanes, *Selections from the Records of the Regality of Melrose*, vol. 1, p. 65.
70 *Register of the Privy Council of Scotland*, 2nd series, volume 2, *1627–1628*, ed. P. Hume Brown (Edinburgh, 1909), p. 530.
71 '[T]his ship was of such great stature and took so much timber that she exhausted all the woods of Fife except for Falkland wood, as well as all of the timber imported from Norway': Robert Lindsay of Pitscottie, *The Historie and Cronicles of Scotland from the Slauchter of King James the First to the Ane thousand five hundreith thrie scoir fyftein zeir*, ed. Æ.J.G. Mackay, volume 1 (Edinburgh, 1899), p. 251.

beams, planks and panels used in his expansion of the palaces at Stirling, Linlithgow, Holyroodhouse and Falkland were imported from eastern Denmark, southern Sweden and Poland-Lithuania.[72] The outflow of bullion from the kingdom that this trade required makes understandable the outrage of Queen Mary's privy councillors in 1564 at what they saw as the wanton plundering of woodland in the Far North that should have been protected to meet the realm's timber needs.[73] Royal parks and forests were still perceived as depleted at the end of the century, when further acts were issued in 1592 and 1594 for their proper enclosure and protection.[74] A third act, issued in 1617, speaks explicitly of inroads made in the king's deer forests for clearings, shielings for summer grazing of livestock, and timber extraction.[75]

Regular felling for building needs was inevitable where the built environment of rural settlements remained primarily of turf, wattle and timber, continuing the cultural traditions originating in at least the earlier Middle Ages and maintained into the nineteenth century.[76] Because very little pre-eighteenth-century timber construction survives, outside of elite contexts, and stone-built structures predominate, there has been unconscious neglect of evidence for large-scale wood use for building across the social spectrum. Even the mainly stone structures we identify conventionally as representative of high-status living, it must be remembered, contained substantial timber elements, many of which required frequent repair or replacement. As we saw in the volume 'Scotland AD 400–1400', throughout the eastern Lowlands, however, demand for building timber had already outstripped local supplies by the early 1200s, stimulating the development of a trade in timber across the North Sea. This trade intensified in the sixteenth and seventeenth centuries, evidenced by port records from Aberdeen especially, but also by surviving Polish-Lithuanian oak and Scandinavian pine beams, planks and panelling in merchant, noble and royal residences.[77] It was to reduce dependence on such externally sourced timber that a new act to promote tree-planting was drawn up in 1661 by King Charles II's Scottish counsellors. But Scotland's large timber needs were never met solely by foreign imports, especially in districts remote from the ports, to which transportation costs were prohibitive.[78]

---

72 A. Crone and R. Fawcett, 'Dendrochronology, documents and the timber trade: new evidence for the building history of Stirling Castle, Scotland', *Medieval Archaeology*, 42:1 (1998), pp. 68–87; Crone and Watson, 'Sufficiency to scarcity', pp. 75, 81.

73 *Register of the Privy Council of Scotland*, 2nd series, volume 1, *1625–1627*, ed. D. Masson (Edinburgh, 1899), p. 279.

74 RPS 1592/4/55, 1594/4/31 [accessed 7 January 2022].

75 RPS 1617/5/32 [accessed 7 January 2022].

76 H. Cheape, '"Every timber in the forest for MacRae's house": creel houses in the Highlands', *Vernacular Building*, 37 (2013–14), pp. 31–50.

77 D. Ditchburn, 'Cargoes and commodities: Aberdeen's trade with Scandinavia and the Baltic, c.1302–c.1542', *Northern Studies*, 27 (1990), pp. 12–22; E.P. Dennison, D. Ditchburn and M. Lynch (eds), *Aberdeen Before 1800: A New History* (East Linton, 2002), pp. 390–391; Rorke, 'Scottish Overseas Trade'; G. Simpson, 'Seeing the wood for the trees: Poland and the Baltic timber trade, c.1250–1650', in A. Roznowska-Sadraei (ed.), *Medieval Art, Architecture and Archaeology in Cracow and Lesser Poland* (Leeds, 2014), pp. 235–254; K. Newland, 'The Acquisition and Use of Norwegian Timber in Seventeenth Century Scotland, with reference to the Principal Building Works of James Baine, His Majesty's Master Wright', unpublished PhD thesis, University of Dundee (2010), especially chapters 3, 9–12.

78 The sources and uses of Scottish wood and wood products are discussed extensively in Smout et al., *History of the Native Woodlands*. Wood use in the post-1600 period is explored by Mairi Stewart in two contributions to contributions to Smout (ed.), *People and Woods*: M. Stewart, 'Using the woods, 1600–1850 (1): The community resource', pp. 82–104 and M. Stewart, 'Using the woods, 1600–1850 (2): Managing for profit', pp. 105–127.

Craigievar Castle, Aberdeenshire. Completed in 1626 by 'Danzig Willie' Forbes of Menie and Craigievar (1566–1627), an Aberdeen-based merchant who had grown rich on the Gdansk and wider Baltic timber trade, the tower's original interiors include panelling and carved work in imported pine.

Demand in regions away from the principal ports saw attention turn to some of the more extensive woodland tracts that had survived the medieval period in remoter districts. With transport costs for this timber still significantly below the costs of foreign imports and their landward transportation, these surviving woods also saw intensifying exploitation. Financial records from the bishopric of Dunkeld reveal inroads already being made by the 1510s into the woods of Rannoch and Glen Lyon in Highland Perthshire for timber required for construction work on Bishop George Brown's many building projects.[79] Sawyers and carters employed by Bishop Brown felled, hauled, processed and carried planks and spars the forty miles from the southern shore of Loch Rannoch to the new bridge he was building over the River Tay, and the sixty miles to another project at the Carmelite friary at Tullilum, just outside Perth.

Mapping the woods

Aside from such references to inroads on previously little-known areas of woodland, what evidence do we have for the location and extent of Scotland's tree-cover in this period? Unlike previous eras, we have detailed maps and topographical descriptions, from the late 1500s, to stand alongside the incidental documentary records of earlier centuries and to reinforce inferences built on pollen and other palaeoenvironmental evidence. Our single most important cartographic and topographical data from this period, dating from the 1590s and early 1600s, are those surviving sketch maps and manuscript descriptions composed by Timothy Pont, referred to earlier.[80] Much has been lost, but Pont's original coverage of the country was perhaps total, for it was material reworked from his originals by Robert Gordon of Straloch (1580–1661) and his son James Gordon of Rothiemay (1617–1686) that formed the basis for the comprehensive maps of Scotland produced in the mid 1600s by the Dutch map publishers Willem Jansz and Jan Blaeu.[81] Between Pont, the Gordons and the Blaeus, therefore, we are blessed with a remarkable suite of map data that reveals the approximate location, nature and extent of some of Scotland's woodland as it existed across a fifty-year period from the 1590s to 1640s. Very significantly, this map data has helped to challenge the deeply embedded belief that the Highlands still had been clothed in a vast wildwood – the mythical descendant of the Classical 'great wood of Caledon' (see the volume 'Scotland AD 400–1400', Chapter 1) – until it fell prey to commercial exploitation and intensive sheep-farming between the late sixteenth and early nineteenth centuries.[82] Pont and the Gordons reveal to us instead that Highland woodland was a rather patchy mosaic, extensive in places like Strathspey but generally broken into small and isolated blocks, very different in character from the blanket of closed canopy woodland filling the glens that has been popularised by ecologists from the early twentieth century

---

79 Hannay, *Rentale Dunkeldense*, 122, 129, 130, 266.
80 J. Stone, 'Timothy Pont: three centuries of research, speculation and plagiarism', in I.C. Cunningham (ed.), *The Nation Survey'd: Timothy Pont's Maps of Scotland* (East Linton, 2001), pp. 1–26 at p. 17 Fig. 7; Smout et al., *History of the Native Woodlands*, pp. 47–56.
81 Stone, 'Timothy Pont', pp. 9–13; Smout et al., *History of the Native Woodlands*, pp. 46–49.
82 The development of this vision of an almost treeless Lowland district contrasting with the heavily afforested Highlands, and the ravaging of those Highland woods by commercialised landlords, foreign industrialists and southern Scottish sheep-farmers is discussed in Chapter 9. For a summary, see Smout et al., *History of the Native Woodlands*, pp. 20–25.

onwards.[83] If we take just Pont's map of the Forest of Atholl we see exactly this character of patches of woodland – often named – in the upper reaches of Glen Garry and its tributaries, densest furthest from the main settlements and penetrating up the courses of the deeply cut streams feeding into the river valley, but with straggles of trees flanking the main valley as far south and east as Blair Atholl. These flanking woodlands are probably the ancestors of the open, mainly birch, woodland that are spread today from Tulach on the south side of the A9 near Blair Atholl to past Dalinturuaine, but the denser woods he depicted north of Dalnacardoch and Dalnaspidal are long gone, largely victims of over-grazing by deer since the later eighteenth century. Equally significantly, however, the Lowland maps also record a subtly different picture to that popular image of a landscape stripped bare of trees by human clearance before the close of the Middle Ages. Pont's maps of Angus, southern Perthshire, the Forth Valley, Clyde Valley, upper Teviotdale and Nithsdale reveal instead scatterings of isolated woodland, much of it protected by enclosures and possibly representing both managed preserves of semi-natural woodland and areas of new planting arising from the earlier sixteenth-century acts of parliament.

Alongside Pont and the Gordons, for those parts of the country for which early manuscript maps have not survived we can draw on the written descriptions composed in the later sixteenth and seventeenth centuries. These accounts, however, are qualitative rather than quantitative and make free use of adjectives to describe the extent, quality and composition of woodland and other natural features without any qualification or explanation of their measure of any characteristic. Thus, the description of Carrick composed c.1700 by the parish minister of Maybole claims that 'no Countrey is better provyded of wood', with the 'great woods' of the Girvan Valley satisfying not only the people of that district but also of Kyle and Cunninghame to the north, providing them with building timbers and wood for making carts, ploughs, harrows and barrows.[84] The woods, however, contained 'birch, elder (alder?), sauch, poplar, ash, oak and hazell', which suggests mainly scrub and softwood species and those hardwoods best suited to coppicing, as the reference to equipment-making perhaps implies. It matches closely what we would expect of woodland composed of coppice-with-standards, where young growth predominated and only a few veterans survived within the fast-growing coppice-wood. The Angus laird John Ouchterlony of Guynd's account of Angus, of similar date, reinforces the Pont record for the condition of woodland in the district a century earlier. Ouchterlony makes frequent reference to new plantings around the greater houses of the shire, but does not name the species planted; his description of the 'great park' at Fothringham reveals that it contained only a birch wood, while at Powrie near Dundee the park

---

83 The woodland cover recorded in Pont is explored in detail in Smout, 'Woodland in the maps of Pont'. Some Lowland areas covered in the surviving Pont sheets correspond well with the treeless vision of later centuries, such as Pont 10 (Buchan), where not a single tree is depicted even to represent an orchard in the enclosures shown attached to some of the houses of the greatest regional noblemen. In contrast to the still extensive woodland to be found in Pont's day in the glens of northern Angus, discussed below (see note 87 below), in southern Angus (Pont 26), trees seem mainly to have been protected within large emparked woods – e.g. the 'Park of Inneraty' (Inverarity) – and smaller areas of tree-cover beside houses such as Panmure and Melgund, with nothing of any significance outside of these. These mapped parks reveal early woodland plantations managed by some of the region's leading families.

84 'A Description of Carrict by Mr. Abecrummie, Minister at Minibole', in Mitchell, *Macfarlane Geographical Collections*, vol. 2, pp. 1–21 at pp. 3.

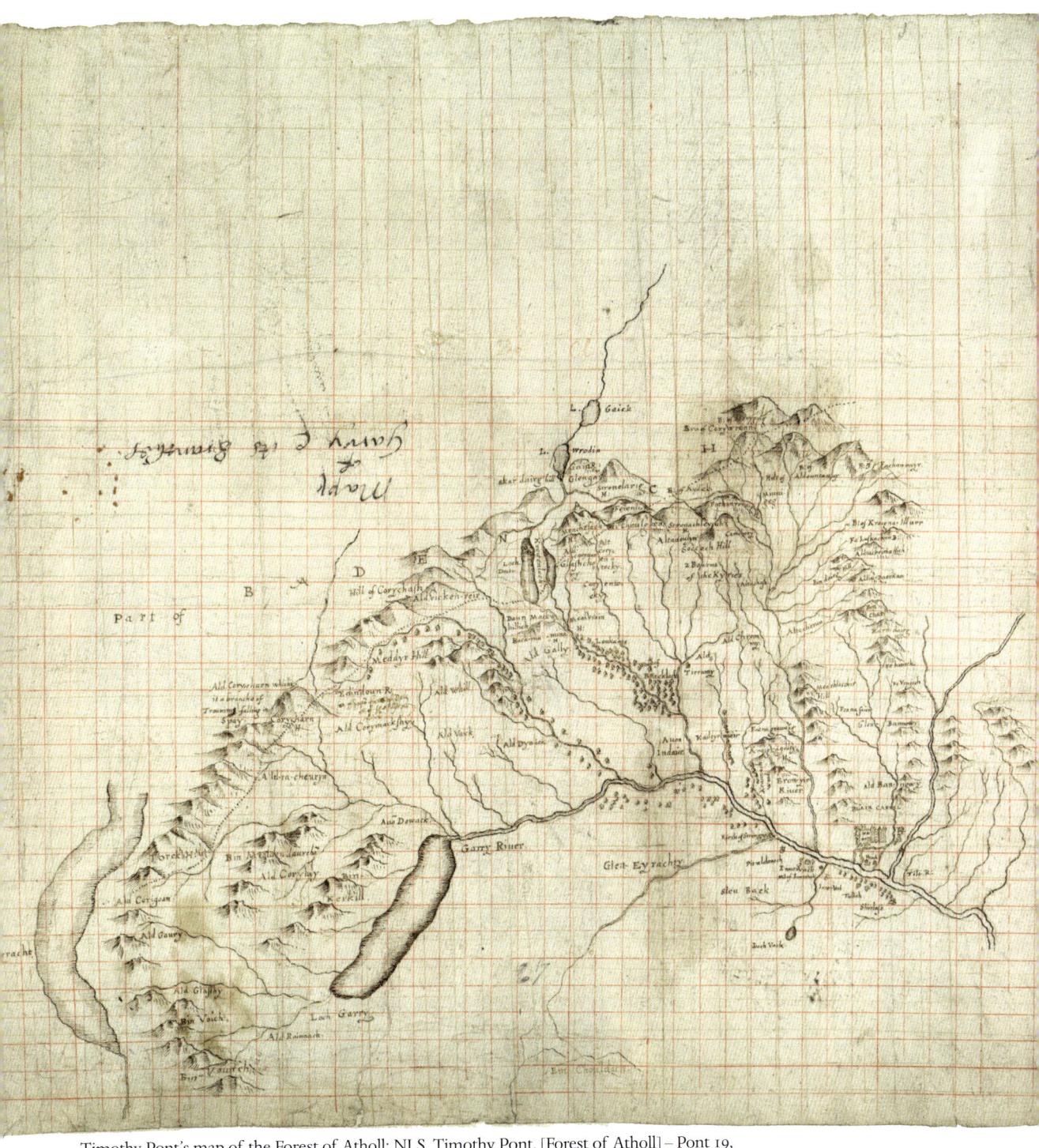

Timothy Pont's map of the Forest of Atholl: NLS, Timothy Pont, [Forest of Atholl] – Pont 19, https://maps.nls.uk/view/00002313. (Courtesy of the National Library of Scotland)

contained 'fir and birk'.⁸⁵ It is interesting that Ouchterlony claims that the timber needs of northern Angus were met from wood cut in the nearby glens and transported for sale in Kirriemuir; he mentions two woods in Glenisla and one at Lochlee in Glen Mark.⁸⁶ Of these, the Lochlee woodland was depicted on Pont (see Chapter 2, Map 1 above) and the woods in Glenisla on Robert Gordon's reworking of Pont's earlier map.⁸⁷ Together, these map sources confirm the survival of extensive wooded areas – apparently of broadleaf trees from the symbol used to represent them – in the upper reaches of these glens and the adjoining hill-country.

Our evidence for other parts of the country is much more mixed than this good data from the Angus and eastern Perthshire glens. For the condition of Hebridean woodland, for example, for which the surviving portions of Pont provide next to no insights and for which there are none of Robert Gordon's manuscript maps surviving,⁸⁸ we are largely dependent on two accounts written around 150 years apart. The first is Donald Monro's 1549 *Description of the Western Isles of Scotland*, which provides the first baseline written record of the general extent of woodland in the Western Isles since the medieval accounts assembled by Fordun and Bower.⁸⁹ Monro presented a picture of isolated and generally small patches of woodland, all of it managed to satisfy local demand. Beginning with the Clyde islands, he recorded how the deer forest of Arran was only 'pairt of woods',⁹⁰ probably in the island's mountainous interior. Jura's hunting forest, by contrast, contained only 'small woods', while adjacent Scarba also had only patches of woodland.⁹¹ Surprisingly, more populous and cultivated Islay, Mull and Skye contained, he said, quite plentiful woodland.⁹² Some insight on the character of these woods is gained from Monro's account of Scalpay, Raasay and Rona, where it is described as 'birkin', that is, of birch.⁹³ Although he is not so explicit, it is likely that these were coppice-woods, from which the islanders obtained the poles they used in house-building as well as wands and withies for wattling and basketry. Throughout the Outer Hebrides, by way of contrast, Monro noted an almost total absence of trees, remarking only that Harris had 'verey faire hunting games without any woodes'.⁹⁴ As discussed in the volume 'Scotland AD 400–1400', this dearth is unlikely to have been solely or even primarily anthropogenic, stemming rather from climatic changes and disease, although human

---

85 'Information for Sir Robert Sibbald anent the Shyre of Forfar by Mr Ouchterlony of Guinde', in Mitchell, *Macfarlane Geographical Collections*, vol. 2, pp. 21–51 at pp. 27, 32.
86 Ibid. pp. 30, 36, 39.
87 NLS, Timothy Pont, [North Esk; South Esk] (front) – Pont 30 https://maps.nls.uk/view/00002327; Robert Gordon of Straloch, 'Brae of Angus, [and] The height of Anguss, M.T.P. Height of Anguss' https://maps.nls.uk/view/00000664 and Robert Gordon of Straloch, 'Glen Yla, Glen Ardle, Glen Shye, out of Mr. T. Pont's papers yey ar very imperfyt' https://maps.nls.uk/view/00000665; Smout et al., *History of Native Woodlands*, pp. 53–56.
88 There is one very rough sketch map of South Uist: NLS, Timothy Pont, [South Uist: Inverkeithing] – Pont 36 (front), https://maps.nls.uk/view/00002338.
89 Donald Munro, 'Description of the Western Isles of Scotland', in Martin Martin, *A Description of the Western Islands of Scotland Circa 1695, by Martin Martin, Gent. Including a Voyage to St Kilda by the same author and A Description of the Western Isles of Scotland by Sir Donald Monro*, ed. D.J. Macleod (Stirling, 1934), pp. 485–526.
90 Ibid., p. 486.
91 Ibid., pp. 488, 489.
92 Ibid., pp. 493, 497, 503.
93 Ibid., pp. 505, 506.
94 Ibid., p. 521.

agricultural and pastoral regimes and wood demand combined to keep the residual woodland as just small patches in isolated, sheltered locations.

Writing 150 years later, Martin Martin observed the same general dearth of trees. On Lewis he noted the presence of (probably prehistoric) tree-roots at the head of Loch Erisort on the island's east side, something he observed also on the west side of Pabbay and in peat-bogs on Skye. Otherwise, he recorded only 'about a hundred young birch and hazle trees' on the south-west side of Loch Stornoway.[95] The same treelessness prevailed in more mountainous Harris, where 'there is not a shrub of wood to be seen',[96] and throughout the Uists and their adjoining islands, except Barra where he saw an unproductive orchard.[97] Martin observed equally few woods on any of the Inner Hebrides or Clyde islands. On Skye and Arran, other than the scattered coppices managed for their fast growth shoots that Monro had observed,[98] extensive woodland was long gone. Even on Rum, formerly described as heavily afforested, the remaining woodland of any significance had been reduced to just its northern end.[99] Here, again, we need to exercise caution, for by the date of Martin's writing the 'little ice age' had also seen the effects of two solar minima. Across the montane regions of western and central Europe, it is known that these episodes helped to lower the tree-line by some 200 metres; we can expect similar impacts on Scotland's upland woodlands but also on those along its exposed Atlantic seaboard.[100] As the climate and weather data reviewed in Chapter 3 serves to underscore, the later seventeenth century in particular saw an extended period in which the altitudinal range of woodland from north-west Sutherland through to the central Highland and Cairngorm glens would have dropped sharply. In the most weather-exposed locations, it is likely that these extreme conditions contributed most to the apparent sudden acceleration of woodland loss that our map and record evidence seems to reflect. Waterlogging and its associated expansion of peat growth during these cold, wet decades added further to natural woodland decline, possibly worsened in areas where there had been felling activity of mature trees, where peat spread prevented natural regeneration. The exceptionally cold conditions of the 1630s and especially 1690s revealed in the summer growth ring record from the Cairngorms perhaps indicate where and why the greatest decline of the seventeenth century occurred. The intensity and duration of the temperature drop and storminess that prevailed across the minima add further complexity to our attempts to understand the processes that brought about the apparent episodes of tree loss illustrated by late sixteenth- and seventeenth-century mapping (discussed below, see pp. 142–146).

Protecting the trees: haining and enclosure

Although the Hebridean birch woods described by Monro and Martin were apparently unenclosed, Pont's maps and subsequent manuscript descriptions, such as Ouchterlony of Guynd's, and numerous court actions concerning illicit wood-cutting, suggest that many Lowland woods were 'hained' or enclosed for protection. Enclosure, too, was employed by landowners in Highland areas, including the Campbells of Glenorchy and Grants of Freuchie, although

---

95 Martin, *Description of the Western Islands*, pp. 92, 122, 199.
96 Ibid., p. 113.
97 Ibid., pp. 156–157.
98 Ibid., pp. 199, 255.
99 Ibid., p. 299.
100 Smout et al., *History of the Native Woodlands*, pp. 56–59.

as a standard practice it was mainly an eighteenth-century development.[101] It has been suggested that the woodbank system on Bowden Moor to the south of the Eildon Hills was established c.1500 by the abbot of Melrose, whose monastery controlled the property,[102] but this proposition has not been tested archaeologically. Similar woodbanks are identifiable in late fifteenth- and sixteenth-century Coupar Angus leases which refer to the haining (enclosure with hedges, fences or banks) of the plantations required of the abbey's tenants and reveal the monks' efforts to enforce on their tenants sustainable management practices within the enclosures.[103] Most of the plantations in this period were of quick-growing species, such as willow, birch and alder specified in the abbey's leases, but included also slower-growing ash. Royal and monastic records also reveal extensive orchard plantings by the earlier 1500s as opposed to just 'barren' trees in timber plantations.[104] James V's 1535 act stimulated planting of garden fruit trees, principally apples and plums, in smaller domestic gardens, which local court records from the later sixteenth and seventeenth centuries reveal became as much the targets of 'plundering' raids by children and youths as mature timber was by resource-hungry neighbours.[105] James VI's 1607 act led to more hardwood planting in the earlier seventeenth century, but it was Charles II's 1661 Act for Planting and Enclosing Ground that seems finally to have stimulated the regular planting of land immediately around elite residences, establishing the practice that was maintained into the twentieth century and which gives much of the wooded character to Lowland districts that is still visible today.[106] This 1661 act required larger estate owners to enclose and plant a minimum of four acres of ground, with a scale of diminishing acreage requirements for lesser property-owners. Some proprietors clearly kept to the absolute minimum, if the later seventeenth- and eighteenth-century maps that depict the planted areas around some houses are accurate, but others, like the Earl of Tweeddale at Yester (discussed below, pp. 119, 202–205), seized the opportunity of the legislation to begin planting on what, for the time, was a prodigious scale and to enclose the plantations within equally prodigious lengths of stone dykes.

Tweeddale was a man ahead of his time and his investment in both trees and enclosures brought him and his heirs a heavy burden of debt. Many parts of our modern rural landscape are still partitioned by the substantial stone-and-mortar enclosing walls of the wooded 'policies' constructed mainly in the eighteenth and nineteenth centuries, but in the seventeenth century, apart from work like the new stone enclosure wall for the King's Park at Stirling or around Tweeddale's mansion, most plantations were bounded in less permanent ways. The earlier hainings erected to enclose coppices and trees for larger timbers could be relatively slight brushwood fences, designed simply to keep out grazing animals, but others could appear as massive physical boundaries that formed as much of an obstacle as stone-built walls. At Doune, when the royal castle was repaired in 1581, the protective dyke around the king's nearby

---

101 Ibid., pp. 158–162.
102 Crone and Watson, 'Sufficiency to scarcity', p. 71.
103 Smout et al., *History of the Native Woodlands*, pp. 157–158.
104 See, for example, Ferrerius for fruits trees at Kinloss Abbey in Ferrerius, *Ferrerii Historia Abbatum de Kynlos*, ed. W.D. Wilson (Edinburgh, 1839), pp. xv–xvii.
105 For example, multiple cases of contested ownership of ash and elm trees, unauthorised and illegal felling of trees, or theft of fruit – mainly from plum trees – are detailed in Romanes, *Selections from the Records of the Regality of Melrose*, vol. 2, pp. 52, 59, 68, 70, 72, 170, 203, 233, 257, 296, 303, 388, 393, 413, 416.
106 RPS 1661/1/348 [accessed 28 December 2021].

Park dyke, Doune, Stirlingshire. The remaining section of the late sixteenth-century enclosure of the royal plantation runs up the slope from the River Teith, immediately to the west of Doune Bridge.

wood preserve was also rebuilt in turf. The works account details the building of 460 roods of dyke (c.3,000 metres), 5 ells wide at the base (c.4.7 metres), roughly 75cm wide at the top and over 3 metres high.[107] The turf volume required for such a barrier, even if only used to revet the face of a heaped earth dyke, is staggering and would itself have meant the stripping of an extensive area of grass or heathland for materials. This was not a deer-park, for three years later when the king pressed for further work to be carried out to maintain the dyke, he expressed fear for the likely damage to the 'ȝoung growth within the samin' which would result from unrestricted access to the enclosure.[108] An absence of any such enclosure, however, need not imply an absence of management or unregulated access. Indeed, the centuries-spanning longevity of areas of alder, birch, willow and hazel woodland in Hebridean, Highland and Lowland contexts strongly argues for the opposite position, particularly in light of the high level of demand which such resources faced.

One remarkable record of the scale of such annual wood need by peasant communities at the turn of the sixteenth and seventeenth centuries survives in two documents from the archives of the Gordon earls of Huntly. These concern the Gordons' dispute with their tenants, the Grants of Freuchie, over the supposed mismanagement of wood resources in Stratha'an in the north-eastern Cairngorms.[109] Dating from the late 1580s and early 1610s respectively, the documents list buildings within the *dabhaichean*[110] of the central area of the Gordons' Stratha'an lordship, noting the number of couples – the paired roof-support timbers – in each and expressing as 'trees' the volume of wood required for maintenance and renewal. The seven-year rolling total of 781,865 trees is breathtaking, even when it is recognised that the Scots term *tree* equates to major tree-limb or branch in modern English usage, and most likely meant a coppice shoot in late sixteenth-century Scotland. It is a remarkable record of the voracious consumption of timber by peasant communities, the systematised coppicing of birch, hazel, willow, and alder to meet that need, and the apportionment and regulation of community access to these resources by the landlord's agents. The litigation to which the documents relate revealed how Stratha'an's lords regulated felling, coppicing and replanting, and pursued transgressors with vigour in the face of the rising challenges of the period. Environmental pressures in the valley, however, worsened local political conflict, which involved competing resource-management agendas at the level of peasant tenants, increased competition for in-demand commodities, and local and superior lordship rivalries. This was occurring at a time of mounting environmental stress when climatic deterioration was altering the growing conditions of woodland across Europe and when the available Scottish resources were already heavily managed to meet the demands of the pre-existing communities. The records contain a richly textured account of the consequences of human choices made in the context of a supply crisis and show how human agency interacting with environmental change delivered unforeseen outcomes, both negative and positive, that affected local land-management regimes into the eighteenth century. As a coda, it should be added that Timothy Pont's 1590s map of Stratha'an shows no woodland anywhere in the upper part of the glen, but carries

---

107 W. Fraser (ed.), *The Red Book of Menteith*, volume 2 (Edinburgh, 1880), p. 421.
108 Ibid., pp. 423–424.
109 A. Ross, 'Two 1585 × 1612 surveys of vernacular buildings and tree usage in the Lordship of Strathavon, Banffshire', *Miscellany of the Scottish History Society*, 14 (2012), pp. 1–52.
110 *Dabhach* (pl. *dabhaichean*): a land division containing a mixture of arable and pasture plus access to all types of natural resource, including fuel and building materials, necessary to sustain a community throughout the year.

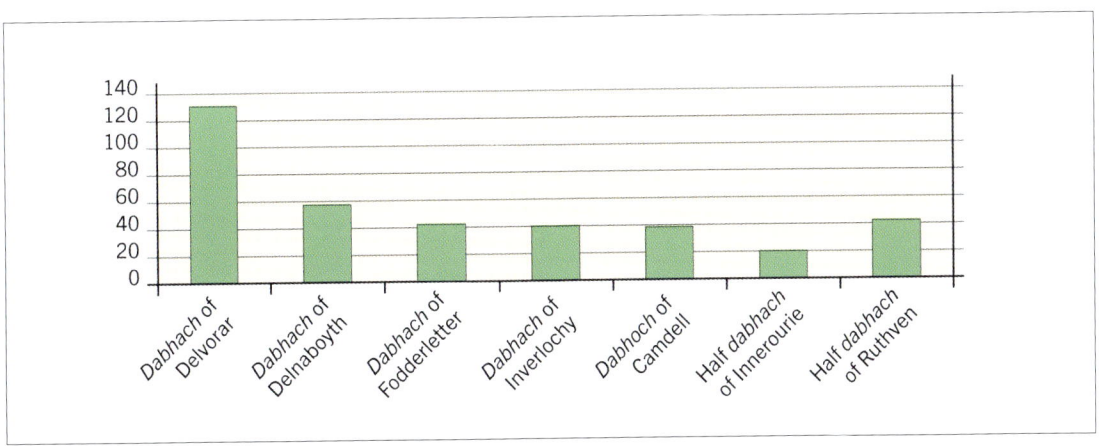

Number of structures per *dabhach* (excluding shielings).
(Reproduced by kind permission of the executors of the estate of the late Dr Alasdair Ross)

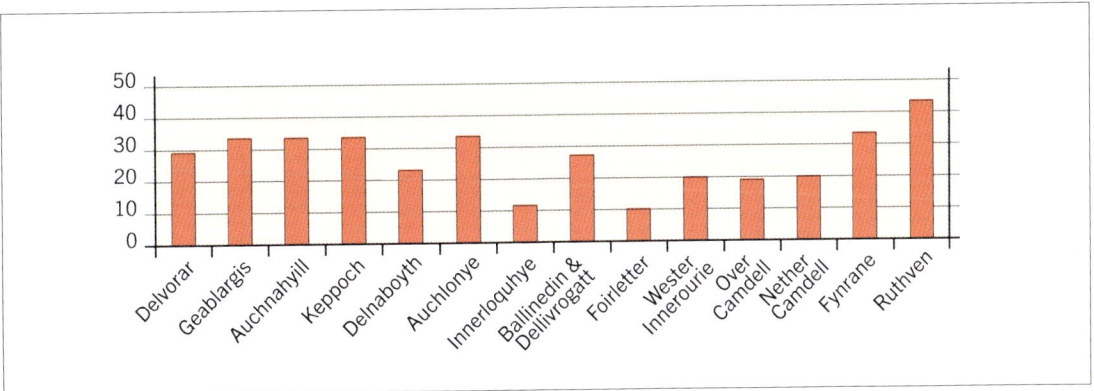

Total number of structures (excluding shielings).
(Reproduced by kind permission of the executors of the estate of the late Dr Alasdair Ross)

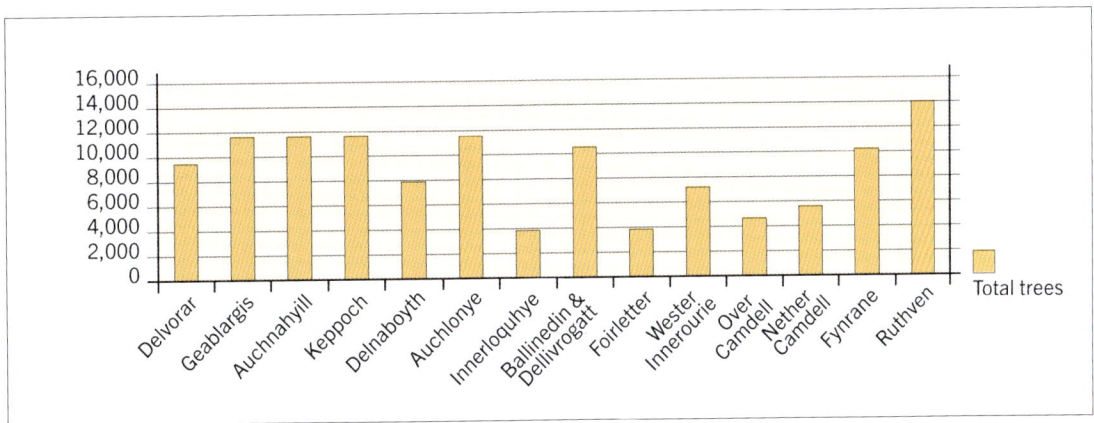

Total number of trees required every two years (excluding shielings).
(Reproduced by kind permission of the executors of the estate of the Late Dr Alasdair Ross)

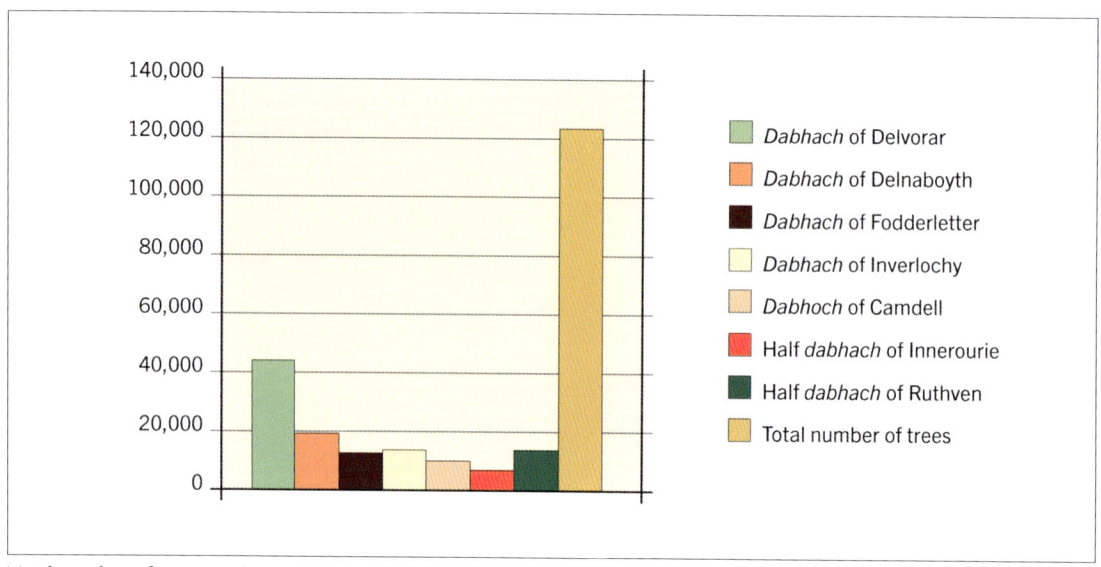

Total number of trees used per *dabhach*.
(Reproduced by kind permission of the executors of the estate of the late Dr Alasdair Ross)

his note that the region around Ben A'an was 'a great wilderness rich in deer'.[111] Unfortunately, we lack his manuscript description of this district, so have no sense of how wooded or denuded the valley seemed to him around the time of the dispute. By way of contrast, Robert Gordon's early seventeenth-century map of Stratha'an that was developed from Pont's sketches and notes, and which was richly embellished by its Dutch engravers, shows the glen upstream of Dalvorar as well wooded, with several separate areas of trees even around Loch A'an but occurring in greater density around Loch Builg in the heart of the Cairngorms.[112] At face value, Pont's map would seem to support the claim that the Grants had systematically wasted the woodland and presided over the devastation of the resource, yet Gordon's map would, on the contrary, indicate that the valley was still very rich in wooded areas. But it has been commented that Pont, Gordon and, indeed, the Blaeus had very different priorities in what they showed on their maps, with Pont more interested in recording the location of settlements than the presence of trees, and the others perhaps exaggerating the extent of tree-cover to 'fill bleak empty spaces with a little more interest'.[113] If we add the testimony of John Farquharson's 1703 map of the Forest of Mar

---

111 NLS, Timothy Pont, [Ben Lawers; Glen Tanar; Strath Avon] – Pont 7 https://maps.nls.uk/view/00002296.
112 NLS, Robert Gordon of Straloch, 'Map of River Avon' https://maps.nls.uk/view/00000355. The remnants of this woodland in the upper strath survive from near Tomintoul southwards along both sides of the glen, past Delavorar and Auchnahyle to around 1km south of Torbain, with a final strip of woodland extending from west of Inchrory for around 1km to Creag Loisgte. Most of this woodland is birch and alder, with some small conifer plantations. Glen Builg, Loch Builg and the 17km of Stratha'an west of Inchrory and the corries around Loch A'an are almost entirely treeless at present, except for straggling survivors in watercourses inaccessible to deer and other grazers. We have no independent historical witness or scientific data to indicate the likely extent of the early seventeenth-century woods there, or the date of their disappearance.
113 Smout et al., *History of the Native Woodlands*, pp. 48–56.

John Farquharson's 1703 map of the Forest of Mar, produced for a dispute over grazing rights in the Forest, shows the wooded areas surviving in districts adjoining the contested district: John Farquharson, 'A Map of the forrest of Mar survey'd &c. / Done by John Farquharson of Invercald, in Anno 1703', https://maps.nls.uk/counties/rec/198.

(Courtesy of the National Library of Scotland)

Upper Stratha'an from Auchnahyle. This upper stretch of the glen, south of Delavorar, was the scene of protracted litigation between the Gordons of Huntly and the Grants of Freuchie over rights in the birch and pine woodland and the inroads of peasant tenants from the fermtouns that flanked the valley.

to the analysis, its omission of any woodland around both Loch A'an and Loch Builg, while the woods of Glenmore are marked, is perhaps more suggestive of imaginative, decorative infilling of these areas with trees by Gordon and the Blaeus.[114] Neither map, then, should be taken as an accurate rendering of the reality on the ground and we must fall back on Ross's assessment of what underlies our written record. If he was correct, it seems that claims of woodland loss were being exaggerated by the complainer to strengthen the Gordons' legal case to expel the Grants. In conjunction with the Pont map evidence, the documents have been read at face value and linked by modern observers to the largely treeless character of the upper glen as evidence of the impact of mismanagement. Woodland loss in this eastern Highland glen has evidently been widespread, even if the density suggested by Gordon's mapping is of doubtful reliability, but that loss occurred long after the litigation brought forward by the Earl of Huntly.

Peat

More credence should perhaps be given to suggestions of a supply crisis affecting peat from the later 1500s onwards. After centuries of exploitation, accounts from all parts of the country indicate that

---

114 John Farquharson, 'A Map of the forrest of Mar survey'd &c. / Done by John Farquharson of Invercald, in Anno 1703', https://maps.nls.uk/counties/rec/198.

the sixteenth century saw the first serious, widespread concerns about the future supply of fuel for domestic consumption from peat sources on common muirs. As observed earlier, although there was a long-running perception of shortages of fuel, it was only in the closing decades of the fifteenth century that perceptions began to become a reality in some districts. We shall never know the scale of extraction that had occurred over the millennia of peat-cutting that has happened in Scotland. Peatland restoration projects, focused on surviving areas of blanket peat, are helping to give a sense of the impact of extraction in the past, but it remains difficult to appreciate fully how transformative the working to exhaustion of peatland has been for both local ecologies and biodiversity, and how it has changed the physical aspect of many areas irreversibly. Around one small burgh – Elgin in Moray – tens of thousands of cubic metres of peat were cut from an area of around 5km^2 and by the eighteenth century the mosses that had once extended far up the valley of the River Lossie and its tributaries there had been stripped to the gravels and boulder clays beneath them. Not all loss had occurred through the actual removal of peat for fuel purposes; damage could be caused by the manner of extraction too. There were repeated efforts, in the burgh council minute books, barony court records and moss overseer accounts, to regulate the manner of cutting to prevent the digging of holes and encourage cutters to work back along a straight cut edge, lift the turf to replace it on top of the deeper layers after the cutting had paused for the season, and so on.[115] Palaeoenvironmental analysis of former areas of peat extraction around Whithorn in Galloway has revealed graphically the impact of such intensive exploitation. What remained there was only the 'foot' of the moss, the mineralised layer at the interface with the sub-peat material, which had developed in the early Holocene, beneath a thin layer of new peat that had formed since extraction finally ended in the later nineteenth century.[116]

Despite the intensity of extraction that occurred, outside of the Highlands and Islands extensive mosses do remain in many parts of Scotland's North-East, central Lowlands and Southern Uplands. The main zones of significant surviving blanket peat, however, are in the central, west and northern Highlands and in some parts of the larger Hebridean and Northern Islands. It is in the non-Highland and non-insular districts that documentary records show that peat was increasingly a source of conflict through the later sixteenth and seventeenth centuries; for it remained by far the most common fuel employed in all places remote from the coal-rich districts fringing the Firth of Forth, and was still important among the poorer peasant and urban-dweller households even in those areas. Mosses that had for centuries been shared as common resources, with easements being enjoyed by communities sharing borders with the mossland or even lying remote from the extraction sites proper, were becoming fields of contention as fears over failing supplies drove efforts to secure exclusive possession.[117] Demand for this once plentiful commodity was, after centuries of exploitation, in many places outstripping the available supplies, and the increasing privatisation and marketisation of a former common resource introduced fuel poverty

---

115 That this was still a perceived 'ill' in the second half of the eighteenth century is illustrated in F. Albritton Jonsson, *Enlightenment's Frontier: The Scottish Highlands and the Origins of Environmentalism* (Newhaven, 2013), pp. 101–102.
116 S. Ramsay, J.J. Miller and R.A. Housley, 'Palaeoenvironmental investigations of Rispain Mire, Whithorn', *Transactions of the Dumfriesshire and Galloway Natural History and Antiquarian Society*, 3rd series, 81 (2007), pp. 35–55.
117 For the allocation of equal shares in fuel and grazing resources attached to communities in Scotland north of the Forth before the tenth century, see A. Ross, *Land Assessment and Lordship in Medieval Northern Scotland* (Turnhout: Brepols, 2015).

to a large segment of the Scottish population.

Indications of mounting concern over control of adequate peat supplies emerge in various districts before 1500, revealing perceptions of an emerging problem in the sub-highland zone. Local tensions over access to peat frequently erupted into physical confrontation across the sixteenth and seventeenth centuries throughout an area extending from central Perthshire, through Angus, and lowland Aberdeenshire to the country between Banff and Inverness. This violence does not seem simply to be an artefact of better record survival from the 1470s onwards, when new centralised law-courts gained oversight of business that had been dealt with previously by local jurisdictions. The steadily increasing volume of evidence and the intensification of efforts to regulate access to and supply of fuel by landlords and magistrates reveal awareness of depletion of traditional sources, rising competition for control over traditional and new reserves, and the mounting desperation of householders to secure adequate and reliable fulfilment of their thermal energy needs. Urban and rural communities, it is clear, experienced shortage of fuel, which contributed to an intensification of conflict over access to and control of peat supplies. Tension is most evident earliest in eastern Perthshire and Angus, where centuries of mossland exploitation for fuel or its draining for cultivation had already exposed social tensions before 1500. Here, urban centres like Perth were competing with rural communities for exclusive use of areas that had formerly satisfied the needs of both. As we have already seen in respect of disputed rights of possession in the medieval period, legally arbitrated settlement between landowners was the preferred route. Their tenants, however, might still resort to physical means to monopolise the indispensable resource.

The following examples are illustrative of the many recorded occasions when dispute escalated into violence before resolution through legal process. The first originated in the early 1500s and records an escalating conflict over peat-cutting on the muirs between Brechin and Arbroath in eastern Angus that had commenced in the late fifteenth century. Attempts to reach a locally arbitrated agreement between the lords of the peasant communities who were in dispute had only achieved temporary compromises, which broke down irretrievably in the 1530s. The resulting lawsuit in 1539 between Lord Ogilvy and Lady Lovat narrates his challenging of the rights of her tenants to cut peat on the Muir of Montreathmont, and her complaints of physical assaults by his men on her people.[118] Local lordship and political power rivalries were at play here; an assertion of control over muir rights during a time of mounting environmental stress provided a vehicle for a very public demonstration of where the balance of might lay. Ogilvy could demonstrate lordship rights over Montreathmont unbroken since 1399, but Lovat exercised inherited rights to common easements of fuel for her dependants in the same ground. These overlapping rights had caused no recorded interpersonal violence before the 1530s but depletion of muirland resources on the Lovat properties had brought increased exploitation of the Montreathmont moss. It was that which triggered Ogilvy's effort to assert exclusive rights for his tenants and Lovat to defend her tenants' easement rights. Ogilvy, however, prevailed and in the late seventeenth century, when tenants on former Ogilvy lands northwest and west of the muir were described as 'abundantly provided of turf' from Montreathmont, no such bounty was enjoyed by the folk of the former Lovat properties.[119]

---

118 NRS GD16/41/9.
119 John Ochterlony of Guynd, 'Account of the Shire of Forfar, c.1682', *The Spottiswoode Miscellany: A collection of Papers and Tracts Chiefly Illustrative of the Civil, and Ecclesiastical History of Scotland*, volume 1 (Edinburgh, 1844), pp. 311–355 at p. 324.

Glenisla, looking south-east from Mount Blair. At the close of the seventeenth century, this heavily farmed and settled mid-section of the valley saw violent conflicts over access to peat-cutting rights as traditional sources of fuel ran low.

Violence and feud arising from peat disputes were the first examples of local dissension cited in 1609 as necessitating the establishment of a system of commissioners and justices of the peace who could resolve such conflict before they escalated to bloodshed.[120] Their conflict resolution efforts, however, were ineffectual in the second example, which dates from 1692. That year saw the north Angus laird, James Ogilvy of Clunie, raise a legal action against seventeen of his kinsmen and neighbours 'for ejecting and dispossessing [him] out of his undoubted right and priviledge, of casting and winning fewell in the moss of muir of pooll and fournouchtie' in Glenisla.[121] The complaint detailed a violent assault on Clunie's servants by an armed band of rival moss-users, who threatened, beat and injured Clunie's men, stole their equipment and drove them away. This attack was part of a wider property dispute arising from Clunie's receipt of exclusive rights over the formerly common mosses of Pool and Fornocht from his superior, the earl of Airlie. Already simmering tensions had grown in the 1690s as a symptom of the prevailing wet and cold weather conditions, where waterlogging rendered some mosses unworkable by traditional means, or the cut peat remained wet and incapable of burning. A failure of traditional tribunals to resolve the localised fuel shortage, attributed by other Airlie tenants to their exclusion by Clunie from a once common supply during a time of mounting social and economic crisis, saw personal grievances escalate into explosive violence.

Such outbursts of violence illustrate how deeply fear of the possible failure of fuel supplies ran in rural communities whose sole sources of thermal energy lay in what they could win with their own hands from woods, muirs and mosses. Urban populations were equally prey to fears over fuel security, and fuel poverty was an ever-present issue for town-dwellers who lived on the social margins. Burgh records have preserved a more continuous narrative for how

120 RPS 1609/4/26 [accessed 9 December 2021].
121 NRS GD16/41/681.

communities attempted to better manage their resources, regulate the behaviour of common users, and maintain the security of their supply against encroachment by fuel-hungry neighbours across the 'little ice age'. From large centres like Aberdeen or Perth, to smaller towns like Kirkintilloch or Banff, we see a range of common responses intended to safeguard the burghs' primary fuel sources from external encroachment. Around the Firth of Forth and in the large east-coast ports, as we have seen, coal bought at market had replaced peat for the main thermal energy needs of the wealthy by the end of the fifteenth century. The rise of coal use in major centres like Edinburgh, Stirling and Dundee, however, has obscured the continuing dependence on peat in towns distant from the mines. Elgin, as mentioned above, was one such town and provides an excellent example of how a small urban community remote from the central Lowland coalfields sought to safeguard its precious and precarious source of thermal energy in a region that had seen clearance of woodland outside a scattering of managed seigneurial parks.[122]

Medieval records reveal that by c.1250 Elgin's common peateries were the only significant local fuel source, but down to the late 1540s there is little explicit evidence of significant supply problems. In December 1549, however, reports of damage caused by unregulated cutting, often by men with no legal right to take fuel from the common moss, forced the councillors to enact protective and punitive by-laws. They reasserted burgesses' exclusive access to the moss but required them to cut their fuel with a council-appointed supervisor present and to transport their peat visibly by horse-load and cart, rather than in back-slung creels which could be moved surreptitiously, on the pain of public shaming in the stocks.[123] This initial decree was strengthened in 1551 when it was ordained that violators caught carrying peat in creels would forfeit the creels, the peat and their outer clothing, pay a 5-shilling fine and suffer a day's imprisonment. Night-time removal of peat was forbidden under threat of banishment,[124] indicating that cutting at night was perceived as probably the work of unauthorised non-burgesses. Unregulated, unsupervised and petty encroachment by individuals were the main issues exposed through these initial acts, but subsequent by-laws reveal continuing disorderly inroads by townsfolk exercising their common rights. Their behaviour led to more radical solutions. In October 1580, the council divided the moss into blocks and leased it out.[125] The intention was for the lessees to manage the moss responsibly, and cut, dry and carry peat to Elgin for sale. There was open hostility to this effective privatisation of a communal resource, with widespread defiance of the new regime by townsfolk who resented having to pay for what had previously been a 'free' commodity enjoyed by right.

The result of the dissent and disobedience was that restrictions on access to the mosses had to be re-enacted in 1581. This time the enactments targeted both individuals who possessed full burgess rights and the 'unfree' townsmen or 'indwellers' who had

---

122 This position is emphasised by the earliest map evidence from the region, Timothy Pont's manuscript map of the Laigh of Moray, produced in the 1590s: NLS, Timothy Pont, Elgin and North-East-Moray, https://maps.nls.uk/rec/302 and the online supporting essay 'Scottish woodlands' https://maps.nls.uk/pont/subjects/woodlands.html. For detailed discussion of the woodland chorography and woodland extent on Pont's maps, see Smout, 'Woodland in the maps of Pont'. The efforts of the burgh council to assert ownership of bog oak recovered during peat-cutting highlights the scarcity of timber on Elgin's common land: W. Cramond (ed.), *The Records of Elgin*, volume 1 (Aberdeen, 1903), p. 156.
123 Ibid., p. 100.
124 Ibid., p. 109.
125 Ibid., p. 158.

no automatic rights in common resources. The burgh court ordained that no one was to cut 'faill [the top, grass layer of turf], fuel or divot' in the moss, or 'hoik' (dig out) or 'fla' (take off the grass layer, literally to flay the ground) in any way, under threat of a much more punitive 40-shilling fine.[126] This extension of restrictions to legitimate as well as unlicensed cutting, and the draconian fines, indicate how ineffective previous regulations had been and how serious the abuses had become in the eyes of the burgh's magistrates. The 1581 ordinance, however, emphasises that inroads into the moss were not caused solely by those seeking fuel, but also by cutters of fail and divot, which were in demand as materials for building dykes and for roofing, as well as for soil-deepening.[127] This specification is indicative of the degree of human impact on local resources in Moray, for records elsewhere in Scotland where woodland was more stringently managed reveal that extensive use was still being made of timber fences rather than fail dykes, and that broom, heather, reeds or marram grass were used for roofing. The act reveals a further concern: that the widespread removal of turf, used to deepen cultivated topsoil, was exposing the underlying peat to weather erosion, further depleting the reserves.

These efforts to regulate access by leasing cutting rights to a few tenants and limiting use of the moss had some effect, with no further evidence for council action being recorded until July 1638, when it was enacted that no burgess should cast peats without a warrant.[128] Strict limits on quantities cut were also imposed, as were measures to ensure that cutting and clearing was regular and orderly to preserve the moss's integrity. Alongside regulation of legitimate use, the council also took moves to prosecute inhabitants of the surrounding countryside who took fuel from the burgh's mosses. Despite such action, local fuel shortages were sufficiently pressing for men from neighbouring fermtouns to risk arrest, fines and imprisonment to secure peat for their domestic needs. Such continuing encroachment led in 1652 to a new by-law against unlicensed cutting on the burgh mosses.[129] This act also reveals the dearth of fuel in adjoining parishes and opportunist action by burgesses to benefit commercially from that shortage. None of the tenants of the moss, i.e. those to whom leases for commercial cutting had been awarded, were permitted to sell fuel from it to anyone except to the townsfolk of Elgin, or face a £10 fine – a very substantial sum – and imprisonment at the will of the burgh court. As with earlier acts, however, the need for such a basic commodity drove inhabitants of the surrounding districts to continue to ignore the burgh's enactments and cut peat, regardless of the consequences.[130]

Despite these successive efforts at regulation, the mosses around Elgin continued to suffer damaging misuse. The moss south of the town was described in 1656 as 'spoiled' by over-cutting, the suspicion being that burgesses were taking more than was necessary for their own use and selling the surplus to non-townsmen.[131] Thirty years later the council fined all those licensed to work the mosses west of Elgin for exceeding quotas and selling peat to non-townsmen.[132] Profit made from supplying fuel to

---

126 Ibid., p. 163. This prohibition was repeated in 1581 (p. 167).
127 For soil deepening in peri-urban contexts using stripped turves, Oram, 'Waste management and peri-urban agriculture'.
128 Cramond, *Records of Elgin*, vol. 1, pp. 260–261.
129 Ibid., p. 188.
130 Ibid., p. 294.
131 Ibid., p. 298.
132 Ibid., pp. 320, 338.

'The Prospect of the House and Town of Elgine': Elgin, Moray, from John Slezer's *Theatrum Scotiae*, published in 1693. Slezer's engraving shows the treeless landscape – other than in urban gardens – and extensive cultivation within which the burgh was located by this date. Peat sources close to the town were by then exhausted. (National Library of Scotland)

neighbouring districts evidently outweighed the risks of detection. The poor state of the mosses closest to Elgin may account for growing concern in the 1670s to secure their legal rights over the more distant mosses that lay within the burgh's control, the whole council visiting outlying sites to seize peat cut there by unlicensed cutters.[133] Actions of this kind became progressively more radical in response to the escalation of unlicensed cutting. In 1680, the council instructed all townsfolk who possessed horses and carts to come in their company for such a visit to the moss, to carry away for the burgh's use any dry peat that had been cut illicitly by non-townsmen, and to throw any wet peat back into the cuttings.[134]

A politicised dimension of the fuel supply question emerged at Elgin in the course of the seventeenth century. The council's response there was threatened or actual litigation with neighbouring landowners and their tenants. As these clashes confirm, however, it was not just Elgin's supplies that were dwindling, for as the record from the 1540s onwards had suggested, most conflict arose where inhabitants of neighbouring districts sought to secure a vital commodity that could no longer be obtained easily within their parish. The councillors were not completely inflexible, however, and as early as the 1640s had made concessions to influential local landowners whose political patronage they courted. Between 1642 and 1655, Sir Robert Gordon of Gordonstoun made successive petitions which

133 Ibid., p. 320.
134 Ibid., p. 324.

indicate that his own usable mosses bordering the wetlands of the Loch of Spynie were exhausted. At first he obtained permission for his tenants to gain unlimited fuel in the burgh's moss for one year,[135] but this was subsequently limited to 200 loads of peat per tenant,[136] the quantity reckoned necessary in late eighteenth-century Scotland to fuel a single hearth per annum, before being lowered to only 100 loads per tenant in 1655.[137] Other disputes in 1657 and 1673 with local landowners were temporarily resolved by negotiation.[138] But that did not bring permanent resolution of the conflict, as the underlying issue of diminishing reserves of accessible fuel for both townsfolk and their landward neighbours could not be addressed without identification of viable alternatives. Fresh proliferation of lawsuits with their neighbours followed a visit to the mosses by a council delegation in 1686 which found that the 1673 agreement had been broken. The burgh responded with steps to police its lands and rights, and during the early 1690s, when the intensely cold and wet weather added to the clamour for access to reliable sources of thermal energy, the council appointed an 'overseer of the mosses' to ensure that all cutters had

135 Ibid., p. 273.
136 Ibid., p. 287.
137 Ibid., p. 297.
138 Ibid., pp. 300, 317. Similar local deals were being negotiated in other areas where formerly abundant peat reserves were becoming depleted in the late seventeenth and early eighteenth centuries, e.g. NRS GD22/1/455 (Offerance, Stirlingshire, July 1703).

licences.[139] But even this step failed to curtail illegal cutting and the council records reveal breaches through the first half of the 1700s.[140] In the late 1600s, it is evident, there were no obvious or innovative legal solutions to the problem of illicit cutting. There was more success in steps to better manage the remaining resources, with efforts beginning in the particularly poor summer of 1693 to improve the quality of mossland that had been described as 'spoiled' in the 1650s and so extend the workable area through improving drainage. This initiative was probably a response to waterlogging and erosion caused by the increased rainfall of the Seven Ill Years,[141] when shortages of adequate supplies of dry fuel added to the general misery of human conditions. The programme of drainage work, however, brought only a temporary respite, as concerns over the adequacy of the supply resurfaced again in the early 1700s and persisted into the late eighteenth century. It was only the increase in shipments of coal northwards from the 1790s onwards that ended the dependence of Elgin and the other northern burghs on the uncertain availability of peat.

• • •

What is evident after the exploration of these core aspects of the environmental history of sixteenth- and seventeenth-century Scotland is absence of passivity in the face of the recurrent blows delivered by the climatic instability of the period. While many saw the hand of God directing the scourges that harrowed the lives of Scotland's people – and prayed and, with no sense of irony, fasted in the hope of a return of Divine favour – they also looked to find physical remedies to address the material shortages they faced. Leadership came from crown and parliament, which began to think in terms of what we would recognise as 'policy' and frame the legislation to deliver outcomes that were intended to be an improvement on the current position. Some we might dismiss as frivolous and imbalanced, like the increased protections afforded to the private hunting interests of the greatest in the land, but others were clearly aimed at betterment of the common weal, such as the repeated efforts to encourage woodland plantations. Through much of that legislation, we can see the emergence of a mindset that is more familiar from the later eighteenth and nineteenth centuries, redolent with the spirit of 'improvement' that inspired some landowners to experiment with ways to increase the value and productivity of their estates. Woodland planting was one key route to that aim, but elsewhere we can see it in the experimentation with a shift towards more commercially minded practices that required new approaches to livestock management, improvement of grazing opportunities, and reorganisation of access to landed resources and the manner in which they were bounded and enclosed. Such changes were scattered, they were uneven in their distribution, scale and effectiveness, and many of them failed, but they marked a faltering first few steps towards one of the most profound episodes of environmental transformation yet seen. And though this was still an era of shocks and transitions, where the environment was an adversary to be confronted, possibly resisted, and hopefully tamed, it was also one where there was a growing belief that this seemingly unpredictable beast which had stalked the lived experience of countless generations could be mastered and reshaped to answer the needs of the human population.

---

139 Cramond, *Records of Elgin*, vol. 1, p. 351.
140 Such pressure was reflected in 1708 in petitions from the landowners in Liggat to permit their tenants to cut peat at Mosstowie in that year (Cramond, *Records of Elgin*, vol. 1, p. 375), and in actions begun by the council against the owners of Pitgaveny, Linkwood and Mayne over their tenants' unlicensed encroachment on Glassgreen Moss (p. 376).
141 Cramond, *Records of Elgin*, vol. 1, p. 354.

CHAPTER 5

Dawning Reason and Improving Practice: Alternative Visions 1700–1790

The villages and the shielings
where warmth and cheer were found,
have no houses save the ruins,
and no tillage in the fields[1]

Most histories whose focus is on early post-Union Scotland, the dawn of the 'Scottish Enlightenment' or the 'Age of Improvement' either divide the eighteenth century around its mid point – frequently represented by the final, failed Jacobite rising of 1745/6 – or adopt a 'long' view that commences in the 1690s and ends possibly as late as 1832.[2] This chapter takes the end of the 'Seven Ill Years' as its beginning and 1790 as its end point. These dates reject one artificial cleavage that is present in much Scottish historiography around the political and religious revolutions of 1689/90, starting instead with the end of the episode of environmental and subsistence crises of the 1690s and early 1700s. Its chosen end point enables us in the chapters that follow to reflect on what change had occurred by the time of the initiation of the great data-gathering exercise of Sir John Sinclair of Ulbster's *Statistical Account of Scotland*, which was undertaken across the 1790s. That national survey, first proposed and started in 1781 by David Erskine, 11th earl of Buchan, but only carried to completion in a new format through a survey that started in 1792 after a further decade in its gestation, marked both a reflection on what had – and had not – been achieved by that point and the continuing wave of transformative 'Improvement' that followed in the first half of the nineteenth century. The nine decades between those points are dominated tradi-

---

1 Extract from Donnchadh Bàn Mac an t-Saoir (Duncan Ban Macintyre), 'Oran nan Balgairean', translated by A. Macleod, printed in R. Watson, *The Literature of Scotland* (Basingstoke and London, 1984), p. 215.
2 For the former, see for example R. Mitchison, *Lordship to Patronage: Scotland 1603–1745* (London, 1983) and B.P. Lenman, *Integration, Enlightenment and Industrialization: Scotland 1746–1832* (London, 1981). For the latter, see the review article by T.C. Smout, 'A new look at the Scottish Improvers', *Scottish Historical Review*, 91 (no. 231 part 1) (April 2012), pp. 125–149, which looks across the 'long eighteenth century'. Fredrik Albritton Jonsson's analysis of the Improvement era in Highland Scotland largely bypasses discussion of the origins of Improvement in Lowland Scotland to focus on the Highlands' experience of the phenomenon and traces its course there from the meetings of the Select Society of Edinburgh during the 1750s and 1760s: Albritton Jonsson, *Enlightenment's Frontier*, pp. 11–12, 16–18.

tionally in mainstream histories by a narrative that is most often distinguished by a 'Whiggish' view,[3] in which there was almost uninterrupted progression from a time of crisis in national self-confidence in the wake of the disastrous 1690s, through financial turbulence in the early 1700s, to the glories of the 'Age of Enlightenment' later in the century. Many historians speak of a 'working out of Union', as Scotland experienced first the shocks of integration into a new economic system that brought as many challenges and threats as opportunities, or they focus on the political and religious divisions most commonly expressed in terms of Jacobitism versus Hanoverian/Government, and Presbyterianism versus all alternative ecclesiastical models. But such simple oppositions mask a complex mix of political, social, economic and religious factors, which makes it difficult to identify any single trend, theme or issue that characterises this period, other than the sense of accelerating change, driven along by enthusiasm and experimentation, that pervaded much of the social, economic and cultural discourse of the time.

Central to the Whiggish narrative is the idea of 'Improvement', itself a somewhat nebulous and contested concept of social and economic betterment.[4] This Improvement was to be achieved principally through development and application of innovative agricultural practices and land-management methods, but also by embracing wider societal and cultural developments aimed at perfecting civilisation. For some recent discussants, the key trigger for this innovation lay in the publicly debated theorising and model construction of the Select Society of Edinburgh from the 1750s, whose 162 members included leading intellectuals like David Hume and

Tomb of David Erskine, 11th earl of Buchan, Dryburgh Abbey, Scottish Borders. A distinguished antiquarian who followed in the footsteps of Scott of Scotstarvit, Sibbald and Clerk of Pennicuik, and a patron of the arts and sciences, Buchan was the political driving force who organised a first attempt at a parish-by-parish survey of the country in 1781, which was left unfinished when Sir John Sinclair started his project in 1792.

Adam Smith, alongside major aristocratic landowners, and whose meetings involved debates that explored the potentials – positive and negative – of Improvement.[5] The Select Society, however, represented a second step in a process that was already underway in the years around 1700, giving greater social and political weight as well as intellectual acuity to older discourse that had been enthusiastic and

---

3 Whiggish historiography in a Scottish context pursues a narrative of inevitable progression and improvement, from past deficiency to present superiority; it judged Scotland's pre-Union past as culturally backward (albeit supposedly with praiseworthy trends towards democracy and social justice), economically impoverished and socially conservative, in contrast to its manifest advance on all fronts within the framework of the British state.
4 See summary discussion in Albritton Jonsson, *Enlightenment's Frontier*, pp. 13–15.
5 Ibid., pp. 16–18.

energetic but had lacked rigour and control. As much philosophy as practice, Improvement, in alliance with 'Politeness',[6] has been presented as the main vehicle for Scotland's escape from its culturally and economically impoverished past into the sunlit uplands of 'Modernity' in all its manifold forms within the British state. The timetable of that escape, however, is as nebulous and ill-defined as the concept of Improvement itself. Most discussion of the broad process identifies Improvement as having its origins in late fifteenth- and sixteenth-century English agricultural innovation, principally in respect of arable cultivation. It also recognises, however, that its articulation as a cultural principle and process through which developments in agriculture, industry, trade and social order would unleash unending prosperity and individual and collective wellbeing was a seventeenth-century phenomenon.[7] Many features of what would later be labelled Improvement were present in the late sixteenth- and earlier seventeenth-century Lowland Scottish (failed) plantations in Lewis and (more successful) plantations in Ulster,[8] and still earlier in the fifteenth- and sixteenth-century parliamentary legislation that promoted agricultural expansion and woodland planting discussed in the previous two chapters. It has been argued persuasively, however, that Scotland's first serious confrontation with an 'Improving' future came only in 1681.[9] Early in that year, James, duke of Albany, younger brother of King Charles II (and future King James VII), held a meeting of a 'committee of trade' of the Scottish Privy Council to discuss why Scotland's economic state was so weak in comparison to its neighbours.[10]

In respect of the agricultural productivity side of Albany's trade question, Scotland was certainly lagging far behind England at this date in the study of key Improvement subjects like agronomics – but not in horticulture or in arboriculture[11] – and a high-level prospection of the contents of private libraries suggests that literature that was in widespread circulation elsewhere in Europe, including England, was still by c.1700 largely unknown to Scotland's landowners or intellectual community.[12] But that does not mean that there was no contemporary understanding of the potential for agricultural improvement, and it is likely that what has been described as the sense of national shame among Scotland's ruling class at their self-perceived backwardness that the duke of Albany's committee had exposed

---

6 L.E. Klein, 'Politeness and the interpretation of the British eighteenth century', *Historical Journal*, 45:4 (December 2002), pp. 869–898. For the Scottish link between Improvement and Politeness, see N. Phillipson, *Hume* (London, 1989), pp. 17–34.
7 P. Slack, *The Invention of Improvement: Information and Material Progress in Seventeenth-Century England* (Oxford, 2014).
8 T. Brochard, 'Plantation: its process in relation to Scotland's Atlantic communities, 1590s–1630s', *Journal of the North Atlantic*, Special Volume no. 12 (2019), pp. 73–94.
9 Smout, 'A new look at the Scottish Improvers', pp. 125–126.
10 A.J. Mann, *James VII, Duke and King of Scots, 1633–1701* (Edinburgh, 2014), pp. 128–132.
11 The prominence of Scottish horticulturalists and arboriculturists in the later seventeenth century is well known (see, for example, Smout, 'A new look at the Scottish Improvers', p. 128) and was one factor that contributed to the rapid development of the large-scale seedsmen and nurserymen businesses in Lothian in the first quarter of the eighteenth century (see below, pp. 160, 202–205, 208).
12 Smout, 'A new look at the Scottish Improvers', pp. 126–127. The author highlights the failure of James Donaldson's *Husbandry Anatomized* (Edinburgh, 1697) and John Hamilton, 2nd Lord Belhaven's *The Countrey-man's Rudiments: or, an advice to the farmers in East-Lothian how to labour and improve their ground* (Edinburgh, 1699) to secure any widespread readership. His identification of Andrew Fletcher of Saltoun as owner of one of the largest collections of agricultural texts alongside his books on gardening, tree-planting and bee-keeping suggests the possibility of a circulation of ideas within the intellectual community with which Fletcher corresponded.

Northern Skye from the Lochmaddy–Uig ferry. Martin Martin's improving manifesto, a 'Brief Account of the Advantages the Isles Afford by Sea and Land and Particularly for a Fishing Trade', constituted the first prospectus for a programme of investment in the farming and, especially, fishing potential of the Hebrides.

had already begun to shift minds in the closing decades of the old century.[13] In language that would not look out of place in any later eighteenth- or nineteenth-century accounts of the economic potential of Britain's colonies around the globe – and, in its tone of exasperation at locals' failure to realise that potential, sharing that air of cultural superiority with which such writing is redolent – the final 'Brief Account of the Advantages the Isles Afford by Sea and Land and Particularly for a Fishing Trade' within Martin Martin's *Description of the Western Isles of Scotland* underscores how widely awareness of Improvement thinking had already permeated before 1700, more than six decades ahead of the excited modelling of Edinburgh's Select Society.[14] But Martin, a Sgitheanach (person from Skye) with personal links to both the MacDonalds of Sleat and the MacLeods of Dunvegan, was a correspondent with and provider of information to the polymath Sir Robert Sibbald, a contributor of papers to the Royal Society of Edinburgh from 1697 and eventually a graduate in medicine of the University of Leiden in the Netherlands.[15] Between Martin's information-gathering journey around the Hebrides in c.1695, his address on that topic to the Royal Society of Edinburgh in 1697, and the first publication of his *Description* in

13 Smout, 'A new look at the Scottish Improvers', p. 125 cites two of Roger Emerson's critiques of late seventeenth- and early eighteenth-century Scottish society as important analyses of this mindset and its results: R.L. Emerson, 'Sir Robert Sibbald Kt, the Royal Society of Scotland and the origins of the Scottish Enlightenment', *Annals of Science*, 45 (1988), pp. 41–72; R.L. Emerson 'Scottish cultural change 1660–1710 and the Union of 1707', in J. Robertson (ed.), *A Union for Empire: Political Thought and the British Union of 1707* (Cambridge, 1995), pp. 121–44.
14 Smout, 'A new look at the Scottish Improvers', pp. 129–130 focuses on Martin's comments on agriculture, but the main thrust of the relevant chapter concentrates on the case for the development of a commercial fishery in Lewis: Martin, *Description of the Western Isles*, pp. 349–359.
15 D.J. Macleod, 'Editorial Note' in Martin, *Description of the Western Islands*, pp. 9–16 at 13.

1703, he had become thoroughly imbued with Improvement thinking. The source of that thought was presumably Sibbald and his network, but the enthusiasm displayed by Martin surely sprang from his own belief in the potential of Improvement in practice to unleash what he was convinced was the unrealised economic wealth of his homeland.

Much modern discussion of Improvement focuses on its accelerating pace post-1750, and Lord Briggs's memorable mid-twentieth-century identification of the period 1783–1867 as the Age of Improvement has helped to cement the late eighteenth-century date of its flourishing into public consciousness.[16] More recently, however, Albritton Jonsson's major re-evaluation of the application of Improvement theories and practices in the Highlands has pushed the 'age' several decades further back to the middle of the century.[17] That, however, is to discount the engagement with Improvement discourse and application of innovative practices by Highland as well as Lowland landowners and their agents that had been underway from at least the 1720s. We cannot underestimate the impact of that influential group of Scottish landowners, lawyers and *cognoscenti* who were already engaging with Improvement and beginning to introduce its thinking in practical application on their properties before the end of the first quarter of the eighteenth century. This group coalesced in 1723 into the Edinburgh-based Honourable Society of Improvers in the Knowledge of Agriculture in Scotland, whose membership list – both current and then deceased – by 1743 included nearly 300 individuals, from great landowners like the dukes of Atholl and Hamilton to medical doctors, merchants and engineers. Although it has been described rather dismissively as 'essentially a coterie of feudal amateurs, aristocrats and lairds, many with an elite legal background' that included no one who described themselves as a farmer, tenant, factor, land-steward or surveyor, and

16 Briggs, *Age of Improvement*.
17 Albritton Jonsson, *Enlightenment's Frontier*.

only three men labelled merchants, one engineer and a handful of 'mathematicians',[18] the membership of the Society merits closer examination. It is important to note that it included two Edinburgh booksellers – James Davidson and John Traill – with premises close to the heart of the city's legal and educational establishments, who were suppliers of books to leading landowners who were members of the Society and to other prominent families engaged in Improvement who were not. Traill, for example, supplied books to the Drummonds of Blair Drummond, the Dundases of Dundas, the duke of Gordon (for whom he obtained multiple books on surveying), the Hamilton-Dalrymples of North Berwick and the Colquhouns of Luss.[19] Three seedsmen or gardeners were listed. The first was Archibald Eagle, described as 'merchant in Edinburgh, seedsman to the Society', a respected and well-connected professional. William Boutcher, the Edinburgh seedsman and nurseryman – as 'Mr Boutcher Gardener' – was the second of these men. His role in supplying tree-seed, seedlings, garden plants and equipment to many of Scotland's leading Improvers is touched on later in this chapter. The final figure was the London-based Stephen 'Sweetzer' or Switzer.[20] Although Switzer is best known for his work in England, he was also involved in Improvement works around some landowners' residences in Scotland, including for the Grants of Castle Grant, and his published works on woodland planting, cultivation of fruit trees and hydraulics were of importance for Scottish landowners long after his style of garden design had become unfashionable.[21]

Published papers by Society members included much that can be dismissed as at best quasi-scientific, often untried and much of it utterly impractical. Much also relied on assertion and unsupported opinion, just like Martin Martin in 1703 in his blind confidence that large parts of the Western Isles could easily be converted to productive arable cultivation through simple effort and improved methods. There was, however, also serious engagement with wider British debates on methods of husbandry and land management, on planting regimes and soil-enrichment, and important papers on new technologies such as more efficient, coal-fired limekilns. Prominent members like Graham of Orchill, Grant of Monymusk, Hamilton-Dalrymple of North Berwick or Maxwell of Arkland were active in the direct farming and management of their own or their employers' properties. Such men shared a perception that agricultural improvement was the key to unlocking a wider array of social, economic and cultural betterment, as propounded with far greater impact in the 1760s by the leading members of the Select Society.

---

18 Smout, 'A new look at the Scottish Improvers', p. 131. The comment on the absence of factors or land-stewards is correct only insofar as none of the members carried those labels, but John Russel, described in the membership list as 'chamberlain to the Earl of Moray', was identified as 'factor' to the Moray estates in the 1730s and David Graham of Orchill, a laird in his own right, was factor to the duke of Montrose. The latter was a major planter and Improver on his properties in southern Perthshire, around Loch Lomond and in western Stirlingshire. It is to be suspected that a number of the lesser lairds and men identified as lawyers in the membership list also held land-management roles on major estates. The engineer, 'Mr Baden, Ingineer', was Lewis Baden, a mining specialist, who became overseer of coalmines in Clackmannanshire.
19 NRS GD24/1/782 (21), GD44/51/465/2 (5), GD75/326, GD110/968, GD248/105/3 (11).
20 Eagle supplied seed to prominent Improvers, like the Earl of Breadalbane (NRS GD112/15/230 (36 and 38), GD112/15/242 (14 and 21)), the Horsburghs of Horsburgh in Peeblesshire (GD178/9/1 (15)), and to other merchants who were dealers with other Improvers (RH15/54/21).
21 W.A. Brogden, 'Switsur, Stephen', *Oxford Dictionary of National Biography* https://doi.org/10.1093/ref:odnb/26855 [accessed 3 April 2022]. E. Clarke, 'Switser, Stephen', *Dictionary of National Biography*, vol. 55 (Oxford, 1898), pp. 241–242. Switzer supplied the laird of Grant with garden flowers in 1728 (GD248/211/14 (3)).

House of Monymusk, Aberdeenshire, drawn by William Taylor, from *Castles of Aberdeenshire* (1887). Sir Archibald Grant's transformation of the sixteenth-century tower into a modern house formed part of a programme of sustained investment in 'improvement' of his estates that followed his expulsion from Parliament in 1732 for fraud.

As with later eighteenth-century Improving tracts, much energy was devoted in papers presented to the Society to unfavourable contrasts between the backward and wasteful practices of the past with the advantages of the new methods; these were generally identified as increased productivity, meaning higher yields and greater profit for the landowners, and market stability and food or materials security for consumers. Unlike the philosophical debates of the Select Society, which engaged in theorising around the most advantageous models and the potential for conflict engendered by unfettered Improvement, the Society of Improvers was concerned with active experimentation, often meeting with failure but equally often with some success. It may have taken decades for the practice of Improving methods in agriculture to catch up with innovation in woodland management and to permeate widely, but on estates spread from the Borders to the northern Highlands, the Society's membership reveals that the fundamental idea of Improvement had already taken firm hold in the first quarter of the century.

It is in the light of this growing conviction of the betterment that was possible through Improvement that we should see the eighteenth-century transitions in perceptions of the wider Scottish environment – and in the character of land in particular – that become evident in contemporary thought and writing. In the early decades of the century, there was a growth in humanist rational or utilitarian views of the land and its potential, albeit often still cast in the language of the Isidorean duality of the medieval period. This new kind of utilitarianism moved far beyond traditional perceptions of humanity and nature in adversarial opposition to one where humanity could be – and should be – unquestionably the master and destined to shape the entire world to suit human needs and desires. In the second half of the eighteenth century, a further shift occurred towards a recast bipolarity between the application of reason and utility on the one hand and the quest for sublimity in nature that formed a central strand within what was later labelled 'Romantic' thought on the other.

Conflict with a visibly and very physically hostile nature, conviction that human ingenuity, labour and

Applying science and reason on the land

Apart from the thousands of kilometres of drystone dykes that still criss-cross the Scottish landscape, one of the most widespread but largely neglected indicators of the progressive introduction and reach of Improvement thinking into every corner of the country is the presence of an estate limekiln or, in fact, multiple such kilns on the larger estates. Some, like the massive example at Tomphubil at the head of Glen Goulandie in Perth and Kinross, are of relatively late date; it was constructed in 1865 but the limestone outcrop where it is located had been exploited for agricultural and building lime already for at least a century by then. As recorded in the correspondence received and responded to by the members of the Honourable Society of Improvers, however, by 1725 major landowners who had commenced with the improvement of their properties were experimenting with liming and the infrastructure to produce it. Some had begun also to introduce the construction of limekilns as a stipulation within the leases of tenants on larger farms where enclosure to enable the introduction of improved agricultural practices was underway. It is difficult to be certain whether the lime produced in these new installations was intended principally for soil-improvement or for the mortar needed for the new style of stone-built house and steading buildings that the leases also demanded. But as the eighteenth century progressed and landlords like Sir Archibald Grant or the earls of Breadalbane enforced the introduction of new agricultural theory and technology on their estates, the bulk of their output was for ploughing in to the acid soils. An excellent example of liming in advance of a general reordering of an unimproved property can be seen at Menstrie Glen on the south side of the Ochil Hills in Clackmannanshire, where in 1754 and 1755 James Wright of Loss was carting in three chalders of lime (between 7,000 and 10,000 litres (dry volume)) per annum (The Royal Commission on the Ancient and Historical Monuments of Scotland, *Well sheltered and watered': Menstrie Glen, a Farming Landscape near Stirling* (Edinburgh, 2001), p. 24). The nineteenth-century example illustrated here, at Whitefield near Kirkmichael in northern Perth and Kinross, was used to make lime for application on the new fields being taken in from the acid grassland pasture and moor in the upland district overlooking Strathardle. It is typical of the single-draw kiln types that had developed from the middle of the eighteenth century, where limestone in the bowl-like depression in the top of the kiln was heated by the fire in the chamber below, fuelled originally by peat but, as transportation infrastructure improved, increasingly by coal.

the appliance of science could make even the most seemingly barren of wastes blossom, and opposing views that on the one hand registered only horror or disgust at its vast awfulness or on the other rapturous wonder at the touch of the sublime in its awesomeness, were most strikingly juxtaposed in alternative perceptions of the Highlands in this era. The equation of topography with the cultural traits and characteristics of a region's inhabitants that derived from the seventh-century theologian Isidore of Seville's *Etymologia* still informed most descriptions of the Highlands and Islands in the first half of the eighteenth century. Martin Martin's 'Brief Account' of 1703, although cast in the language of Improvement and of unrealised potential, is redolent with that Isidorean view where the physical state of the land is reflective of the cultural state of its inhabitants. However, the islands were, he argued, 'of all

North Uist machair looking north-west from Cairinish to Baleshare. Martin Martin was convinced that investment in knowledge and equipment was all that was required to transform the islands into a new source of bountiful grain for all of Europe.

others most capable of improvement by sea and land'.[22] It was, he opined, their remoteness and the Gaelic language of the natives that presented the greatest obstacles to attainment of an informed understanding of the possibilities for development of their homeland, other than in the handful – no doubt himself included – who had acted 'by the force of their natural genius' to learn from the example of their neighbours. They had 'not yet arrived to a competent knowledge in agriculture, for which cause many tracts of rich ground lie neglected, or at least but meanly improved in proportion to what they might be'. Certain that it was just ignorance that prevented this potential from being unleashed and that the Islesmen had somehow failed for millennia to adapt to the circumstances that nature permitted, he advocated sending just 'two or more persons skilled in agriculture' from the Lowlands to open islanders' eyes to opportunity and turn their homelands into a new European bread-basket supplying grain 'to furnish the opposite barren parts of the continent with bread'. Claiming an already rich natural grassland and cultivable soil state but also pointing to the abundance of seaweed available for manuring the land, Martin advocated a programme of enclosure, soil-improvement, drainage, development of grass and hay production, and the planting of wheat, peas, orchards and kitchen gardens that 'they might have as great plenty of all things for the sustenance of mankind as any other people in Europe'.[23] The blind faith that change, i.e. Improvement, would bring transformative social, economic and cultural advance, sits like a fine veneer over a deeper belief that 'native' indolence and intransigence were

22 Martin, *Description of the Western Islands*, p. 349.
23 Ibid., p. 350.

keeping both the land and those natives themselves locked in a primitive state.

It is still a very Isidorean perspective of upland environments and an imperfect or incomplete natural state that pervades the letters of the early Georgian military engineer Edmund Burt sent from the Highlands in the 1720s, albeit with the occasional glimpse of an alternative where 'fortune and education' and 'letters and converse with polite strangers will render all mankind equal'.[24] Such passing awareness of the principles of Improvement apart, as Charles Withers has expressed it, Burt's writing generally carries no sense of potential for betterment and instead speaks of the land in terms of horror, inconvenience and malformation: 'this is not a landscape of aesthetic majesty. It is one of "huge naked rocks", of "monstrous excrescences" and of an all-pervading "horrid gloom" where even the colours are disagreeable'.[25] The 'otherness' of Scotland and of the Highlands in particular permeates almost all of Burt's correspondence, but Letter XV's descriptions of the physical state of the region – the mountains' 'stupendous bulk, frightful irregularity, and horrid gloom' – with the barriers it presented to profitable economic exploitation and Letter XVI's description of bogland and rivers are the most detailed.[26] Burt, indeed, questioned (rhetorically) the utility of the 'monstrous excrescences' of the mountains and responded that their value probably lay in the minerals within them, their suitability for the breeding and feeding of cattle, wildfowl and other 'useful animals' at little or no cost, and their service in collecting water 'for the use of mankind in time of drought'.[27] In no place did he see the rock, bog and heath as land simply still awaiting 'completion' through conversion into arable fields or pasture, nor did he see it as possessing any kind of beauty or attraction.

If anything, this bleak view of the Highlands, in sharp contrast to the supposed softer, more refined civility of the Lowlands, became more entrenched in the economic-intellectual discourse around Improvement as the eighteenth century progressed. This entrenchment came despite the emergence of an alternative vision among those attracted to what would become the Romantic literary, cultural and philosophical movement. It figured prominently in the evolving models of Scottish stadial history – the discussion of the stages of social development in the different parts of the world – where the Classical tradition of drawing equivalence between the civility and refinement of people and the physical form and agricultural state of the land they inhabited provided a theoretical schema of a gradation of social and cultural development.[28] The Highlands' supposedly mountainous, intractable and desolate landscape, caricatured as the home to unsophisticated hunters and pastoralists, was suited to the feudal tribalism of the clan system. By way of contrast, the Lowlands, with their preponderance of arable agriculture, expanding urbanism and engagement in commerce, had enabled their inhabitants to advance beyond that primitive stage and, capitalising on the time freed by these more productive and profitable means of economic existence, Lowlanders had developed in civility and evolved in their democratic political sophistication.[29] This contrast is the theoretical pivot

---

24 Edmund Burt, *Burt's Letters from the North of Scotland*, ed. A. Simmons, with 'Introduction' by C.W.J. Withers (Edinburgh, 2012). Quotation from Letter XV, p. 154.
25 Ibid., p. xiii.
26 Ibid., pp. 153–161 at 157.
27 Ibid., p. 158.
28 P. Fielding, *Scotland and the Fictions of Geography: North Britain 1760–1830* (Cambridge, 2008), pp. 2–3.
29 The stadial view of historical development was exemplified by the writings of Henry Home, Lord Kames, especially as expressed in his *Historical Law Tracts*, 1st edition (Edinburgh, 1758).

Liathach, Glen Torridon, looking east from Mullach an Rathain over the Am Fasarinen pinnacles to Spidean a' Choire Leith. For Edmund Burt, 'monstrous excrescences' such as this lacked any beauty or utility, beyond the minerals their rocks might contain, the water they might store, or the game that lived there.

around which, for example, the Enlightenment thinker Edmund Burke's delineation of the world into graded zones of barbarity and civility turns.[30] In this construct, the physical landform might determine the cultural, social and economic state of its inhabitants and, without the introduction of different ideas of how to live and how to exploit the land and its resources by those of greater civility from regions of advanced socio-economic development, it would present the insurmountable barrier to their betterment and 'improvement'. Highlanders, in this distinctly colonial philosophical view, were equivalent to other 'savages' with whom Britain's colonial agents were engaged, in North America or Africa and, like them, they required to be educated in the most appropriate methods for the optimum development and exploitation of their land.[31] Thus, as well as having an economic imperative driving them,

---

30 G.H. Guttridge (ed.), *The Correspondence of Edmund Burke*, volume 3 (Cambridge, 1961), pp. 350–351, letter to William Robertson, 9 June 1777.

31 This attitude is encapsulated in Samuel Johnson's memorable description from 1773: Highlanders, even educated and relatively cultured ones 'were content to live in total ignorance of the trades by which human wants are supplied, and to supply them by the grossest means. Till the Union made them acquainted with English manners, the culture of their land was unskilful and their domestick life unformed; their tables were as coarse as the feasts of Eskimeaux, and their houses filthy as the cottages of Hottentots', Samuel Johnson, 'A Journey to the Western Islands of Scotland', in Samuel Johnson and James Boswell, *A Journey to the Western Islands of Scotland* and *The Journal of a Tour to the Hebrides*, ed. P. Levi (London, 1984), p. 51.

North Berwick Law across the cultivated landscape of East Lothian, from near Athelstaneford. As a contrast to Burt's 'monstrous excrescences', the rolling farmland of East Lothian represented proof positive of the stadial theories espoused by Edmund Burke, where the state of a people's civilisation and culture was reflected in the form of the land in which they lived.

Improving landowners had a moral obligation to lead the inhabitants of their properties out of the ignorance and primitivism of their traditional lifestyles into releasing and realising the full potential of the land and themselves. It is no surprise that many of the practices employed in the North American colonies in the second and third quarters of the eighteenth century were introduced, with widely variable degrees of success, into the estates of people who had personal experience of colonial methods or who employed those who had such experience as their factors and change agents.[32]

As Edmund Burt's descriptions from the 1720s and, later, Samuel Johnson's views from the 1770s illustrate, there were those who considered parts of the Highlands and wider upland districts of Scotland to be beyond any greater economic or social betterment due to the absence of any means to cultivate the deformed 'monstrous excrescences' or miry sloughs encountered there. Johnson was both attracted and repelled by what he believed he saw in the expanses of 'dark heath' and general absence of varied vegetation, describing the alternative as 'nakedness, a little diversified now and then by a stream rushing down the steep'. He immediately contrasted what he saw with the 'flowery pastures and waving harvests' of southern districts and claimed astonishment and repulsion at 'this wide extent of hopeless sterility' which had the appearance of 'matter incapable of form or usefulness, dismissed

---

[32] For a detailed case study, see A. Ross, 'Improvement on the Grant estates in Strathspey in the eighteenth century: theory, practice and failure', in R.W. Hoyle (ed.), *Custom, Improvement and the Landscape in Early Modern Britain* (London: Routledge, 2011), pp. 289–311.

by nature from her care and disinherited of her favours, left in its original elemental state, or quickened only with one sullen power of useless vegetation'.[33] Throughout his accounts, he expressed revulsion at the lack of development or even the potential for development, dismissing the whole as lacking in any utility through which its inhabitants might achieve social, economic or cultural betterment. Even at Loch Lomond, which the Welsh-born topographer, observer of nature and early ethnologist Thomas Pennant had so recently extolled, he found much that on closer inspection failed to please him: 'the islets, which court the gazer at a distance, disgust him at his approach, when he finds, instead of soft lawns and shady thickets, nothing more than uncultivated ruggedness'.[34] It is not that there were no attempts to achieve the desired transformation of these supposed wastelands, from the successful, like the drainage and clearance efforts from the late 1740s to create arable directed by Hugh Graham of Arngomery on the southern edge of Flanders Moss in Stirlingshire, to the fruitless attempts to drain parts of Rannoch Moor in the 1760s by government soldiers under the direction of the administrators of the forfeited Perthshire estate of Alexander Robertson of Struan.[35] Supreme confidence that all that was needed to turn Rannoch Moor into productive farmland was the application of the superior knowledge of the Improvers and the guiding example of non-Highland practitioners, however, could not overcome the challenges of geology and climate. Even with the advanced, steam-powered agricultural technology of the second half of the nineteenth century, this elevated expanse of rock, water and blanket bog remained beyond the ambitions and capabilities of any Improving landlord.

Even as the government soldiers toiled in the saturated peat of Rannoch Moor an alternative vision of the land and its inhabitants was gaining popular currency. Debate surrounding the authenticity and authorship of the poetry published by James Macpherson between 1760 and 1765 as the works of Ossian, a Gaelic poet of the third century AD, is irrelevant to the impact his works had in changing perceptions of the Scottish Highlands.[36] Within five years of the publication of the first of Macpherson's 'translations', travellers from the south of Britain were heading for the Highlands to experience for themselves the exhilaration and awe that the combination of 'so much beauty and so much horror' that the 'ecstatic' mountains generated within them.[37] This quest for the 'sublime' through experience of

---

33 Johnson, 'Journey to the Western Islands', p. 60.
34 Ibid., p. 149.
35 For the work on Flanders Moss and its context, see J. Harrison, *A Historical Background of Flanders Moss*. Scottish Natural Heritage Commissioned Report No. 2 (Edinburgh, 2003), pp. 49–51. This example is also discussed in Albritton Jonsson, *Enlightenment's Frontier*, pp. 1–2, 33 (where the eighteenth-century spellings 'Graeme' and 'Ardgomery' are used). The so-called 'Soldiers' Trenches' or Lub nam Buth can still be traced at the eastern edge of Rannoch Moor. See https://canmore.org.uk/site/341143/lub-nam-buth-soldiers-trenches#references-fragment and https://www.scottish-places.info/features/featurefirst19006.html. This is the unspecified 'bog near Rannoch' referred to in Albritton Jonsson, *Enlightenment's Frontier*, p. 33.
36 J. Macpherson, *Fingal, an ancient epic poem, in six books: together with several other poems, composed by Ossian the son of Fingal. Tr. from the Galic language by James Macpherson* (London, 1762); J. Macpherson (ed.), *The Works of Ossian the Son of Fingal*, 2 volumes, 3rd edition (London, 1765).
37 The full quotation containing these descriptions is drawn from the correspondence of the English elegiac poet, Thomas Gray, that forms the introduction to T.C. Smout, 'The Highlands and the roots of Green consciousness, 1750–1990', *Scottish Natural Heritage Occasional Paper* No. 1 (1990), published in its revised, fully developed form in T.C. Smout (ed.), *Exploring Environmental History: Selected Essays* (Edinburgh, 2009), pp. 21–51 at 21.

Reaching the end of the road, looking west from Kinloch Rannoch at the maximum reach of the attempted penetration of Improvement into the uncultivated expanse of Rannoch Moor. Here, finally, science and reason ran hard up against the realities of climate and geology in the failed efforts to drain and cultivate the moorland.

and within the land, which would characterise much Romantic thinking as it unfolded in the later eighteenth and earlier nineteenth centuries, was perhaps an emotional response to the emotionless pragmatism of the rational Improvers, but it also constituted a profound philosophical challenge to the stadial historical model. Rejecting the detachment of the 'aerial' or external position adopted by Edmund Burke, who stationed himself as an outside observer and, thus, able to comment from a position of superior elevation and dispassionate disassociation, the Romantics sought immersion in place and local culture to secure an affective engagement with and understanding of the land, its people and the interrelationships of both. Over time, that philosophical view became almost the antithesis of the principles that drove Improvement.

It would be wrong to caricature views of the land and environment in the later eighteenth century as simply polarised between the utilitarianism of the

Falls of Acharn, Loch Tay, Perthshire. Planted with larch and beech for aesthetic effect, the gorge of the Allt a' Chilleine was 'improved' as a place of wonder and delight, to be viewed from the comfort of the Earl of Breadalbane's tea-pavilion in the Hermit's Cave overlooking the Falls.

Improvers and the emotion of the Romantics. Despite their later high public profile and unquestioned influence in matters of aesthetics, the dissenting voices of the Romantics made little impact on the advance of Improvement as a general management strategy within the Highlands in particular. Nevertheless, as we shall see in a later chapter, although the greatest of the Improving landowners of the era, men such as John Murray, 3rd duke of Atholl, and John Campbell, 3rd earl of Breadalbane, made hard-nosed business decisions that determined the means by which they transformed the physical environment of their estates on the grand scale, they were inspired also by the Romantic landscape vision stimulated by Macpherson's *Ossian*. Amid the commercial planting of his estate in Strathbraan near Dunkeld, Atholl constructed Ossian's Hall and the Hermitage overlooking the gorge and falls of the River Braan, while Breadalbane laid out walks and built the 'caves' and grottoes of his own Hermitage overlooking the Falls of Acharn on the south side of Loch Tay. But these were for their personal pleasure and the amusement of their guests rather than reflections of a wider tension in their personal philosophical view of their property. Atholl and Breadalbane were demonstrating that other facet of Improvement, their superior culture and refinement which enabled them to create the perfect landscape in which beauty and horror could be admired and experienced from tastefully designed settings rather than through the immersion of the soul in the lived experience of place sought by the new wave of Romantics from the 1790s onwards. It was in the following century that an opposition that was founded as much in politics and economics as in philosophy or cultural theory generated a more heated conflict over the form and fate of Scotland's land and its inhabitants.

• • •

Most modern discourse focuses on the novelty, energy and ambition with which eighteenth-century Improvement thinking was imbued. As a foundational element in those Faustian chimeras of continuous economic growth and unending socio-cultural betterment, the relentless positivism of the Improvers' message has been repeated and re-presented endlessly as an exemplifier of the unleashed potential of post-Union Scotland and her people. Here, writ large in their own words and with its applied experiment visible in the physical legacies of transformation wrought at landscape scale across the face of the land, we come face to face with the manifestation of a decisive step being made in the stadial theory so favoured by Scottish philosophers. Freed from the hostile constraints of environment, and unrestrained by the shackles of their traditional agricultural practices, all Scots could leave the stage of primitive cultural and social disorder to ascend to a higher level of civilisation. Proponents of the new philosophy were quick to point to the self-evident truth laid out to see across Scotland, where the contrasts between the settled, economically advanced, agriculturally experimental and culturally precocious eastern Lowlands and the archaic, disordered, economically underdeveloped and socially inferior Highlands and Islands were easy wins cheaply taken. Deeply grounded in inherited cultural prejudices formulated in the later fourteenth century and ingrained by the seventeenth, reinforced by deep suspicion of the cultural 'other' with which Lowland Scots shared this land, and redolent of cultural models formulated in the post-Roman Latin West, the theory that equated human characteristics and the geomorphology of their land was repackaged to demonstrate how the endemic political instability and economic backwardness of Highland society was rooted in the undeveloped state of the region. Intellectually lazy and philosophically questionable, this determinedly prejudiced, deeply skewed model fixed an image of a land of two halves in the minds of the political and economic leaders of the British state. What were perceived as the endemic problems of the one came to be seen as resolvable only by making it more like the other, in essence presenting Lowland Scotland as the 'norm' and Highland Scotland as an aberration ripe for reforming. It was in just such terms that James VI had viewed his 'Highland Problem' in the early 1600s and Martin Martin had seen opportunity for advance at the end of the seventeenth century. The notion was thus already hard-wired into Lowland Scottish and wider British mentalities by the mid 1700s, when everyone from the Forfeited and Annexed Estates Commissioners to travellers like Dr Johnson saw the 'answer' in terms of economic development and cultural assimilation. Despite the failure of so many Improvement schemes and the rise of what Albritton Jonsson has presented as a new negativism in perceptions of the potential for economic growth and Improvement in the Highlands in the early decades of the nineteenth century, it was a contrast still being made and a solution still being sought for by politicians and economists at the beginning of the twenty-first century.

CHAPTER 6

Changeable Times: Climate and Weather in the Eighteenth Century

---

> [It] began to rain below, but it was snow above to a certain depth from the summit of the mountains. In about half an hour afterwards, at the end of near a mile, there arose a most violent tempest. This, in little time began to scoop the snow from the mountains, and made such a furious drift, which did not melt as it drove, that I could hardly see my horse's head. The horses were blown aside from place to place as often as the sudden gusts came on, being unable to resist those violent eddy-winds; and, at the same time, they were nearly blinded with the snow.[1]

Of all the things in the dawning 'Age of Improvement' that perhaps showed the least change from what had come before, the prevailing patterns of winter and summer weather were the most significant. Bizarrely, it is one of the greatest peculiarities of the historiography of earlier eighteenth-century Scotland that, until the late 1990s, the first stages of the profound transformation of Scottish rural societies and economies were discussed in abstract or technical contexts quite divorced from any consideration of the lived experiences through this continued era of climatic volatility. Instead, the focus was strongly on social and economic factors, and there was little discussion of the constraints and opportunities that climate change placed on landlords and tenants beyond off-the-cuff references to the 'combination of natural and human deficiencies' that meant 'the soil made niggardly returns'.[2] A reader of that work would be forgiven for thinking that the period saw no weather extremes to compare with what had come in the 1690s. Indeed, apart from sweeping – and quite inaccurate – generalised statements that 'the first four decades [of the eighteenth century] were broadly favourable',[3] most historians adopted a Whiggish vision of positivity, progress and 'Improvement' that focused on the post-1750 period. Yet, recognition by some historians of the potential socio-economic impact of the extreme weather and

---

1 Burt, *Burt's Letters*, p. 177.
2 W. Ferguson, *Scotland: 1689 to the Present* (Edinburgh, 1968), pp. 70–71.
3 T.M. Devine, *The Transformation of Rural Scotland: Social Change and the Agrarian Economy 1660–1815* (Edinburgh, 1994), p. 75.

harvest failures of the early 1770s and early 1780s, and its constraining effect on the abilities of landlords or tenants to invest in Improving activities, renders the failure to explore conditions through the 1700–50 period all the more inexplicable.[4] Even the most cursory of examinations of the primary sources reveal that although the worst extremes of the 'Seven Ill Years' had indeed ended in 1699, unfavourable weather continued into, and regularly if episodically throughout, the new century; it was a further 150 years yet before the 'little ice age' drew finally to a close. Tracing even general trends beyond the headline events, however, remains difficult due to the sporadic nature of surviving record sources – most of which are very localised in their reporting and concentrate on those headline extreme events – with some parts of the country lacking even such basic material.[5] Striking though individual accounts of great storms are, and we can relate the profound social and economic disruption that they caused to our own contemporary experiences of such events, they were short-term episodes and provide little insight on the less immediately obvious processes of broader trends in weather patterns across annual or decadal periods.

Extending the Seven Ill Years: the cold continues

Although a growing volume of weather observation data accumulates as the eighteenth century progresses, including from the 1770s the first systematically collected instrumental records, climate proxy data remains essential for providing insights on prevailing weather conditions regionally, especially in much of the Highlands and Islands. As in earlier periods, it is the Traligill speleothem record that at present provides the most consistent data across this span. The evidence there for water percolating into the limestone caverns indicates that the eighteenth century started with a period of unusually dry winter conditions for the north-west mainland, pointing to an episode of low winter NAO and the likelihood of northerly and easterly airflow that brought cold, drier conditions elsewhere over Scotland.[6] There seems also to have been an extended spell of low summer NAO, for the northern Cairngorm tree-ring record indicates that while cold summer temperatures had reached their nadir in the late 1690s they still remained significantly below the mean throughout the 1700s and only staged a brief recovery late in the 1710s.[7] What such conditions meant for the people of Scotland is revealed in the details offered by parliamentary records, private correspondence and diaries, which chronicle a continuing atmosphere of popular anxiety amid still-unstable weather conditions.

The weather manifestation of the negative winter and summer NAO of the proxies stalk contemporary written reports from locations spread across the country. Accounts from the Lothians, for example, noted a mild winter in 1699/1700 and an early but windy and dry spring, cold but again windy and dry summer and a delayed harvest that was barely in before snow and frost started in early November.[8]

---

4 Ibid., pp. 75–76.
5 For the most part, we are reliant on the blunt instrument of grain prices, which from 1723 were set annually by the fiars courts in each shire.
6 Proctor et al., 'A thousand year speleothem proxy record', p. 818 Fig. 4.
7 Rydval et al., 'Reconstructing 800 years of summer temperatures in Scotland', p. 2959 Fig. 4B.
8 George Turnbull, 'The diary of the Rev. George Turnbull, Minister of Alloa and Tyninghame, 1657–1704', ed. R. Paul, in *Miscellany of the Scottish History Society* (Edinburgh, 1893), pp. 295–445 at 389–390, 393, 395, 396; Francis Masterton, 'Masterton Papers, 1660–1719', ed. V.A. Noël Paton, *Miscellany of the Scottish History Society*, volume 1 (Edinburgh, 1893), pp. 449–493 at 475.

Glen Derry and Derry Cairngorm, Cairngorms, Aberdeenshire. Cooler summers persisted from the late seventeenth into the early eighteenth century and helped to suppress pine growth and the setting of new trees from seed, an episode evident in the trees' annual growth rings.

This unexpectedly early start to winter, however, did not bring the prolonged and bitter cold and snows for the whole country that had been experienced in the 1690s. Instead, there was greater regional difference. In highland Perthshire, for example, the early heavy snowfall in November only delayed the already very late grain harvest on the Breadalbane estate, which was completed shortly before 25 November 1700.[9] The snow apart, however, this milder start to winter 1700/1 in Highland Perthshire raised hope that the incessant summer and autumn rains and bitter winters of the 1690s had passed. That optimism was dashed when the latter part of the winter reverted to severe cold and poor weather that continued into May and affected the sowing of the spring grain in arable districts of East Lothian.[10] Anxiety again rode high for the likely condition of the grain yield, but despite the poor start, the crop flourished and was harvested in full in Lowland districts before October.[11] That success was provident, for weather conditions deteriorated sharply in late autumn, with reports of sea-floods in the inner reaches of the Forth from Kincardine to Clackmannan in November pointing to easterly gales and storm surges up the narrowing estuary, perhaps combining with spate run-off from the western hills.[12]

The easterlies of autumn 1701 prevailed still in spring 1702, but now they brought cold and dry

9 NRS GD112/39/182/22.
10 Turnbull, 'Diary', pp. 398, 399, 401, 402.
11 Ibid., p. 406.
12 'Masterton Papers', p. 476.

Culbin Forest and Findhorn Bay, Moray. The modern conifer plantations cover the towering sand dunes on the west side of the bay, which are spread over the long headland that formed the eastern end of the estate of Culbin.

conditions to the eastern Lowlands. Widespread drought had taken hold by May, leading to a withering of the pastures before the end of the month and shortages of early season grass for the already malnourished livestock.[13] And then the summer turned wet, resulting at the end of August in a saturated harvest with much of the crop brought in wet and in need of drying in kilns to prevent it from rotting or mildewing.[14] The weather continued to deteriorate through the autumn, culminating in a great storm with winds initially west-south-westerly then veering to north-westerly on 11 October. That wind direction brought sea-floods, this time to parts of the East Lothian coast, with further flooding inland that destroyed some of the spring-sown grain crop. In the North-East, the same gales wrought extensive devastation along the Moray coast, with further sand-drift over the fields at Culbin and intensified long-shore drift and spate conditions on the Findhorn creating a new shingle barrier that blocked the river mouth, which then forced a new passage to the sea, destroying part of the old village of Findhorn.[15]

In light of the traumas of the previous decade, it is easy to understand why famine still haunted the memories of Scots through these disturbed weather episodes in the first decade of the new century. Petitions to parliament regularly recalled the horrors of

---

13 Turnbull, 'Diary', pp. 413–414, 415.
14 Ibid., p. 420.
15 Ibid., pp. 421–422. H. Lamb, *Historic Storms of the North Sea, British Isles and Northwest Europe* (Cambridge, 2005), p. 59; Dawson, *So Foul and Fair a Day*, pp. 128–129 misdates these events to 1703.

the 1690s and spoke in fearful terms of their possible return.[16] In February 1700, a 'solemn national fast and humiliation' was proclaimed, to seek Divine release from 'the continued pinching Dearth ... the great and unusual Sickness and Mortality, which hath gone over all the Land'.[17] The General Assembly of the Church of Scotland and the Scottish Parliament again agreed in July 1703 to a further solemn national fast and humiliation to seek God's continued mercy after the end of the famine and to pray that the hardships of the previous decade were not repeated.[18] Interestingly, in the 1790s, one Midlothian minister looking back on the Seven Ill Years described the famine period as extending from 1695 to 1702, which tallies better with the long tail of poor harvests and food shortages that staggered on into the new century.[19] In fact, the economic repercussions of the successive failures of the grain harvest, especially the most severe failure in 1698, were still playing out in the second half of the first decade of the eighteenth century. Men who had held the tack of the excise collection from the late 1690s, for example, still pursued compensation in 1704 for their losses caused by spoiled and failed harvests, while sub-collectors used claims of reduced excise payments which had 'the same effect as famine' as late as 1707.[20] Some Highland estates, where rents had not been collected during the worst years and where the loss of income could not easily be made up, rarely managed to balance their accounts for many years after the famine's end as tenants still struggled to subsist.[21]

Alongside the added trauma of the financial ruin brought by the failure of Scotland's attempt to found a trading colony at Darien in the Isthmus of Panama, the indebtedness of some of Scotland's greatest landowning families caused by the catastrophic loss of income across this period helped to polarise political divisions for and against parliamentary Union with England or a restoration of the exiled senior male line of Stuart monarchs.

And after that, rain and snow

The death of the childless King William in March 1702 and accession of his sister-in-law, Queen Anne, helped to further deepen these divisions. Continuing adverse weather conditions added worsening economic hardship to the complex mix of tensions. Weather-related impacts on noble incomes persisted and were a further factor that fuelled discontent with the ruling regime, and their contribution to the prevailing climate of disharmony should not be underestimated. For a culture in which poor weather and harvest failures were still widely seen as evidence of Divine displeasure with the ruling regime, the poor start to Anne's reign did not bode well. Perhaps fortunately for the queen, after a cold late summer and autumn in 1703, which saw snowfall in the Ochil Hills in late August, then an indifferent harvest followed by a stormy winter,[22] relatively good summers with early and satisfactory harvests from

---

16 RPS A1700/10/34 [accessed 24 March 2022].
17 NRS GD150/3381.
18 RPS 1703/5/127 [accessed 24 March 2022].
19 Rev. Dr John Walker, 'Parish of Collington', *Statistical Account*, vol. 19 (Edinburgh, 1797), p. 586.
20 RPS 1704/7/90, A7O16/10/45 [accessed 24 March 2022].
21 NRS GD112/39/200/15.
22 'Masterton Papers', p. 478. Reports from Caithness in January 1704 speak of stormy weather over an extended period: NRS GD112/39/192/4. Scotland escaped the devastation of the storm of early December 1703 which swept across southern England and the southern North Sea and through into the southern Baltic, but still experienced high winds and rain: Lamb, *Historic Storms*, pp. 59–72.

1704²³ to 1706 eased some pressure. The relief was short-lived, however, for deteriorating conditions from March 1706 saw 'excessive raines' in eastern Scotland that continued throughout the year and down to February 1707.²⁴ The result by the end of 1707, with its unsuccessful harvest, was a spike in grain prices to levels unseen since the 1690s.²⁵ Poor weather continued through 1708 and into 1709 and helped to stoke up discontent that was already riding high in the wake of the Treaty of Union. The exceptionally poor winter of 1708/9 – one of the European 'great winters' – struck a people whose resilience was already fragile;²⁶ the resulting delayed spring and unsettled summer of 1709 brought a poor harvest across the whole country.²⁷ Local dearth in even otherwise prosperous trading ports like Leith, where the poor European conditions removed any possibility of imported grain to make good shortfalls at home, stimulated fear of a return of the conditions of the 1690s and prompted the burgh council to organise relief for the poor in anticipation of the starvation and deaths to come.²⁸ Clearly, the experience of the 1690s' famine years had left deep scars.

At Traligill, speleothem growth signals an increase in winter precipitation in the north-west Highlands that extended through the 1710s into the 1720s, indicating a higher winter NAO and a phase of wetter and milder conditions.²⁹ This short, mild episode appears to have coincided with a generally strong warming trend in the north-western European climate down to c.1730, although scattered references to extreme frosts and snow in Caithness in 1710–11, to severe winters and poor summers through the 1710s in Shetland, and to bad summers with poor harvests in Islay from 1714 to 1721 serve as reminders that such general trends can mask greater instability regionally than the proxies can reveal, especially but not exclusively in those districts most directly exposed to fluctuations in Atlantic surface-water temperatures.³⁰ Major one-off storms such as that in early March 1724 which devastated Portsoy on the Banffshire coast represent headline events that stand out against the broader record of prevailing conditions, but they perhaps also signposted the beginning of a transition into the next phase of the LIA.³¹

23 'Masterton Papers', p. 480.
24 Ibid., p. 482. Masterton reported that the dam at 'Gartstank' (Gartmorn?) burst in December 1706 and several houses were said to have collapsed because of the rain.
25 Ibid.
26 The snow was reportedly so deep around Loch Tay by 24 November 1708 that the roads were impassable to men on horseback, while in January 1709 severe stormy conditions were reported around Edinburgh: NRS GD112/39/222/25, GD124/15/942.
27 In Caithness, rain was identified as the main factor holding back the harvest: NRS GD112/39/231/15.
28 NRS RH9/14/83.
29 Proctor et al., 'A thousand year speleothem proxy record', p. 818 Fig. 4. Reports of heavy snowfall on the Banffshire coast, coupled with severe storms at sea in the Moray Firth, in February 1724 suggest that the higher winter rainfall coming in from the Atlantic experienced further north and west had encountered colder air coming from the north and north-east, typical of negative winter NAO conditions, as it crossed the country: NRS GD248/563/68 (41). The short, mild episode had definitely ended by the mid 1720s, with low precipitation/negative winter NAO prevailing for around fifty years thereafter.
30 NRS GD112/39/250/2. Dawson, *So Foul and Fair a Day*, p. 129.
31 NRS GD248/563/69 (18). The harvests and grain supply of 1723/4 and 1724/5 appear to have been especially poor, yet there is little reference to this episode in studies of this period. The exception is P.R. Rössner, 'The 1738–41 harvest crisis in Scotland', *Scottish Historical Review*, 90 part 1 (2011), pp. 27–63 at 29, where he calls also for research into the crisis of 1724 as one of a series of under-researched eighteenth-century subsistence crises.

Portsoy, Aberdeenshire. Prosperous through seaborne trade in the seventeenth century, when its harbour was first developed, the impact of the 1724 storm on the outer harbour wall proved to be a significant economic setback that was not put wholly right until the new harbour for the herring fishery was completed in the 1820s.

The reversal of the brief trend towards increased precipitation visible in the speleothem record suggests that the north and west Highlands perhaps experienced some colder and drier winters, as delivered by low winter NAO. This appears to be the position reflected in the comments of Sir Archibald Grant of Monymusk's estate factor, Thomas Winter, who in early May 1726 wrote to his employer to bring him news of work underway on his Donside estate. Winter referred to the unseasonably hot and dry conditions at that time that he was concerned would impede the growth of the barley crop. The drought was perhaps exacerbated by Monymusk's location in the Cairngorms' rain-shadow, but his observations on the very low level of the River Don at this date is suggestive of low winter and spring precipitation over the mountains and most likely no significant snowfall to bring the water-release of a spring melt spate.[32] This pattern of weather persisted through the central decades of the century. We have evidence for such conditions on Speyside in the central Highlands in late winter 1727/8, where long-lasting and severe frosts delayed building work at Castle Grant. Such conditions are suggestive of extended periods

---

32 H. Hamilton (ed.), *Selections from the Monymusk Papers (1713–1755)* (Edinburgh, 1945), p. 103.

of sluggish anticyclonic circulation around a high-pressure cell, but the winters in the years to either side saw significant precipitation. On the nearby Moray Firth coast in winter 1725/6, nearly three weeks of heavy snow was reported before a thaw set in from mid January; in the west Highlands, deep snow was reported in December 1728. Both reports appear to indicate the collision of intensely cold Arctic air with moisture-laden Atlantic fronts.[33] The colder and perhaps drier winters stemmed from wider changes in North Atlantic conditions, which saw the extent of sea-ice increasing around Iceland in the late 1720s and especially from winter 1729/30. The plunging temperatures brought a collapse in the fisheries off Iceland and the Faroe Islands but, as species like cod and herring were sensitive to cold conditions and moved southwards, the sea-fisheries around northern Scotland should have benefited. Oceanic systems, however, seem to have experienced wider disruption, for the white fish catch around Shetland in 1731 was apparently very poor and the Forth ports' North Sea herring fishery was in steep decline already in the 1720s.[34] Stormy weather, however, made capitalising on any movement in the cod to compensate for those declines a dangerous venture, as the losses of boats and men from communities along the Kincardineshire coast in a great windstorm in February 1730 attest.[35]

That pattern of possibly cold/dry winters and warm/damp summers suggested by the Traligill speleothems can be seen perhaps also in the tree-ring record from the northern Cairngorms. There, it is a rising trend in summer temperatures that can be posited from the dendrochronolgical evidence to have prevailed through the 1720s, accompanied by the damper conditions that favoured the trees. Such significantly wetter weather over the mountain massif is visible in the reports of excessive rainfall and consequent flooding that affected the Spey basin in June 1727. The rain over the mountains brought hardship in the lowlands, with accounts of widespread inundation of farmland on the floodplains in the lower reaches of the river apparently following intense and prolonged precipitation over the central Highlands.[36] Similar increased precipitation seems to be reflected in reports from August/September in the following year from parts of Mid Argyll. There, devastating storms, including hailstorms with stones large enough to injure cattle and flatten the unharvested crops, punctuated the late summer and early autumn of 1728.[37] Although warmer summers may have continued into the next decade, the winter conditions deteriorated sharply and the 1730s witnessed a return to the cold and wet of earlier years.

Volcanic forcing and the icy 1730s

As perhaps happened in the 1250s with the Samalas eruption of Gunung Rinjani in Indonesia, the 1730–6 Timanfaya eruption of Lanzarote in the Canary Islands and its massive release of aerosol particulates into the atmosphere contributed to a pronounced episode of volcanically forced cooling globally.[38] Incidental references to extreme cold suggest that low NAO winters with cold anticyclonic airflow became established down the eastern side of the country and,

---

33 NRS GD112/39/289/22, GD248/47/2 (65), GD248/564/72 (5).
34 Smout and Stewart, *Firth of Forth*, p. 33.
35 NRS CH1/2/62 f. 5.
36 NRS GD44/43/16 (59).
37 NRS GD112/39/288/25.
38 J.C. Carracedo, E. Rodriguez Badiola and V. Soler, 'The 1730–1736 eruption of Lanzarote, Canary Islands: a long, high-magnitude basaltic fissure eruption', *Journal of Volcanology and Geothermal Research*, 53 (1992), pp. 239–250.

Glen Almond, Perth and Kinross, frost and snow on the hills between Glen Almond and Loch Tay, 9 November 2008, illustrate the conditions experienced in the late autumn of 1732.

with low air circulation, the cold pooled in valleys in the southern Highlands and across the Lowlands. The year 1732 appears to have been especially bitter, with ice on the Nor' Loch at Edinburgh thick enough to take the weight of a man on horseback being reported in early May.[39] The summer was short and by early November 1732, the minister of Killin was writing to the earl of Breadalbane's factor to request repairs to his manse before the expected sharp frosts prevented the mortar used to point the stonework from hardening properly.[40] The following winter again brought extreme conditions that lasted into early spring, with around one metre of snow reported as having fallen at Careston in the lowlands of Angus in the second half of March 1733.[41] This poor weather affected much of the central and southern Highlands, with tenants on the Loch Awe portion of the Breadalbane estates petitioning for relief of

39 Dawson, *So Foul and Fair a Day*, p. 131.
40 NRS GD112/27/46 (20).
41 NRS GD237/20/11 (4).

rent due to storm losses experienced in the early part of the year, and on the Montrose estates in Stirlingshire numerous bridges were reported to have been lost in spate floods as the snows melted when the delayed spring finally began to warm up.[42]

Winter 1738/9 brought an already poor decade of generally unfavourable weather conditions to a violent close, with cold northerly anticyclonic systems clashing with cyclonic westerly airflow from the Atlantic. John Clerk of Penicuik recorded a three-month period of intense frost in Lothian from December through to February, including a six-week period of deep snow cover, with streams and rivers freezing over and high mortality among wildfowl and game.[43] There were, however, violent episodes within this phase of intense cold, the worst of which was a storm on 24/25 January 1739 (Clerk's 13 January date is 'old style'), which swept Scotland from Arran in the west to the Angus and Berwickshire coasts in the east, causing widespread damage to buildings and woodland and sinking or capsizing ships on both coasts.[44] Clerk noted that he lost around 1,400 trees on his properties, many snapped through the trunks or blown over completely. Modern analysis of this event suggests that a combination of the funnelling effect of the midlands valley and convergence of atmospheric cells – a cold front moving south and a wet and warmer front moving east – intensified the wind strength around the pressure systems.[45] Destructive and disruptive though this storm was, however, it is the wider trends that helped to create it that are more revealing. The anticyclonic winds that delivered the cold air from the north were symptomatic of the current low winter NAO, as was the more southerly track of the westerlies with which they collided. As the Traligill proxy data implies, northern and western Highland winters were probably drier but also significantly colder than they had been even in the 1690s, as were conditions through the north-east Highlands and more eastern districts of the country. Unlike the Seven Ill Years, however, it seems that these cold 1730s winters, though severe, were short and the summer trend was towards warmer weather, but not necessarily meaning more favourable crop-growing conditions.

While the winters of the 1730s were intensely cold, those years constituted one of the warmest decades in the central Highlands for summer temperatures since the early sixteenth century. But warm summer conditions do not necessarily mean benign weather. Indeed, summer weather appears to have been as relatively unstable as that of the 1720s, with violent but short storms and downpours affecting the more mountainous districts. Such episodes, however, constituted the brief, socially disruptive and eye-catching headline events beloved of earlier chroniclers and later diarists, for the decade more generally was one of the peak periods in the last eight hundred years for tree establishment in the central Highland region. Thus, although precipitation was bad for those eking out a subsistence arable farming existence in the valleys that radiate from the Cairngorm massif, those same warm and damp conditions were favourable to tree growth and woodland regeneration through seed-setting.[46] But what was good for trees was certainly not good for grain and the end of the decade saw the start of one of the worst runs of failed or poor harvests of the eighteenth century, extending from 1738 to 1741.[47]

An indifferent summer and bad harvest led into

---

42 NRS GD112/10/1/3 (56), GD220/5/884 (20).
43 J.M. Gray (ed.), *Memoirs of the Life of Sir John Clerk of Penicuik, 1676–1755* (Edinburgh, 1892), pp. 149–150.
44 Ibid., pp. 150–151; Lamb, *Historic Storms*, p. 82.
45 Lamb, *Historic Storms*, p. 82.
46 Rydval et al., 'Reconstructing 800 years of summer temperatures', pp. 2957, 2958 Fig. 3B.
47 Rössner, 'The 1738–41 harvest crisis in Scotland'.

a winter in 1739/40 that was much worse than the poor winter of 1738/9 recounted by Clerk of Penicuik and which epitomised the contrasts that low winter NAO could bring to the British Isles generally. In November, the weak westerlies had given way to a strengthening cold, north-easterly airflow coming down from Scandinavia across the North Sea. It was this that eventually delivered in mid January 1740 the intense windstorms and sea-ice that affected England's east coast from Northumberland round into the English Channel,[48] but the winter had already commenced with a phase of extreme cold and hard frosts that started the development of ice on major river-courses. Scotland has usually been omitted from discussion of the impact of this event, but eastern Lowland districts suffered from the same extreme cold, heavy snowfall and blizzard conditions as affected England. The most extreme phase started around 11 January 1740, when a first casualty – a man travelling from Cupar on the hill-road to Ceres in central Fife – was reported as a consequence of the dreadful conditions.[49] By 13 January the snow was deep enough to impede movement of mail coaches from the south and west towards Edinburgh, and between 18 and 25 January it was reported that the River Forth was entirely frozen west of Alloa and the Tay was frozen upstream from around its confluence with the Earn, 8km below Perth.[50] Interspersed with further substantial snowfalls, the latter part of the month saw the intensification of the frost, with reports of the outer Firth of Forth carrying some three inches of ice shore to shore at the Queensferry Narrows, and various journals narrated the widespread inability of mills to grind grain on account of their lades and ponds freezing solid.[51] By the start of February, Edinburgh newspapers carried stories of cattle and sheep starving in the open because they could not access grazing beneath the snow and farmers were unable to bring what meagre feed they had stored out to them because of snow-drifts. But the broadsheets also carry tales of human suffering due to shortages of fuel either to keep warm or to heat food or drink.[52] Further snow and intense cold followed and by 8 February news reports from the northern Highlands spoke of the Dornoch Firth frozen over from Meikle Ferry west, and, from further south, of the Tay being frozen right across only 4 miles above Dundee on the tidal estuary, where it is over 1.5 miles wide.[53]

It is likely that the freezing of the Dornoch Firth was part of the 'great storm' in northern Scotland reported in 1740 that wrought widespread devastation through Ross and northern Inverness-shire.[54] In the same week as that event was reported, accounts from the Kincardineshire coast speak of the snow at Stonehaven being so deep that householders had to tunnel up from their front doors to get out.[55] It was not just the east that was affected, however, for Glasgow was blanketed in snow throughout the event and even coastal towns like Greenock reported how their poorer townsfolk suffered extremely in the cold. Hopes of a thaw came with a brief, mild phase when the wind switched to the north-west around 15 February, but the cold soon returned. By the second half of the month it was reported in the Edinburgh newspapers that sheep-farmers in the western South-

---

48 Lamb, *Historic Storms*, p. 82.
49 M.G. Pearson, 'The winter of 1739–40 in Scotland', *Weather*, 28 (1973), pp. 20–24.
50 Ibid., p. 20.
51 Ibid.
52 Ibid., p. 21.
53 Ibid., p. 22.
54 NRS GD248/97/4 (42).
55 Pearson, 'Winter 1739–40', p. 22.

Dornoch Firth and Meikle Ferry from the Struy Road, Highland. In January 1740, the tidal waters of the Firth from Meikle Ferry – which ran from the headland projecting into the centre of the view below the bridge to the smaller headland projecting into the Firth at the left of the image – up to where the River Oykel discharges into the Firth were frozen from side to side.

ern Uplands had begun to drive their flocks down to lower-lying districts where they could still get pasture – but at great cost in grass-rents.[56] Data on human mortalities during this period has not been worked through systematically, with most concentrating on large urban centres like Edinburgh – which saw a sharp spike in excess deaths doubling the January and February mortalities over those reported for December – but it seems that higher numbers of winter deaths and more that arose from the food shortages of this period were averted through local famine relief efforts organised by parish authorities and local landowners.[57] With the diversion of funds to provide such relief, it is unlikely that many among even the greatest landowners were in any position to invest heavily in the development of their properties across these years.

Further pulses of cold with more snowfall continued through February and March, with reports of the ground still being so frozen in some places in early May that peats for the following winter could not be cut. This added inability to secure the most basic of necessities for daily life repeated the crisis conditions of the later seventeenth century and brought further instability to communities that were already struggling in the face of the challenges to subsistence agriculture posed by the weather. Nevertheless, replicating the trend of the 1730s, the summer warmed rapidly from May and raised hopes of a respite. Although some newspapers reported good crop-growing conditions and a good harvest, contemporary diarists continue to speak of dire weather, food shortages and generally poor growth in the new season's crops or grass. John Cockburn of Ormiston in Midlothian noted on 6 May 1740 that:

> I never did see such a Winter and Spring as this has been. Nothing green to be got of any

---

56 Ibid.
57 M.W. Flinn (ed.), *Scottish Population History from the Seventeenth Century to the 1930s* (Cambridge, 1977), p. 222.

kind and few roots at any price, except onions. Money could not buy what was not. . . . We have been so used to open Winters that little care was taken to provide against a hard one, and this cold dry weather brings in very small supplies. . . . No Grass can rise in this Weather and want of rain will raise unequal Crops of what is in the ground.[58]

The poor growing conditions continued through the year, with North Sea storms reported in mid September and mid November affecting the east-coast harvest and sea-fishery.[59] Although these seem to have inflicted the greatest damage across southern parts of England, severe north and north-easterly gales with rain, snow and hail recorded in Northumberland during the November storm suggest that the east coast of Scotland was also affected. Modern meteorological reconstruction of this event points to a deep low-pressure system that tracked north-eastwards across the south Midlands of England and into the central North Sea, combined with already cold air that had persisted throughout the year, further driving down temperatures. Although the storm did not lead into another winter as bad as the two preceding, the poor conditions brought further suffering and extended the crisis years into 1741.

Our speleothem proxy record from western Sutherland indicates that the 1740s saw a decline in winter precipitation in the north-west Highlands to the lowest levels experienced since the first quarter of the seventeenth century.[60] The persistent low winter NAO conditions point to prolonged episodes of anticyclonic Arctic airflow that delivered generally cold and dry weather but led to heavy snow or long-lasting snow cover when it met moisture-laden systems moving north-east from the Atlantic. While the far north of Scotland may have seen below-

58 J. Colville (ed.), *Letters of John Cockburn of Ormistoun to his Gardener, 1727–1744* (Edinburgh, 1904), p. 64.
59 Lamb, *Historic Storms*, p. 83.
60 Proctor et al., 'A thousand year speleothem proxy record', p. 818 Fig. 4.

Garmouth and Kingston, Moray, looking east across the pre-1829 channel of the Spey to Tugnet on the eastern shore of the 'new' channel. Garmouth, an important timber-processing, shipbuilding and trading harbour on the west side of the river estuary, was the location of important estate operations like the Grants' sawmill, where timber that was floated down the river could be cut for onward trade. The 'Great Spate' of 1829 damaged its harbour and shifted the river's main channel further east, starting a long era of decline.

average rainfall or snowfall, the central Highlands and southern parts of the country experienced above-average poor winter conditions throughout this period, albeit not of the severity of 1738/9 and 1739/40. The northern Cairngorm tree-ring record for the later 1730s and 1740s reveals through a run of narrow rings, which signal limited growth, that there was an episode of pronounced cooling of summer temperatures in this region. That cooling in the montane region is consistent with the poor summers and adverse harvest conditions experienced in lower-lying areas across these years.[61] It is frustrating that the historical records for the 1740s are less informative in terms of environmental data than those for the period down to 1740, but incidental references in estate correspondence are generally indicative of continued weather-related difficulties across the decade. For example, on the Grant estates on Speyside and in the northern Cairngorms, early 1744 saw a prolonged period of intensely cold and stormy weather that left unusually deep snow. This was seen, however, as presenting a potential opportunity to use the inevitable spring thaw as means of floating logs felled in the woodland flanking the river down the Spey for processing at Garmouth, so delivering a much-needed boost to estate income. But summer 1744 seems also to have remained unseasonably cold even on the Moray coast, for by 31 August it was reportedly too cold for the sawmiller at Garmouth to work at night.[62] Dispatches from military units stationed in the north of Scotland indicate that similar poor weather continued into the second half of the decade. Highland mountains were described as under widespread snow cover by late

61 Rydval et al., 'Reconstructing 800 years of summer temperatures in Scotland, p. 2959 Fig. 5B.
62 NRS GD248/168/3 (7 and 25).

October 1747, but this gave way at first to milder conditions with heavy rains that caused spates throughout the region, before turning again to snow that blocked the military roads west of Glenmoriston and south of Fort Augustus in late December.[63] A slight improvement in conditions set in across the first years of the next decade but it was not until the mid 1750s that a more general change in weather patterns established itself.

The winds from the west

This new shift was announced by a major Atlantic storm that swept across the whole of the British Isles, southern Scandinavia, the Netherlands and northern Germany in early October 1756.[64] Storm surges were reported at the head of the Solway Firth, where salt spray was blown inland to the district around Carlisle, blasting vegetation.[65] It is likely that the low-lying merse zone around the Nith, Annan and Esk estuaries on the Scottish side were also affected by this event, as they had been in similar episodes in the thirteenth century. On the exposed Atlantic coast of the Western Isles, this storm is reported to have wrought widespread destruction in machair districts, especially down the highly vulnerable western side of the Uists. The catalogue of damage included loss of agricultural land through tidal surges and immersion beneath sand through dune deflation. The storm seems to have had a profound impact in particular at Baleshare off the west side of North Uist,

---

63 NRS GD113/3/265 (8), GD113/3/269 (10).
64 Lamb, *Historic Storms*, pp. 85–86.
65 Ibid., p. 85.

The weather observed: 1782–1783 from the pages of Janet Burnet's diary

Remarkable survivals, like the diary kept by Janet Burnet, wife of the Aberdeenshire laird George Burnet of Kemnay, first at Kemnay House and then, following her husband's death in 1780, at Disblair House near Newmachar, 19km north-west of Aberdeen, provide rich sources of locally observed detail of weather impacts on Scotland's agriculture. Much of our evidence for the poor harvest of 1782 and the resultant dearth of 1783 comes from upland districts on properties administered by the Commissioners of Forfeited Estates. Burnet's diary provides us with insight on the effects in a district where Improvement was well established by the last quarter of the eighteenth century (M. Pearson (ed.), *More Frost and Snow: The Diary of Janet Burnet 1758–1795*. Sources in Local History 2 (Edinburgh, 1994)).

1782
January
Unsettled with frost and snow, but little on the ground.

February
- 14 The Snow computed 8 Inches over all, after a fortnight of snowfall, which lasted for another week
- 22 Thaw began; a good deal of rain, but cold
- 28 Snow in the Hills little melted

March no entries

April
- 1 Sowing of oats begun at Aberdeen – also some field peas
- 8 Left Aberdeen for Disblair. Very little ploughing done and hardly any seed sown in Disblair district.
- 18 Sowing proceeding very slowly – Snow and cold
- 20 No sign of growth – all as winter
- 21 A great deal sown during the past week. No leaves on the trees – not even the gooseberry bushes showing any sign of leaf.
- 24 All low ground sown, and most of the sowing completed around Disblair. In many other places little done
- 30 The peas which had just been showing above ground on 8 April were not yet half an inch above ground. Not one gooseberry bush in full leaf, nor had any of the buds on the trees opened. Blossom ready to burst given one warm day.

May
- 11 In many places oat seed not sown
- 12 Some cherry plum and apple blossom out. Some larches looking green and the grass looking better. Gooseberry bushes not fully leafed yet.
- 21 and 22 Bear seed being sown
- 23 not a Bud burst on any Forest Tree, but the Larch
- 26 Plain Tree buds opened & Some Leaves on the Hathorn – little Flowrish on the Fruit Trees
- 28 Sowing of bear seed finished in most places, though hardly begun in others

June
- 9 No tree yet in full leaf, though plane, elm, birch and laburnum had some leaves. No blossom on standard fruit trees. A little more blossom on the pears and apples on the wall than there had been on 28 May. Cherry blossom fully out, but nothing else making progress. Some grass very poor and natural grass not growing at all. Sowing of bear not nearly finished
- 15 Gean in full bloom. Strawberries, pears and apples fully blow; also plums on the wall and one standard pear tree. Spinach ready
- 22 All trees now in leaf except the ash. Some cherries set and some pears on the wall. Some gooseberries the size of peas. Standard apples not in blossom yet, except the Oslen.
- 26 Standard trees in full blossom. Peas in full flower.

July
- 1 Laburnum, rowan, lilac and hawthorn in full flower. Grass much improved.
- 14 Hay crop very good.
- 24 Plenty of strawberries and green peas ready. Also turnip.
- 29 Cabbage at table.

August
- 30 After a cold and tempestuous month the leaves of some of the ash and plane trees were quite withered. They had been fully out for only two months.

September
- 6 A few stooks of bear seen in Hatton of Fintray. Oats all quite green.
- 14 Began to cut a small park of bear. Potatoes pretty large.

October
- 6 Bear into the yard, but not very dry. Peas cut down.
- 13 Much of the oat crop still green. Bear not all cut and very little gathered in most places.
- 20 Snow most of the day. Oats hardly changed colour. Very little of the crop in general cut down and gathered in throughout the whole country.
- 24 After three dry days everybody cutting down and taking in as quickly as possible – ripe or green.

November
- 19 Oats cut and some taken in.

December
- 8 Some hill farms above Monymusk still not even cut. Little ploughing done.

1783
January
- 18 and 19 A great deal of Snow with Blowing & sever Frost, which Stopt all traveling with wheel carriges for some days
- 24 Thaw till end of month

February
- 2 Some ploughing got don, but much of the Snow lying & Bad traveling
- 9 Snow and rain in Aberdeen, but all Snow in the Country
- 23 Dry with strong wind; so sever a Frost as to stop Plughing
- 28 Mild and snow melting a little during the day

March
- 18 Oats sown at Disblair

April
- 1 Great activity sowing everywhere. Grass greener than at end of April 1782. All sowing, except the low ground, finished at Disblair. Some spring flowers out and the gooseberries showing leaf
- 9 Low ground sown
- 12 Began to sow the Bear park. The grass looking green and some fields of oats green. Plum blossom out. Gooseberry bushes in leaf. Sowing of oats finished in most parts.
- 19 Larch quite green. Gooseberries, currants and raspberries in leaf. Pear blossom fully out. Buds on the birch green
- 20 At Fintray young Latice [lettuce] of this years sowing & Radish at Table. Peas of this year Stringd & will be soon in Blosom, French Beans far advanced. Cherrie Trees in full blossom – the buds of the Plain Tree Burst & Showing there Leaf tho not Expanded
- 29 Bear seed finished in the park

May
- 1 Larch fully out. Hawthorn green and leaves of many plane trees quite expanded. Pears, plums, apples and cherries on the wall in blossom. Grass and braird [first shoots of grain] looking fine. Gooseberries and currants in leaf and flower
- 9 Bear seed finished at Disblair, with grass seed sown and rolled
- 21 Large Sallet at table. Blossom on peas, pears and apples. All trees, except ash, in full leaf and glory. Gooseberries and currants netted
- 24 Severe frost took off tops of potatoes

Burnet's entries for the year record the effects of the cold, delayed spring on all plant growth on her property and the lateness of the harvest, disrupted as it was by early snow and the late ripening of the all-important oat crop. The bleak picture presented for this eastern Aberdeenshire property is further sharpened by the passing reference to some of the crops on upland farms in central Strathdon remaining unripe and unharvested into the second week of December, which provides us with a hint of how communities in the hillier interior of Scotland had fared through this exceptionally bad year.

Baleshare from Carinish, North Uist, Western Isles. The long, low strip of land on the skyline is the southern, Eachkamish, part of Baleshare, which in the storms of autumn 1756 was temporarily separated from the township in the northern part of the island, to the right in this view.

temporarily separating the northern part of that barrier island from its southern Eachkamish portion and inundating the main Baleshare township beneath several feet of sand from the deflated coastal dunes.[66] The timing of this disaster in the islands was particularly bad for the unharvested grain crop, which was already showing signs of a likely poor yield nationally, and 1756 and 1757 returned the poorest harvests of the mid-century period, triggering yet another but thankfully short-lived subsistence crisis for many districts.[67]

A relative absence of references to extreme events or generally poor weather through the 1760s should not be interpreted as signalling any return to 'good' conditions. Indeed, 1761/2 saw another exceptionally poor harvest with food shortages in 1762.[68] Although the later decades of the century saw a gradual return to higher winter NAO and the milder conditions of a westerly airflow, at no point before the 1790s did mean winter precipitation in the north-west Highlands exceed average levels experienced in the seventeenth century, nor did summer temperatures in the Cairngorms regain the highs seen in the earlier 1730s.[69] Instead, there was a see-saw between cold and mild winters, with an extremely cold winter in 1769/70 that saw hard frosts in January delaying ploughing by several weeks on Speyside. The fluctuating conditions extended across the decade, with 1777/8 seeing hard frosts into March recorded in normally mild Easter Ross, so severe that the spring ploughing had to be delayed as the ground was too hard to work.[70] Winter 1779/80 was again generally

---

66 Gilbertson et al., 'Sand-drift and soil formation', p. 443.
67 Rössner, 'The 1738–41 harvest crisis in Scotland', p. 29.
68 Ibid.
69 Proctor et al., 'A thousand year speleothem proxy record', p. 818 Fig. 4; Rydval et al., 'Reconstructing 800 years of summer temperatures in Scotland, p. 2959 Fig. 5B.
70 NRS GD199/88, GD248/81/1 (36–37).

severe in the northern Cairngorms, with reports of 'knee deep' snow and deer starving in Glenmore; the duke of Gordon and earl of Fife resorted to providing feed from their straw-store for their game animals in an attempt to prevent the total loss of their future hunting.[71] The infrequency of truly extreme winters affecting the whole country, however, at least in comparison to the pre-1750 period, and an apparently downward trend in storminess – or at least its reporting – possibly encouraged hopes of a general improvement. Any optimism, however, was dashed when full-blown famine afflicted parts of the Highlands in 1782–3, with the Commissioners of Forfeited Estates paying for the purchase and distribution of grain and potatoes for tenants in the worst-affected areas from the income from these properties.[72] In Easter Ross, the local authorities attempted to prevent distillation of whisky – licit and illicit – from November 1782 in the hope of averting famine locally due to 'the severity of the season and consequent poor state of agriculture'.[73] Tree growth in the northern Cairngorms was seriously affected in this year, which saw the coldest summer temperatures since the 1230s and the second coldest summer of the last 800 years creating exceptionally hostile conditions for the setting of new trees.[74]

Laki and its aftermath

One potentially significant contributor to the poor conditions from June 1783 was the eruption of Laki

---

71 NRS GD44/43/232 (10 and 68).
72 NRS E704/8, E713/3, E730/30, E732/23, E732/24, E777/281.
73 NRS GD427/183.
74 Rydval et al., 'Reconstructing 800 years of summer temperatures', p. 2960 Table 2.

Caithness coastline looking east past Dunnet Head, Highland. The plume from the eruption of the Icelandic volcano, Laki, in summer 1783, deposited ash over grain fields in Caithness, severely damaging the crop and earning the name of 'the ashie year' in local folk memory.

in southern Iceland. The whole of Britain, along with much of northern Europe, was affected by the physical fallout of the volcano in terms of the particulates from the ash-pall and sulphurous, acidic gases released into the atmosphere and carried south and east by airflow and pressure-system movement.[75] Geologists since the nineteenth century have been aware of local traditions of the ash fallout from the eruption affecting harvests in north mainland Scotland, referring to Caithness lore which named 1783 as 'the ashie year'.[76] Recent analysis of weather diaries from the time of the eruption, kept in locations spread from Fochabers in Moray to Dalkeith in Midlothian, indicate that the first ash-rich portion of the volcanic plume was over Scotland just five days after Laki's eruption and that the sulphurous haze

---

75 J. Grattan and M. Brayshay, 'An amazing and portentous summer: environmental and social responses in Britain to the 1783 eruption of an Iceland volcano', *Geographical Journal*, 161 (1995), pp. 125–134; T. Thordarson and S. Self, 'Atmospheric and environmental effects of the 1783–1784 Laki eruption: a review and reassessment', *Journal of Geophysical Research: Atmospheres*, 108 (2003), pp. 1–29.
76 A. Geikie, *Textbook of Geology*, 3rd edition (London, 1893), p. 217.

was evident a further nine days later.[77] Although it has for long been argued that the Laki ash-plume led to a volcanically forced cooling episode over the following three years, the Scottish record evidence does not support the notion of such impacts here. That experience perhaps suggests that the cooling and wettening seen in the mid 1780s was as much a consequence of wind flow and synoptic air pressure conditions as of the atmospheric impact of the eruption. We cannot be sure that Laki worsened what was an already deteriorating position, but the declining storminess of the period was also reversed in December 1783 in a severe event that swept in on easterly winds, with the first snow reported on 23 December in Aberdeenshire.[78] Accompanied by plunging temperatures, worsened by wind-chill, snow continued into early January 1784, resulting in drifts up to 2 metres in depth, while the storm-force winds wrought widespread destruction to woodland and buildings and losses of vessels in ports and at sea. But an early thaw and milder spring conditions prevented the 1782/3 subsistence crisis continuing into the following year, 1784, which delivered an acceptable harvest. Nevertheless, general conditions remained poor from summer 1783 through to winter 1786/7, keeping the threat of famine alive in the minds of people across the social spectrum. Although early autumn 1786 brought an intensely cold northerly airflow south over Scotland, it was southern Britain and mainland Europe that experienced the worst weather,[79] and spring 1787 hinted that the worst had passed. But less bad does not mean good, and as our present period drew to a close in 1790, there was nothing to suggest that the future held any prospect of a general improvement in conditions. Indeed, as the sun moved into another episode of solar minimum, the 1790s and 1800s – as will be discussed in Chapter 10 – saw the weather worsen severely.

• • •

Although the tone of weather reporting across the nine decades reviewed here was almost unremittingly negative, it would be wrong to see the human population of Scotland as passive victims of conditions beyond their control. As in earlier centuries, landowners and their tenants had options as to how best to respond to the ebb and flow of conditions and, once the immediate memory of the Seven Ill Years faded from their consciousnesses, most of them probably reverted to the patterns of subsistence agriculture or the entrepreneurial quest for profit that their ancestors had pursued. But as the eighteenth century progressed, there was a new dynamic added to the mix of responses, with the spread of innovative thinking on how to achieve social, cultural and economic betterment through new, more rigorous and 'scientific' approaches to the exploitation of natural resources, new methods of farming, and new approaches to the management of the land itself. Through such 'Improvement', it was believed, humanity could escape from the figurative shackles of what were presented increasingly as the old, inferior and wasteful practices of the past and the exposure to the unpredictability of the weather and its impact on harvests that were worsened by such practices. By the application of reason and through enquiry, experimentation, observation and transmission of new ideas, Scotland was taking its first steps into what we would now consider to be the Anthropocene.

---

77 A.G. Dawson, M.P. Kirkbride and H. Cole, 'Atmospheric effects in Scotland of the AD 1783–84 Laki eruption in Iceland', *The Holocene*, 31:5 (2021), pp. 830–843.
78 Lamb, *Historic Storms*, pp. 87–88.
79 Ibid., pp. 88–90.

CHAPTER 7

Improvement's First and Greatest Child: Woodland and Plantation 1700–1790

> [T]he glen has the appearance of incorrigible sterility: yet on more attentive observation, thousands of young pines are seen forcing their way through the chinks of the rocks, and rising amidst layers of loose stones and gravel, where one would be apt to imagine, vegetation would be denied.[1]

Where the philosophy and practice of Improvement took hold first and wrought some of the greatest environmental transformations in this period was in the composition and exploitation of the nation's woodland. Here, as the eighteenth century progressed, the traditional anxieties voiced over shortages of woodland within Scotland and the consequent dependence on foreign suppliers and outflow of money to buy it, along with the ad hoc enactments by parliament that aimed to address those ills, gave way to concerted efforts to develop something approaching a national strategy. In the arguments surrounding this, we can recognise emergent notions of patriotic duty to develop security of supply in what were viewed as materials of outstanding naval and military as well as economic significance, as well as the age-old drive to improve landowner incomes. By the middle of the century politicians, political and economic theorists, men of business and landowners were engaged in wide-ranging debates over the extent to which the State should be a key actor in driving private plantation activity and the nature of any role for it in delivering expansion of domestic availability of such a key, strategic resource as timber; over the types of tree that should be planted to provide the most suitable, in-demand timber; where those plantations should be located; and whether there should be any financial recognition for this patriotic investment. While there were sharply divergent views about the process of Improvement generally and the likely outcomes of such transformative change – positive and negative – most agreed that there was an urgent need for action in respect of woodland.

By 1700, cover of what we would describe as 'native' woodland in many parts of Scotland was approaching its lowest ebb, but in some other areas it was only the eighteenth and early nineteenth

---

1 C. Cordiner, *The Antiquities and Scenery of the North of Scotland in a Series of Letters to Thomas Pennant* (London, 1780), p. 23.

century that saw its first sharp decline in extent from its late medieval level. Pressure on woods through encroachment or clearance by land-hungry peasant farmers in Lowland districts had been easing since the mid seventeenth century, but even in the mid eighteenth century woodland loss in those regions continued, albeit at a diminishing rate, despite a more general shift towards woodland protection and plantation among the greater landowners.[2] At the same time, different pressures were emerging in Highland areas, where commercial interest in mainly oakwoods developed rapidly in the early decades of the eighteenth century, leading to far greater intensification of exploitation than had occurred under traditional management regimes. There, pulses of felling and regeneration were to characterise the experience of woodland across the region and provide us with some of the most memorable anecdotes of the devastating impacts of commercial exploitation on native woodlands. But it was not simply a story of unthinking and wanton unsustainable practice triumphing over traditional methods, for most of the new woodland exploiters were in fact acutely aware that the resource they were felling was finite and, if mismanaged, fragile. Despite Frank Fraser Darling's graphic and very memorable portrayal of the rapacity of the commercial ventures with interests in Highland woodlands, most of these businesses had long-term plans and were making investments in infrastructure that underscore their intention to be more than simply fly-by-night asset-strippers.

As we have already seen and contrary to long-held tradition, investment in woodland planting and its more commercial management had already commenced, several decades before the parliamentary Union with England in 1707. Nevertheless, it was Scotland's progressive integration into the economic and regulatory regime of the new British state and its mercantile empire that provided one of the greatest stimuli to large-scale exploitation and subsequent investment in sources of timber and wood products and then, ironically, rendered the original purpose intended for that wood redundant. The needs of the emerging eighteenth-century fiscal-military state, especially during the protracted series of foreign wars that disrupted many of Britain's overseas supply routes for essential materials – from iron and timber for weaponry manufacture and shipbuilding to tanbark for curing leather for clothing, boots, shoes and harness – created both government concern for security of supply in these materials and demand from producers and processors for more secure domestic sources of these essential commodities. The potential of Scotland's woodland to supply charcoal for iron processing and tanbark for leather production had been recognised in the later sixteenth and earlier seventeenth centuries in ventures such as George Hay's Loch Maree ironworks, discussed in Chapter 4 above, but it was firstly through the burgeoning demand for processed leather goods in the eighteenth century that their more systematic – if episodic – exploitation commenced in earnest, followed later in the century by the quest for security of shipbuilding resources. Britain's early imperial wars and the loss of the well-established trade in leather goods to the American colonies after 1776 triggered cycles of more intensive investment and then retrenchment in Scottish woodland, both the exploitation of the existing semi-natural mixed broadleaf and pine resources and the creation of new broadleaf and conifer plantations. Perhaps more than any other element of Scotland's environment in the eighteenth century, the visible transformation of the already heavily managed woodlands reflected the State's integration into the new social and economic networks of a rapidly reconfiguring world system and its pulses of supply-and-demand driven change.

2 Smout et al., *History of the Native Woodlands*, p. 47 for what seems to be an isolated example of woodland eradication on a west Fife estate in the 1740s.

A land bereft of trees?

If we look first at the Scottish Lowlands, there is a persistent perception of that region for most of the 1700s as a nearly treeless zone until large-scale planting was commenced by Improving landlords in the last decades of the century. Much of that view, as we shall explore below, stems from a conscious nineteenth-century and later misrepresentation of the words of that most self-opinionated and critical of later eighteenth-century travellers, Dr Samuel Johnson. His prominence in the historiography of the age, founded on his literary stature and the consequent widespread dissemination of his travel journal, has overshadowed the contributions of earlier – and often better-informed – travellers than the metropolitan urbanite Johnson. The bleak picture has been worsened further by the weight placed on the national scope of the maps of General William Roy, which suggested that by the middle of the century as little as 4 per cent of Scotland's land surface was under woodland cover of some form. Comparison with estate maps and records, however, indicates that Roy's surveyors missed substantial areas of woodland, in some cases possibly because the open structure did not conform with their notion of what constituted 'woodland', and in others because the mapping exercise was cursory and probably involved observation from a distance rather than close inspection. Indeed, it is likely that the surveyors missed almost half of the wooded areas, meaning that somewhere nearer 8 per cent of Scotland still carried woodland. To balance that criticism, however, it is also evident that Roy's maps show wooded areas that escaped comment in other mid-eighteenth-century sources, much of it in its manner of representation clearly semi-natural woodland rather than plantation. Contemporary with Roy, travel journals kept by English visitors to Scotland also present a broader and more varied picture of the woodland in those parts of both the Highlands and the Lowlands through which they passed than is suggested by the General's map sheets. From their frequently contradictory reporting of the condition of woods in the same general localities, however, we are confronted with the fruits of southern English views on landscape aesthetics, imperfect understanding of the differential impacts of climate and physical environment on woodland development, and knowledge of management practices that ranged from the fanciful to the professional. It is through a synthesis of these sources that we can build a picture that is at once positive and at the same time often dismissively critical.

One of the earliest surviving and published of such journals is that written by Sir William Burrell, a member of a Kent and Sussex gentry family. In 1758, he undertook one of the most extensive – and earliest journalled – tours through Scotland, penetrating Galloway and the western Lowlands, the central Highlands as far as the Great Glen, the northeast, the eastern Lowlands and the Borders.[3] Since his primary interests were in art and architecture, material culture and social and economic conditions, Burrell's observations on landscape are limited in comparison. Nevertheless, he is a rich source of information on the spread of Improving practices into areas of Scotland seen conventionally as late recipients of the new thinking, on agricultural management (especially manuring practice and new crop types), and on the introduction of field drainage and enclosure. Of particular value here, however, is his reporting of what he regarded as especially noteworthy areas of woodland and especially on the number of plantations around the new houses of the gentry and nobility. Looking just at his northward route to Inverness, there is through the first stages of his journey in Dumfries and Galloway and

---

3 Dunbar, *Burrell's Northern Tour*.

Caerlaverock Woods, Dumfries and Galloway, coppiced oak and sycamore. The plantations of oak that Sir William Burrell noted at Drumlanrig in Nithsdale had been present for centuries at smaller scale around many of the greater houses of the region, exemplified by the enclosed oak coppicing adjacent to the Maxwells' original chief residence at Caerlaverock.

southern Ayrshire an absence of comment on woods, seemingly corroborating the Johnson-based view. This reticence, however, seems at odds with Roy's maps of the region, which record extensive plantations around the mansions of the local nobility in the district between Annan and Dumfries and then an increasing number of areas of open and extensive woodland in central Nithsdale and its tributary valleys. These clearly failed to interest Burrell, for whom the first plantation that was deemed worthy of mention lay around Drumlanrig, where he saw 'very noble and extensive' woodland covering around 700 acres.[4] This plantation appears on Roy as a dense block divided by radiating avenues, a geometrically planned example of land management that contrasts sharply with the random tree symbols of the semi-natural woodland around it.

Moving west from Drumlanrig through Galloway, Burrell saw nothing that was considered worthy of note until entering Glen App in southern Carrick, where he observed what he described as 'bushes' on the valley sides. By this he seems to have meant scrub species like willow and hazel rather than substantial broadleaf trees, and he noted extensive alder groves in the Girvan Valley, further north.[5] The Glen App 'bushes' are marked on Roy's map as open woodland of indeterminate composition. At Eglinton, he encountered his first evidence for what was an eighteenth-century Scottish preference for conifers, when he remarked on the 'remarkably beautyfull [sic]' and spacious plantations of what he labelled 'furs' that surrounded the otherwise unremarkable castle. If Roy's mapping of these plantations is reliable, they comprised a remarkable expanse of avenues and rides, groves, roundels and thickets laid out with geometrical precision over many hectares around the Montgomeries' mansion. 'Fur' was a species that he again saw forming plantations at Kelburn, and, distinguishing between avenues of 'pine' and plantations of 'furs', surrounding Hamilton Palace, and again at Drummond Castle near Crieff.[6] It was not until he had passed out of the Lowlands into the Highland district beyond Aberfeldy that he saw what seems to have been (semi-)natural woodland in the trees that cloaked the hillside from top to bottom behind Castle Menzies. That scene, however, was dismissed as 'fine, but it owes little to judgement or taste';[7] for him, the regularity of planned woodland plantations was superior to the disorder of naturally regenerating managed native woods. It was still in the Highlands, at the earl of Breadalbane's residence at Taymouth, that he first saw woodland plantation that truly impressed him in its extent, planning and species composition, providing an endorsement of the planting policy that had been pursued by the Campbells here since the late seventeenth century. Around Taymouth, he observed 'fur, oak, birch, mountain ash, broom, beech, laurel, spirea, larix and many others', including well-established sycamores, covering the hillsides and forming extensive groves and walks around the house.[8] Moving on to Blair Atholl, he was less

---

4 Ibid., p. 20. His use of 'noble' seems to have been a sparingly used shorthand for oak, occurring again in respect of the woods at Dalkeith Palace (p. 42).
5 Ibid., pp. 25, 26.
6 Ibid., pp. 27, 34, 35. He did not enter the park at Cadzow, so did not see the oakwood there. Drummond's 'extensive plantations of furs, and other trees' were notable in an otherwise 'barren, mountainous country'. These 'furs', which he refers to as trees, are apparently distinct from 'furze', or gorse, which was being planted in volume as a soil-improver on sandy or gravelly ground and as a source of fuelwood throughout many Lowland British areas in the later seventeenth and eighteenth centuries.
7 Ibid., p. 36.
8 Ibid., pp. 36–37.

impressed by what the Murrays had achieved there by that date. He does, however, give insight on the success of their larch planting from the later 1730s (discussed below), noting that estate-sourced larchwood was being used already in building work there before the time of his visit.⁹ Although these few examples from Burrell hardly project a picture of a well-wooded landscape, they demonstrate that through a broad transect from the south-western Lowlands into Highland Perthshire, large-scale planting had already created mature woodland alongside the surviving, more limited expanses of semi-natural native woodland. Most was contained within the walled enclosures around the houses of the nobility and gentry, but some of the most enterprising landowners had clearly been investing for decades in more extensive planting across their property.

Burrell's matter-of-fact writing style and relatively short journal form, coupled with the fact that it remained unpublished until 1997, lacks both the literary weight and published reputation of Samuel Johnson's voluminous and bitingly critical travelogue. Since its first publication in 1775, his *Journey to the Western Isles of Scotland* has been mined for data on the condition of the land, the economic (mis)management practices that he witnessed, and the cultural backwardness of the Highlands and Islands, often in a distortion of what Johnson had intended or expressed in his text. For those of a Whiggish and Improvement mindset, it was an especially valuable source that could be exploited to illustrate their arguments for the benefits that Union had brought to a land languishing in the shackles of its primitive agricultural and forestry practices. Foremost among these was that shocking vision of a land devoid of woodland, which became an article of faith in much of the past historiography of Scottish woodland decline produced in the nineteenth and twentieth centuries and is still cited in popular accounts of Scottish woodland history. That view is founded on Johnson's somewhat dyspeptic and often repeated observation, made on his journey north from Berwick in 1773, that 'from the banks of the Tweed to St Andrews I had never seen a single tree' and that 'there is no tree for either shelter or timber'.¹⁰ Presented thus, it is a devastatingly negative view that still pervades much contemporary discussion of human impacts on Scotland's woodlands from prehistory to the Age of Improvement. Regardless of the contrary evidence that our other sources tell us, including the witness of other travellers like Burrell or Thomas Pennant, it is deployed as the trump card when presenting the damning indictment of anthropogenic woodland loss everywhere in Scotland, not just the Lowlands. But the quotation is rarely given verbatim and is usually incomplete, for Johnson continued with the important qualification that it was not that he had seen no trees at all but that he had seen no trees 'which I do not believe to have grown up far within the present century'. He went on to add that most gentlemen's houses he passed had small plantations around them of very young trees. These further observations are of vital importance, for they change the tone and meaning

---

9   Ibid., p. 38. For the first larch plantations on the Atholl estate, see Albritton Jonsson, *Enlightenment's Frontier*, p. 150. The Atholl planting of the Tummel and Tay valleys from Blair south to Dunkeld does not appear to have made significant advances at the time of Burrell's visit, the sides of the valley down to Dunkeld being covered mainly with birch. Interestingly, at Dunkeld, he noted that Birnam Hill, famous for the wood that shielded Mael Coluim mac Donnchada as he advanced to defeat Macbethad at Dunsinnan, was 'now a barren mountain' (p. 39), suggesting that the late medieval demands for oak for Bishop Brown's work at the cathedral, the bridge over the Tay, and in Perth, had seen the woodland nearest to his see stripped and not replenished. Again, relating to Atholl and Menzies traditions about larch, Burrell noted that at Dalkeith there was an avenue of larches a mile long that linked two of the demesne parks (p. 43).
10 Johnson, 'Journey to the Western Islands', p. 39.

Haddington in the 1690s (from John Slezer, *Theatrum Scotiae*). The almost treeless landscape depicted by Slezer's engraver outside the orchards in the town's backland gardens seems to reinforce Dr Johnson's 1773 observation of the form of woods in this same district. (National Library of Scotland)

of what Johnson was saying. Yes, he was decrying the absence along the route he followed of the mature standards that he was used to seeing in southern England, but he recognised that there were many apparently immature and certainly smaller trees in addition to those planted within the 'policies' around lairds' mansions. And here he may be revealing his lack of interest in, or ignorance of, the nature and purpose of much of the woodland he observed, both semi-natural and recently planted. These were intensively managed woods, the broadleaves usually coppiced on long rotations that produced a coppice-with-standards structure where most growth was barely more than twenty-five years old, albeit springing from stools that were at least many decades – and in some woods, centuries – older. In both the managed ancient woodlands and the newer plantation trees, he was witnessing both the products of medieval parliamentary efforts and also the first fruits of the Improving spirit that had been pervading Scotland since the later 1600s.

To gain a clearer picture of the realities of eighteenth-century woodland management and planting – and of the extent of woodland cover, both semi-natural and planted – it is necessary to set aside for the present the retrospective projection of late eighteenth- and nineteenth-century views of Scottish practice, founded on these travel journals, and explore the contemporary evidence for landowner action in this regard across the 1700s. Even before the end of the seventeenth century, the cumulative effect of legislation – including the older fifteenth- and sixteenth-century acts to encourage tree-planting – coupled with changing social fashion and economic land-management models, had seen a steady expansion of new woodland planting, especially by Lowland estate owners. It had also seen experimentation, with some major landowners trialling the introduction of tree species that were previously rare in or wholly alien to Scotland. In a virtuous circle where a new breed of specialist tree-growers were patronised by landowners, who in turn were inspired

by the work of those specialists, arboriculture – the study of how trees grow and how they are affected by both cultural practices and the environment in which they are grown – was developing rapidly as a new science in this period. A major advance in arboricultural research came with the publication in 1664 of John Evelyn's *Sylva; or, A Discourse of Forest Trees and the Propagation of Timber*.[11] Intended as a stimulus to the planting of trees to ensure English self-sufficiency in timber for warship construction, *Sylva* provided a rationale and a blueprint for planting that was equally relevant to Scottish landowners. Among the earliest Scottish adopters of Evelyn's ideas was John Hay, 2nd earl of Tweeddale, whose plantations around his family seat at Yester in East Lothian, which we encountered in Chapter 5, became a model for other would-be Improvers and a source of seed and seedlings for distribution around his friends and connections.[12] Social networks became the primary mechanism for the dissemination of such ideas through the first and second quarters of the eighteenth century.

Evelyn, however, was far from being the sole originator of Scottish interest in arboriculture, for the

---

11 John Evelyn, *Sylva; or, A Discourse of Forest Trees and the Propagation of Timber in His Majesty's Dominions* (London, 1664). Evelyn had first presented his thoughts to the Royal Society in London in a 1662 paper. It went into four editions by 1706 and was among the most widely circulated books on the propagation and cultivation of trees throughout the first half of the eighteenth century.

12 S. House and C. Dingwall, '"A nation of planters": introducing the new trees, 1650–1900', in T.C. Smout (ed.), *People and Woods in Scotland* (Edinburgh, 2003), pp. 128–157 at 131. For woodland management in East Lothian more generally in the seventeenth and eighteenth centuries, see T.C. Smout, 'Managing the woodlands of East Lothian, 1585–1765', *Transactions of the East Lothian and Antiquarian and Field Naturalists' Society*, 26 (2006), pp. 41–53.

The panoramic 'North Prospect of the City of Edenburgh', from John Slezer, *Theatrum Scotiae*. On the extreme left is the Palace of Holyroodhouse, the original location of the Physic Garden, which in 1676 was moved to the gardens of the former Collegiate Church of the Holy Trinity at lower centre of the right-hand half of this image. In these gardens, specimens of exotic plant and tree species were propagated for study. (National Library of Scotland)

establishment in 1670 by Sir Robert Sibbald and Andrew Balfour of the Physic Garden at the Palace of Holyroodhouse to educate trainee medical doctors of the newly-founded Royal College of Physicians revealed an already thriving interest in the study of trees and plants.[13] Following its relocation in 1676 to a new site at the former Collegiate Church of the Holy Trinity (now under the tracks and platforms of Waverley Station), the Physic Garden began to broaden its focus from the potential medical properties of plants into botany more generally, from which Edinburgh's Royal Botanical Gardens began their development. It has been noted that the first catalogue of the garden's contents, produced in 1683, included the European larch (half a century before its supposed first introduction on the Perthshire estates that claim to possess the surviving earliest trees of the species), an array of other European trees, and a wide range of species native to the eastern seaboard of North America (possibly obtained through the connections of James, duke of Albany).[14] That same year saw the publication of possibly the most influential book for landowners seeking to develop their properties, written specifically for Scottish climate and soil conditions, John Reid's *The Scots gard'ner*.[15] Reid, who was heavily influenced by John

---

13   F.P. Hett (ed.), *The Memoirs of Sir Robert Sibbald (1641–1722)* (London, 1932), p. 65.
14   House and Dingwall, 'A nation of planters', p. 132.
15   John Reid, *The Scots gard'ner in two parts, the first of contriving and planting gardens, orchards, avenues, groves, with new and profitable wayes of levelling, and how to measure and divide land: the second of the propagation & improvement of forrest, and fruit-trees, kitchen hearbes, roots and fruits, with some physick hearbs, shrubs and flowers: appendix shewing how to use the fruits of the garden: whereunto is annexed The gard'ners kalendar / published for the climate of Scotland by John Reid* (Edinburgh, 1683).

Evelyn, used Evelyn's translations of French gardening treatises, and was also a close friend of John Sutherland of the Physic Garden, had worked at several important houses, including Niddry Castle in West Lothian, Hamilton Palace in Lanarkshire and Drummond Castle in Perthshire, before moving in 1680 to work for the Lord Advocate Sir George Mackenzie of Rosehaugh on his estate at Shank in East Lothian. Reid's book, which ran to multiple editions and circulated widely at first among Improving landowners and then their tenants,[16] was explicit in recognising that the development of landed estates had profit as much as pleasure as the driving motivations behind it. Containing advice on obtaining tree-seed and its propagation, on selection of the most suitable species for growing, on woodland care and disease and pest management, it became a handbook whose guidance supported generations of Improvers through the following century.

The first experiments

As observed by the English novelist and journal-writer Daniel Defoe in his *Tour thro' the whole island of Great Britain*, published in 1727, the fruits of Evelyn's and Reid's writings were to be seen gracing the land around the great houses of Lothian.[17] Defoe noted especially Tweeddale's work at Yester, describing him as having taken from King Charles II 'the love of managing what we call forest trees, and making fine vistas and avenues' and that the earl had

---

16 It is, for example, among the books of David Drummond, 3rd Lord Maddertie, which formed the basis of the Innerpeffray Library collection, which were borrowable by farming tenants throughout central Strathearn.

17 Daniel Defoe, *A Tour thro' the whole island of Great Britain*, volume 3 (London, 1727), Letter 11.

planted over 6,000 acres around his house.[18] Although still modest in extent in comparison to what was being planted by the end of the eighteenth century, these were for the time large-scale trial plantations that contained several species, of which beech (*Fagus*) was to establish itself as the most successful of the introductions. The prominence of references to Tweeddale's work in modern historiography has tended to promote an assumption that he was the sole promoter of this species and the instigator of its introduction to Scotland, but we have already seen that beech was established in many parts of the country and formed mature woodland by the middle of the eighteenth century. Tweeddale's contemporaries, the 2nd and 3rd earls of Panmure, at their new house of Panmure in Angus, were also planting beech at scale. When the plantation close to the mansion was seen by Sir William Burrell in 1758, these trees represented the greatest concentration of maturing beech that he had yet seen in Scotland.[19] The trialling was also not purely driven by commercial motivation, although that was always to be a prime consideration; most of Tweeddale's trees were grown both for visual impact and for future profit. Yester and the estates of several of his close political and family connections provided models which were being replicated at large and small scale across Scotland in the early decades of the eighteenth century.

Fuelled by John Reid's advice on how and where to obtain the best seed, the importation of tree-seed expanded rapidly in the years following the Union and reached its apogee after 1765, the year in which the Society for the Importation of Foreign Seeds was founded in Edinburgh.[20] The city, however, had by then long been established as a centre for the import and onward sale of tree-seed and saplings, with the Edinburgh-based merchant Arthur Clephane becoming a field leader in the quarter century after 1705.[21] Clephane's accounts show him importing a rich diversity of foreign tree-seed and seedlings, mainly from merchants in Rotterdam and Amsterdam, including *Pinus pinaster* (maritime pine) from the western Mediterranean and Iberia, *Abies alba* (silver fir) found across northern Europe from the Pyrenees to Carpathia, 'great pine', 'Spanish brown seed' (perhaps carob or *Ceratonia siliqua*) and *Tilia x europaea* (common lime), yew, myrtle, horse chestnut and 'pitch fir'. By the 1710s, Clephane was supplying clients throughout Lothian, southern Perthshire, Fife and Kinross, mostly among the small landowner community but providing them with volumes of seed that imply large-scale planting ambitions. Larger estate concerns were more commonly early direct importers using English and European suppliers. One of the most enthusiastic of these aristocratic early planters and seed-importers was James Graham, 1st duke of Montrose, who had been developing woodland for commercial exploitation on his estates since the last quarter of the seventeenth century. The duke's probable engagement with Reid's treatise can be seen in the seed-buying activities of his then estate factor, Mungo Graham of Gorthie, whose responsibilities extended across the entire Graham property complex from Loch Lomond to the Angus coast. In

---

18 Beech was a trial tree at Yester before its large-scale adoption, and sale of 'haggs' of wood there in the mid eighteenth century indicates that they were part of mixed planting alongside oak, elm, ash, 'firr' and miscellaneous smaller species: Smout et al., *History of the Native Woodlands*, pp. 70–71, 74.
19 Dunbar, *Burrell's Northern Tour*, pp. 71–72.
20 House and Dingwall, 'A nation of planters', p. 138. Although focused on the English experience, see M. Thick, 'Garden seeds in England before the late eighteenth century. II: The trade in seeds to 1760', *Agricultural History Review*, 38 (1990), pp. 58–71 for discussion of the commercial development of this trade.
21 T. Donnelly, 'Arthur Clephane, Edinburgh merchant and seedsman, 1706–1730', *Agricultural History Review*, 18:2 (1970), pp. 151–160.

Beech woodland, the Knock, Crieff, Perth and Kinross. The beech was the most successful of the early plantation woodland in the first half of the eighteenth century, valued both for its aesthetic qualities that changed by the season and for its timber.

Veteran hornbeam (*Carpinus betulus*), Murthly Castle, Perth and Kinross. Although its status as a plantation tree declined through the later eighteenth and nineteenth centuries, English hornbeam was initially among the most widely planted broadleaf species favoured by Scottish improving lairds.

1710, for example, Gorthie purchased a bushel of beech mast, 2–3 bushels Spanish and English acorns, several bushels of hawthorn berries for growing 'quicks' for hedging, 4,000–5,000 young 'quicksets' or hawthorn plants from a nurseryman near Chester, and a mixture of 100 wych and Dutch elms 'for a variety', to be planted around the duke's house at Buchanan near Drymen.[22] The best beech mast and acorns were here being sought from English sources and seed-importers to sow in what would become tree nurseries from which the wider Montrose estate would in future be provided, with the hawthorns planted around them to form protective enclosures against grazing animals. It was through such efforts

22 NRS GD220/5/806 (4). Gorthie made regular purchases of beech mast, holly berries and acorns for estate purposes. In 1723, he bought beech mast, yew seed, filbert nuts and lime-tree seed from Edinburgh seedsman William Miller for use at Buchanan (GD220/6/1015 (27)). In 1727, among other trees, he enquired about 'seed' for 'evergreen oaks' (GD220/5/1051 (3)) and in 1728 he bought hornbeam, beech and English elm for estate planting, plus standard cherry, plum, apricot and peach trees for the gardens at Buchanan from William Boutcher of Comely Garden, Edinburgh (GD220/6/1025 (5)).

that southern English beech made its rapid advance throughout Scotland and the pedunculate or English oak (*Quercus robur*) became more widespread alongside the previously more common sessile variety (*Quercus petraea*).

As will be explored further below (see pp. 206–209), there was diversity in the species whose seeds and saplings were being introduced, with both fast- and slow-growth varieties and ornamental or more utilitarian types planted. In 1712, for example, the marquis of Huntly purchased Dutch elm and beech trees for the new French-style block-and-vista plantations being laid out around Gordon Castle, but his purchase also included small numbers of crab and 'codlin' (green cooking-apple) trees for his new orchards and kitchen gardens, again as recommended by Reid.[23] The following year, laburnum, beech mast, yew, lime, hornbeam, chestnut seedlings and acorns were purchased from Arthur Clephane, along with a further similarly diverse batch including hornbeam from another seedsman, John Turner, marking the start of new plantation and diversification of species on the Grant family's lands around Castle Grant on Speyside.[24] The 1713 purchase from Turner was among the earliest references to trees for large-scale hornbeam planting in Scotland. Although its especially hard timber was in some demand in England for construction and furniture-making – and the commercial potential of such demand perhaps encouraged Scottish woodland experimentation with hornbeam – this species was rarely planted on the almost industrial scale associated with beech, oak and pine, with numbers more often similar to the 500 obtained in 1739 by the earl of Breadalbane for the plantings around his house at Taymouth. One of the earliest largest investments in hornbeam is recorded in May 1726, when Sir Archibald Grant at Monymusk in central Aberdeenshire had in excess of 50,000 hornbeam seedlings germinated from 'exterordinary (sic) good seed' for future planting around his property.[25] Although this mainly southern British species did not especially thrive as a plantation tree in Scotland, improving landowners persisted with it through the eighteenth century for more ornamental purposes, and numbers of exceptionally large hornbeams from this period still survive in the gardens of mansion-houses throughout the country.[26]

23 NRS GD44/51/498/16/12. The layout of the blocks of trees separated by broad rides that provided vistas to and from the house can be seen on William Anderson's *c.*1768 plan of the castle and grounds produced just before their remodelling: NRS RHP2379/6. Orchard design and the propagation of fruit trees were significant elements of both parts of Reid, *Scots gard'ner*, while the book's appendix contained advice on how to use the produce.

24 NRS GD248/104/2 (4), GD248/104/8 (7). This planting policy continued throughout the century, with larch and cedar seeds and 'enough beech to plant all the waste land in Scotland', acorns, horse and Spanish chestnuts obtained in 1765 (GD248/672/6 (16)); large quantities of hedge hawthorns, crab apples, beech, oak, elm, sycamore, lime, chestnuts, walnut trees, New England pines, silver firs bought in 1766 (GD248/530/1 (8–10)) and another substantial purchase of seven different species in 1772 alone, including 24,000 hawthorns for quickset hedging, 'Scotch' and English elms, lime, hornbeam and beech and crab apple stocks for the kitchen garden (GD248/179/1 (77)).

25 NRS GD112/15/266 (41); Hamilton, *Monymusk Papers*, p. 105. The ecologist, historian and woodland management expert Oliver Rackham viewed hornbeam planting as an early nineteenth-century fashion but it was evidently a well-established practice through the eighteenth century. Rackham expressed some bemusement as to why anyone would want to plant hornbeam: O. Rackham, *Trees and Woodland in the British Landscape: The Complete History of Britain's Trees, Woods and Hedgerows*, revised paperback edition (London, 2001), pp. 129–130.

26 For example, on the Innes of Stow property, hornbeam was planted in 1787/88 along with horse chestnuts, walnuts, limes, laburnum, holly, hazel, elders, whitebeam, crab[apples] and Spanish chestnuts, all obtained from Dickson of Hassendeanburn, one of the biggest commercial tree nurserymen in Scotland at that date: NRS GD113/4/167 (337). One of the finest surviving later eighteenth-century hornbeams in Scotland is that at Murthly Caste in Perthshire.

But experimentation with non-native species was not ubiquitous, and elsewhere in Scotland landowners were often planting native species whose qualities were well understood and whose timber remained in demand by Scottish craftsmen.[27] There are records of plantations of mixed birch and 'firr' (largely but not exclusively Scots pine (*Pinus sylvestris*)) being made on the Nairne estate between Perth and Dunkeld in 1709, with Lady Nairne obtaining the 'firr' from her neighbour, the earl of Breadalbane.[28] The Nairne mixed plantation evidently had as much aesthetic as economic reasoning behind it, Lady Nairne commenting on how 'very beautiful [a] plantation it is, the streight lines of birch, through the irregular firrs, makes a very agreeable diversity'. Located on the interface between Highland and Lowland Perthshire, the choice of trees by the Nairnes perhaps reflected the local market for their timber and reveals the existence of an important trade in native species saplings by landowners with estates where such species were abundant in their managed woodlands. The Breadalbane estate from which Lady Nairne had obtained hers continued to be a source of pine seedlings and young trees for other landowners alongside its better-known role as a major introducer and planter of non-native species. In 1734, for example, Breadalbane's woods at Ardmaddy on the Argyll coast south of Oban were seen by the laird of nearby Lochnell as a potential source of 600–800 young trees that he wanted to complete a new plantation behind his house in Benderloch.[29] These trees were part of the mature woods seen at Lochnell House by Burrell in 1758 (depicted on Roy as enclosing the house to north and west, which appears on the map under its original name of Ardmucknish), which he described as covering the hills, thriving, and 'disposed with taste and judgement'.[30] This early development of pinewood plantations serves to remind us that the rapid expansion of Scots pine and other conifer woodland planting later in the eighteenth century had its origins in early estate Improvement policies.

It was, however, non-native species (including the introduction of non-native varieties of species native to Scotland) that most interested Improving landowners. Although it is now argued that beech should be considered a 'native' species, as genetic analysis of growing trees has established its origins in the south-eastern British reservoir of European beeches that existed after the end of the last ice age,[31] the trees in Scotland are the product of an anthropogenic acceleration of the likely natural northward spread of this active woodland coloniser. Favoured for its visual impact – height and striking colour-change through the seasons – as much as its versatility as a plantation tree, shelter-belt former or managed hedge-species, and for its high-quality timber, it was perhaps the first major species introduction since the arrival of sycamores in the sixteenth century. Certainly, the liveliest internal trade in tree-seed in late seventeenth- and early eighteenth-century Scotland seems to have been in beeches, giving them a distribution from Galloway to Sutherland before the 1790s. The earl of Tweeddale's Yester House estate in East Lothian was, as we have already seen, an important early source for beech mast from which seedlings were propagated and for established seedlings, derived from the trees introduced here in

---

27 See Smout et al., *History of Native Woodlands*, pp. 90–92.
28 NRS GD112/39/234/13.
29 NRS GD170/772.
30 Dunbar, *Burrell's Northern Tour*, p. 58.
31 M.J. Sjölund, P. González-Díaz, J.J. Moreno-Vellena and A.S. Jump, 'Understanding the legacy of widespread population translocations on the post-glacial genetic structure of the European beech, *Fagus sylvatica* L.', *Journal of Biogeography*, 44:11 (2017), pp. 2475–2487.

Scots pine (*Pinus sylvestris*), Glen Derry, Aberdeenshire. This most iconic of Scottish trees, whose surviving expanses of semi-natural woodland are often labelled as remnants of the mythical 'Great Wood of Caledon', was planted and felled on an almost industrial scale by central Highland lairds through the eighteenth century.

the last quarter of the seventeenth century.³² In 1724, for example, Tweeddale supplied the Hamilton-Dalrymples of North Berwick with mast from Yester.³³ Tweeddale was not alone, however, with George Gordon, 1st duke of Gordon, planting beeches on the grounds around his house at Fochabers in Moray before 1712.³⁴ The duke of Montrose was a more prodigious importer of English seedlings but especially beech mast, with which he intended to plant much of the low ground of the Buchanan estate, and by 1732 the duke's own nursery was supplying relatives, tenants and neighbouring lairds for plantations on their properties.³⁵ Commercial nursery- and seedsmen, successors to Arthur Clephane, such as William Boutcher whose nursery at Comely Garden below Abbey Hill in Edinburgh was well established by the later 1730s, were also supplying Improving landowners across Scotland. In 1739, Boutcher supplied the earl of Breadalbane's factor with beech seedlings as well as beech mast and also holly and thorn plants for hedges to protect the newly planted trees.³⁶ Two years later, another Edinburgh seedsman, William Miller, supplied Breadalbane with more beech mast and young hornbeam.³⁷ The scale of planting on the Breadalbane estate of which these purchases were part, especially around the principal residences at Taymouth and Finlarig at the east and west ends of Loch Tay respectively, as witnessed by William Burrell, can be gauged from the accounts for work undertaken in 1742 alone. Beech seed – along with fir (probably Scots pine) and elm seed that had been collected at Taymouth – as well as 3,000 hornbeams and an unspecified number of lime trees were planted at Finlarig in spring 1742, while at Taymouth there was additional expense in taking up about 16,000 firs, 2,000 oaks and 60 big lime trees for relocation.³⁸ Further quantities of beech, elm, 'fir' and sycamore were planted on the lands around Finlarig in 1746.³⁹

Wych elm was native to Scotland from its post-glacial (re-)establishment some 8,000 to 8,500 years ago, occurring in both lowland mixed broadleaf woods and upland mixed ashwoods.⁴⁰ As mentioned above, wych elm was bought for estate planting at least as early as 1710, being prized as second only to oak for all manner of wood-use from large structural timbers to water-pipes.⁴¹ Planting of this native species continued but other elm species – occasionally labelled 'Dutch' but most likely English elm and European field elm – were also being introduced in the early 1700s. Commonly occurring with oak in mixed broadleaf plantations but also used in avenues, in shelter-belts and for screening purposes, it was another species whose seed was exchanged between estates, spreading elms into parts of the country where they had been rare or absent previously.

---

32 Beech trees reach nut-producing (mast) maturity at between forty and sixty years, so Tweeddale's trees were likely planted by c.1680.
33 NRS GD110/892. The Hamilton-Dalrymples were themselves supplying beech trees by the 1780s: NRS RH4/189 (19).
34 NRS GD44/51/498/16 (12). The beech, along with Dutch elms, were purchased from the merchant Arthur Tate.
35 NRS GD220/5/882 (21–22); GD220/5/1018 (4). In 1732, John Graham of Grahamshall requested 200 beech or hornbeam to plant in a row round a new yard he had built: GD220/5/1221 (2). Walter Macfarlane of Arrochar requested beech trees from Buchanan to complete the avenue that he had laid out at his principal residence, New Tarbet at Arrochar, in March 1735: GD220/5/1342.
36 NRS GD112/15/265.
37 NRS GD112/15/275/22.
38 NRS GD112/15/284/16.
39 NRS GD112/15/301/27.
40 Smout et al., *History of the Native Woodlands*, pp. 2, 25, 26, 28–29.
41 Ibid., p. 92.

Bog o' Gight or Gordon Castle, as it was renamed, beside Fochabers, Moray (mistakenly labelled in John Slezer, *Theatrum Scotiae*, as 'The Castle of Inverero'). The engraving shows the formal broadleaf planting that had already begun around the castle by the last quarter of the seventeenth century. (National Library of Scotland)

Lowland estate owners enjoyed demand from upland proprietors. Among these were the Cummings of Altyre near Forres, who in 1787 offered as much elm seed as was wanted for planting at Castle Grant to increase the broadleaf presence in the policies around the mansion there.[42] Lady Grant was seeking at that time to diversify what seems to have been mainly pinewood planting on the core of the family's estate, where Burrell in 1758 had observed extensive 'fir' planting around the castle designed to make the landscape 'as agreeable to the eye as possible',[43] looking also to obtain beech mast, acorns and larch seed or seedlings.

Pine's readvance

In the previous section, we explored the changing patterns of woodland exploitation and planting, starting in the Lowlands and extending into Highland districts. Here, we will examine the mainly Highland phenomenon of the re-expansion of conifer woodland – chiefly Scots pine – but we should note that the fashion for experimentation among Improvers also saw the introduction of various species of conifer in plantations in the Lowlands and Southern Uplands. The new Grant plantations of the 1780s were being established in a district where pine was

42 NRS GD248/61/1 (51).
43 Dunbar, *Burrell's Northern Tour*, p. 61.

Allt a' Chonnais pinewoods, Glen Carron, Wester Ross, Highland. Surrounded by modern commercial forestry, the Scots pines clustered in the glen of the Allt a' Chonnais are the regenerated successors of the trees taken by the Admiralty in their timber procurement operations.

still dominant as the large-timber source among smaller species like birch, geanies (wild cherry), hazel, willow, alder and aspen. Contrary to modern pseudo-historical traditions, as explored in the volume 'Scotland AD 400–1400', this was not a land of dense and extensive pine forest – supposed remnants of the Great Wood of Caledon of Classical Roman and more recent popular mythology – but one littered with a 'wood of clumps and granny pines, open-grown and park-like in character', comprising trees of varying sizes and ages among which were managed, reserved stands of tall and densely close-grown examples.[44] Such close-growth reserves must have existed on many Highland estates through the sixteenth and seventeenth centuries, supplying the long, straight-grained pine timbers used in the floor and roof structures of lordly houses like the Grants' main residence at Freuchie (later renamed Castle Grant) or the Doune of Rothiemurchus in Strathspey and Invergarry Castle and Achnacarry in Lochaber.[45] Some reserves of such tall, straight trees were exploited briefly for masts for Royal Navy and Commonwealth warships in the 1630s, early 1650s and later 1660s, but especially after 1707, when Strath-

---

44 T.C. Smout, 'The pinewoods and human use, 1600–1900', in T.C. Smout (ed.), *Exploring Environmental History: Selected Essays* (Edinburgh, 2009), pp. 71–85, quotation from p. 73.

45 Smout et al., *History of Native Woodlands*, pp. 84–86. The surviving sixteenth-century roof structure at Castle Grant employs lengths of timber up to 18 feet long by 14 inches square, while the span of internal spaces in the now-gutted shell of Invergarry indicate that lengths of 24–25 feet were used there.

carron in Wester Ross became an important source for a short time.[46] In such areas, these episodes of felling probably saw only the most mature, tallest and straightest trees of suitable circumferences for naval purposes taken, leaving the younger, smaller and twisted trees to continue to seed and regenerate the woodland. More generally outside such reserves, however, unregulated access to woodland areas for the timber and wider wood needs of local populations and established practices like wood pasture for cattle, especially in the winter and early spring months, had led to a thinning of trees into a more park-like character.[47] That more open form can still be seen in places like Glen Lui and Glen Derry in the southern Cairngorms or Strathcononish in Breadalbane. Mapping from the early 1700s, like John Farquharson's 1703 survey of the Forest of Mar, shows areas of pinewood with approximately the same distribution as the existing wooded areas, but with distinct zones of denser cover or thinner, more park-like growth.[48] These different zones probably represented areas of a more protected character and those where grazing pressure and local wood consumption had made significant inroads. In a 1760 court case between the earl of Fife and the Farquharsons of Invercauld over the use of land within the Forest, Farquharson claimed that 'there is hardly any Pasture on the Moor adjacent to the Woods, and the Cattle are all grazed in the Forrests, where there is the best Out-pasture for all Kinds of Cattle to be found any where in Scotland'.[49] The intensity and regularity of that cattle pasturing, while beneficial to the herds, was the mechanism by which the open woodland with lawns characteristic of much of the Highland forests was created and maintained.

Commercial exploitation versus sustainable use

At the root of the Fife–Farquharson dispute was an attempt by the earl to exclude other users from the wooded areas to protect the trees from further damage or loss. This was not an action driven by any environmental awareness or recognition of the need for sustainable wood-use; it arose from Fife's intention to exploit the pinewoods commercially, principally as timber, and to safeguard its future value by excluding any other user whose actions could diminish its monetary value. His drive towards maximised profit must be set against a backdrop of attempts at intensified commercial extraction throughout the central and west Highlands that commenced in the 1710s and expanded in scope and scale from the second quarter of the eighteenth century.[50] This development constituted a new departure for many of the landowners whose properties included still substantial blocks of woodland, for its previous exploitation – although still potentially episodically

---

46 Smout, 'Pinewoods and human use', pp. 74, 77.
47 A.L. Davies and F. Watson, 'Understanding the changing value of natural resources: an integrated palaeoecological–historical investigation into grazing–woodland interactions by Loch Awe, Western Highlands of Scotland', *Journal of Biogeography*, 34 (2007), pp. 1777–1791.
48 National Library of Scotland, John Farquharson, 'A Map of the forrest of Mar survey'd &c. / Done by John Farquharson of Invercald, in Anno 1703', [accessed 13 April 2022]. Most of the current woodland is a result of replanting and regeneration following a late eighteenth-century felling episode: see, RCAHMS, *Mar Lodge Estate, Grampian: An Archaeological Survey* (Edinburgh, 1995), p. 6.
49 J.G. Michie (ed.), *Records of Invercauld* (Aberdeen, 1901), p. 153.
50 For detailed discussion of the eighteenth-century exploitation of Highland pinewoods, see Smout et al., *History of the Native Woodlands*, pp. 202–224. Alongside other wood and wood by-products being explored as viable commercial commodities, Breadalbane had been experimenting with turpentine extraction from pine on his lands as early as 1692: NRS GD112/16/11/2 (12), GD170/629.

destructive – had been on a quite different scale, primarily for local wood needs, and subject to strict oversight and regulation by the laird's officials and local agents. Commercial exploitation held up a prospect of more rapid, extensive and systematic operations that could, potentially, change the physical character and ecological profile of forested areas in dramatically short periods of time.

Indications of a likely significant impact on existing pinewoods can be seen in the proposal from the Inveraray merchant James Fisher, who offered in January 1717 to make a trial to extract turpentine from pine trees and roots on the earl of Breadalbane's lands.[51] This does not seem to have been financially successful and the venture was not carried forward. Instead, among the first large-scale operations on the earl's lands was that in the Glenorchy core estate, where the establishment in 1723 of the Glenorchy Firwood Company was a spill-over from commercial oak-felling on the Campbell estates there for tanbark and charcoal by mainly Irish contractors involved in the leather tanning industry.[52] The Irish, left unsupervised by the earl of Breadalbane's officers and – according to the earl's personal account – failing to observe the terms of their contract, appear to have clear-felled a large portion of the Glenorchy pinewoods within two years (see below, pp. 223–224 for their impact on the Glenorchy oakwood). Closer analysis of this case, however, suggests that traditional management of these woods as reserves for the tall pines that were in demand for high-status structural use on the Campbells' lands had perhaps contributed to a failure of natural regeneration. The earl, indeed, was forced to recognise that the age structure and physical character of his pinewood was such as had created a uniformity that the contractors were able to exploit legitimately – but devastatingly – to their advantage. Here, close-growth had also prevented successful re-reseeding beneath the high canopy, meaning few younger and slenderer trees that would be spared to replenish the woodland before natural regeneration from already fallen seed took place.[53] Occurring during a time of poor climatic conditions when, as the Cairngorm climate proxy pine data suggests, few new trees set naturally and natural regeneration of woodland generally largely failed, however, the damage was irrecoverable without a policy of new plantation. But this devastation in Glenorchy appears to have been an extreme case and there is little evidence from elsewhere throughout the Highlands that commercial exploitation in the pre-1790 period had similarly devastating consequences.[54] Indeed, some of the most notable failures of pine woodland in this period, such as the often-referenced loss of the 'mechtie parck of nature' as described by Timothy Pont in the 1590s around Loch na Sealga in the Strathnasheallag Forest district of Wester Ross and still present in a much-reduced state in General Roy's map of c.1750, were more probably natural rather than entirely anthropogenic processes.[55]

One of the most extended accounts of the interplay between human and natural forces in the decline of woodland is that provided by the late eighteenth-century Banff Episcopalian minister, traveller, antiquarian and artist, Charles Cordiner, of the forests around Braemar that had been earlier in contention between the earl of Fife and Farquharson of Invercauld.[56] Cordiner's account commences with the report of the systematic felling of the pinewoods of

---

51 NRS GD112/39/275/4.
52 Their activities are analysed in depth in Smout et al., *History of the Native Woodlands*, chapter 13.
53 Smout, *Exploring Environmental History*, p. 83.
54 Ibid., p. 78.
55 Ibid., p. 81. For further examples of natural failures and the underlying processes, see pp. 81–84.
56 Cordiner, *Antiquities and Scenery of the North of Scotland*, pp. 22–30.

Scots pines, Glen Quoich, Aberdeenshire. These trees may be among the 'thousands of young pines' seen by Charles Cordiner as the glen's woodland regenerated following the felling operations of the Duff estate between 1695 and the 1770s.

Glen Quoich in association with a sawmill that was operative there from 1695 on the haugh where the glen debouches into the northern floodplain of the Dee. This operation, he reported, cleared the glen of all but the most inaccessible trees growing on the rocky mountainsides that were beyond the capability of the woodsmen to reach. Most often, discussion of his report ends here, and we have the impact of the sawmillers presented as another example of utter devastation of a supposedly ancient woodland. The following page, however, presents an altogether different picture, for he reports that 'the rest of the glen has the appearance of incorrigible sterility: yet on more attentive observation, thousands of young pines are seen forcing their way through the chinks of the rocks, and rising amidst layers of loose stones and gravel, where one would be apt to imagine, vegetation would be denied'.[57] Together with the already fallen seed from the felled trees, the few remaining veterans were the most likely source of the seed for the regeneration of the pinewoods of Glen Quoich which Cordiner saw in progress. He gives no indication of any effort to enclose the sprouting seedlings for their protection from grazing animals, but reference to the ending of wood pasture practice and the management of the deer for hunting perhaps gives us some idea of how conditions favourable to this regeneration were achieved.[58]

57 Ibid., p. 23.
58 Ibid., p. 27 for the former use as wood pasture and the removal of the population.

At the time of Cordiner's visit to Braemar and the Forest of Mar, the felling operations had moved west of Glen Quoich and were active in the pinewoods upstream of Linn of Dee and in Glen Lui, with the felled wood floated downriver to the Inverquoich sawmill.[59] Again, Cordiner presents us with a scene of devastation, in which 'on the sides of the hills bordering on the valley, are many thousand stumps of trees, the remains of woods which have been floated down Lui water to the Dee'.[60] While Glen Lui's woods had largely been felled, the mouth of Glen Derry was still filled with an established pinewood, whose trees of 'irregular size' speak of a woodland of variable age in which natural reseeding and conservation of preferred timber trees had created diversity in height, density and shape. Here, however, Cordiner also noted a large quantity of fire-damaged pines which he believed had been burned deliberately but, importantly, he could see that the fire had spared many trees (or those still-growing trees represented regeneration post-conflagration).[61] The final section of his account is a remarkable record of a growing, maturing, ageing and regenerating woodland, where he observed all stages of the predominantly Scots pine woods' lifecycle. The oldest trees were, to him, remarkable for their 'amazing size, and venerable appearance', especially the isolated examples that stood 'in a kind of solitary majesty' on the high ground separate from the denser woods below. He lamented that great trees, 70 feet or more in height and 13 feet in girth, victims of age, infirmity and winter storms, lay rotting because they had fallen too far from places where they could be transported to the sawmill.[62] Here he also saw trees still standing but dead, others still alive but with their trunks beginning to hollow and opened as shelters for the woodsmen. In a description redolent of better-known accounts of dying woods in north-western Scotland, where the waterlogging of the ground and growth of peat in the colder and wetter conditions of the seventeenth and earlier eighteenth centuries killed the growing trees and prevented regeneration from seed, he painted a vivid description of natural decay. Here were trees 'immured in moss: their perished leaves and dissolving branches, strewed around, constitute the present soil, and aiding to corrupt and soften the timber, are in a great measure turned into one general mass of vegetable earth'. Waterlogging had accelerated loss and created a morass in which the trees fell and decayed. But, again, amid that decay Cordiner saw growth and regeneration, with gorse, heather, blaeberry and other small shrubs thriving on the rotted wood, from which the next generation of the greater forest trees was already rising. It is altogether a vivid, closely observed account, in which human action and natural processes can be seen interacting to create something other than the blasted wasteland of modern popular tradition. Even amid the zones of near clear-felling, we are presented with a picture of solitary survivals, swathes of regenerating seedlings, woodland edges from which new growth was beginning to recolonise felled areas and former grazing lawns, leading again to the development of a woodland of widely varying form, tree-size and age.

Reconfiguring the woodland

Among the biggest impediments to the attempts at more intensive exploitation of the Highland pinewoods was that variable physical constitution described by Cordiner. Centuries of mixed usage for

---

59 Ibid., p. 25.
60 Ibid., p. 27.
61 Ibid., p. 28.
62 Ibid., pp. 28–29.

wood pasture, fuelwood and timber needs had created a structure in which, despite the extravagant claims of lairds made to attract buyers and contractors, there were limited numbers of large, mature trees that met the expectations and requirements of timber dealers. The woodland around Loch Arkaig in Lochaber provides an illustration of this diverse age structure in a survey of 1756, completed following a fire in the pinewoods in 1746. The surveyor identified some 162,000 trees in the woods, of which a little over 32,000 were suitable for felling and sale immediately, a further 15,000 would be sufficiently mature within twenty years, while over 50,000 remaining young pines would take significantly in excess of that time to mature.[63] This is a healthy age structure for a naturally regenerating woodland but it rendered such woods unattractive to outside buyers of felling rights – mainly eastern Scottish and English merchants – who were seeking to maximise profit and who wanted mature, tall and large-circumference trees to predominate. Many contracts appear to have been agreed sight unseen, based on the descriptions provided by the sellers. The Grants of Rothiemurchus, for example, ended up facing litigation in 1742 when their promised 300,000 large trees measuring 12 inches square 6 feet above the ground materialised as fewer than 500 at those dimensions.[64] As estate records from Rothiemurchus following the collapse of this contract reveal, it was not English or even Lowland Scottish dealers who exploited the pinewoods intensively and profitably but mainly local men who were catering for traditional wood needs in the Highlands, where short and slender trees continued to be in demand for house-building and fencing.[65] Cultural as much as environmental factors created a continuing regional demand for sizes of timber that were unattractive and uneconomical for extraction for sale in southern markets.

Recognition among some of the greater Improving landowners of these different markets for wood and the need to ensure greater consistency in tree dimensions if they were to access the lucrative English timber trade led to increased investment in new plantation rather than continued attempts at exploitation of naturally regenerating conifers. Small-scale experimentation with planting new species from the late 1730s gradually gave way to the large plantations seen as one of the main characteristics of the Improvement era. Among the better known of the Improvement planters who introduced these methods is a man whose estate instructions and letter-books constitute one of the best-known sources for the rate and nature of change in central Aberdeenshire, Sir Archibald Grant of Monymusk. He started woodland plantations on his demesne land as early as 1719, but introduced also a requirement on his tenants that they plant specified numbers of 'barren' trees for wood needs. By 1754, he had planted or directed plantation of around 2 million trees on his Donside properties in west-central Aberdeenshire.[66] But the greatest of these new plantations dating from before 1790 were made by the 2nd, 3rd and 4th Dukes of Atholl on their central and northern Perthshire estates from the later 1730s onwards.[67] Between them, these three had over 21 million trees planted on around 15,000 acres of their

---

63 T.C. Smout, 'Cutting into the pine: Loch Arkaig and Rothiemurchus in the 18th century', in T.C. Smout (ed.), *Scottish Woodland History* (Edinburgh, 1997), pp. 115–125 at 116.
64 Ibid., pp. 116–117.
65 Ibid., pp. 118–119.
66 Hamilton, *Monymusk Papers*, pp. xlv–xlix.
67 House and Dingwall, 'A nation of planters', pp. 139–140; J. Fowler, *Landscapes and Lives: The Scottish Forest Through the Ages* (Edinburgh, 2002), chapter 7.

Dunkeld, Perth and Kinross. The cluster of larches in autumn colours at the centre of this view include the survivors of the supposedly original trees planted here in 1738 – adjacent to the cathedral – by the 2nd duke of Atholl.

property, of which nearly 10,000 acres was almost entirely of European larch.

The Atholl dukes' investment in larch was an undertaking that constituted the first large-scale effort at planned, monoculture conifer plantation anywhere in Europe. Larch itself, however, despite deeply ingrained popular tradition and more recent academic discussion,[68] was not an eighteenth-century introduction; trees of this species were being grown as specimens in Scotland by the 1680s, when they were among those listed in Edinburgh's Physic Garden and were included in the catalogues of English seedsmen and nurserymen before 1700 and those of Scottish dealers by around that date. Tradition, however, claims that the Atholls' relationship with larch dates from the gift in 1738 of seventeen trees to James, 2nd duke of Atholl, by a 'Mr Menzies of Glenlyon', probably Sir Robert Menzies of Menzies, five of which were planted adjacent to Dunkeld Cathedral and the remainder at Blair Castle.[69] The most extensive planting on the Atholl estate was undertaken in the time of the 4th duke (1774–1830), who identified the tall, straight and rapidly growing larch as ideally suited for shipbuilding as well as for general construction needs, and believed that his plantations could become major suppliers to the Royal Navy.[70] Even before inheriting the title in 1774, the 4th duke had been an enthusi-

68 Albritton Jonsson, *Enlightenment's Frontier*, chapter 6, especially p. 150.
69 Fowler, *Landscapes and Lives*, p. 87.
70 House and Dingwall, 'A nation of planters', p. 140; Albritton Jonsson, *Enlightenment's Frontier*, pp. 147, 153–156.

astic follower of the works of James Anderson, who from 1771 had mounted an energetic advocacy of larch as the key to securing Britain's wood needs for shipbuilding and for domestic timber supply, and also as the key to delivering financial security to Scotland's upland landowners.[71] Atholl's contribution to efforts to achieve what has been labelled 'larch autarky',[72] that is, self-sufficiency of supply in this timber tree, was the product of both his patriotic ambition to safeguard the future supply of timber for constructing the warships necessary to defend Britain's overseas empire and a personal determination to increase the value of the Murray family's sprawling estate, whose upland expanses otherwise supported little of economic value. Despite the 4th duke's enthusiastic proselytising for the advantages of larch silviculture, including observations of the efficacy of its deciduous needles as soil-enrichment and a means of improving the quality of wet, acidic soils and peatland, political opposition to reliance on a domestic source meant that larch lost out to timber from Canada and especially to teak from southern Asia.[73] It was not that failure to secure the Navy timber contracts, however, that signalled the end of the brief

71 Anderson's collected papers and general argument on this topic were published under the pseudonym Agricola as, *Miscellaneous Observations on Planting and Training Timber-Trees; Particularly Calculated for the Climate of Scotland. In a Series of Letters*, by Agricola (Edinburgh, 1777).
72 This is the title of Albritton Jonsson, *Enlightenment's Frontier*, chapter 6.
73 Ibid., pp. 156–164.

mania for larch in Scotland, nor any environmental factor, but advances in shipbuilding technology that moved first towards steam-engine propulsion from sail and from there to iron frames and hulls. The development of iron-hulled ships in the second quarter of the nineteenth century and, from 1859 onwards, the Royal Navy's transition from wooden to iron-hulled and steam-powered warships, finally ended any hopes of securing even a share of that market. For a brief moment, however, the owners of larch plantations profited from naval demand.

Quickening the pace

The Atholls were not alone in seeing the potential value in larch, and the species featured prominently in plantations of both native and non-native types, often running into hundreds of thousands of trees, that were being developed on some of the larger estates from the 1780s. These plantations were often driven by a combination of those same motives that inspired men like Atholl but they were also a response to Anderson's active promotion of the larch as a means to 'improve' the quality of poor ground and deliver better profits from it. Archibald Douglas of Douglas, for example, included major plantation work as part of his improvement of the land around Douglas Castle in the peaty and waterlogged uplands of southern Lanarkshire. He entered contracts in 1780 with the Nottinghamshire-based nurseryman Thomas Retford for the provision of 5,300 trees for each acre of a plantation of 100 Scotch acres (around 52 hectares) there, each one containing 500 birch, 2,000 larch, 800 Scots pine, 500 ashes, 500 alders, willows and Lombardy poplars, 500 oaks and beech, and 500 sycamore (530,000 trees in total of which 200,000 were larch).[74] Douglas was evidently investing at a large scale in valuable timber-yielding trees, both fast and slower growing species, as well as in more specialist varieties in smaller numbers. Retford also contracted with the Montrose estate in 1787 to provide over 700,000 mixed trees for planting in a 136 Scotch acre enclosure at Buchanan, each acre to contain 1,500 larch, 500 Scots pine, 300 spruce, 1,100 oak, 300 beech, 500 ash, 500 sycamore and 500 elms.[75] The emphasis on oak and larch – again over 200,000 of the latter alone – provides a clear illustration of where the chief timber and bark value was believed to lie by these particular investors. In 1786, George Innes of Stow instructed new planting on his property and ordered trees from the leading tree-supplier Robert Dickson of Hassendeanburn that were to be delivered in 'matted bundles'.[76] The new plantations were of mixed native and introduced species, including beech, oak, larch, sycamore, 'silver firs', 'balsam poplar', weeping beech, 'spruce fir, common and white American ditto', rowans, ash and elm, with a second batch comprising alders, Egyptian and black Italian poplars, ash, common and weeping birch, oak, beech, and a mixture of four- and one-year-old Scots pine.[77]

Conifer plantations, of Scots pine and other pine species as well as of larch, saw the spread of significant new areas of pinewood out of the Cairngorm and central and west Highland redoubts to which it had been largely confined since prehistory. On the earl of Moray's Darnaway estate – famed in the Middle Ages as a source of oak – in excess of 10 million Scots pine seedlings, many probably grown

---

74 NRS GD220/6/585 (1).
75 NRS GD220/6/585 (17) for the plantation at Ballindorin.
76 Dickson's nursery at Hassendeanburn in Roxburghshire was established in 1729. Modern calculations suggest that he supplied trees for around 48,000 acres of plantation: House and Dingwall, 'A nation of planters', p. 142.
77 NRS GD113/4/167 (191 and 213).

Douglas Castle, Douglas, South Lanarkshire. Only a handful of the over 500,000 trees planted in the 1780s by Archibald Douglas on the land around his house have survived the break-up of the estate, the timber needs of two World Wars, and extensive coal-mining and gravel-extraction operations.

from seed collected in the upland district of the earl's lands, were planted in the four decades after 1767.[78] Much of the surviving Scots pine woodland in these areas dates from, or is composed of seedlings of, these plantations. Consisting of densely planted trees of the same species and roughly the same age and growth dimensions, such new plantations as these were future cash-crop investments intended for clear-felling, wholly different in character to the more complex age, species diversity and spatial structures of the largely self-regenerating Highland woods. Property-owners whose lands contained tracts of native pinewood were deeply concerned by these newcomers. In enterprising estates like those of the dukes of Atholl and earls of Breadalbane they recognised commercial rivals with superior products who were often better placed to secure lucrative contracts, such as those to supply the Royal Navy. By the early 1780s, as the scale of larch and other conifer plantation was becoming apparent, landowners on Speyside were fearful of the loss of value of woods like Abernethy, Glenmore and Rothiemurchus, which had been the principal bulk, commercial sources of pine in Highland Scotland in the middle of the century.[79] Those who could moved quickly themselves to enter this new type of woodland management, exemplified in the North-East and in Banffshire and Moray by Farquharson of Invercauld (whose 16 million Scots pine and 2 million larches dwarfed the 2 million trees overall planted by Grant of Monymusk) and the earls of Fife, Moray and Seafield.[80] From the time of Sir Francis Grant (d.1726), who began the development of his lands around Cullen which formed the core of the Seafield estate, down to the later nineteenth century, successive Grant lairds planted over 20,000 hectares of land with Scots pine – mostly on land with no pre-existing native pinewood – between them planting almost 200 million trees of different species,[81] but also 'weeding out' those native species that were viewed as unprofitable and undesirable. It is in this context of investment in plantation trees that Lady Grant's efforts in the 1780s to secure the newly fashionable larch seed through the Cummings of Altyre, mentioned earlier, should be seen.

Later eighteenth- and nineteenth-century observers generally extolled the Improved conifer plantations and the value that they gave to what had previously been land of little profit to its owners. By the twentieth century, the negative impact of these first blanket coverings of conifers was better understood. Planted, enclosed and managed in ways that effectively excluded other forms of use, especially as wood pasture, and with a density that prevented subsequent seedling generation or development of the complex understoreys of semi-natural woodland – dense planting was intended to provide a variety of materials, including straight lengths of pole for use in fencing or as supports – these new plantations had an adverse impact on habitat and ecological diversity. 'Weeding' of pine plantations, when the young trees were thinned to encourage growth, yielded poles of usable lengths that were in demand for local needs, but the discarded lateral branch trimmings – or brash – and the rapid growth of the remaining trees ensured that there was no regeneration of other ground-cover or scrub species in planted areas. Some impacts from such practices might seem minor in the grander scheme, as on the Atholl estate and the Speyside properties of the dukes

---

78   House and Dingwall, 'A nation of planters', p. 140.
79   NRS GD44/43/241 (34–6). Sir William Burrell described the felling operations on the Grant estate around Abernethy in 1758, commenting that he had been told locally that a deliberate fire in the 1740s – intended to drive up the price of the wood – had destroyed over 2 million trees (Dunbar, *Burrell's Northern Tour*, p. 61).
80   House and Dingwall, 'A nation of planters', p. 140.
81   Smout et al., *History of the Native Woodlands*, p. 71.

of Gordon and the laird of Grant, where changes in land-use from rough pasture to woodland plantation quickly had visible, unintended consequences. Most arose from the removal of grazing animals or exclusion of local communities who previously had made use of the scrub of such areas for fuel, thatching and fencing, which had managed its growth and spread sustainably while preventing the build-up of old and dead material that posed a significant fire risk. Enclosure ended that form of management but, with no grazing load, it enabled scrub to proliferate beyond its natural limits. On the Grant estate, for example, a pine plantation in a moorland enclosure known as Jackson's Park required regular cutting back of the broom which grew even faster than the young trees once the inroads of cattle and materials-hungry tenants had been excluded.[82] The fear was that the broom would choke the more valuable trees,[83] but a greater threat was probably fire in the dense and highly combustible scrubwood underlayer of dead and dry branches and leaves.

Oak – for bark and charcoal

Conifer species were not the only trees to be planted in high numbers when demand for wood and wood products soared as the eighteenth century progressed. Oak was in particularly high demand, for its timber, the high-quality charcoal obtained from processing of its smallwood trimmings, and the high tannin content of its bark.[84] In 1758, Sir William Burrell noted that the cyclical felling of the oak coppice on the island of Inchtavannach in Loch Lomond netted its Colquhoun of Luss owners 7,000 merks with each cutting.[85] Oak bark from woods within the Tay basin was being traded into Perth from the late Middle Ages and continued to be brought there for sale or onward shipment in the early 1700s.[86] Ironmasters and tanners from England, Ireland and Lowland Scottish districts had since the later sixteenth century been turning their attention to the ancient but long-used Atlantic oakwoods – of sessile oak – that fringed the west coast's sea-lochs and inner Hebridean islands and formed the denser inland forests of Argyll and Lochaber.[87] From as early as 1716, the Campbells of Breadalbane were selling timber and bark rights in their Glenorchy lands, starting with the oakwoods on the east side of Loch Awe, a six-year lease of the woods on Innis Chonnain, and the woods of Glen Noe, all sold or let to Scottish interests.[88] Woods closer to Glasgow had already been supplying the burgeoning industrial processing of the rapidly growing city with both charcoal and bark in the late seventeenth century and continued to do so through the eighteenth. On the eastern shore of Loch Lomond, for example, the duke of Montrose sold the oak bark from the woods around

---

82 NRS GD248/47/4 (1).
83 NRS GD44/43/270 (41): in April 1782, John Gordon of Cairnfield asked the duke of Gordon's factor for the cut 'weeding' of the plantation on the 'red moss park' for use as hop poles, since his own plantation was not yet mature enough to provide poles of that length.
84 The most detailed modern analysis of oakwood (and other broadleaf species) for charcoal and tanbark needs is given in Smout et al., *History of the Native Woodlands*, chapter 9.
85 Dunbar, *Burrell's Northern Tour*, p. 48.
86 See, for example, NRS GD112/75/163, dated 28 July 1706, for the transport of oak bark from the Menzies properties in lower Glen Lyon to Perth.
87 P. Sansum, 'Argyll oakwoods: use and ecological change, 1000–2000 AD – a palynological-historical investigation', *Botanical Journal of Scotland*, 57 (2005), pp. 83–97; Smout, *Exploring Environmental History*, p. 88.
88 NRS GD112/16/11/1 (1 & 2), GD112/16/11/2 (14–16).

Oak-dominated woodland, Glen Kinglass, Argyll and Bute. Coppiced on a rolling basis from the 1720s to 1870s to provide materials first for the adjacent Galbraith and Murphy iron furnace and subsequently for the Bonawe furnace, a few kilometres down Loch Etive, the regenerated woodland illustrates the fallacy of Frank Fraser Darling's vision of the all-destructive ironmasters.

Sallochy in July 1744 to a Glasgow tannery and continued to supply tanneries in the city with materials through the 1760s.[89]

Irish interest in Argyll and Dunbartonshire oakwoods had surged in the second half of the seventeenth century and they remained major consumers of south-west Highland oak into the 1730s.[90] One of the more notorious operators was the Dublin-based business of Roger Murphy and Captain Arthur Galbraith, who secured timber and bark contracts on the Campbell of Glenorchy lands from 1721 (for their impact on the earl's pinewoods, see above, p. 212).[91] Some of this oak-felling and processing operation was linked to an iron furnace that Murphy, Galbraith and their partners had established close to the mouth of Glen Kinglass, on the south-east side of Loch Etive, which had a demand for a nearby charcoal supply.[92] Much was obtained from the proprietor of Glen Kinglass, Campbell of Lochnell,[93] but adjacent landowners such as Breadalbane were also approached. Breadalbane was to regret his deal with them bitterly, writing in August 1725 that:

> I never was so out of humour as yesterday I went to see all the oak timber the Irishmen have cutt. They have not left one standing oak tree in the countrie ... The countrie looks like a desolation and will be more, nor as there is already not one oak tree in the countrie for any use, so there will not be or they [are] done one fir tree, to help a house, nor the mans to the minister here, nor a bridge on a burn in the countrie whereof there are a great many ... I would not for double the value have the bargain hold. It is ane everlasting ruine to us, and a 15 year slaverie, and the tennents gets nothing by them, and we may easily make double that monie by it.[94]

Among the alleged misdeeds of the Irish operators was failure to adhere to the contract terms that stipulated proper hagging, to maintain coppice stools for tree regeneration, and to fell only trees that had been marked by the earl's local officers.[95] For those failures, as has been observed, Breadalbane and his senior estate officers were partly culpable but, at least, it was a mistake from which they learned and which they did not repeat once they were free of the Irish contracts.

Despite the devastation supposedly wrought by Murphy and Galbraith, the Breadalbane estate and the adjoining Lochnell properties continued to be a significant source of oak bark. Indeed, contrary to the earl's despondent view of the likely consequences for the region's woods in 1725, even the oak-dominated mixed broadleaf woods of Glen Kinglass recovered from their depredations and by the 1750s were being recorded by Roy's surveyors as an extensive woodland on both sides of the glen near its mouth. The Glen Kinglass woods went on to supply the Bonawe furnace with charcoal down to the later 1870s, and they still cover the lower hillsides for the

---

89 NRS GD220/5/1584 (20), GD220/6/837. The commercial exploitation of the Loch Lomond oakwoods is discussed in R.M. Tittensor, 'History of the Loch Lomond oakwoods. 1: Ecological history', *Scottish Forestry*, 24 (1970), pp. 100–118.
90 Smout et al., *History of the Native Woodlands*, pp. 202–203. For an offer made to the earl of Breadalbane by the Belfast-based merchant Peter McNaughton in September 1716, for cash purchase of oak timber or bark, see NRS GD112/39/273/32.
91 Smout et al., *History of the Native Woodlands*, chapter 13 for the activities of this rapacious company.
92 For the furnace, see Historic Environment Scotland, Canmore database . The Argyll furnaces and their charcoal needs are discussed further in Chapters 8 and 13.
93 Smout, *Exploring Environmental History*, p. 90.
94 NRS GD112/16/11/1 (26).
95 Smout et al., *History of the Native Woodlands*, p. 203.

bottom 6km of the glen's north side. Greater care, however, was taken in future contracts to ensure sustainable extraction of timber and bark, with a rotational arrangement of enclosed haggs set in place and a system of wood-officers instituted to maintain oversight of and dialogue with the contractors. By the early 1750s, over a decade after the end of the contract with Murphy and Galbraith and nearly twenty-five years since the earl had been forced to intervene to put a stop to their mismanagement, the oakwood of his Argyll properties was deemed to be in sufficiently good condition for new contracts to be negotiated.[96] In 1758, in advance of a further episode of bark contracting, a full valuation of the oakwoods across the Breadalbane estate was carried out for the earl by a Dunkeld-based merchant, providing a clear record of the location as well as the financial value of the woods on his estate.[97] Further contracts were negotiated with the owners of the Bonawe iron furnace on Loch Etive in 1775, indicating that the rotational hagging and coppice management that had been enforced in Breadalbane's woods continued to deliver sustainable crops of wood for charcoal-making to feed the ironworks.[98]

As revealed by the broadleaf planting undertaken for the earl of Breadalbane around his castles at Taymouth and Finlarig, discussed earlier, this major landowner was not relying solely on natural woodland regeneration of native sessile oak and other hardwood species. As on other large estates, existing areas of traditionally managed mixed broadleaf woodland had been brought into a rotational coppicing arrangement, enclosed and internally hagged to help to protect the regenerating shoots from the coppice stools and to facilitate 'weeding' of less financially attractive tree and scrub species. It should be emphasised that there were still blocks of semi-natural oakwood that had not yet been brought into systematised, commercial exploitation in the late 1770s and beyond. At 'Achnacross' (Ardnacross) on Mull's east coast, for example, it was only in autumn 1779 that the duke of Argyll instructed that the 'wood of Sloich' there be divided into haggs for a rotational felling and enclosed before cutting commenced.[99] Alongside such now intensively managed woods, however, new plantations were made dominated by lucrative oak – often of English and European pedunculate oak rather than native sessiles – to capitalise on the steadily increasing demand for timber, bark and charcoal, both for estate use and for sale. On both sides of Loch Tay on the Breadalbane estate, zones of heavy population and agricultural activity, new reserves of oak had been established in the second quarter of the eighteenth century and were coming to first growth maturity in the 1780s.[100] These reserves included some that had been planted for ornamental or leisure purposes, like those around the Hermitage in the glen above Acharn, as well as those intended purely as slowly maturing cash crops. Whether ornamental or commercial, however, all of these woods were disposed of in 1787/8 at an advertised roup or public auction, where the earl obtained the best available market value through open bids rather than privately negotiated contracts.[101] Even in the Highlands, however, these oak and other broadleaf plantations

---

96   NRS GD112/16/11/2 (32–33).
97   NRS GD112/15/353 (72).
98   NRS GD112/16/11/2 (35–36). As early as 1752, the proprietors of the Bonawe furnace had been looking to obtain a ten-year lease for making charcoal in the duke of Argyll's oakwoods on the west side of Loch Awe: NRS GD224/388/5 (6).
99   E.R. Cregeen (ed.), *Argyll Estate Instructions – Mull, Morvern, Tiree – 1771–1805* (Edinburgh, 1964), p. 111.
100 NRS GD112/16/11/2 (40).
101 Ibid. (43). For the development of roups in woodland sales, see Smout et al., *History of the Native Woodlands*, p. 164.

Atlantic oakwood, Ardtornish, Morvern, Highland. The woodland clinging to the steep slopes rising from the shore of the Sound of Mull was, like the Sloich woods on the opposite side of the Sound, divided into enclosures and cut on a rolling basis from the late eighteenth century.

remained on a significantly smaller scale than those of conifers, principally because of their slower growth rates and consequently deferred financial contribution to estate incomes. Throughout most of the central and south-eastern Lowlands, the norm was for the smaller oak coppice-with-standards plantation formula that Reid had proposed in *The Scots gard'ner*, whose regenerated descendants can still be seen in places like the Glamis estate in east-central Angus or the Melville estate in central Fife.[102]

Further south and east, from the 1650s the Grahams of Montrose were already practising extensive oak coppicing to yield bark and timber from the 400–500-acre Wood of Kincardine on their southern Perthshire lands near Auchterarder and on woods they owned in Menteith.[103] By the early 1700s, the Montrose estate had a well-established operation at Kincardine, with a twenty-four-year coppice arrangement running from 1705 to 1728 and feu agreements with their local tenants requiring them to furnish horses to carry bark from the woods to Perth, from where it could be shipped to consumers down the east coast of Britain or exported or purchased by local tanneries.[104] The woods were later disposed of and in 1772 the then owner, Duncan Campbell of Glenure, sold the rights to all timber, bark and bough of what was described as 'natural wood of all kinds' at Kincardine, excluding that which was in haining for commercial hagged rotation, to the local landowner Robert Carrick of Kildace.[105] Even after more than a century of hard-nosed business management of this wood, it evidently maintained, in parts, a mixed broadleaf character that was closer to the older, semi-natural Lowland woods that had predominated into the early 1700s. Importantly, this was also a structure that afforded commercial interest to contractors.

It would be wrong to assume, as shown by the mix of species – chiefly oak, larch or pine – being planted on the Atholl, Douglas, Montrose or Seafield estates, that it was exclusively high-value, slow-growth timber trees that were being planted as cash crops. The mixes being planted on both Lowland and Highland estates, exemplified by the variety on Innes of Stow's property in 1786, obtained from Dickson of Hassendeanburn's nursery, underscore that there was experimentation with the commercial potential of other species, interest in the quicker returns of faster-growing species like larch (whose annual growth is six times faster than oak's), and continued investment in what are often viewed as inferior native species but which were in high demand by Scottish wood-users.[106] Willow, whose whip-like branches and fine-grained trunk wood were valued for wickerwork and wood-turning, was a component in mixed willow, alder and poplar planting at Douglas Castle in the 1780s, but it was also grown earlier in monoculture plantations. In 1766, for example, a willow plantation at Colinton in Midlothian was a subject of proposed commercial partnership,[107] perhaps reflecting the continued demand for this important source of basketry and treen materials from craftsmen in Edinburgh. Improvement was not just about innovation in the name of science and reason or the supplanting of native species by more profitable exotic introductions, but also about finding better

---

102 House and Dingwall, 'A nation of planters', p. 141.
103 A contract of December 1710 sold the standing timber, bark and boughs of various woods in Menteith to John Graham of Mackeanstoun near Doune: NRS GD220/6/579.
104 Smout et al., *History of the Native Woodlands*, p. 179. For the tenants' obligations, see: NRS GD220/1/H/2/1/2, GD220/1/H/2/1/3, GD220/1/H/3/1/1, GD220/1/H/5/1/1, GD220/1/H/6/3/1, GD220/1/H/6/3/5.
105 NRS GD170/431 (18).
106 House and Dingwall, 'A nation of planters', p. 141.
107 NRS GD30/1879.

ways to grow and exploit the native trees that provided the materials that traditional craftsmen and emerging industries needed to manufacture essential products upon which the everyday lives of most Scots depended.

Native and non-native woodland: continuities and innovations in (un)sustainable management

It is important to acknowledge that despite a proliferation of management of woods for commercial purposes – often for charcoal-making or peeling of tanbark – many parts of the northern and western Highlands and Hebrides also continued to rely on often very small patches of native woodland that had been exploited for centuries for traditionally mixed purposes. Martin Martin's descriptions of the limited areas of woodland on most of the Inner as well as the Outer Hebrides at the end of the seventeenth century are strongly suggestive of careful regulation of what little woodland remained, and they point to locally negotiated arrangements for access to more abundant sources, possibly located at a distance on the mainland. As late as 1786, the duke of Argyll's tenants on treeless Tiree, for example, were exercising a customary right to cut timber for house- and boat-building and making tools in the Atlantic oakwoods on the duke's property along the shores of Loch Sunart.[108] From the 1770s on more wooded Mull, however, which had also supplied the wood needs of Coll, Tiree and smaller islands such as Ulva and Gometra, Argyll had been instructing his officers to exercise more stringent control on access to woodland and to ensure that negligent or wasteful damage to it was punished.[109] Community rights were recognised and safeguarded but abuse of those rights was not to be tolerated.

Despite the exhortations of the most widely circulated Improving treatises on woodland planting, enclosure to protect new plantations or regenerating coppice was not ubiquitous or even standard practice for most of the eighteenth century. Even by the last quarter of the eighteenth century, as the duke of Argyll's instructions to his Mull chamberlain reveal, neither existing areas of managed woodland nor new plantations were automatically enclosed within stone dykes or fences as defence against cattle and sheep grazing or casual thefts, possibly because of cost. As William Boutcher's 1775 *Treatise on Forest-Trees* laments, failure to enclose was one of the 'idlest ways of throwing away money'.[110] Neighbouring estates might take opposite approaches that were possibly but not necessarily determined by financial concerns. Thus, at Drummond Hill on the Breadalbane estate immediately north of Taymouth Castle, 60 acres of new planting was enclosed immediately within a purpose-built stone dyke for the significant sum of £50.[111] Just a few miles further south-east near Meikleour around the same date, however, a young plantation owned by the Mercers of Aldie was left unenclosed and was ravaged by the cattle of a neighbouring tenant, destroying its future timber value.[112] It was not, however, just the threat of grazing animals against which enclosure and appointment of woodrangers was intended to protect; it was also to limit unregulated or illicit inroads by the local human population. On the earl of Eglinton's property at

---

108 Cregeen, *Argyll Estate Instructions*, p. 7. The duke's chamberlain of Mull, who had oversight of these woods on the adjacent mainland, had reported his fear that the 'abuses' committed by the Tiree men in their voracious demand for wood was likely 'in a few years [to] utterly destroy the woods'.
109 Ibid., pp. 103, 105, 106, 111.
110 W. Boutcher, *A Treatise on Forest-Trees* (Edinburgh, 1775), p. 243.
111 NRS GD112/74/370.
112 NRS GD132/674 (3).

Out-grown beech hedge, Laggan Hill, Strathearn, Perth and Kinross. The beech trees on the right grow out of the top of an earthen bank and originally formed a bank-and-hedge haining enclosure to protect the plantation on the left from livestock grazing on the open grassland which surrounded the woodland.

Benshie in Ayrshire, woodland continued to suffer illicit felling into the 1760s, when his factor petitioned the local justices of the peace for action against those responsible for felling oak and elm trees there.[113] Rigorous enforcement of the law and harsh penalties imposed on lawbreakers were deterrents, but especially during times of weather-related crisis the pressing need for access to woodland materials outweighed the threat of fines or imprisonment.

There has been a considerable volume of research undertaken on the management of Scottish woodland – principally semi-natural woods – in the seventeenth and eighteenth centuries.[114] The coppice-management practices employed in the Improvement era were clearly not innovations, as the names of enclosures, forms of enclosure and materials all have an Anglo-Scandinavian derivation: for example, haggs (subdivisions within a wood for cutting), stake-and-rice (vertical stakes with wattle (rice) panels forming fencing), haining (a hedged or otherwise protected area).[115] As explored elsewhere in these volumes, although we lack the explicit detail of the

---

113 NRS GD3/2/115.
114 See especially Smout and Watson, 'Exploiting semi-natural woods'.
115 Ibid., p. 88.

Coppicing in a Lowland Scottish oakwood in the mid eighteenth century

An illustration of coppice-management and timber-cutting arrangements in practice can be seen in a wood contract for a small, Lowland semi-natural broadleaf wood. This document is a remarkably short-term contract of March 1750 for three years' access to timber at Headswood beside Denny in Stirlingshire, made between John Stirling of Herbertshire and two dealers, Thomas Johnston and William Laing (NRS GD86/930). Johnston and Laing paid 2,200 merks (£1,466) to Stirling for all his green wood and growing timber inclusive of trunk, boughs and bark at Headswood, of all species (oak, ash, elm, plane, birch or others are listed). The woodland was enclosed by fences and banks, and Stirling prohibited Johnston and Laing from cutting the planted oak growing on the woodbanks, all hazel trees, plus three of the 'spare' trees or 'old maidens' and twenty young oaks that measured 9 inches in circumference 5 feet above the ground. These would become mature, acorn-producing standards for future replenishment of the woods, distinct from the young-growth coppice that regenerated from the stools of the felled trees. Johnston and Laing were obliged to fell and remove the timber and clean the remaining brash from the clearance from the enclosures by 1 March 1753, erecting a barrier between each of the two haggs that they were felling in successive years before they were permitted to start cutting the second. Stirling bound himself to carry yearly for two years 120 bolls of bark (1,200 stone) to Linlithgow (or an alternative location at a similar distance) for supply to tanneries. Johnston and Laing received the right to the grass in the woods for pasture before they started felling, but all tree stools were to remain Stirling's property, with clear instructions that they were to be left in a good state for coppice growth. To ensure that they complied with these requirements, the felling was to take place under the supervision of Stirling's burleymen (minor local officers who arbitrated and settled local disputes and oversaw implementation of legal agreements). Far from being a free-for-all of clear-felling and ruthless, unsustainable extractive practice, as even the arrangements on this minor Lowland estate emphasise, by the last quarter of the eighteenth century Scotland's woodland was possibly being better and more sustainably managed than it has been at any point since the mid nineteenth century.

eighteenth- and nineteenth-century records, the scattered references to woodland management practices in monastic and secular estate documentation and the remaining examples of ancient – mainly oak – coppice make it clear that division of woods into hained haggs for coppicing was practised in many parts of Scotland throughout the medieval period. That this was sustainable practice and not just 'slash-and-burn' exploitation is emphasised by Smout and Watson's analysis of these practices and especially the careful regulation of how coppicing was to progress systematically through woods (possibly with a stipulation to leave a certain number of seed-grown 'maiden' trees to create a wood of coppice-with-standards). Other stipulations included the height and shape of stools to be left after cutting; the seasonality of the timber and bark harvest; the staggering of cutting and duration of the rotation; and measures to be taken to enclose and protect the woods and the individual haggs, especially during the regeneration period.[116] As the example of Murphy and Galbraith's inroads on the woodland of the Breadalbane estate illustrates, there were instances of what look like gross mismanagement and wanton despoliation. Contrary to modern popular mythology, however, these were extreme and very rare cases for, as discussed in more detail in Chapter 8, even the supposedly unscrupulous English ironmasters of the mid and later eighteenth century needed a guaranteed supply of materials for charcoal manufacture to

---

116 Ibid., pp. 88–89. For woodbanks and enclosures, see Rackham, *Trees and Woodland*, pp. 114–118; Smout et al., *History of the Native Woodlands*, pp. 165–171. For haggs and the management of hagg enclosures, pp. 171–172.

ensure the long-term viability of their furnaces. Rather than the eighteenth century seeing the remaining large areas of Scottish woodland ravaged by commercial exploitation, the rigorous management to which they became subjected in this period might, counterintuitively, have caused a stimulus to their growth and ecological dynamism that has again begun to revert to a baseline that we regard as a 'decline' since that intensive management ended.

• • •

With the acceleration of planned woodland planting that occurred after 1700, the vague and generalised mandates of the fifteenth- and sixteenth-century legislation gave way to an increasingly technical or scientific approach to the management of trees and Scotland's timber resources, accompanied by intense public debate in political and intellectual circles over the why, how and what of woodland improvement. The rapid development of arboriculture and the circulation of published work on the whole lifecycle of trees from seed selection, through thinning and form-shaping, to felling and regeneration, provided landowners with the knowledge and information necessary for them to begin to contemplate planting at scale, and the confidence to experiment with introduced species. The rapidity with which specialist providers of services and materials established themselves as commercial agencies in the decades either side of 1700 underscores the burgeoning demand and the confidence of these entrepreneurs that they were catering for something that was more than simply a passing fashion. As the volumes of investment in new woodland emphasise, the buyers were planning at more than just ornamental specimen scale and their suppliers needed to be able to meet their demands immediately. As we have seen throughout the first half of this volume, the seedsmen and nurserymen were in place by the first quarter of the eighteenth century to service the growing market in trees. Investment in the nursery facilities that enabled them to supply seed, seedlings and saplings at volume and so rapidly after their first establishment is one of our least well-understood dimensions of the environmental reconfigurations that formed part of Improvement.

As we have seen in this chapter, Improvement meant new introductions in many but not all cases, with pedunculate oak, beech, English or Dutch elm, and hornbeam being the most prominent of the new broadleaf trees and larch the predominant conifer. But they were not the only species being introduced (or extended, in the case of beech, as its slow postglacial northward spread was accelerated by Improving planters), for with them came a growing band of more exotic trees as seedsmen penetrated the interiors of the North American and African continents and visited the forests of south Asia. By the end of the century, the specimen gardens and policy planting of many great houses were beginning to fill with non-native trees introduced as curiosities, for ornament and as talking-points, and as visible symbols of the reach of the owning families into the expanding network of British imperial power. Such trees, while representing the extent of Britain's global dominion and the wonders it contained, were also symbolic of the new belief in the ability of humans, through their application of science and reason, to master nature and harness it wholly to the needs of human society. How better to demonstrate the taming of wild nature than to confine the greatest of natural organisms – trees – to the gardens of the families who controlled that empire?

Amid the novelty and introductions, however, we have also seen that the inherited natural woodland resource was subjected to an unprecedented intensity of use. Much of it was woodland that had been managed intensively for centuries, supplying the timber needs of landowners and their dependants and providing valuable wood pasture opportunities for grazing domesticated herds throughout the year. Its structure and ecological diversity had developed through that complex interplay of human and environmental factors. But, within this era, it experienced a profound change in how it was managed and perceived, shifting from a resource that was exploited sustainably and whose use was well regulated and overseen, to one where its commercial value supplanted all else. Nevertheless, it was no commercial free-for-all that ensued, despite deeply ingrained popular mythology of the devastation wrought by ironmasters and tanbark suppliers, for these operators needed to ensure access to a sustainable resource for decades to come and, some few horror stories apart, the continued existence of the woods where they extracted their materials speaks of the generally good practices they pursued. Importantly, we also need to recognise that the demand for the native tree types that they were exploiting helped to ensure the spread of a species like Scots pine back from the redoubts into which it had retreated as its post-glacial optimum conditions faded and environmental and human pressures had grown. Much of the semi-natural native woodland that we can still see across Scotland today is the product of this phase of intensive investment and exploitation. In the mid eighteenth century, however, even with this active investment, the native woodland areas amounted at most to around 7 per cent of the land cover of Scotland and that proportion was to diminish sharply as non-native species became ever more popular, as offering faster – and greater – returns on investment. But the comparative rarity of native woodlands (around 1 per cent of modern land cover today),[117] amid the tides of Sitka and Norway spruce that dominate most modern commercial forestry – a product of twentieth-century and later planting – conceals a last efflorescence of Scotland's iconic native tree species under the auspices of the individuals traditionally stigmatised as their nemesis.

---

117 Smout et al., *History of the Native Woodlands*, p. 76.

Huntingtower Castle, Perth, Perth and Kinross. The lower of the two fireplaces in this room was reduced in size in the eighteenth century as fuel fashions evolved. Part of the early sixteenth-century building, it originally had a broad and deep hearth necessary for burning wood and peat. As the packing to the left of the fireplace shows, this was reduced by over a third to create the narrower form of hearth required for a coal fire.

CHAPTER 8

Fuel: Improvement Fashion versus Practicality and Access 1700–1790

[W]e have neither coal, nor limestone nor freestone nor any wood considerable, except planting about Gentlemen's houses . . . Those that live near the coast side, may if they please, furnish themselves with coales from England, but for the most part, the countrey is, except towards the sea, well furnished with Mosses, from whence, in the summer time they provide themselves with peits . . .[1]

Against the backdrop of the woodland plantation and its management explored in the previous chapter, and late seventeenth-century observations on the abundance of peat in some districts, like that of Andrew Symson above for western Galloway, it might seem odd that the eighteenth century saw recurring anxieties expressed about the availability of fuel for thermal energy needs. Across Scotland, the surviving records convey a sense of heightened anxiety that had ratcheted up progressively from the first emergence of a sense of incipient crisis in the later sixteenth and seventeenth centuries (as discussed in Chapter 4). In many urban and rural areas away from the Lowland coalfields that were centred principally on the Forth estuary and, of increasing importance, central Ayrshire,[2] shortages of available fuel worsened through changes in land management, including division and enclosure of commonties, as much as through depletion of traditional reserves. Fuel poverty might seem a very modern issue, with questions of price, access and choice dominating contemporary politics and public discourse, but it was already a perennial matter that faced the bulk of Scotland's population in the 1700s, despite the almost invisibility of the topic in most modern discussions of fuel in the Age of Improvement.[3] Ironically, in a

---

1 From Symson, 'Large description of Galloway', p. 100.
2 For the local coal trade from the colliery at Bargany in Carrick in 1758, see Dunbar, *Burrell's Northern Tour*, p. 26.
3 R.D. Oram, 'Social inequality in the supply and use of fuel in Scottish towns c.1750–c.1850', in G. Massard-Gilbaud and R. Rodger (eds), *Environmental and Social Inequalities in the City, 1750–2000* (Banbury, 2011), pp. 211–231; Oram, 'Arrested development?'. Albritton Jonsson, *Enlightenment's Frontier*, p. 101 discusses an example of abuse of cutting rights for commercial provision of peat at the Moss of Delnies in Easter Ross, but does not set this into a context of depletion of supplies in a region of centuries-old arable intensification and rising population.

land that was relatively fuel-rich and which had a diversity of sources of high-calorific thermal energy available in different regions – mainly wood, peat and coal on the mainland but also dried seaweed and cattle dung in parts of the Northern and Western Isles – access to those fuels was not guaranteed, even if the labour to obtain it was 'free' in the sense that most Scots outside the main urban centres were still directly employed in securing their own supplies. For those with wealth, who were not gleaning wood or cutting peat for themselves, cost was not an issue, but fashion became of mounting concern in the choices of what they burned on their hearths.[4]

Burning questions

Interlinked with questions of societal progression, politeness, social status and fashion in influencing what fuel you burned in your domestic hearths was a trend in 'progressive' thinking aimed at the wider economic, social and cultural betterment that lay at the heart of much Improvement literature. The authors of such tracts presented the use of supposedly inferior fuels – principally peat – as a vice of those locked in ignorant cultural conservatism, socio-economic backwardness and personal indolence; the accusation of indolence, though, was strikingly incongruous alongside arguments that the high personal labour cost involved in securing an annual peat supply was a misuse of effort that should be better spent on agricultural improvement. In most contexts, these vices were linked explicitly with Highland Scots and the Irish in part of a wider discourse of the supposed social, cultural and moral inferiority of the Gael that had developed aggressively since the Middle Ages. This view had received a significant boost from the fear of 'other' within Britain stoked by the prominent support for Jacobitism among certain Highland Scottish and Irish clans. Within that thinking ran that deep strand of self-confident but usually wholly misplaced or misconceived cultural superiority with which much Improvement and later colonialist literature is redolent.

A central tenet of the Improvement literature was the superiority of specific fuel types over others and that the marketisation of fuel supply was innately superior to traditional sourcing and individual provision at domestic level. It presented as wasteful inefficiency the use of one's own personal labour to obtain domestic fuel supplies rather than devoting that energy to profitable productivity in agricultural activities. Likewise, the reservation of areas of land as fuel sources which had potential agricultural or woodland plantation value was most commonly dismissed as a profligate misuse of an asset. From that thinking came a slackening of interest among some landowners and burgh magistrates in protecting peatland, a trend that saw mosses that had been curated as fuel sources for many users converted for other purposes for the profit of a few. In the last quarter of the eighteenth century, there was a counter-movement in some upland districts of the country where the main alternative, coal, remained prohibitively expensive to obtain, to promote peat as an alternative, but with extraction on a commercialised basis to ensure that tenants who could be more profitably employed on other activities were not frittering away their labour on winning fuel.[5] For those dependent on traditional fuel sources and the use of their own labour, this was to be a painful transition that introduced new social and economic dependencies as energy became marketised and a growing number of Scots moved from subsistence agriculture into waged employment. It also brought a decisive switch to another new dependency – on coal – which led to

---

4 Oram, 'Social inequality in the supply and use of fuel', pp. 217–220.
5 This counter-movement is discussed in Albritton Jonsson, *Enlightenment's Frontier*, pp. 172–175.

one of the most profound episodes of environmental change in Scotland. That new dependency in turn helped to trigger the wave of industrialisation whose high carbon legacies are among the principal drivers of climate change globally.

Peat in retreat?

In 1700, coal dependency was still a thing of the future. At the beginning of the eighteenth century, for most Scots living outside Scotland's central Lowlands and main east-coast ports, peat remained the primary, if not quite the only, fuel available. Urban and rural dwellers in many Lowland districts faced the same problems of obtaining secure and conveniently located supplies of this domestic staple as their ancestors had done since the 1500s. As the tensions that had become evident through the sixteenth and seventeenth centuries demonstrate, there was a widespread belief that lack of a secure supply of this staple posed an existential challenge to communities the length and breadth of the country.[6] Despite the recorded evidence for such tensions, much published historical research into the nature of the early modern Scottish economy has assumed that fuel resources were one commodity that was both abundant and relatively accessible.[7] Most of that research, however, is focused on the experience of Edinburgh and the Forth estuary ports, which sat adjacent to what were then Scotland's most intensively exploited and extensive coalfields. Beyond that zone, the experience was often radically different. Fuel for all thermal energy needs – on domestic and industrial scales – was a hard-won commodity for many communities. That position only changed with the mid-nineteenth-century development of rail and steamship transportation that brought cheap coal from central Scottish mines to coal depots in northern and western Scotland and the islands. Until then, fuel security and fuel poverty were a constant concern for the many.

A perception of shortages and the reality of the situation throughout the country are complex matters. Extensive peatland still survives in many parts of Scotland where record evidence indicates that before the end of the seventeenth century peat had already become a common subject of litigation concerning access rights or control measures that attempted to regulate (ab)use of what were believed to be rapidly diminishing sources. In some areas, like the hinterland of Perth, where fuel supplies had become a concern as early as the 1200s, the current peat-stripped zones extending far to the west of the city, where fuel was once cut, seem to bear out the reality of the eighteenth-century sense of crisis. Yet, in much of this district, the most extensive peat-stripping occurred in the later eighteenth and nineteenth centuries, when alternatives had become available and agricultural technology advances encouraged farmers to convert such areas of mossland to cultivable ground. Elsewhere, however, like the country around Elgin in Moray, fuel shortages were a reality and, as we saw earlier, mosses that for centuries had been shared as common resources became sites of conflict as locally dwindling supplies stoked fears of the exhaustion of this essential commodity. After centuries – in some areas, millennia – of peat extraction, accessible and viable sources of supply were increasingly scarce. It is through palaeoenvironmental studies, such as that conducted on Rispain Mire outside Whithorn in Galloway, that we are presented with the unequivocal reality of past over-exploitation and the consequent likely shortages locally. There, where large-scale peat-cutting commenced only in 1763 after the exhaustion of the medieval peateries

---

6 Oram, 'Abondance inépuisable?'.
7 A.J. Gibson and T.C. Smout, *Prices, Food and Wages in Scotland 1550–1780* (Cambridge, 1995), p. 349.

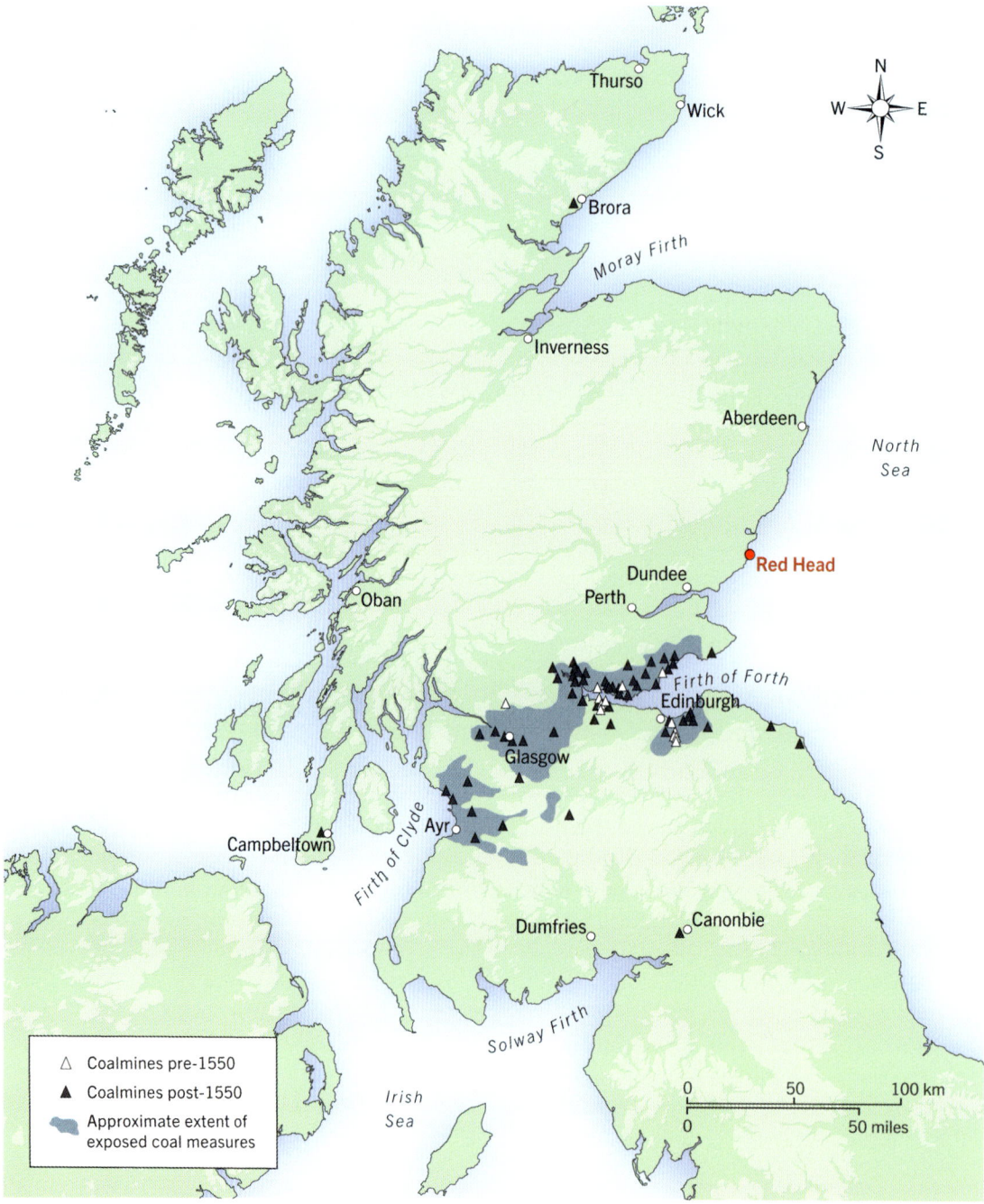

Scottish coalfields: the shaded area marks the primary coal measures, as exploited down to the present. The red dot marks the position of Red Head, north of which the duty on carriage of coal increased markedly. The small black triangles mark the location of individual mines, including the outliers at Brora in Sutherland, at Canonbie in Dumfriesshire, on the East Lothian coast and in Kintyre. (Map copyright R. Oram; redrawn by Helen Stirling)

Rispain Mire, Whithorn, Dumfries and Galloway. The willow and alder carr in the centre of this view is growing on the remnant of the otherwise stripped-out peat moss, from which the burgesses of Whithorn cut all but the stony 'foot' at the interface of the peat and the subsoil.

in the Ket Valley closer to the town of Whithorn,[8] research has shown that what remains is a thin sequence of peat growth dating from the late Holocene, i.e. comparatively modern growth, sitting on top of late glacial and early–mid Holocene sequences.[9] In effect, eighteenth-century and later cutting had removed by the mid 1800s most of the 12,000-year accumulation of peat from this shallow, wet basin over an area of around 700 square metres. This experience was far from unique; such a scale of impact can be identified at other locations throughout Galloway and distributed widely around the country from Shetland to the Borders.

Much of our evidence for supply difficulties relates to the experience of the smaller burghs like Whithorn, where population levels were too low to generate the demand to make coal imports commercially viable and where the common resources of burgh muirs and mosses were exposed to intense pressure during an era of generally deteriorating climate. Records of rearguard defensive actions from burghs, such as were explored in Chapter 4 in respect of Elgin, may give the impression that these communities were fighting legitimately to defend their rights and preserve their resources from encroachment by fuel-hungry predators. Often, however, we can see

8 NRS GD455/27 (10): this records the prospection of an area of peat moss within the Horse Isle Park of Rispain for suitability as a source of fuel for the minister of nearby Whithorn.
9 Ramsay et al., 'Palaeoenvironmental investigations of Rispain Mire, Whithorn', p. 35.

that it was burgesses who were the predators on their landward neighbours' rights. One such case occurred at Inverness, where the common fuel reserves were in a poor state by the later seventeenth century, leading the burgesses into opportunistic but illegal action; in 1701 they started cutting peat on land west of the town belonging to the underage Lord Lovat. They may have believed that neither the child nor his guardians would respond effectively, but Lovat's grandfather the marquis of Atholl initiated legal proceedings by having a notary record the encroachment and serve the burgesses with a notice to desist. At the same time, he instructed his grandson's Fraser kinsmen to go with their own servants to the moss and remove the cut peat.[10] His prompt action forced the burgesses to abandon their attempt and the threat of Atholl using his political influence at national level against the burgh ensured no future repeat of their fuel raid. For the more affluent Inverness burgesses, shipments of coal from the Forth soon provided an alternative if expensive source of fuel, with local merchants engaged in the coal trade by the 1710s.[11] For those unable to afford the high retail price of the imported commodity, however, their failure on Lord Lovat's property was a crushing blow and accessing adequate fuel supplies remained for them a constant struggle into the nineteenth century.

As with its sixteenth- and seventeenth-century travails, Elgin's experience of fuel supply issues in the eighteenth century again provides a detailed account of the interleaved factors in play. There, the main indication of escalating shortages after c.1700 can be seen in councillors' attempts to regulate prices charged for peat by those who held cutting licences.

With no viable fuel alternatives and faced with growing consumer demand, these licensed suppliers had a monopoly upon which they capitalised. Council minutes from 1729 record how the peat-dealers' actions were denounced as extortionate and daily visits to the peat market by councillors were organised to ensure that only the price set under the licence agreements was charged.[12] Subversion of these arrangements through forestalling – selling goods before they had been displayed openly in the market – led to threatened legal action against any burgess buying fuel other than in the public market-place and for the publicly agreed prices.[13] It was understood, however, that rising instances of forestalling reflected the dearth of peat remaining in the burgh's reserved muir and so Elgin's councillors launched further attempts through the 1730s and 1740s to assert or enforce the burgh's exclusive rights in various mosses where their claims were at best tenuous. The consequence was mounting conflict with their landowning neighbours and their tenants which intensified as shortages worsened after 1750.[14]

In a seeming paradox, despite acknowledgement that Elgin's traditional peat sources were nearing exhaustion, there was an acceleration in losses at these locations rather than attempts to conserve supplies. Not all the losses were from cutting for fuel; some were related to demand for land for conversion primarily to arable uses but also for woodland plantations. In July 1749 a council inspection of the moss found new evidence for such threats as well as the expected evidence of illicit cutting;[15] some moss ground had been prepared for cultivation. This first identified encroachment was apparently small-scale,

---

10 NRS GD347/91.
11 W. MacKay (ed.), *The Letter-Book of Bailie John Steuart of Inverness 1715–1752*, Scottish History Society 2nd series (Edinburgh, 1915), p. 10.
12 Cramond, *Records of Elgin*, vol. 1, p. 431.
13 Ibid., p. 435.
14 Ibid., pp. 443, 452, 459.
15 Ibid., p. 189.

Looking across the bed of the Laich of Moray towards Kintrae, near Elgin, Moray. By the later eighteenth century, most of the mossland on the north-facing slopes had been stripped by the local communities, forcing them to seek fuel in Elgin's protected peat reserves on the south side of the ridge of hills that separates the Laich from the valley of the River Lossie.

but visits in 1752 and 1762 revealed an accelerating rate and expanding area of loss. Elgin's subsequent recourse to legal action against non-burgesses to prevent further agricultural encroachment was part of an escalating trend across the country. As early as 1725 at Whithorn, attempts by the council to meet their statutory duty to find the parish minister a suitable peat supply had been baulked by William Agnew of Wigg. Despite the best efforts of the council, he refused to surrender mossland on the former commonty north of the burgh in the Moss of Wigg, which he had converted at his own expense into what was, for him, profitable pasture.[16] It took a judgement of the Court of Session to force Agnew to assign part of the moss to the minister but he and his successor at Wigg continued to convert the remaining moss into profit-yielding agricultural land and deny access to the underlying peat to the local population. During a further dispute in 1763, this conversion was accused of having hastened the exhaustion of the minister's portion of the moss, but the council eventually found that it 'was not by the Heretor (the laird of Wigg) his casting Peats, Draining, Burning or other Improvements made by him exhausted sooner than it could have been . . .'.[17] Although the findings exonerated the then proprietor, they identified that by the

---

16 NRS GD455/27 (5). Agnew stated that he had 'by my industry Converted to Ground property of emeyr Nature as affording me good pasture wherefore in no event ought my project to be disappointed and the said ground be condemned to be used as a moss for ever'.

17 NRS GD455/27 (10). Conflict between Improving landlords and neighbours who claimed fuel easements on their land proliferated from the middle of the century and occurred widely around the country. See NRS GD1/221/52 (Lanarkshire); GD77/221 (Dumfriesshire).

middle of the century draining, moss-burning and 'other Improvements' were making substantial inroads on mossland in the Machars district of Galloway.

Whithorn and Elgin were not alone in experiencing loss of fuel resource to agricultural encroachment; conversion into agricultural land was worsening supply issues and adding new dimensions to a now perennial problem. One response, as exemplified by Elgin in 1752, was to employ a paid 'overseer and moss-grieve' to monitor peat-cutting and prevent agricultural encroachment.[18] This move met with initial success but the system broke down by the early 1760s. In September 1762, the council brought charges against twenty-seven men for illegally cutting and selling peat from the burgh's moss.[19] In a late example of a practice that the Scottish Parliament had sought to proscribe in the later seventeenth century, some of the men confessed to removing peat and turf to mix with material from their dunghills to make richer soil for garden cultivation. In most cases, however, it was alleged that the accused were breaking areas of moss for cultivation on that ground rather than for transportation of the mossy materials elsewhere.[20] The very public action against these individuals was clearly no deterrent, for in 1764 a further nineteen men faced similar charges and were fined and imprisoned for destroying the moss, building houses on it, enclosing parts of it, and planting potatoes and other roots and vegetables in it.[21] The echo of Whithorn council's findings from the Moss of Wigg are loud in the factors driving encroachment at Elgin. Despite the enthusiasm for agricultural improvement which was contributing to the inroads onto mossland, the threat to fuel supplies meant that no burgh could permit its resources to be eroded in this way without viable alternative fuels being found.

As the rate of Improvement operations undertaken by landowners in the rural districts around Elgin increased in the 1770s and 1780s, further areas of moss were lost to the plough. Estate plans from the 1780s show extensive inroads had been made in the wetlands and mosses of the Laigh of Moray around Spynie Loch, from which the populations of the parishes of Spynie and Drainie had formerly satisfied their fuel needs and in some of which the burgh had claims.[22] As well as new arable fields, however, here the lairds of Gordonstoun, Findrassie, Pitgaveny and Westfield among others had planted woodland shelter-belts and larger plantations for timber. Similarly, it was in the 1780s on another peat muir that had been a zone of conflict over access to fuel since the 1500s, Montreathmont in Angus, that the interested landowners were securing clarification of their respective interests and dividing their common muir area.[23] What then followed was the beginning of woodland plantation that has remained the dominant land-use on this ground down to the present. That this was a logical and, in Improvement terms, the only 'rational' development permeates modern discussion of the change, for although 'surrounded by fertile arable ground, Montreathmont was an unproductive wet Lowland heath, impossible to drain for arable crops or even permanent grass. The obvious thing . . . to do to make it productive was to plant trees'.[24] That same perception of peat areas in general as useless for anything other than cutting for fuel or

---

18 Cramond, *Records of Elgin*, vol. 1, p. 461.
19 Mitchell Library MS.246130.
20 For this practice, see Oram, 'Waste management and peri-urban agriculture'.
21 Cramond, *Records of Elgin*, vol. 1, p. 199.
22 See, for example, NRS RHP427, RHP31465/4, RHP94440.
23 NRS GD45/16/2194.
24 House and Dingwall, 'A nation of planters', p. 141.

'The Great Cut', Gilston, Moray. This central drainage channel cutting east–west across the bottom of the former bed of Spynie Loch gathers in the water from a capillary network of ditches feeding in from north and south. It forms the main element of the final and successful attempt to drain and improve the wetland between Duffus and Drainie on the north and Findrassie on the south.

planting with trees has lingered with immensely damaging consequences into the early twenty-first century.

By the 1780s, those who could afford coal in the district south of Montreathmont Muir could be supplied from the port of Arbroath. Duty on coal being shipped north of Red Head on the Angus coast between Arbroath and Montrose, however, made the importing of alternative fuel through Montrose more expensive, prohibitively so for those who either lay at the bottom of the economic spectrum or who had high demand for bulk fuel supplies. Accordingly, those communities further inland and north of the muir, who had depended on it for their peat, were pushed to the margins. Similarly Elgin, as an inland burgh, could not shift smoothly to importing coal in the manner of Inverness, for carting bulk supplies just the 8km from the port at Lossiemouth to the market in the town could double its cost on top of the high duty charged for its shipment north of Red Head.[25] Instead, the council sought to secure control over disputed areas of moor south-east of Elgin,

25 For fuel transportation costs, see Oram, 'Social inequality in the supply and use of fuel' and Oram, 'Arrested development?'.

commencing litigation in June 1767 and in February 1768 restating claims from 1638 to exclusive use of these more distant mosses.[26] The aim was to obtain a judgement through which they could pursue encroachers and halt further agricultural development of that land. Improvement to local road networks in the later eighteenth century eased access to these more distant peat sources, but travelling time to them meant a consequent loss of the opportunity value for this labour for other activities, effectively driving up the price of obtaining fuel, making the effort of obtaining it unattractive to individuals and tightening the grip of the cartel of licensed cutters. *Prima facie*, these efforts show a council fulfilling its duty to protect the burgesses' interests but they masked deeper, more complex and often contradictory local political manoeuvring. In 1751, for example, the council took simultaneous action to prevent William Duff Lord Braco's tenants closest to Elgin from removing peat from a disputed area while also working to persuade Braco to permit his tenants on his peat-rich estates at Pluscarden and Coxton to sell cut peat to Elgin burgesses.[27] Such contradictory stances arose from the lack of local alternatives to peat and the leverage that control of this precious staple gave to the council and cutting licence-holders and to neighbouring landowners. Continued local dependence on peat meant that the few licence-holders and rural suppliers could make high profits, despite annual price-setting. For Elgin's wealthier inhabitants, first glimmers of a long-term solution came in 1774, when, on account of another excessively wet summer and a lack of dried peats threatening to drive prices to unaffordable levels, the council bought 50 tons of English coal for sale to burgesses.[28]

Coal's day comes

Scotland's coal-mining industry had developed little since the sixteenth century and, even into the 1760s, there had been little investment in the new technologies that were transforming the scale and quality of mining operations in parts of England.[29] Lewis Baden, a mine engineer and member of the Society of Improvers in the Knowledge of Agriculture in Scotland,[30] was one of the first of a new breed that remained scarce until well after 1750. Most Scottish mines – still located mainly around the Forth estuary, in lower Clydesdale and in central Ayrshire (see Map 8.1) – continued to operate as they had done in the sixteenth and seventeenth centuries, largely as adjuncts of landed estates and most often associated directly with saltpan operations, which consumed the 'panwood' or coal dross that was unattractive to domestic buyers. The increasing commercialisation of the industry in the eighteenth century was linked to growth in demand from buyers among the urban elites, who saw coal use as a highly visible social signal of their cultural sophistication, from Improving landowners – mainly for fuel kilns that produced lime for building work and soil-dressing – and from expanding industrial operations. Increased demand, however, was not simply down to fashion or to changes in farming and industrial practice. Declining reserves of accessible peat, as we have seen, had since the 1500s encouraged a transition towards coal-burning in domestic contexts in areas where it could be obtained with comparative ease and, crucially, cheaply. More generally, however, it has long been recognised that climatic deterioration stimulated further demand for coal, especially from the later seventeenth century, when a succession of wet

26 Cramond, *Records of Elgin*, vol. 1, p. 201.
27 Ibid., p. 459.
28 Ibid., p. 482.
29 B.F. Duckham, *A History of the Scottish Coal Industry*, volume 1, *1700–1815* (Newton Abbot, 1970), pp. 14–15.
30 See above, Chapter 5, p. 160, note 18.

Peat versus coal: a question of efficiency and opportunity

Given the passion and intensity of current arguments surrounding energy sources and specific forms of fuel, we can perhaps gain a better understanding of the vehemence with which some Improvement literature denounced peat use. But, with peat, while our modern preoccupations are with securing zero carbon emission and safe, clean energy, the denunciations were driven by ideology and cultural prejudice rather than any questions of sustainability or environment. Peat was, in the eyes of ardent Improvers, a symbol of retrograde and culturally inferior practice, where individual effort was expended pointlessly on the procurement of a fuel supply rather than invested in betterment of one's economic position through increased agricultural output delivered by the new methods advocated by those same Improvers. Yet, in terms of thermal efficiency, peat was more than adequate for the needs of both domestic and small-scale industrial consumers. Indeed, as Andrew Symson had noted in Galloway at the end of the seventeenth century, salt-masters in Glenluce and Kirkmaiden parishes used peats to fire their pans rather than the 'panwood' coal used in Ayrshire and on the Forth estuary (Symson, 'Large description of Galloway', in Mitchell, *Macfarlane Geographical Collections*, vol. 2, p. 100). Although the Improvers stigmatised domestic and commercial consumers of peat alike, however, it was not their repeated criticism of the wasteful adherence to inferior practice and inferior fuel that won the day. Ultimately, it was the greater thermal efficiency and economic viability of coal for the larger manufacturing concerns that clinched the argument. Good-quality peat delivers 20–23 megajoules of thermal energy per kilogram, a thermal output capable of smelting iron, as opposed to the 28–33 megajoules from coal. But a far greater volume of peat was needed to deliver that output than the equivalent of coal, rendering it uneconomic to use at the scales needed by the new industries of the later eighteenth and nineteenth centuries, compounded further by the limitations of the available furnace technologies. Leaving aside the manifold negative environmental consequences of the adoption of coal on a large scale, access to coal avoided the intensification of peat extraction through the nineteenth and twentieth centuries that has left such a negative environmental legacy, as can be witnessed in the denuded peatland of the Irish Midlands and West, or the Meuse Valley in northern France.

summers and autumns and bitterly cold winters first prevented the cutting and drying of adequate quantities of peat to satisfy household requirements and then left those households with inadequate fuel to meet thermal energy needs in the midst of some of the coldest winters of the 'little ice age'. In such contexts, it has been noted, there was a direct correlation between the weather conditions and demand for coal in regions normally reliant on peat.[31] Much of that demand was met by English suppliers, mainly from the Northumberland coalfield, but by the second half of the eighteenth century Scottish collieries were receiving the capital investment necessary to transform their essentially late medieval physical and commercial structures into producers of bulk fuel that was fed into the British imperial trading network.[32] All that held back the large-scale shipment of coal from the Forth Valley coalfield to the north-east and north of Scotland was the heavy duty of 3 shillings 8 pennies per ton levied on it, which became the target of intense lobbying in parliament through the 1750s to early 1790s by mine-owners, burgh councils and would-be Improving landowners, who viewed it as an unjustifiable impediment to more widespread coal use and to straightforward security of their fuel needs.[33]

---

31 Duckham, *Scottish Coal Industry*, vol. 1, p. 17.
32 Oram, 'Social inequality in the supply and use of fuel', pp. 225–226.
33 Duckham, *Scottish Coal Industry*, vol. 1, pp. 230–233; Oram, 'Social inequality in the supply and use of fuel', pp. 225–226; Oram, 'Arrested development?', p. 22.

Moat Mine site, Culross, Fife. The low mound in the centre of the picture marks the position of the off-shore pithead of the Bruce estate's famous mine, from which coal could be loaded directly onto ships for transport to local users or to more distant markets.

Landowners and councils in districts north of the Red Head or where peat supplies were significantly depleted were not simply parliamentary lobbyists seeking repeal of coal duty. In the 1770s, Sir James Grant of Grant and several other prominent northern Improvers entered ultimately unproductive contracts with English 'viewers and borers of coal' to prospect for mineable coal in Ross and Cromarty and Sutherland.[34] Only the mine at Brora on the Sutherland coast, where coal to supply local saltpans was mined from the late sixteenth through to later eighteenth centuries, yielded any exploitable reserves in the whole of northern Scotland. On the extensive Gordon estates in the north-eastern Highlands, the duke's factor in 1771 mused on the 'need to get a proper miner to search for a vein' of coal.[35] Tacks to tenants on the earl of Breadalbane's estate in Strath Tay in 1760 gave them the right to make trials for coal and other minerals, reflecting the earl's deep personal interest in Improvement of his property.[36] In Galloway, supreme confidence that the coal beds known from Cumberland and the Isle of Man to the south and east, and from Ayrshire to the north and west, *must* extend beneath Wigtownshire led to extensive prospection in the southern Machars district, where peat reserves were nearing exhaustion after millennia of use by consumers and inroads by Improving landlords. No coal was found. At Perth, to which coal had been imported since the Middle Ages, despite its geographical location south of the boundary of the coal duty, shipment costs still restricted its use to the wealthiest in the community.

34 NRS GD248/160/12 (14).
35 NRS GD44/43/50 (15–16).
36 NRS GD112/10/1/4 (2–3).

Red Head, Angus, viewed from the site of the Scotts of Dunninald's great limekiln operation at Boddin, 5.5km to the north. The coastal headland was the landmark north of which a higher duty was paid on the seaborne shipment of coal.

Here, peat had been in increasingly short supply locally from the fifteenth century and, by the eighteenth century, was being obtained mainly from mosses 6km west of the burgh. The Commissioners of Forfeited Estates had undertaken trials for coal on the Drummond estates north of the town in the period 1763–1774, while in 1779 the provost and council were approached by a would-be mining speculator to support coal prospection on land belonging to the burgh's King James VI Hospital at Scones Lethendy, 4.5km north of the town.[37] Again, like almost every other trial in the country north of the Forth and in Galloway, both efforts yielded nothing. A major longer-term consequence of this discovery that Scotland's geology was far more complex than first assumed and that most districts outside the Midland Valley were devoid of coal-bearing strata was continued reliance by the vast majority of the Scottish population on traditional fuel types for their domestic and industrial needs.

Woods, ironmasters and charcoal burners

John Campbell 3rd earl of Breadalbane's interest in finding coal on his estates was in large part driven by his and his father's experience of dealing with English and Irish tanbarkers and charcoal burners. As discussed in Chapter 7, over-generous contracts and a failure to recognise the latitude given to the contractors by generations of management that had produced woodland with fairly uniform age structures had made serious inroads on their woodland in the 1720s and 1730s. The 3rd earl maintained

37 NRS E777/139; Perth and Kinross Archives, A.K. Bell Library B59/24/15/13.

contracts with charcoal burners after the failure of Murphy and Galbraith's business, along with his neighbours and kinsmen the duke of Argyll and Campbell of Lochnell; in the 1750s they entered arrangements with two north Lancashire iron-manufacturing companies who established furnaces for the production of pig iron at Bonawe on Loch Etive in 1753 and Craleckan on Loch Fyne in 1755.[38] These were longer-term, economically successful ventures developed by companies with established experience and vested interests in maintaining a regime of sustainable management of woodland to provide the charcoal that was essential to their operation. Earlier operations, such as the 1727/8 establishment of a short-lived furnace at Invergarry by the Rawlinsons of Graithwaite, had been sited close to the wood sources and charcoal-production locations in Glen Garry but failed largely because of the high costs involved in the transportation of ore and other materials from Corpach on Loch Eil to Invergarry, and not because of any inadequacy in the fuel supply.[39] Indeed, the Rawlinsons knew that the issue was the furnace's location in respect of its sources of materials other than fuel and the transportation of its product to market, not any deficiencies in the availability of an energy source. Further proposals by the company to develop new furnace operations further up the west coast were founded on the reliability of access to the coastal woodlands of the Macdonalds of Clanranald and Kinlochmoidart, but they failed to materialise when the mineral sources the Rawlinson company expected to find close to their new site proved to be non-existent.[40] Murphy and Galbraith in Glen Kinglass[41] and the operations at Bonawe and Craleckan did not make that mistake, all being sited adjacent to the landing-places for the importation of raw materials and shipment out of the pig iron, and with access to nearby woods for charcoal or its easy transportation in by water from more distant production sites.[42]

Contrary to the still oft-repeated assertion of Frank Fraser Darling that the decline of West Highland woodland was a consequence largely of the unsustainable practices and voracious charcoal consumption of these iron-producing operations,[43] close analysis of their activities instead supports late

---

38 Smout, *Exploring Environmental History*, p. 91.
39 NRS GD1/168/11; Smout et al., *History of the Native Woodlands*, pp. 238–239.
40 Smout et al., *History of the Native Woodlands*, p. 239.
41 Despite the poor reputation of these Irish operators, the furnace at Glen Kinglass was a sophisticated and quite technologically advanced type of blast furnace and might have enjoyed the longevity of operation of its nearby successor at Bonawe had Murphy and Galbraith been more competent and less dishonest in their business activities. See Smout, *Exploring Environmental History*, p. 90; Smout et al., *History of the Native Woodlands*, p. 354. For the Glen Kinglass furnace, see Lindsay, 'Iron industry in the Highlands', pp. 56–57; J.H. Lewis, 'The charcoal-fired blast furnaces of Scotland: a review', *Proceedings of the Society of Antiquaries of Scotland*, 114 (1984), pp. 445–463 at 457–460. The site is described in *The Royal Commission on the Ancient and Historical Monuments of Scotland. Argyll: An Inventory of the Ancient Monuments*: volume 2: *Lorn* (Edinburgh, 1975), pp. 280–281, No. 358.
42 J.M. Lindsay, 'Charcoal iron smelting and its fuel supply: the example of the Lorn Furnace, Argyllshire, 1753–1876', *Journal of Historical Geography*, 1 (1975), pp. 283–298; Smout et al., *History of the Native Woodlands*, pp. 240–248. For a description of the Bonawe site and its history, see G.P. Stell and D.H. Hay, *Bonawe Iron Furnace* (Edinburgh, 1995).
43 He had expounded this thesis, based on mainly nineteenth-century antiquarian studies, in a short essay: F. Fraser Darling, 'History of the Scottish forests', *Scottish Geographical Magazine*, 65:3 (1949), pp. 132–137, and repeated it without further reference to sounder historical or ecological evidence in studies such as F. Fraser Darling, *West Highland Survey: An Essay in Human Ecology* (Oxford, 1955), pp. 4, 167 and F. Fraser Darling and J.M. Boyd, *The Highlands and Islands* (London, 1964), pp. 54–59.

Loch Etive, Argyll and Bute, looking south-west down the loch from Invernoe and the Glen Noe oakwoods towards the location of the Bonawe furnace in the centre of the view. The sea-loch allowed the easy import of raw materials to sites close to the prized oak charcoal sources and the shipment out of the iron pigs for processing at foundries in north-western England.

eighteenth-century claims of good management and the healthy condition of much regional woodland.[44] Modern calculation of the extent of woodland necessary to support Bonawe and Craleckan suggests that somewhere in the order of 20,000–24,000 acres coppice-managed on a twenty- to twenty-four-year rotation would have been required to keep the furnaces supplied.[45] The ironmasters were active in seeking contracts for access to woodland for their charcoal needs, with Breadalbane being one major partner from the early 1750s onwards, along with Campbell of Barcaldine and, later, the Maclaines of Lochbuie on Mull.[46] As is clear from contracts such as that drawn up in 1772 between the Macdonalds of Clanranald and the Cumberland ironmasters of the Netherhall furnace, the woodland was not to be diminished through their charcoal-burning operations and would revert to the landowner in as good a state as it had been in when the contract had been entered.[47] Collectively, the contracts cover much of

---

44 This had been postulated as long ago as the early 1900s: A. Fell, *The Early Iron Industry of Furness and District* (Ulverston, 1908). For the more modern analyses, see Lindsay, 'Charcoal iron smelting'; Lindsay, 'Iron industry in the Highlands'; and J.M. Lindsay, 'The commercial use of woodland and coppice management', in M.L. Parry and T.R. Slater (eds), *The Making of the Scottish Countryside* (London, 1980), pp. 271–289. The reassessment is summarised in Smout et al., *History of the Native Woodlands*, pp. 240–248.

45 Summarised in Smout, *Exploring Environmental History*, p. 91.

46 NRS GD112/16/10/6, GD112/16/11/2 (35–39), GD112/16/11/3 (18–33), GD170/438, GD174/878.

47 NRS GD201/1/281.

Coppiced oak, Glen Kinglass, Argyll and Bute. This oak was probably last cut for charcoal production between 1850 and 1875, for use at the Bonawe furnace, whose owners had secured coppice rights to the woodlands of the glen following the failure of the Galbraith and Murphy furnace located there.

the broadleaf woodland of coastal districts from Kintyre to Moidart and the adjacent islands. The age profile and form of the still existing woodland there is a legacy of their successful long-term management of the resource, and belies traditional views of the voracity and malpractice of the iron companies. Certainly, the Craleckan furnace was maintaining its rotational management of the woods under its contracts until c.1813, when the duke of Argyll's decision to increase rents due to him from the ironworks led to its closure, while Bonawe continued to operate down to 1876, sustained on the basis of the favourable contract terms they had negotiated in the 1750s and their evidently careful conservation of the fuel source upon which their production relied. Indeed, as mentioned in Chapter 7, it has been posited that the active management of the region's broadleaf woodland for such industrial concerns may have brought these semi-natural woods to a modern apex of health and ecological diversity that has declined since the ending of these practices.[48]

The Bonawe furnace, like several others, contin-

---

48 Sansum, 'Argyll oakwoods'.

ued to use charcoal down to the end of its productive life in 1876. Ready access to its chosen fuel and the favourable terms on which that access was founded, however, were not an advantage enjoyed by many other ironworks and, elsewhere, the trend from the 1760s was towards increasing reliance on coked coal in furnaces. Although coke production had a history in England from the sixteenth century, it was the development of a large-scale coke-fired blast furnace in 1709 by Abraham Darby that accelerated high-volume production of coke and, after the 1768 development of larger coke-ovens, the fuel was being produced at a volume that eclipsed what could ever be yielded by charcoal production in Scotland's Atlantic-facing oakwoods. It was partly with an eye on this expansion in the industrial uses of coal and partly because of the need to find a local alternative to dwindling peat supplies that Breadalbane had turned his mind to prospecting for coal on his lands in the 1760s. Although Breadalbane and his successors failed in every prospection for coal and almost every trial for other minerals, the development of the new coal-using technologies of the second half of the eighteenth century signalled the beginning of a coal rush and the massive expansion in output from mines across Scotland that powered the new iron, steelmaking and shipbuilding heavy industries of the Clyde Valley and inner Firth of Clyde. That, however, is a subject for the volume 'Scotland 1850 to COP 26'.

・ ・ ・

By shifting the perspective from the Improvement-driven discourse of the later eighteenth and nineteenth centuries and modern social and economic historians' focus on the roots of the nineteenth-century coal-iron-shipbuilding economy of lower Clydesdale, in this chapter we have gained a clearer picture of the continuing diversity of approaches to securing domestic and industrial thermal energy needs across Scotland. While the eighteenth century is rightly recognised as a period of transition which saw a step-change in fuel demand by domestic and industrial consumers and a pivot towards the increased consumption of coal, we have seen that peat, wood and charcoal remained the primary thermal energy sources for the bulk of Scotland's people – including for industrial processing – in most areas distant from the mines. It was not until the later nineteenth century that coal use became an almost universal choice, aided by expansion of production and new means for its bulk distribution, and limited only by personal financial position and transport opportunities. But abundance of peat and its relative ubiquity did not mean that fuel for heating, cooking and processing was not a pressing concern for many Scots, especially those in urban communities and on society's economic margins. The securing of an adequate supply of peat across a century noted for its poor weather and recurrent periods of bitter cold further intensified pressure on both the sources of fuel and those who needed it. Persistent wet summer conditions rendered it difficult to dry adequate amounts of peat to heat homes across the longer winters, while waterlogging and poor management of the mosses brought deterioration of the vital resource. Consequently, in some areas where peat remained the only available form of fuel, despite the reality that Scotland was a fuel-rich country in comparison to other parts of northern Europe, there were persistent deficits in fuel security for the bulk of the local population. The position in such places was worsened by a combination of privatisation of once common resources, poor management of the remaining areas of mossland, and a progressive programme of conversion of mossland into arable by lairds whose fuel needs were met from other sources. In environmental terms, this era witnessed an acceleration in the rate of loss of wetland and moss as human priorities shifted from their fuel value to arable expansion, and alternative solutions offered themselves.

It was not just peat, however, that witnessed an

intensification in pressure, for we have also seen that woodland – especially but not exclusively oak – likewise experienced rising demand as Scotland's integration into the wider British economy and industrial complex accelerated after 1707. Some of the great expansion of woodland discussed in Chapter 7 was intended to meet that growing demand, but most of the pressure fell on the already ancient areas of managed woodland around the country. Although it has been seen traditionally as an era of devastating inroads that caused the wasting of woods that had survived the voracious wood consumption of the medieval period, the recorded evidence suggests instead that this period probably witnessed the peak of sustainable management practice for timber and fuelwood needs. Headline instances of bad management have long made for powerful polemic, but even in what have been presented as extreme cases of rapacious extraction by asset-stripping contractors – for example, on the Glenorchy and Lochnell Campbell estates – the supposed clear-felling that occurred seems rather to have been extensive and unregulated cutting of un-hagged woodland, for regeneration from the stools provided more considerate operators with a further 150 years of material for charcoal manufacture. In the case of the woods, as with peat moss, while the impact of fuel demand intensified in the eighteenth century and brought more regular and systematised coppicing, these activities constituted the more rigorous and larger-scale application of methods that, as discussed in the volume 'Scotland AD 400–1400' and earlier in this present volume, had maintained those woods successfully for centuries already. What changed was the frequency and extent of the operations, moving from management for local needs to management for the commercial requirements of businesses whose reach had already extended from local operations to global markets. Abandonment of those older sustainable practices later in the nineteenth century and a consequent deterioration in the ecological health and fundamental structure of the woods still lay a century ahead.

CHAPTER 9

Achieving Improvement amid a 'Sea of Waste': Agricultural Change to 1790

> Glasgow is 6 miles distant, through the most beautyfull country in Scotland. A spirit of industry reigns in the neighbourhood and a de[sire] to set off everything to the best advantage. In every part you see gentlemen's houses neatly built, with large plantations to adorn their estates. Near the town grow wheat and turnips, the first we saw in Scotland. They sow them in rows with a small space between each row. This being left unsown serves instead of a summer fallow, so that every year the same quantity of ground is cultivated without it being necessary for the whole field to lye fallow.[1]

The changes in the choice, procurement and general availability of thermal energy sources that we examined in the previous chapter became key facilitators of the great expansion of Scottish industrialised society through the later eighteenth and nineteenth centuries. It is accepted generally that they served as the primary motors of the profound environmental changes that the coming era of industrialisation was to bring. It was, however, fundamental changes in agricultural practice that enabled some of the most profound social restructuring and triggered the dramatic environmental reconfigurations that became increasingly evident as the nineteenth century progressed. As with past considerations of woodland planting and management explored in Chapters 2, 4 and 7, traditional historiographical presentation of agricultural change as a profound and rapid process running from the middle of the eighteenth century onwards has given way to a more nuanced understanding that what we would recognise as Improved practices had begun to be introduced in the seventeenth and, in some areas, possibly sixteenth centuries, and continued into the early twentieth century.[2] At the start of the period under review in this chapter, areas of 'Improved' agriculture could still be characterised as islands in a 'sea of waste',[3] but by the early nineteenth century these islands had grown and in some areas Improvement

---

1 Dunbar, *Burrell's Northern Tour*, p. 29.
2 Whyte, *Scotland Before the Industrial Revolution*, p. 143.
3 The quotation is from Whyte, *Scotland Before the Industrial Revolution*, p. 146.

had spread to engulf what had once been regarded as waste. In this chapter we shall examine the acceleration of these processes and the associated acceleration in the application of new agricultural methods and the technology to deliver them, and the introduction of new crops, grasses and soil-improving plants, in a steadily mounting wave that transformed ecological systems across the whole of Scotland. It was not just a question, however, of landowner trial-and-error experimentation persisting in the quest for improvement in yields, grassland quality, and livestock. Their effort was also founded upon greater access to funds to borrow and capital to invest in development of their estates, much of it siphoned from plunder and profits accrued through participation in Britain's expanding mercantile empire and slave-worked plantations in its colonies. It was also aided by employment of legal authority to consolidate their properties, to divide and enclose commonties, and physically re-constitute the social organisation of the agricultural communities upon their land. Environmental change was thus the result as much of political and economic factors as it was of the spread of Improvement agriculture and the introduction of the new plant species and animal breeds that accompanied it.

Reshaping rural structures

Environmental history blurs into social and economic history when we start to look at the process of agricultural improvement through the eighteenth century. Changes in the socio-legal framework of rural settlement and society as much as in the physical organisation of farming communities, extending from lease terms through to building types, went hand-in-hand with the intensification of anthropogenic change at landscape scale. Although we can easily overstate the degree to which development and adoption of new tenancy arrangements (from multiple-tenant fermtouns on short or annual leases to single tenancies on long leases), new arable technologies or approaches to livestock management had caused widespread change before 1700 (as discussed in Chapter 6), few would question that in some parts of the country there had already been an appreciable switch in practice achieved in the last quarter of the seventeenth and first quarter of the eighteenth centuries. It is now widely accepted that the organisational changes represented by consolidated farms under long-term single-tenancy leases prefigured and enabled the introduction of new agricultural practices. These came as a result of both lease conditions (many of which stipulated prescribed crop rotations and annual liming of fields) and incremental introduction by tenants as and when their finances permitted. A growing number of locally focused studies, such as the detailed historical analysis of agricultural practices in and around Flanders Moss in Stirlingshire, has revealed the steady penetration of these Improvement ideologies and practices, from adoption by landowners to application by tenants in the century down to 1750.[4] As the example of Hugh Graham of Arngomery referenced in Chapter 5 illustrated, underscored by the enthusiastic take-up of his thinking by Lord Kames and his son-in-law, mossland drainage was seen as one of the new 'frontiers' of Improvement efforts.[5] Travel journals, moreover, such as that written by the keenly observant and sharply enquiring Englishman Sir William Burrell, also reveal how widely these practices had spread before the middle decades of the century.

There is, furthermore, now a broad recognition that the attendant processes of 'clearance', whereby

---

4 Harrison, *Historical Background of Flanders Moss*, especially chapters 3, 4 and 7.
5 Albritton Jonsson, *Enlightenment's Frontier*, pp. 11, 33–34.

Boundary between the Abercairney and Monzie estates near Foulford, Perth and Kinross. The late eighteenth-century turf dyke and its successor twentieth-century fence line are representative of the new compartmentalisation that was introduced to the Scottish rural landscape during the first great tide of Improvement.

the human population was removed from the land in the name of increased agricultural efficiency and economic exigency, long represented as an almost uniquely northern and western Highland and Hebridean experience, was equally characteristic of regions spread from Moray and north-east Highland Aberdeenshire to Galloway. In central and southern districts as well as the north-eastern Lowland fringe, clearance probably commenced earlier than elsewhere as Lowland lairds embraced Improvement principles from the 1670s onwards. In Lowland Moray, for example, some multiple-tenancy fermtouns were already being consolidated under single-tenancy arrangements in the last quarter of the seventeenth century, and in central and west Aberdeenshire clearance was underway by the 1720s.[6] Although the pace of Lowland clearance accelerated in the second half of the eighteenth

---

6 P. Aitchison and A. Cassell, *The Lowland Clearances: Scotland's Silent Revolution 1760–1830* (East Linton, 2003); A. Watson and E. Allan, 'Depopulation by clearances and non-enforced emigration in the North East Highlands', *Northern Scotland*, 10:1 (1990), pp. 31–46.

century, the spread of Improvement agriculture through Lothian and the broader south-east, the south and east of Dunbartonshire, south-west Stirlingshire, and northern parts of Lanarkshire and Renfrewshire before 1750, recorded graphically in the maps of General William Roy, indicate that agricultural enclosure and new plantations and the attendant processes of division of commonties, displacement of population and economic degradation from tenant to landless labourer status were already well established in those districts.[7] What is hinted at in the cartographic evidence, however, is that Improvement first happened on the already heavily invested-in land, where gains would be easiest and quickest to achieve, but which delivered a patchwork of Improvement amid a sea of unimproved land. It meant that islands of Improvement could exist remote from the supposed early pioneering regions of such activity, like Lothian, if the local laird was an early adopter of Improvement principles and practices and had the means to implement innovations. It was often only once these scattered blocks of better land had been Improved, consolidated and able to demonstrate their evident success and advantage over traditional practices that the spread of new agricultural methods out from these nodes into the less favoured land could commence.

It is difficult nowadays to gain any true appreciation of the scale of transformation represented by these processes. They were the triggers of a first phase of modern depopulation of some rural districts that accelerated in the later eighteenth and nineteenth centuries. Depopulation, however, did not mean abandonment; it most often signalled intensification of exploitation, but with the driving economic consideration being production for the market rather than subsistence to sustain the extended family networks of multi-tenant peasant communities. They also marked the beginning of the transition away from a landscape of open fields to the compartmentalised one with which we are familiar today. With it came species changes, of domesticated livestock and crops, which themselves signalled wider change in the ecology of the farmed landscape as different patterns of selective grazing began to change the composition and structure of herbage. New crop species and the technology of growing them introduced or destroyed niches occupied by specialist biota, from land snails to rodents, ground-nesting birds to fungus varieties. It was a revolution in agriculture that affected every dimension of the Scottish environment.

Most discussion of this 'agricultural revolution' tends to be socio-economic in its focus or more aligned to exploration of the technological developments and experimentation in crop rotations and the use of nitrogen-fixing legumes, selective breeding of livestock and plants to obtain bigger and better animals and yields. Yet, the reconfiguration of patterns of settlement and the modes of exploitation employed by rural populations at this time, which laid the foundations of the lowland and upland landscapes that we recognise today, triggered the greatest and probably fastest episode of anthropogenic environmental transformation experienced in Scotland before the later twentieth century. Improvement agriculture is usually represented in terms of both an increase in the quality of arable farmland and pasture (through the application of new methods of management and use of new and increasingly cheaply available agricultural technology like the cast-iron ploughshare) and the quantity or extent of ground so improved. It is viewed conventionally as enabling the optimisation and maximising of the potential of

---

7 Whyte, *Scotland Before the Industrial Revolution*, pp. 145–146. As an example, see the new patterns of fields around Duns in Berwickshire on Roy's map, where the recently enclosed fields form large blocks of regularity amid surrounding zones of unenclosed agriculture and land that was only exploited extensively for pasture or fuel: NLS General Roy Military Survey https://maps.nls.uk/geo/explore/#zoom=12&lat=55.77451&lon=-2.33862&layers=4&b=1.

the land to support cultivation and livestock, yielding the surpluses that brought profit to producers and cheaper food to consumers.[8] Although there were areas – like Rannoch Moor – that were beyond the capability of would-be Improvers to tame and 'improve' for profitable agricultural use, elsewhere this period saw the draining of coastal and inland wetlands, breaking into at least rotational cultivation and fallow of formerly occasionally cropped outfield and pasture, alongside the beginning of a new retreat from some high-altitude marginal areas.[9] Through mineral enrichment and intensive manuring of the soil, changes to ploughing technology and technique, modifications to drainage, the introduction of crop rotations, sowing of new cultivated plants (including the cropping of clover and new grass species and increased use of lucerne, peas and beans as nitrogen-fixing fodder crops), and removal of peat blankets, the spread of Improved agriculture initiated an ecological sea-change whose consequences we live with today.

Enclosing and compartmentalising the land

Although only a few landowners had acted on any large scale to take advantage of the legislation available from the 1660s onwards that had encouraged enclosure for arable cultivation (see above, pp. 102–103), woodland plantations and livestock management, realignment of land boundaries or division of commonties, that legislation remained in place post-1707 and provided the legal authority upon which an increasing number of proprietors chose to act in the following decades.[10] Division of commonties, those areas where multiple landowners held shared rights of access for grazing, fuel and other organic materials, with their rights usually exercised through their tenants, is often presented as a phenomenon primarily of the post-1760 period and mainly in the Lowlands. It was, however, a process already well in train before 1750 and throughout Scotland.[11] Change was certainly slow and most often linked to the availability of the capital needed to fund development. Access to capital was itself linked to market conditions that enabled landowners to convert the rents which were still received largely in kind – mainly grain – into the cash needed to pay not only for new equipment, breeding stock, plants and seeds but also for the legal costs involved in bringing about enclosure and the costs of then constructing the new enclosed fields. Capital flows increased in the later eighteenth century as landowning families also drew profit from colonial and foreign trade investments or the prize money and plunder of foreign wars, appropriating the natural capital of overseas territories to fund the development of their estates at home. None of this, however, was in any sense systematically planned in the earlier 1700s and, indeed, it was still proceeding piecemeal in the middle decades of the century. As Roy's maps record, the result (or at least the intention) was a patchwork of blocks of enclosure and improved farmland dispersed through a still largely unimproved landscape. These blocks were most often immediately adjacent to the homes of the Improvers and representing land under their direct control as demesne, which formed the 'Mains' or 'Home' farms of the estate in the later eighteenth

---

8 As expressed, for example, in M.L. Parry, 'Changes in the extent of improved farmland', in M.L. Parry and T.S. Slater (eds), *The Making of the Scottish Countryside* (London, 1980), pp. 177–199.
9 Ibid., pp. 181–184. That retreat was slow and in some areas not evident in the palynological record for upland cultivation until after 1800: see Tipping, *Bowmont*, pp. 199, 201–203.
10 Whyte, *Scotland Before the Industrial Revolution*, p. 144.
11 The scale and differential rates of commonty division around Scotland are evident from the detail provided by I.H. Adams (ed.), *Directory of Former Scottish Commonties* (Edinburgh, 1971).

East Lothian centred on Haddington, showing the mixed landscape of Improved agriculture with regular, enclosed fields and regimented new woodland planting set amid wider expanses of unenclosed rig-and-furrow cultivation, such as had prevailed since the later Middle Ages. (From William Roy, *Military Survey of Scotland 1747–55*)

and nineteenth centuries. They were, however, islands of difference embedded within zones of unenclosed rig-and-furrow cultivation or pasture that were still leased to multiple tenants.[12]

A few examples illustrate the spread of such actions around the country, with most 'lowland' districts from Galloway to Caithness witnessing division and enclosure of former commonties progressing at an accelerating pace through the first half of the century. In 1713, for example, the town council of Cupar in Fife and the Clephanes of Carslogie, west of the burgh, entered negotiations for a division between them of the muir in which they had shared rights, and in the same year the first divisions took place on the commonty at Kirk Yetholm in Roxburghshire.[13] Moves for division of commonty in Annandale in Dumfriesshire, in Dunsyre in Lanarkshire and at Scone near Perth were underway in the 1720s.[14] At Wick in Caithness in 1739, the council was canvassing George Sinclair of Ulbster, their provost, for his assistance in resisting Sir William Dunbar, who had raised a court action for division and enclosing of the commonty enjoyed by burgesses and Dunbar's tenants alike.[15] Most of these divisions were achieved through agreement, at least between the principal landowners involved, but it

---

12 I. Whyte, 'The emergence of the new estate structure', in M.L. Parry and T.S. Slater (eds), *The Making of the Scottish Countryside* (London, 1980), pp. 117–135 at 130.
13 St Andrews University Library Department of Special Collections, B13/22/57; NRS GD6/826.
14 NRS CS226/362, CS226/6414, GD5/267, GD5/510, GD5/511, GD5/513.
15 Caithness Archives Centre, B73/2/1.

only needed one of the joint possessors of commonty rights to raise a court process for the division to be enforced. Wick's experience, where the burgh wished to retain the land in commonty in the face of Dunbar's wish to divide and enclose, and various challenges brought in the name of lesser property-owners and ejected tenants across the country, suggest that the larger landed interests and those of an Improving bent could be aggressively proactive in securing their desired result.

Parks for cattle

Regional divergences in the speed and nature of enclosure were in part influenced by the market opportunities to which the enclosers were responding. For example, the letter-book of the Inverness bailie and merchant John Steuart reveals him operating as a dealer in salted beef in the 1710s, trading with Rotterdam.[16] Steuart was acting as a middleman in the trade in cattle from within the northern Highlands and northern parts of the Hebrides, where numbers of cattle had been increasing from the later sixteenth and seventeenth centuries as a response by tenants to the increasing demand from clan chiefs for cash rents rather than rents in kind.[17] The tenants' animals were brought to local markets for cash sale, providing the cash with which the tenants paid their landlords, and then slaughtered there either for local consumption or salted for onward trade. For men like Steuart, operating remotely from the centres of highest demand in southern Scotland, shipment of already processed carcases rather than live animals was the best option. Further south, however, the long-established trade to Lowland towns and increasingly into England saw engagement by landowners in new opportunities to maximise profit from live cattle.

In Galloway, the development of large-scale cattle 'parks' – in the Scottish meaning of that word as simply an enclosed piece of ground rather than its more modern recreational or leisure function – that had begun in the later seventeenth century (see above, pp. 123–125) accelerated in the first quarter of the eighteenth century as the trade in and movement of beef cattle to English markets expanded rapidly in the 1700s. The trend begun in the eastern Machars district of Wigtownshire by the Dunbars of Baldoon – whose great park was described in the 1680s as measuring some 4.5km in length by 2.5km wide and containing 'excellent grass'[18] – was adopted by the Stewart earls of Galloway and most of the major landowning families throughout the region's lowland districts. By the middle of the century, parks were common throughout the region although the land they enclosed might not necessarily have been much improved, reflected in William Burrell's dismissive observation that 'the people have a method of styling all ground enclosed parks, though perhaps some of those parks produce nothing of benefit of (sic) man or beast'.[19] Using the seventeenth-century enclosure enabling acts and the associated division of commonty legislation, landowners had taken the opportunity to clear multiple-tenancy fermtouns from their property and divide the land into large, drystone-walled grass parks to provide grazing for herds running into thousands of heads.[20] By 1721 in the eastern Machars, Sir John Clerk of Penicuik could report that around 20,000 acres of land had been given over to cattle enclosures, to the great profit of the local landowners – who he estimated made over

---

16 MacKay, *The Letter-Book of Bailie John Steuart*, pp. 6–8.
17 Dodgshon, *From Chiefs to Landlords*, p. 113.
18 Symson, 'Large description of Galloway', p. 78.
19 Dunbar, *Burrell's Northern Tour*, p. 21.
20 Livingston, 'The Galloway Levellers'.

10,000 guineas profit per annum – but at a loss to the displaced farming population.[21] Clerk, indeed, described the former arable land as 'laid waste'. The process was often heavy-handed and insensitive, triggering an unprecedented degree of local resistance from those who had been dispossessed or who feared their likely dispossession. Simmering tension erupted into disorder and resistance in 1724, when groups of tenants and dispossessed families began to throw down the newly built walls, earning them the name of 'the Galloway Levellers'.[22] Although the large-scale enclosures had developed first in Wigtownshire, the Levellers' revolt began with incidents mainly in the districts around Kirkcudbright. Here, surprisingly, their first targets were cattle parks that had been in existence already for decades rather than the more recent constructions.[23] It took military intervention in 1725 to suppress the rising, but punishments were generally lenient, possibly because it was recognised that the landowners' enclosing actions had been provocative.

Despite the resistance encountered in Galloway in the 1720s, enclosure for livestock management was viewed as a positive development in much pro-Improvement correspondence in the next half century. It was seen as enabling among other things supervision of herds to improve the quality of animals through controlled breeding.[24] It was, however, viewed largely as a means of asserting exclusive use of a grazing area, through which improvement of the quality of pasture within it became a worthwhile venture for owners. Some were capitalising on the developing drove trade in live cattle, creating enclosures seeded with the new high-quality grass species advocated in Improving literature and available from the growing number of specialist seed-merchants. Animals could be fattened in these enclosures after being transported by ship or driven down from Orkney, the Western Isles and the Highlands, before sale in nearby urban centres. Moves to develop land for grazing at the point of origin and to create secure areas where new breeds of 'Improved' stock were kept separate from inferior native varieties may have underlain efforts to develop grass enclosures on isolated headlands in Orkney, such as that created in 1703 at Snelsetter and Cantick Head in South Walls in the Orkney Islands by James Moodie.[25] These, it seems, were not being created out of divisions of former commonties or commons, which some Orkney proprietors were only beginning to divide and enclose in the 1760s with varying success,[26] but from land within their sole ownership. Although Moodie's experiment failed due to want of good grazing for the sheep he imported from his relatives' lands on the Scottish mainland, an even larger operation for cattle at Park of Cara at the north end of South Ronaldsay was more successful. Park of Cara, extending to around 170ha within drystone walls like those of the Galloway parks, was the creation of Sir James Stewart of Burray, a second cousin of the earl of Galloway. We cannot be absolutely certain, but these had probably been inspired

---

21 J.W. Leopold, 'The Levellers Revolt in Galloway in 1724', *Journal of the Scottish Labour History Society*, 14 (1980), pp. 4–29; A.S. Morton, 'The Levellers of Galloway', *Transactions of the Dumfriesshire and Galloway Natural History and Antiquarian Society*, 3rd series, 19 (1933–5), pp. 245–254; W.A.J. Prevost (ed.), 'A Journie to Galloway in 1721 by Sir John Clerk of Penicuik', *Transactions of the Dumfriesshire and Galloway Natural History and Antiquarian Society*, 3rd Series, 41 (1962–3), pp. 186–200 at 194–195.
22 Mitchison, *Lordship to Patronage*, pp. 155–156.
23 Livingston, 'The Galloway Levellers', pp. 60–74.
24 NRS GD14/17.
25 S.M. Gow, *James Moodie Younger of Melsetter: An Orkney Laird at the Time of the Union* (Kirkwall, 1977); Thomson, *History of Orkney*, p. 201.
26 Thomson, *History of Orkney*, p. 202.

'Baldune Parks', Wigtownshire. The great cattle park of the Dunbars of Baldoon, delineated in red ink, covered land previously occupied by a number of small, multi-tenant fermtouns. (From William Roy, *Military Survey of Scotland 1747–55*)

by the parks of his southern relatives on their Garlies estate north of Newton Stewart.[27] The construction of this Orkney cattle park is a reminder of the strong role that social networks played in the circulation of new ideas and an illustration of why many early Improvements seem like inexplicably random and isolated developments remote from other examples of change. But Moodie and Stewart were also seeking to capitalise on local economic conditions and to become important players in the movement of live-

---

27 Ibid., p. 201; W.P.L. Thomson, *The New History of Orkney*, 3rd edition (Edinburgh, 2008), pp. 336–337. The Park of Cara is shown as a large, rectangular enclosure at the northern point of South Ronaldsay on William Aberdeen's manuscript map 'A Chart of the Orkney Islands' (1769) in the National Library of Scotland https://maps.nls.uk/coasts/chart/829. The park still appears on the map of Orkney in John Thomson's 1822 Atlas of Scotland: https://maps.nls.uk/atlas/thomson/476.html and was still included on Johnston's c.1850 map of the islands: https://maps.nls.uk/counties/rec/7256. That the enclosure does not appear on the 1st edition OS one-inch to the mile Sheet 117 (Hoy), which was surveyed in 1877/8, suggests that at least the c.1850 map simply plagiarised the information recorded on earlier coverage.

Cara, South Ronaldsay, Orkney, looking from the nineteenth-century Grutha Farm across the site of the former cattle park.

stock southwards out of the Orkney Islands. Both men had interests in the Orkney to Caithness trade, Moodie owning a ferry and Stewart having landed and commercial interests on both sides of the Pentland Firth. Moodie's park at Snelsetter and Cantick Head lay just 3km east of Brims, from where his ferry operated, while Stewart's Park of Cara lay only 1km from the port at St Margaret's Hope. Here, it is suggested, Stewart fattened cattle that he had bought from tenant farmers in the district before selling them on to drovers who used the nearby harbour to ship them out.

Parks on similar scale to those in Galloway and Orkney were also built at the southern end of the drove routes from the Highlands, where animals were fattened before being brought to market in Edinburgh or the Clyde burghs. Most of these appear to have been private enterprises carved from estate land rather than from divided commonty, but the legislation that enabled the creation of regular enclosures by exchange of property with adjoining proprietors was a key mechanism in encouraging their creation. One important investor in such cattle enclosures in central Scotland was the Edinburgh lawyer George Smollett, younger son of the Improving baronet Sir James Smollett of Bonhill and himself a member of the Society of Improvers. In the 1730s, he leased the parks he had constructed and seeded in the 1720s on his estate at Ingliston, west of Edinburgh – now the Royal Highland Society showground beside Edinburgh Airport – to drovers bringing stock to the city.[28] He had invested significantly in the improvement of the grazing quality of the land to be rented, buying ryegrass and clover seed in bulk to sow on his property from the prominent merchant-seedsman, Arthur Clephane.[29] Through to the second quarter of the nineteenth century, such parks served as recovery ground on which cattle

28 NRS RH15/31/88, RH15/31/91, RH15/32/136A. The parks are shown clearly on General Roy's map of 1752–55 https://maps.nls.uk/geo/explore/#zoom=14&lat=55.94350&lon=-3.36400&layers=4&b=1.
29 Donnelly, 'Arthur Clephane', p. 156. Smollett is here wrongly labelled as 'a gardener from Ingleston'.

driven on foot from northern and western districts for sale in the major Lowland cattle markets or trysts at Crieff and Falkirk were re-fattened before being brought in to the city slaughterhouses.

Highland cattle and the drove trade

On a grander scale than most of the divided Lowland commonties, shared grazing rights and traditional grazing practices in areas still designated as 'forest', including wood pasture within the forested area, also became subjects of increasingly bitter conflict. In most cases, these arose when some landowners sought to assert exclusive use, as we have already seen in respect of the disputes concerning the Forest of Mar. In west-central Perthshire in the 1730s, for example, the earl of Breadalbane's assertion of exclusive rights to grazing and the introduction of enclosures resulted in a protracted dispute with neighbouring lairds and their tenants. In May 1738, despite a House of Lords judgement in favour of Breadalbane, Menzies of Culdares, Macdonald of Kenknock and their tenants drove their cattle onto land assigned exclusively to the earl's ownership within the Forest of Mamlorn.[30] The correspondence concerning the case reveals a steady escalation in the levels of violence between the herdsmen and Breadalbane's men, with reports of the one being armed with dirks leading to requests to carry guns from the other. The case details also expose the consequences of enclosure and exclusion for the excluded; Alexander Macdonald of Kenknock's cattle were starving as he had depended on access to the common grazing in the forest to sustain his animals when his grass elsewhere was depleted.[31] For Kenknock, the choices were stark but involved mainly intensifying grazing on poorer-quality land or on summer pastures outside the traditional spring grazing areas, which threatened degradation of grassland where growth was at best just resuming after the winter. Breadalbane, however, was seeking to improve the quality of the grazing from which he was excluding others. His aim was to lease this to drovers who were assembling herds for driving south and east through his lands heading to the central Lowland markets.

Smollett of Ingliston's parks and Breadalbane's drive for exclusive possession of pasture resources in the west-central Highlands were both responses to the increasing numbers of cattle within the north and west Highlands and the steadily rising demand for beef in external markets. The Campbells of Glenorchy, the earl of Breadalbane's ancestors, had been marketising their estate's cattle-based economy since the end of the sixteenth century and had reorganised its internal structures in the seventeenth to enable the movement of herds between grazing areas under their control.[32] The scale of their landholding made this development possible and the size of their herds gave them considerable leverage in direct business with the drovers and cattle-dealers who completed the supply chain to the Lowland and English markets. Similar trends were evident in large landholdings like those of the Mackenzies of Seaforth, extending from Easter Ross to Lewis, where by the mid seventeenth century the earl of Seaforth was able to instruct the assembly of 300 cattle annually for five years at Stornoway to satisfy a contract with an Edinburgh dealer.[33] Just as in the Middle Ages the movement of cattle on this kind of scale had meant negotiation of arrangements for watering and

---

30 J. Menzies, *Answers for James Menzies of Culdares, and Angus Macdonald of Kenknock, to the petition of John Earl of Breadalbane* (Edinburgh, 1738).
31 For the progress of the dispute between May 1738 and July 1739, see NRS GD112/59/20.
32 Dodgshon, *From Chiefs to Landlords*, p. 113.
33 NRS GD201/1/54.

Inveroran, Loch Tulla and the Bridge of Orchy hills, Argyll and Bute. Inveroran, at the west end of the loch (centre right), lies in an area of rich pasture on the alluvium washed down from the surrounding hills, at the point where routes from Glen Kinglass and Glen Etive meet the Military Road from Fort William and the roads leading to Glasgow, Stirling and Perth.

overnight grazing en route to their destination, so the increasing commercialisation of this trade in the later seventeenth and eighteenth centuries created profitable new opportunities to supply these needs in return for cash payment for landowners along the major droving routes.[34] Breadalbane was permitting the establishment of extensive so-called droving stances, where the herds could be pastured overnight, at various stages across his estate – one of the most important being at Inveroran – and the securing of exclusive grazing rights in areas where drove routes converged can be seen as a response to this market opportunity.

As livestock numbers grew in the late seventeenth and eighteenth centuries, with the southern and central Highland Breadalbane and Atholl estates focused on sheep as much as cattle, grazing pressures increased and greater control over and regulation of pasture is evident in records and literature. The Gaelic poet Duncan Ban Macintyre, for example, in a lament for the spread of sheep-farming, spoke of the grassy glens that had once been grazed by cattle and supported a dairying culture, unconsciously recording the already transformed landscape of his youth; that landscape had been the result of intensified pasturing and was then undergoing a transition into something emptier but even more intensively grazed under the new sheep-dominated regime.[35]

---

34 A.R.B. Haldane, *The Drove Roads of Scotland*, new edition (Edinburgh, 2021), p. 52.
35 Duncan Ban Macintyre, 'A Song of Foxes' (trans. John Stuart Blackie, in Dunlop and Kamm (eds), *Scottish Collection of Verse*, pp. 83–85.

This trend may be one of the key factors behind significant changes that occurred in the use, location and oversight of shieling grounds; tenants saw access for their animals to some traditional summer pasture reduced as landlords sought to preserve the grass for their own or drovers' cattle.[36] But potential profit from pasturing was also attractive to tenants and it was evident from the 1720s that some were renting their grazing to 'strangers', thereby adding to pressure on remaining expanses of common pasture. It has been suggested that this activity was one of the drivers behind the progressive subdivision of some of the lower-level hill-grazing into still large blocks defined by new, mainly drystone dykes from the 1750s onwards and the gradual shift away by the 1760s from common herds and flocks grazed for dairying purposes on the higher-level pastures. One Breadalbane innovation of the 1680s, the requirement for shieling sites to be relocated every five years, since this caused 'a great deall of good to the tennents and causes more grass to grow', had perhaps stimulated this trend and added to the pressure, for it has been noted on the Atholl estate that quinquennial re-siting was often a first step towards the breaking of pasture into arable cultivation.[37] On these estates, it can be seen that the progressive reconfiguration of their economies to capitalise on the expanding commercial opportunities in the supply of meat, wool and leather to Lowland and external processors and consumers had had a cumulative effect in bringing about a reorganisation of land-use and more rigorous exploitation of the natural grass resources. In Atholl, Breadalbane and across much of the southern and south-western Highlands, it was this shift that laid down the grid of stone dyke-defined land-division and mode of exploitation that prevailed into the mid nineteenth century. It was a phenomenon whose physical legacy is still to be seen in the compartmentalisation of the modern landscape.

New seeds, new plants

Lowland grassland, especially at first within the great landowners' cattle parks but progressively extending over tenanted pasture, experienced one of the most profound but, in popular perceptions, least recognised changes of this era. Sowing of better-quality fodder grasses, clover and legumes like lucerne and various pea varieties – which also had benefits as nitrogen-fixers that helped to boost soil-fertility and productivity – was a central element in Improvement practices in England in the seventeenth century and spread into Scotland with those ideas. Research into the take-up of grass-sowing remains limited, with much of what has been published focusing on the few well-documented cases like John Cockburn of Ormiston on his East Lothian estate in the 1730s or George Dundas of Dundas in West Lothian in the 1740s, and suggesting that the rapid expansion of the practice occurred only after c.1760. It seems, however, already to have been current in Lothian and within the network of Improvement-minded landowners in the first quarter of the century.[38] George Smollett at Ingliston, for example, was sowing grass and clover provided by the seedsman Arthur Clephane by 1723 and it is likely that his father and brother were likewise active on the family's Dunbartonshire estates,

---

36 J. Atkinson, 'Ben Lawers: an archaeological landscape in time. Results from the Ben Lawers Historic Landscape Project, 1996–2005', *Scottish Archaeological Internet Reports* 62 (2016), pp. 211–212.
https://archaeologydataservice.ac.uk/library/browse/issue.xhtml?recordId=1137495&recordType=MonographSeries.
37 Quoted ibid., p. 211.
38 Whyte, *Scotland Before the Industrial Revolution*, pp. 146–147; Devine, *Transformation of Rural Scotland*, pp. 54–55. For Dundas, see I.D. Whyte, 'George Dundas of Dundas: the context of an early eighteenth-century Scottish improving landowner', *Scottish Historical Review*, 60 (1981), pp. 1–13.

(a) Red clover (*Trifolium pratense*) and (b) white clover (*Trifolium repens*) were among the plant species most favoured by the Improvers as both a fodder crop and a mechanism for soil-enrichment. Clover's prolific sowing around the country in the eighteenth and nineteenth centuries has made it ubiquitous in what were managed pastures.

where droving routes converged at the south end of Loch Lomond.[39] Clephane's accounts, however, suggest that demand for grass and clover seed remained low among his circle of buyers, mainly from those seeking to specialise in provision for cattle, until well into the second quarter of the century.[40] Orders for red and white clover seed grew with the spread of Improvement practice to the management of grassland, and unquestionably accelerated after 1750, with landowners from the Hebrides and Orkney to Lothian and throughout the Highlands seeking the manifold advertised benefits of introduced grass, clover and legume species. These well-documented benefits ranged from increased milk yields and beef weight-gain, or abundant winter feed, to increased soil fertility for subsequent cultivation, as the growing number of bulk purchases of grass and clover seed attest.[41]

39 NRS RH15/32/136A.
40 His purchasers included Robert Hay of Naughton in Fife (1718), Lord Dalmeny for Dalmeny in West Lothian (1723), and George Aytoun Douglas of Strathendry in Fife (1732), suggesting that the small demand was perhaps more a reflection of his narrow business network than limited interest among Scottish buyers: see NRS RH15/32/77, RH15/32/96 and RH15/32/131. In the 1730s, the earl of Rosslyn, for his lands around Dysart in Fife, was importing clover seed from Rotterdam: NRS GD164/621 (25) and GD164/628 (17), while in 1740 and 1741, Sir Robert Menzies of Weem was obtaining clover seed in Perth and the earl of Breadalbane was importing direct from Rotterdam: GD1/369/137 and GD112/15/275 (11).
41 For an illustration of such improvement on Burray in Orkney from the late 1760s, see Thomson, *History of Orkney*, pp. 202–203.

Ryegrass (*Lolium perenne*) was sown in abundance on pasture across Scotland, favoured both as an excellent hay crop and as a nutritious fodder for grazing animals. It quickly established itself as a dominant grass across much of Lowland Scotland.

Former outfield areas, previously cropped only occasionally and usually after the folding of livestock there to give a nutrient boost prior to ploughing, were prime land upon which to experiment with grass, rapeseed, novel legumes like sainfoin and lucerne, and clover, the latter also contributing to enriched soil fertility for future cultivation.[42] Estate and business accounts show a well-established commercial trade in grass seed in existence by the middle of the century. The Grants of Castle Grant, for example, used their connections and access to credit arrangements to buy grass and clover seed in bulk from long-established Edinburgh merchants, nursery and seedsmen by the mid 1760s. Smaller landowners like Alexander Anderson of Strichen in north-east Aberdeenshire, who lacked such network links and financial flexibility, ordered smaller quantities of seed more locally but at higher prices from the Aberdeen merchant James Elder.[43] By the 1780s, experimentation with ryegrass – the most common species used in the eighteenth century and the prevalent lowland grass type today – and other species was widespread.[44] The duke of Gordon purchased ryegrass and other grass seeds in 1782 for sowing on his Speyside properties.[45] In the Borders, the Inneses of Stow trialled various species in 1787, including

---

42 Rapeseed was being planted as a soil-improver on the Donside estate of Sir Archibald Grant of Monymusk by 1726 and sainfoin (St Foin) by 1746: Hamilton, *Monymusk Papers*, pp. 105, 170.
43 NRS GD248/530/1 (8–10), RH15/27/87.
44 The novelty of grass monoculture is perhaps indicated by the naming in 1765 of one of the parks on the Sandilands' estate in West Lothian as the Ryegrass Park: NRS RH15/44/209.
45 NRS GD44/51/358/5.

red, white and yellow clover seeds, ribbed grass and ryegrass. Gilbert Innes instructed his estate officers to try the yellow clover and ribbed grass on different parts of the property, noting that 'as they are not much in use in the country a trial of that kind will more easily ascertain their utility'.[46] Ryegrass was, by then, the tried, tested and proven fodder grass but Improvers were always keen to experiment with new and potentially more productive varieties.

Deliberate cultivation of grass, usually as an almost mono-species crop, as opposed to exploitation of the naturally occurring mixed grass and wildflower herbage, marked the beginning of a profound change in the composition and appearance of lowland grasslands. The creation of these cultivated grasslands also signalled the first steps in widespread, large-scale habitat loss. The inclusion of clover, especially the vigorous red variety, which was widely recognised as beneficial for the fertility of the underlying soils even if its role as a nitrogen-fixer was not discovered scientifically until 1838, was perhaps even more important. Wherever it was sown, its nitrogen-fixing quality affected the soil chemistry over areas and on a scale far beyond any soil-improver employed in the past. It was, however, just one tool in an Improving toolkit that was deployed to transform agricultural productivity and whose evident utility was endorsed by satisfied experimenters the length and breadth of the country.

Lime and marl

Word-of-mouth networks and eyewitness evidence from visits to trial areas ensured that the successes of these new plant types, but also of other Improvement practices like liming, were broadcast nationwide. Among the scientific discoveries of the seventeenth century had been the value of liming: using pulverised burnt lime – either from seashells or limestone – for reducing soil acidity and enhancing nutrient availability to crops. Used on both the regularly cropped infields and the infrequently cropped outfields, it led to notable improvements in crop quality and yields, and delivered a new viability to marginal areas as permanently cultivated fields. That new viability led in turn to a dramatic increase in the extent of land under regular if not continuous tillage in the Lowlands especially by 1790.

Among the many observations of that discerning traveller, Sir William Burrell, was the nature of soil-improvement being practised in the country through which he travelled and the diversity of forms that it took. Of these, liming and marling were becoming common, where the materials were available. Liming had spread as a practice from those areas of Fife and the Lothians where it had already become common in the later seventeenth century into neighbouring districts where water or road transportation allowed or where local limestone sources had been identified. On the agricultural land flanking Flanders Moss in Stirlingshire, for example, liming using materials brought in from sources in eastern Stirlingshire is first recorded in the 1690s, local lime sources were being identified and exploited by 1707, and by the 1720s annual liming was a standard lease condition.[47] Bearing in mind that a very high proportion of the lime for which we have early eighteenth-century documentary evidence was intended for building work rather than agricultural use,[48] records show

---

46 NRS GD113/4/167 (216 and 263).
47 Harrison, *Historical Background of Flanders Moss*, p. 47.
48 As, for example, in the shipping of limestone in 1718 to Barcaldine in Argyll for building operations at the castle there and the burning of lime at Lix in Glendochart in 1726, probably for use at Finlarig: NRS GD112/15/161 (12), GD160/229 (8).

Boddin limekilns, Boddin, Angus. The massive estate limekilns of the Scotts of Dunninald at the northern end of Lunan Bay supplied lime for building and agricultural purposes to most of the surrounding district from the 1690s, but in increasing volume from the 1730s, when the existing complex began to be developed.

that limestone was being imported for lime-making or searched for locally all around Scotland from the 1700s. This traffic established a network of supply and production that was quickly redirected towards providing lime for agricultural use as the demand from Improvers grew. In 1707 in Angus, the earl of Airlie was obtaining limestone shipped to Dundee and carted north from the burgh for burning principally for work at Glamis Castle but also for application to his demesne fields.[49] Supplies to Dundee at that date were already coming largely from the lime-quarrying and processing operations controlled by the Buce family and the earls of Rosslyn on the Fife coast of the Firth of Forth, the former of whom were to dominate the Scottish lime trade in the second half of the century. Further east on the Angus coast, however, limestone was being quarried in the 1690s at Boddin on the estate of the Scotts of Rossie, Dunninald and Usan, with Patrick Scott of Rossie but especially his son Robert Scott of Dunninald (both members of the Society of Improvers) investing heavily in limestone burning for agricultural purposes.[50] Demand burgeoned in the 1720s and 1730s, triggering prospection for exploitable limestone sources and experimentation with new kiln technologies for its more efficient burning. Scott of

---

49 NRS GD16/26/1/206.
50 For Robert Scott's major limekilns at Boddin, see https://canmore.org.uk/site/36303/boddin-point-limekilns.

Dunninald was a leading developer of kiln technology in the later 1730s. Among the thirty-eight queries, responses and advice notes received and given between 1723 and 1743 and published in the 1743 *Transactions of the Honourable the Society of Improvers in the Knowledge of Agriculture in Scotland*, twenty-six were concerned with methods of arable and grassland improvement. Most of these were concerned with soil conditions in accordance with the Society's aim of 'directing the husbandry of the different soils for the most profitable purposes'. Two in particular, one provided by Robert Scott of Dunninald, gave details of the construction of new and more efficient forms of limekilns, fired with coal, and how best to process and apply their product.[51] The majority of the submissions from members indicated that liming was either already practised or soon to be introduced on their land or near where they lived.

By the 1730s, lime was being applied to Improvement operations on acid soils all around Scotland using both bought-in supplies and locally sourced materials. In 1737, for example, the Breadalbane estate contained at least six limekilns, utilising the limestone outcrops that had in previous centuries been exploited for building-lime.[52] One prominent Society of Improvers member with lands distant from the limestone-rich south-east was Sir Archibald Grant of Monymusk in Aberdeenshire, who was promoting its use to his tenants and neighbours in the mid 1730s. Grant made liming a requirement in his tenancy rules in the 1740s and initiated local searches for limestone outcrops to reduce the deterrent impact of the cost involved in transporting it to his inland properties.[53] Alongside such individual actions, however, must be recognised the role of broader government initiatives in promoting the use of lime across large parts of both Lowland and Highland Scotland. Most prominent of these as an influential agency for the introduction of liming from the 1750s was the Forfeited Estates Commission, especially through its actions in respect of the Annexed Estates (those properties forfeited by attainted Jacobites and annexed inalienably to the British crown), among whose members were several influential Improvers like Sir John Clerk of Penicuik. The Annexed Estates Commission was seeking to safeguard the pacification of formerly Jacobite-supporting areas through their economic development, with agriculture regarded as a key lever for achieving these aims. On the forfeited Robertson of Struan estate in western Perthshire in 1769/70, the Commissioners oversaw the establishment of a water-powered limestone crushing mill at Rannoch, in part with an eye to continuing with the fruitless efforts from 1763 to drain and Improve the wetlands of Rannoch Moor (see pp. 167–168 above).[54] Through such developments, limestone quarries, crushing mills and kilns became by 1790 almost ubiquitous features of large estate infrastructures across those areas of mainland Scotland where stone and fuel sources were present or materials easily transportable.[55] We can only guess

---

51 Robert Maxwell of Arkland, *Select Transactions of the Honourable the Society of Improvers in the Knowledge of Agriculture in Scotland, 1743*, reprint (Edinburgh, 2003), nos 29 and 29, pp. 185–195.

52 NRS GD112/15/255 (6).

53 Hamilton, *Monymusk Papers*, pp. 130, 141, 174, 177. Sourcing limestone in the north-eastern Highlands was at first problematic. In the Moray uplands, however, limestone sources described as 'inexhaustible and would supply all Scotland' were being quarried in the 1770s at Glen Rinnes between Dufftown and Aberlour and supplying Improvers as far west as lower Strathdearn and throughout Highland Banffshire: NRS GD44/43/101 (20).

54 NRS E783/93. For farm improvement plans generally, see E777/313 and E783/98.

55 Despite this importance and the ubiquity of agricultural limekilns, these symbols of Improvement receive little notice in most secondary literature on eighteenth-century landscape change.

Site of Restenneth Loch, Angus. The spire of the medieval priory rising over the trees indicates the location of the peninsula that once projected into the loch from its south-west side, before the water was drained and the loch-bed excavated for marl to a depth of 15 metres. The area of reeds, and the willow and alder carr between it and the priory, represent the wetlands that re-formed once marl extraction ended.

at the volume of limestone quarried and introduced to agricultural land throughout Scotland in the course of the eighteenth century. We should not, therefore, underestimate the scale of the change that it had probably begun to wreak even before the yet greater application of lime and other chemical agents to the soil in the nineteenth and twentieth centuries. Contemporaries might not have noticed or remarked upon the changes in the wild flora and the biota which depended on specific plant species as the soil pH moved from acid to alkaline, for many of these were disparaged as unproductive wildflowers or plants dismissed as weeds and pests that had plagued pre-Improvement agriculture. These, however, were among the first unintentional casualties of the drive to escape from the millennia-old knife-edge upon which Scottish subsistence agriculture had teetered.

Lime was the first additive introduced on a large scale, but others followed close behind. The principal of these was marl, a carbonate-rich mineral soil usually derived from the decomposed shells of aquatic molluscs that could be found as a sediment on loch-beds and fen-infilled wetland hollows. Although widespread use of marl was a mid-eighteenth-century development, reference to water-filled former marl pits in a seventeenth-century Orkney ballad suggests that its value was appreciated in some places from much earlier.[56] From the 1750s,

---

56 Thomson, *History of Orkney*, p. 200.

'The face of the country . . . has undergone a very wonderful change'

Where it was available, marl was regarded as a soil-improver of the greatest quality. Its extraction from loch-beds and kettle-holes around the country was undertaken on an almost industrial scale by Improving lairds, who exercised remarkable ingenuity in devising new techniques for maximising the amounts removed from the watery depths and for speeding its carriage to the land that was awaiting its miracle-working addition. The *Statistical Account* entry by the Rev. Mr Halliday of the Parish of Kelton, in which marl-rich Carlingwark Loch was located, provides us with insights on the operation and its effects.

> *Lakes and Marl* – Near the north corner of the parish, there is a lake, commonly called *The Carlingwark Loch*, along the west side of which runs the military road. The extent of this loch, before it was partly drained in the year 1765, was 116 square acres. Ten feet of water were taken off from it by a cut, or canal, to the water of Dee. Now it is only 80 square acres in extent. The loch is one great source of improvement to Kelton and the neighbourhood. It contains in itself, and the mosses adjoining, an inexhaustible fund of the very best shell marl. The marl is taken out of the loch by means of boats and ballast bags, wrought with a wheel; a mode that is pretty expeditious; and it is taken out of the mosses, in the way of throwing, usually practiced in other parts of the country.
>
> *Improvements* – Since the loch was drained, the face of the country, all around, has undergone a very wonderful change in point of improvement. Not only Kelton, but the parishes of Buittle, Crossmichael, Balmaghie, Parton, Balmaclellan and Kells, reap the benefit of the marl, from the Carlingwark Loch and its vicinity. Marl is carried from the Loch in flat bottomed boats, along the canal to the Dee, in large quantities, for the improvement of the lands on each side. It is conveyed up the river, by the means of these flats, as far as New Galloway, to the distance of 15 to 16 miles. Before the late improvements in husbandry, the crops in these places were, in general, very light; and the grain, in quantity and quality, inferior, by far, to what is now
>
> (Rev. Mr Thomas Halliday, 'Kelton, County of Kirkcudbright', *Old Statistical Account*, Volume VIII (1793), pp. 303–304).

Halliday's account gives us great insight on the geographical range over which the marl was transported and applied, on the transformation of the water bodies and wetland areas from which it was being extracted, and the soil-enrichment that it delivered to land being brought into intensive arable cropping across this area. Although he does not mention it, the areas of drained loch-bed at Carlingwark and in the area between the loch and the Dee which the Carlingwark Lane or canal helped to drain, were also brought into cultivation or developed as pasture. His claims of the marl's 'inexhaustible' abundance were never tested, for although it was still being extracted in the 1840s new forms of fertiliser, especially guano imported from southern Africa and South America, replaced it as a cheap, easily transportable and readily available commodity.

Detail from John Thomson's 1821 map of Kirkcudbrightshire, showing Carlingwark Loch and the canal or 'Carlingwark Lane' leading from the loch to the River Dee, along which barges carried the marl to areas of Improved cultivation upstream. (National Library of Scotland)

Map excerpt — place names listed roughly top-to-bottom, left-to-right:

DaneVale P. · Hardyard · Loch
Oldcroft · Highpark · Ernminnie · Whiteneuk · Gerranton
Camp · Midtown · Woodside · Pyetthorn · Castlehill · Greenhart
Douglas · Glentocher Br. · Greenlaw · Hillotown
Creoch · Supd Seat of an · Chapmanton · Blackadams · Blackera
Boat Ho. · Abby Lt Mains · Mavisbank · Redbrae
Boreland · Wheatcroft · Drap · Boghead
Old Greenlaw · Dungary · Pirnespie
School Ho. · Springfield · Townhead · Millda
Blackbridge · Goose Black pk. · Wheatcroft · Burnside
Thrave Grange · Isles · Burgh of Cast. Torrs · Leathe
board · Castle · Fuffock · Douglass Torrs
Evergreen · Canal · Isle Whitepark · Loch
Barmboard · Nr. Hall · Kelton Ho. · Black
Hill · Carlingwark Inn · Ih · Crogo Gorri
White Hill · Kelton Lodge Hitae · Floors · Black
Long Is. · Kelton Outr · Lochbar · Doonend · Craig
Hardhill · Middtown · Newark
Bridge end · Lamb Isd · Hillhead · Kirkland · Cool
Keltonhill · Keltonhill · Manse
Grany · Granyford · Kirk · Braoch
Bridge · Rhonhouse · Kirkland · Cootraw
Asterown · Slack · MidKelton · Burnstick
Birkhill · Crosthill · Hallmyre
Brownhill · Auchlane Mill · Gelston Mill · Castlegower
Dilldown · Craigley · Supd a
Poland · Bridge · Slagnar · Earl
sh · Auchlane · White Cairn
holes · Low Arkland · Gelston · Doach
D Bank · Howyard · Ingleston
crosh · High Arkland · Airylind · KELTON
Billie's · Bridge · Keltonbush · Gelston Kirk in ruins · Glen
Maxfield · end · Lit Lochdaugan · Dongayle
ormock · M Lochdaugan · Benkreek · Killdow · Strand
Millthird Supd a Lead Mine · Green Lane · Whitehill · Hol
Hillhead · Moat · Gels
Third · Castle
amp · Barr · P^H · Linkens
Hill

discussion of the efficacy of marl and where to source it increased and estate records begin to record the presence of marl pits as well as limestone quarries or limekilns. At Kirkhill farm in Dalton parish in south-western Dumfriesshire, a region of small kettle-hole lochs and fenny hollows, a marl pit was part of the Improved farm's infrastructure by c.1756, while at the opposite end of the country on the low-lying properties of the earl of Moray's Darnaway estate on the gravel terraces of the River Findhorn west of Forres, marl pits and limekilns were recorded on a 1760 estate map showing the extent of Improvement.[57] Since marl recovered from watery contexts did not require burning prior to application, it was sought as a substitute for limestone, especially in districts where fuel sources were not abundant or where, as Scott of Dunninald found, the duty on coal made kilns expensive to operate. In Angus, the geology that rendered it devoid of coal seams had created landforms where marl-rich lochans proliferated and these were being drained and dug by the 1760s.[58] The impact of marl extraction is another little-appreciated aspect of the environmental consequences of Improvement agriculture but, as can be seen at Restenneth in Angus, the process was utterly transformative of local land and water features. There, the loch that had once enclosed the medieval priory's peninsular site was drained and then its bed was excavated to a depth of 15 metres below its former surface under the direction of the Improving laird George Dempster of Dunnichen. The effect was devastating. Restenneth's once near-island site is now a hilltop above a deep, scrub-filled marshy hollow, an abandoned anthropogenic landscape which has developed as a species-rich habitat in its own right. Elsewhere, kettle-hole lochs were deepened by dredging of marl from their beds, as occurred at Carlingwark beside Castle Douglas in Galloway, where in the 1760s Sir Alexander Gordon of Culvennan began operations that included the cutting of a canal or 'lane' from the loch to the River Dee 2.4km distant. The lane was used for barges to transport loads of marl from the loch to Gordon's farmland up to 21km upstream.[59] Physical landform and water systems were thus modified and 'Improved' to meet the demands of the new agriculture.

Sea-ware, midden waste and animal dung

New forms of fertiliser did not mean that traditional, organic methods were abandoned. Some, indeed, saw expansion in their use. Sea-ware, the wave-dislodged fronds of seaweed washed up on shores, had been used for centuries in parts of the Northern and Western Isles, as illustrated by Martin Martin's early 1700s account. It also, however, formed a major component in soil-enrichment practices around the mainland coasts, where it remained the principal non-chemical fertiliser into the twentieth century. Although sea-ware use has been identified in palaeoenvironmental analyses of anthrosols datable to the twelfth century in Orkney, as we have already seen it was only when landowners began to appreciate its value in the late fifteenth century that references to its management and use start to appear in documentary sources.[60] By the mid sixteenth century its application was recorded in almost all coastal

---

57 NRS RHP1743, RHP3371.
58 See, for example, the maps of the Earl of Panmure's estates in the parishes of Edzell and Lethnot in north Angus, where marl was being dug out of former wetland or fenny areas on the gravel beds of north-eastern Strathmore: NRS RHP1665/10, RHP1666/4, RHP1666/6, RHP1666/7.
59 https://canmore.org.uk/site/217628/carlingwark-lane-canal.
60 Oram, 'From "Golden Age" to Depression', pp. 209–211; Davidson and Simpson, 'Soils and landscape history', pp. 69–71; Simpson, 'Relict properties of anthropogenic deep top soils'.

Sea-ware, Birsay Bay, Mainland Orkney. The accumulations of seaweed on the shores of the islands have provided a rich source of fertiliser and, when dried, low-calorific fuel for millennia.

districts, and its efficacy was still being lauded and its use encouraged in the *Transactions* of the Society of Improvers in the 1740s.[61] During William Burrell's 1758 'tour', he noted that most coastal communities in the parts of the country he visited from Galloway northwards to the upper tidal reach of the Firth of Clyde, on the coast of Lorn, around Banff, and in Fife, used seaweed in their mix of soil fertilisers and dressings.[62] Sea-ware's importance as an abundantly available fertiliser led to competition and dispute over locations where it accumulated or where still-growing seaweed could be cut, as occurred on the duke of Argyll's Mull and Morvern properties, some accounts revealing also recognition of a risk of over-exploitation.[63] Although rich in nitrogen and potassium, seaweed is deficient in calcium and phosphates, which perhaps explains the 1775 observation concerning the condition of Orcadian 'black earth' anthrosols which, despite annual application of sea-ware, were said to lack volume and have a shallow

---

61 Fenton, 'Seaweed as fertiliser'; Smout and Stewart, *Firth of Forth*, pp. 18–23; Maxwell, *Select Transactions*, 'Number XIV, Scotstarvat's Letter with Respect to Sea-ware', pp. 114–117. For a discussion of the efficacy of sea-ware in Mull in the 1790s, see Cregeen, *Argyll Estate Instructions*, pp. 189–190.
62 Dunbar, *Burrell's Northern Tour*, pp. 25, 27, 29, 53, 64.
63 Cregeen, *Argyll Estate Instructions*, pp. 101, 122, 135, 182. Erskine Beveridge recorded the methods of harvesting and transporting sea-ware to the cultivated ground still being used at the end of the nineteenth century in North Uist and its adjacent islands: Beveridge, *North Uist*, pp. 325–326.

Greenock, West Renfrew. The farmers of Dunbartonshire, across the estuary to the north-east from Greenock, were ready customers for the waste from the port's thriving herring fishery. The dung and midden waste of the Clyde fishing ports was considered superior to other town dungs, its quality being described as greatly improved on account of the waste materials from the herring-gutting yards, listed as 'refuse salt, putrid fish, blood, and other animal substances' (A. Whyte and D. Macfarlan, *General View of the agriculture of the County of Dumbarton* (Glasgow, 1811), pp. 197–198).

tilth that produced poor, thin crops.[64] Sea-ware alone, therefore, was not a 'silver bullet' for fertiliser needs and long experience by farmers saw it being mixed with other materials to secure optimal growing conditions.

Manuring using the dung of livestock, human faeces, domestic and farmyard midden and shambles waste was ancient practice, but in the eighteenth century the ingredients, accumulation, storage, transport and application of manure became the almost fetishised subjects of extensive Improvement discourse.[65] Archibald Grant of Monymusk's farm regulations of *c.*1744, for example, outline the recommended constituents and procedures for developing

---

64 J. Fea, *The Present State of the Orkney Islands considered* (Edinburgh, 1884), p. 37. This deficiency may, in part, have been addressed through mixing lime with the seaweed to provide calcium and release other nutrients: Smout and Stewart, *Firth of Forth*, p. 19.

65 J. Shaw, 'Manuring and fertilising the Lowlands 1650–1850', in T.C. Smout and S. Foster (eds), *The History of Soils and Field Systems* (Aberdeen: Scottish Cultural Press, 1994), pp. 111–118; D. Woodward, '"Gooding the earth": manuring practices in Britain 1500–1800', in T.C. Smout and S. Foster (eds), *The History of Soils and Field Systems* (Aberdeen, 1994), pp. 101–111; Oram, 'Waste management and peri-urban agriculture'.

the best manure for different soils, including the recycling of old farm-buildings, which were primarily of turf, wattle and heather-and-straw thatch construction.[66] Scotland's expanding towns were looked to by their landward farming neighbours as important sources of manure, as fewer burgesses were actively cultivating their own ground and so no longer wished to preserve their own middens for use as fertiliser.[67] As early as 1715 at Elgin, council statutes required townsfolk to clear middens weekly and the council assumed responsibility for clearing street dung, all of which was then sold for agricultural use to farmers in the district around the burgh.[68] Street-cleansing and midden removal became a commercial proposition, rights being auctioned at roup and the successful bidder organising weekly collections for accumulation at private midden-heaps outside the town and then making a profit from the on-sale of the waste to farmers.[69] There was particular demand for the dung and waste from fishing communities like Newhaven on the Forth and Greenock on the Clyde, whose fish-processing waste was deemed especially efficacious as a fertiliser.[70] Through the dispersal of their waste over quite extensive areas – Perth, for example, was supplying manure within a zone radiating some 16km from the town – urban communities were significant contributors to the spread of Improvement agriculture in their immediate hinterlands. And burgh authorities were well aware of the growing demand for manure and other soil-improvers and the potential profit to be derived from it; at Stirling, whose waste was already being traded to farmers 5km distant, the provost was engaged in discussion with Lord Elgin over the commercial supply of lime from the latter's kilns on the Forth estuary through the burgh's market to these same customers.[71]

Reshaping the land

How significant were these developments in the period before 1790? Although the significance and impact of the Society of Improvers, its membership and the propositions floated in the *Transactions* have been questioned, what is at once striking from the answers offered in its pages is the geographical distribution of the landowners seeking advice and the diversity of the land and soil types with which they were concerned. The queries received and answered by members ranged over coastal wetland, lowland bogs, wet and dry upland areas, clay, gravel and sand, underscoring the ambitions of would-be Improvers to convert low-yield and marginal land to productivity and profit. The *Transactions* editor, Robert Maxwell of Arkland, was especially fulsome in his praises of Archibald Campbell, earl of Ilay, singling out his experimentation with mossland which was hitherto 'almost everywhere in this kingdom neglected' and which Ilay saw as a most suitable soil for improvement for growing everything from grain to trees.[72] The selected material in the 1743 *Transactions* reveal that all land types from floodplain gravels and seasonal water-meadows to upland moss and whin-covered moorland were being seen as ripe for Improvement for arable purposes. From Sir George

---

66 Hamilton, *Monymusk Papers*, p. 141.
67 Oram, 'Waste management and peri-urban agriculture', pp. 13–14.
68 Cramond, *Records of Elgin*, vol. 1, pp. 390, 469.
69 This arrangement was in place at Perth by 1738: Perth and Kinross Archive, B59/25/3/39. It continued at several Scottish burghs into the mid nineteenth century: Oram, 'Waste management and peri-urban agriculture', pp. 14–15.
70 Oram, 'Waste management and peri-urban agriculture', pp. 14–15.
71 Ibid., p. 14; NRS E727/10.
72 Maxwell, *Select Transactions*, p. vi.

Dunbar of Mochrum's enquiries about improvements to the arable potential of grassland on his Wigtownshire estates, to Lord Rollo's drainage of a peat moss in lower Strathearn, or Lord Reay's efforts to improve the cultivable ground around his seat in north-west Sutherland, the questions and the answers given show an active network that reached into every corner of mainland Scotland and which was already engaged in applying experimental methods on varying scales from single meadows or fields to whole landholdings. Although the scale of investment might yet have been limited, the correspondents leave little doubt that Improvement was already a nationwide endeavour in the second quarter of the eighteenth century.

As Maxwell's praise of Lord Ilay's efforts implies, peatland was certainly at the forefront of much Improvement activity before 1750. Much of the encroachment on mossland recorded in examples like those at Elgin, discussed above, was small-scale and for domestic or subsistence agriculture, but larger initiatives directed by proprietors of significant mossland areas were making substantial inroads on lowland and coastal mosses. In Benderloch, Sir William Burrell clearly had a detailed discussion with his host, Campbell of Airds, about improvement of peatland locally, where shell sand was used to improve the peat either to support good-quality pasture or, if ploughed into the moss in bulk, to create rich arable. The caveat in this improvement was that the land needed first to be drained, and Burrell reported how the gentlemen of the district were draining 'unfathomable bogs' and modifying the soil so that in the space of three years it supported cereal crops.[73] Lord Rollo's previously mentioned drainage of the moss on his Duncrub estate in south-ern Perthshire is typical of such actions. It involved the cutting of ditches to enclose an area approximately 1.6km in circumference, with further channels cut through it to drain the interior.[74] Until Rollo had started his work, the moss, which was described as between 2.75m and 3.7m deep, supplied local fuel needs and his plan was to sell off the remaining fuel-quality peat to defray the cost of his Improvement work. The Society's answer, however, was for him not to remove the peat but to plough the drier and less compact upper 1m or so, burn half of that to produce what was believed to be mineral-rich ash, and then plough that into the remaining area before sowing it with black oats or grass and clover. For the next century across most peatland areas of Britain, this process – described as 'paring and burning' – became one of the principal means of removing the unwanted peat, destroying with it the surface vegetation and creating an ash that was then ploughed into the sub-peat soil as fertiliser.[75]

Far larger mossland transformations than those undertaken by Lord Rollo were the ambition of the Society's leading members, who saw such land as representative of the underdevelopment and general economic backwardness of Scotland. Unlocking the arable potential of the soil beneath the peat and, indeed, of the peat itself, was seen as affording the opportunity to transform the economic condition of the nation. By greatly expanding the area under cultivation, their belief was that they could end the threat of a general failure of cereal crops such as had been experienced as recently as the late 1730s. Perhaps the grandest such scheme proposed, but thankfully only ever partly implemented, was the draining of the vast Lochar Moss in south-western Dumfriesshire, where the peat was recognised as an

---

73 Dunbar, *Burrell's Northern Tour*, p. 53.
74 Maxwell, *Select Transactions*, pp. 46–47, with answer following.
75 Woodward, 'Gooding the earth', p. 105. On the Ogilvy of Findlater estates in Banffshire, a paring plough designed to strip the peat layer preparatory to burning was purchased in 1769/70: NRS GD248/680/2.

Blair Drummond Moss, Stirling, looking east towards Stirling. The level fields that extend up the broad valley of the Forth to the west of Stirling were formed by the clearance of up to 5 metres depth of peat that had grown over the blue clays of the former shallow marine gulf that stretched from the Firth of Forth to the Trossachs, in operations directed by Lord Kames and his son-in-law in the second half of the eighteenth century.

infilling of what had once been an arm of the sea and where it was believed that fertile marine sediments were locked beneath the deep mossy overburden.[76] The Society had hoped that drainage and Improvement here would be carried out under the direction of the largest proprietor, the duke of Queensberry; it was instead smaller proprietors, acting as resources allowed, who carried out most of the operation piecemeal across the next century. This incremental erosion of the wetland area was also what occurred on the mosses flanking the Forth west of Stirling. Here, however, the process was ultimately on an altogether different level and, contrary to popular traditions of clearance operations attributed to Lord Kames starting in 1766 but undertaken chiefly by his son-in-law after 1790, also built on smaller-scale removal of peat that had commenced long before 1700.[77]

Physical and chemical change to cultivated land was just one side of the coin; accompanying such changes was the introduction of new crops, both improved strains of cereals and new types of root crops and legumes as both animal fodder crops and human food. Prominent among these were potatoes and turnips, which were planted and sown in increasing numbers from the first quarter of the eighteenth century, as seedsmen's stocklists, bills and receipts indicate. In Arthur Clephane's stock c.1710 there

---

76 Maxwell, *Select Transactions*, pp. 63–72. For a summary of the evolution of Lochar Moss, see I.A. Morrison, 'Galloway: locality and landscape evolution', in R.D. Oram and G.P. Stell (eds), *Galloway: Land and Lordship* (Edinburgh, 1991), pp. 11–16 at 10–11.
77 Harrison, *Historical Background of Flanders Moss*; Smout and Stewart, *Firth of Forth*, pp. 127–128.

were parsnips and what were labelled as 'Best' and 'Yellow' turnips,[78] the latter possibly swedes, but over the following two decades he seems to have specialised more in garden flowers and vegetables, trees and herbs. Interestingly, turnips were not listed among the garden vegetables, probably reflecting their use from the outset as winter feed for livestock rather than for human consumption. This distinction may be reflected in the purchase from him in March 1721 of half a pound of turnip seed by the laird of Pitmilly in north-east Fife, too much for a domestic garden but sufficient for a field trial.[79] Burrell in 1758 noted the by-then still limited take-up of large-scale turnip growing, seeing his first fields planted with the crop between Paisley and Glasgow,[80] and observing elsewhere that turnips could be sown to improve the soil and increase fodder opportunities. There was, however, clearly mounting interest in the potential of the yellow turnips, with clearing and improving lairds like Henry Home, Lord Kames, engaged in correspondence with Thomas Boyes in Edinburgh in 1762 over the success of turnip crops as an experiment in the improvement of 'muir ground'.[81] By the 1780s, turnips – most likely swedes – were being grown at scale as winter fodder across most districts of the country where Improvement was progressing. Cereal crops and experimentation in varieties lagged still further behind. Although a shift from cultivation of poorer varieties of oats to bere and to more valuable wheat is discernible in many areas, especially once enhanced soil-enrichment methods had been adopted and the soil pH changed to favour growth of these species, the biggest transition in this regard still lay far in the future. Here, though, cultural and dietary preferences as much as environmental conditions might have contributed to the slow uptake in planting of what were exotic novelties in the eyes of the majority of the Scottish population.

Among the newer crops, the slow transition from exotic novelty to domestic commonality is best illustrated by potatoes, which even in 1790 were yet to gain the dominance and dietary staple importance that they held just fifty years later. Early eighteenth-century references to potatoes suggest that they were still grown mainly as culinary novelties in aristocratic kitchen gardens but, from the 1720s, they were also becoming established in the repertoire of lowlier kailyards.[82] That they were already a significant component of some Hebridean diets in the first quarter of the century is indicated by their inclusion within a 1722 rental from Tiree, which suggests that they were being grown on a scale sufficient to attract the notice of the laird, and by 1743 they were of such importance as to have regulations concerning their planting proclaimed to tenants on the Campbell lands on Jura.[83] In 1735, Sir Archibald Grant of Monymusk was lamenting that planting of potatoes was 'too much neglected' by tenants on his land and by other proprietors.[84] By 1744, he had introduced regulations requiring the planting of potatoes on wetter areas of 'dead rigg' ground improved with manuring, with the intention that this would prepare that land for

---

78 Donnelly, 'Arthur Clephane', p. 154.
79 NRS RH15/32/127. Garden turnips at this time seem mainly to have been the white-fleshed varieties, referred to as 'Dutch': see, for example, the seed purchases for the earl of Breadalbane and Innes of Stow in 1737 and 1738 NRS GD112/15/22 (56) and GD113/5/161e (20 and 21).
80 Dunbar, *Burrell's Northern Tour*, p. 29.
81 NRS GD24/1/553 ff. 292–293.
82 NRS GD44/51/739 (50–52), GD112/29/66, GD190/3/285, GD220/6/1624 (12); Dodgshon, *From Chiefs to Landlords*, p. 165. For early potato recipes, see NRS GD248/238/5 (14), for baked potato pudding and GD132/871 for 'white potatoes'.
83 Dodgshon, *From Chiefs to Landlords*, p. 165; NRS GD64/5/17.
84 Hamilton, *Monymusk Papers*, p. 130.

Dalmore, Glen Almond, Perth and Kinross. The enclosure on the north side of the late eighteenth-century single-tenancy farm formed the 'kailyard' and potato ground for the household, separated from the adjacent arable fields by an enclosing dyke to protect the vegetables it contained from straying cattle.

future planting with bere.[85] As a draft letter of 31 December 1754 from Sir Archibald to the editor of the *Aberdeen Intelligencer* reveals, however, although potatoes carried good prices at market, they were seen by him largely as a fodder crop for cattle and a supplement – along with that other new fodder crop, turnips – for the diets of his poorer tenants and crofters.[86]

It is in the post-1750 period that we can see most evidence for the rapid spread of potato-planting across central and western Highland estates from Breadalbane to Barrisdale and Coigach but perhaps still largely as a kailyard rather than a field crop.[87] Slower introduction occurred in Orkney, where they seem to have been an introduction of the middle years of the century, with increased planting after the harvest failures of the 1750s, but with widely variable take-up in different islands until after c.1780.[88] The heaviest dependence on potatoes for human subsistence, however, seems to have been in the Outer Hebrides, where a transition from oats and bere to potato cultivation occurred across the second half of the century.[89] As in western Ireland, this increased reliance on a single crop was to threaten

85 Ibid., p. 141
86 Ibid., p. 159.
87 Dodgshon, *From Chiefs to Landlords*, pp. 185, 187–190.
88 Thomson, *History of Orkney*, p. 200.
89 Dodgshon, *From Chiefs to Landlords*, pp. 192, 195.

catastrophe when environmental conditions proved perfect for the epidemic spread of the *Phytophthora infestans* (potato blight) pathogen in the mid 1840s (see below, pp. 307–309). In the later 1700s, however, the potato was a saviour that allowed communities to escape their historic dependence on more weather-reliant cereal crops and weather-dependent harvests,[90] albeit with the consequence that population levels in some of the most marginal arable districts, buoyed artificially by the new staple's availability, began to climb to levels that were unsustainable in the longer term and introduced new stresses to an already pressurised environment.

Fishing

One naturally occurring environmental resource that was most commonly held up as a panacea for the ills of the precarious rural economies of much of Atlantic-facing Scotland was the sea-fishery. Published at the beginning of the eighteenth century, Martin Martin's confident assertion of the unrealised richness of the Hebridean fishery rested on firmer foundations than his claims for the islands' arable potential.[91] Foreign fishers had long visited those waters and, despite brief episodes of indigenous large-scale exploitation in the later fifteenth, sixteenth and seventeenth centuries, they maintained that dominance almost to the close of the eighteenth century. Martin's was just one voice among those whom modern research has styled 'piscatorial enthusiasts' or 'piscatorial optimists',[92] who believed that the off-shore fisheries far more than agricultural Improvement represented the greatest potential source for the economic betterment of the nation. Intermingled with such opinion was a further and more overtly political view that saw in the development of an economically powerful west-coast fishery that delivered prosperity and employment locally a 'silver bullet' effective against the powerful hold of Jacobitism among the Highland and Hebridean clans. Among men of that mindset, it was also seen as a potentially vital contributor of finance and skilled manpower to the continuing growth of Britain's emergent maritime empire.[93]

In 1727, the newly established Board of Trustees for Fisheries, Manufactures and Improvements in Scotland, which was charged with developing an economy that was complementary to and not competitive with England's, started a long, intermittent and ultimately largely unsuccessful effort to stimulate development of a large-scale fishery and fish-trade to rival that of the Dutch. Despite its title, however, the efforts of the Board were weighted heavily towards the promotion of a Scottish flax-growing and linen industry and trade rather than a fishery.[94] Half a century later, when the earl of Buchan was beginning his lobbying for production of the survey of the agricultural and industrial state

---

90 The devastation of the potato crop in the severe frosts of September 1782, however, revealed to many mainland proprietors and their tenants that even this wonderful root-crop was not impervious to extreme conditions. That year's potato failure was a significant contributor to the near-famine of 1783, which required the Forfeited Estates commissioners to authorise the purchase of emergency food supplies for tenants on the properties they administered. The petition of March 1783 from the tenants of Sligarow on the Robertson of Struan estate illustrates the distress: A.H. Millar (ed.), *A Selection of Scottish Forfeited Estate Papers 1715; 1745* (Edinburgh, 1909), p. 249.
91 Martin, *Description of the Western Islands*, pp. 352–354, 357–359.
92 T.C. Smout, *Scottish Trade on the Eve of Union* (Edinburgh, 1963), p. 220; R. Harris, 'Scotland's herring fisheries and the prosperity of the nation, *c*.1660–1760', *Scottish Historical Review*, 79:1 (2007), pp. 39–60.
93 Harris, 'Scotland's herring fisheries', p. 40; J.R. Coull, 'The herring fishery', in J.R. Coull, A. Fenton and K. Veitch (eds), *A Compendium of Scottish Ethnology*, volume 4: *Boats, Fishing and the Sea* (Edinburgh, 2008), pp. 208–235 at 210–211.
94 A.J. Durie, 'The markets for Scottish linen, 1730–1775', *Scottish Historical Review*, 52:1 (1973), pp. 30–49.

of the nation that Sir John Sinclair would deliver in the *Statistical Account of Scotland*, ardent supporters of investment in fisheries were still advancing the old, ringing claims to encourage the Board of Trustees into new efforts. David Loch, for example, argued that the sea-fishery was 'on the whole an inexhaustible source of riches to the whole nation',[95] through which Britain could secure its maritime hegemony. Despite further similar and even more impassioned pleas for government investment in a west-coast fishery, however, the position had changed little in terms of coherent policy and intensive capital investment before the end of the century. It was principally in the period after 1790 that the British Fisheries Association (founded in 1786) undertook the large-scale investments that were to transform Scotland's herring fishery and lay the foundations of its later nineteenth-century dominance of the European and colonial trade.

It is true that the scale of the sea-fishery was locally significant in terms of numbers of men and women involved and in numbers of fishing-boats, even if only seasonally or more intermittently, but in operational, technological and commercial terms it lagged far behind that of Scotland's foreign competitors. The major concentrations of what could be classed as commercial fishing as opposed to estate or subsistence fishing activity by the first quarter of the eighteenth century were located in the Moray Firth, Firth of Forth and Firth of Clyde, with the last extending into the sea-lochs of Argyll.[96] Although it was claimed to be over 2,000 boats strong at its height in the seventeenth century, the vessels of the largest of the 'herring buss' fleets – the Forth fishery – were a fraction of the size and sophistication of the 500-strong Dutch fleet in its heyday and probably a fraction, too, of the number of smaller, open boats operating in the Clyde fishery.[97] Operations, however, had contracted hugely since the mid seventeenth century, when Fife and Lothian boats had plied their trade in fishing grounds in Orkney (for cod) in March; the central North Sea and outer Firth of Forth (for herring) from June to late July and the latter also in August; the Minch, Sutherland and Wester Ross coasts (for herring) in September to late December; concluding with late-winter fishing for all species in the Firth of Forth through into the following March.[98] Firth of Forth involvement in the Lewis fishery in the Minch had declined in the later seventeenth century and was taken over largely by Clyde and Moray Firth interests. The Forth fishermen, however, seem to have stepped up their North Sea and Firth activities, more than doubling catches traded to Baltic ports by the early 1700s and probably contributing the bulk of the roughly 2,900 tonnes of herring exported from Scotland in 1707.[99] The Clyde operation – centred on Greenock[100] – was particularly important for trade with western England, Ireland and after 1707 with the Caribbean, where processed Scottish herring formed a staple in the diets fed to the enslaved workforces on the plantations, while the merchants behind the Moray Firth fishery displayed remarkable opportunism in seeking out a diverse European trade. For example, Bailie John Steuart oversaw shipment of herring and cod to France, cod to Bilboa in northern Spain and herring to Gdansk

---

95 D. Loch, *Essays on the Trade, Commerce, Manufacture and Fisheries of Scotland*, volume 2 (Edinburgh, 1778), p. 133, cited in Harris, 'Scotland's herring fisheries', p. 39.
96 Harris, 'Scotland's herring fisheries', pp. 41–42. The Forth fishery is explored in detail in Smout and Stewart, *Firth of Forth*, pp. 31–45.
97 Coull, 'Herring fishery', p. 210.
98 Smout and Stewart, *Firth of Forth*, pp. 31–32.
99 Ibid., p. 33.
100 Dunbar, *Burrell's Northern Tour*, p. 29.

St Monans, Fife. The coat-of-arms of the former burgh of barony, as granted in 1962, depicting four fishermen with a draw-net in an open boat, was based on the design of a seventeenth-century burgh seal, which highlighted the community's relationship with the herring fishery in the Forth.

from Inverness in the mid 1710s.[101] The cod fishery which supplied him in Inverness was operating from west-coast bases. Steuart was one of a syndicate that had boats based in Gairloch in Wester Ross that were catching fish, processing them ashore and then trading the dried cod with markets spread from Bilboa to Hamburg.[102]

One of several striking aspects of these earlier eighteenth-century fisheries is the absence of significant indigenous commercial operations in some of the areas dominated by the Clyde, Forth and Moray Firth fishermen. There were certainly fishing communities all along the coast from the mouth of the Tay to the Deveron, but they were still relatively small-scale operations at this time in comparison to the much larger Forth and inner Moray Firth fleets. The north and west mainland and Western Isles fisheries, which were the focus of much of the British Fisheries Society efforts from the 1790s onwards, are well known as a region where the fishery in adjoining coastal waters was exploited by boats from distant harbours, but less attention has been given to the pre-1790s situation in the Northern Isles. In Orkney, for example, a consortium of Kirkwall-based merchants had been prominent in an ultimately futile effort in 1721 by the Convention of Royal Burghs to promote fisheries. Their failure to generate support meant that it was not until they became involved in the activities of the London-based Society of Free British Fishery between 1750 and 1771 that local employment in fishing reached significant numbers.[103] The loss of the Society's parliamentary charter in 1771, however, led to the near-total collapse of this activity and the skilled fishermen drifted away into other employment. It was said in 1774 that the islands' indigenous cod fishery, based in Walls, had also collapsed, supposedly through the 'laziness' of the local men.[104] There was more success with a lobster fishery in the inshore waters around the islands, where there was no competition from the better-equipped Lowland outsiders; from 1775 to the end of the century, annual exports (mainly to London) climbed to staggering levels of between 100,000 and 120,000 lobsters. It is little wonder that the numbers of lobster caught began to decline

---

101 MacKay, *The Letter-Book of Bailie John Steuart*, pp. 4–6, 8, 20–21, 26, 27–28. Steuart was always looking for the best destination for his fish, seeking intelligence in September 1716, for example, as to whether Marseilles or Livorno would be better ports offering good prices than northern French and Flemish markets.
102 Ibid., pp. 20–21.
103 Thomson, *History of Orkney*, pp. 216–217.
104 Ibid., p. 217.

sharply in the nineteenth century. In Shetland, the crofter-fisherman of the *haaf* tradition that had evolved during the MCA prevailed through the eighteenth century. The fishermen were tenants of the major Shetland landowners and were obliged to fish as part of their tenancy arrangement using open boats and deploying long, baited lines for catching cod and ling.[105] This organisational basis constituted a significant limiter on the development of a large-scale commercial fishery in the islands, with the potentially lucrative trade in cod continuing to be dominated by foreign merchants into the nineteenth century. While we might question the sustainability of the Orkney lobster fishery, the numbers of cod and other species caught by *haaf* fishers off Shetland or by the Pentland Firth fishermen in Orkney were tiny and are unlikely to have made any impact on the fish populations around the islands. Fluctuating catch sizes that are recorded through the eighteenth century in these fisheries seem instead to have been a consequence of environmental changes related to the successive switches in ocean conditions during the LIA rather than human action. The impact of changes in ocean water temperature in particular on migratory fish behaviour and on the more static species in the waters around Scotland were to become increasingly evident across the eighteenth century, but especially during the final extremes of cold in the early nineteenth century.

Changing oceanic conditions, principally temperature in both shallow surface levels and deeper waters, affected fish migration, feeding and spawning behaviour of marine pelagic and demersal species and also of anadromous species like salmon, which are spawned and develop in inland freshwater systems, migrate to the mid-Atlantic to mature, then return to their hatching waters to themselves spawn.

Episodes of warming and cooling might take some years to cause significant changes in the movement of fish, but the colder North Atlantic conditions of the 1720s and 1730s affected both the cod and herring fisheries. These conditions were visible especially around the Northern Isles, but also, apparently, the herring migration and spawning in the North Sea was affected at this time. There, the fish seem to have moved from their once-favoured late-summer spawning grounds in the Firth of Forth into deeper waters, ending the bounty of the so-called 'Lammas Drave' in the ports of north-east Fife and Lothian.[106] Poor weather also affected the ability of the smaller boats that still dominated the Scottish fisheries to operate safely in deep, open waters, as the losses suffered in communities on the Kincardineshire coast in the storms of February 1730 mentioned in Chapter 6 attest. It was these changes rather than over-fishing that at this date threw the Scottish fishery into sharp decline in the second quarter of the eighteenth century, for even at the peak volume of around 2,900 tonnes per annum of the catch being exported, such numbers would have made little impact on the still vast herring shoals.

For the Forth fishermen, the change in herring behaviour brought decline in other species, for other white fish – like bass, cod, haddock and pollock – fed on herring spawn and fry, and a decline in their availability drove them, too, to seek their prey elsewhere. It was not, however, a permanent decline and there were periods when the fish returned to their former migratory patterns, as in the 1770s when the Forth fishery in particular returned to boom conditions.[107] But these episodes were unpredictable and, while the fishery had been strong through into the 1780s, by the 1790s it had again collapsed, probably once more affected by changes in the circulation and tempera-

---

105   J.R. Coull, 'White fishing', in J.R. Coull, A. Fenton and K. Veitch (eds), *A Compendium of Scottish Ethnology*, volume 4: *Boats, Fishing and the Sea* (Edinburgh, 2008), pp. 253–276 at 259–260.
106   Smout and Stewart, *Firth of Forth*, pp. 33–34.
107   Loch, *Essays*, vol. 2, pp. 232–241.

ture of North Atlantic waters entering the North Sea that influenced herring behaviour. When the herring shoals entered a different migratory pattern there were widely ramifying repercussions affecting other fish species, as well as porpoise, seal and bird species, that fed on herring eggs and fry and on mature fish, and also for human populations that depended on them all directly or indirectly for subsistence or commercial livelihood.[108] The domino-effect that it triggered was not visible at the time, nor were the causes understood of the disappearance of the herring and of its impact on the main food-species – herring and haddock – exploited by the human population. Although the impact of the potato failures of the mid nineteenth century was far greater than anything caused by the collapse of the herring fishery in the 1780s – certainly in terms of human suffering and longer-term cultural changes – the experience of the east-coast fishing communities presents us with the first example of the consequence of over-reliance on one natural resource that was at the centre of a whole series of interlocking dependencies.

• • •

As the above discussion has illustrated, we can endlessly debate what is meant by 'Age of Improvement' and the date range to which that label can and should be applied. Splitting hairs over when exactly this era commenced, however, serves little purpose in an exploration of the environmental ramifications of Improvement more generally. What matters more for this exploration is that there is little doubt that the processes wrapped inside the term – both long-established and novel – were key agents of change in Scotland's most profound episode of environmental transformation in the historic era. Compartmentalisation into temporally defined and artificially labelled periods serves only to obscure both the longer run of trends that created the circumstances for change and the playing out of the consequences that were unleashed. Traditionally, the transformative dimensions of Improvement have been – and continue to be – most commonly discussed in social and economic terms and explored through the physical reconfiguration of patterns of settlement and land-use. But the debates surrounding the legacies of Improvement can also be usefully informed by the more recent recognition of the deeper and not simply negative environmental effects wrought by the ambitions of the nation's Improving landowners to secure their own financial positions and deliver wider social, economic and cultural betterment to the Scottish people. That framing of perspectives is much needed, for we are burdened with the legacy of mid-twentieth-century historical ecologists who traced the roots of large-scale environmental degradation to what they projected as the baleful impact of Improvement philosophies and the spread of commercialism. Despite decades of more recent research that has challenged that view, the negative framing of the 'Age of Improvement' maintains a tenacious grip especially in respect of current so-called 'rewilding' activities or restorative projects.

What is perhaps most striking in most explorations of the eighteenth-century era of Improvement is how little consideration is given to the effects – positive or negative – of the practices being introduced. Debate, where it occurs, is generally over the rate and method of the spread and adoption of Improvement thinking and practice. It has most often focused on the success or failure of individuals and associations, stressing the important influence of network and connection, economic imperatives and constraints, and on the actualising of political and social theory. Couched in the language of progress and cultural advancement, it is still viewed most

---

108 Smout and Stewart, *Firth of Forth*, pp. 40–43.

commonly through a Whiggish lens of positive social development. As this chapter has explored, however, that development came about largely at great cost in environmental terms, but also with some significant benefits to certain plant or animal communities, woodland especially. It is important that we do not lose sight of the balance of these positives as well as the negatives in our present movement towards reversal of centuries of human influence in shaping the land and environment in which we live, supplanting one form of 'Improvement' with another that is more in tune with current concerns. Perhaps yet more importantly, as some of the material explored in this chapter has illustrated, while Improvement most certainly entailed disjunction, discontinuity and dislocation, it could also see continuity of long-established practice and even acceleration of trends that are visible from centuries earlier. Where the focus of future research needs to fall is not so much on the process and the 'why' behind these instances of continuity or change, but on how it was achieved and not just on the direct impacts but on the manifold ramifications across ecosystems that resulted. Failure to do so risks another episode of well-intentioned, scientifically rational but potentially irreversible human manipulation of our environment.

CHAPTER 10

The Sting in the Tail: Rain, Snow, Frost and Drought at the End of the 'Little Ice Age' 1790–1850

The storm, which has been increasing since the evening of Sunday last has turned out a hurricane that I imagine has scarcely a parallel. The tempest is so extreme that it is with difficulty that we can go over the doors, and to go against the wind is suffocating. It is only in intervals of the gust that I can see above a hundred yards out of my shop window. I do not remember to have seen wreaths [of snow] of the same height. There is one collected in my back close twelve or fifteen feet in height. It stands alone and has the appearance of a house. There is another runs along the front of the house, which may be calculated at eight or nine feet deep. . . . There may be expected great havoc amongst the coast shipping.[1]

The six decades covered in this and the following three chapters are most commonly explored in terms of the profound and rapidly moving changes to the fabric of Scotland's social and economic life that they witnessed. It can be argued, however, that their environmental-historical significance was at least as great, if not greater, for they spanned the protracted period of climatic turbulence that can be identified in retrospect as the violent closing episode of the 'little ice age'. Whereas in earlier periods such extreme conditions might have resulted in subsistence crises, bringing famine and high mortality that kept human population numbers suppressed and economies weak, these decades of climatic disturbance were a time of rising population and – for some – growing prosperity. This was the period which witnessed Scotland's fuller integration into the mercantile web of Britain's spreading overseas empire. That integration brought access to colonial trade and investment in slave-worked plantations, from which income was channelled back into expenditure at home; some of this was intended to stimulate estate economies but more often it was simply to transform and beautify ancestral properties with exotic planting of alien species and landscaping that physically altered the character of the land, purely as a demonstration of their owners' new wealth and connection. The flipside of this access was the consequent exposure of Scotland's still largely rural, agricultural regime to the economic rollercoaster of a new, global trade network.

The profits and plunder of full participation in

---

1 Diary entry for 7 February 1823, from D. Stevenson (ed.), *The Diary of a Canny Man 1818–1828. Adam Mackie, Farmer, Merchant and Innkeeper in Fyvie* (Aberdeen, 1991), p. 121.

the exploitation of a growing overseas empire, already trickling back to Scotland by the mid eighteenth century, became a flood as the new century dawned. Landowners and institutions who speculated in and exploited the colonies and grew rich on the trade in human lives and the plantations built on slave labour were provided by empire with the capital to undertake ambitious remodelling of their property at home. It was the era that also saw Scotland contributing increasingly to Britain's foreign wars, offering a further route to prosperity through military service, plunder and 'prize' money, or the provisioning and equipping of armies and fleets. Trade, armed service and government office brought income, much of which joined the homeward flow of capital to be invested in ancestral estates and bankroll the betterment of their infrastructure. It was not just smart new mansions and public monuments that were funded from the questionable profits of empire, but also the schemes for large-scale land improvement, woodland planting and industrial infrastructure. Indeed, much 'Improvement' undertaken by landowners (as opposed to their tenants) was funded through the profits of war and slavery. Underwritten by the asset-stripping of Britain's overseas empire, the tide of Improvement that had been rising through the eighteenth century continued its relentless spread. It was accompanied by the steady expansion of industry and the accelerating growth of urban centres in areas close to the main sources of fuel for the thermal energy that powered the new, steam-powered mechanised factories and mills. This was the era when Improvement went from being a philosophy that looked to the social, cultural and economic betterment of the people generally to a dogma that saw undeveloped nature simply as raw material that could be improved upon. In this new prescription, there was nothing that Man could not transform through the appliance of reason and science. For the many still trying to eke an existence from the land, however, the bitterly cold and saturated years of the early 1800s presented a stark contrast between the ideal of Improvement and the reality of their precarious existence with the continuing, age-old struggle to subsist.

Checking reality: climate versus Improvement

As with the histories of eighteenth-century Scotland, it is a striking aspect of most general accounts of the post-1790 period that they give almost no notice of the continuing impact of weather events as the 'little ice age' drew towards its close. Suggestions of negativity, hardship and failure, even when a result of poor weather conditions, sit awkwardly in historiography dominated by the continuing Whiggish narrative of socio-economic improvement and progress, at least for Scotland outside the Highlands. That, however, is not to suggest that the spread of new agricultural techniques and land-management practices was anything less than remarkable, it succeeded generally despite the adversity of the times and the hostility of the environmental conditions in which it was achieved. Where 'environment' is given any prominence in narratives of agricultural advance, however, is in respect of the potato blight and consequent Highland famine of the late 1840s and 1850s, discussed later in this chapter. Yet, even potato blight's place within the broader context of the decades of poor weather that characterised the general Scottish experience of the early nineteenth century is rarely considered. The famine years are presented most often as a case study in socio-economic breakdown, demographic unsustainability and political failure; few historical analyses consider the climate context and cool, damp weather conditions that encouraged the spread of *Phytophthora infestans*, the micro-organism that causes blight. And this absence of broad discussion is despite the steady proliferation of both weather observations set out in personal correspondence and more formal weather records, including the keeping of instrumentally gathered data from locations spread around Scotland. A further internal bias of the historiography is that alongside the generally positive presentation of agrarian change, albeit tempered with acknowledgement of the often traumatic social and cultural consequences for traditional farming communities across the country, the balance in the narrative switches from rural and agricultural to urban and industrial. And there is generally an assumption that the central Lowlands' experience of industrialisation was the norm, from which other areas deviated for a variety of ingrained, mainly cultural, occasionally geographically determined reasons. Within discussion of urbanisation and the rise of manufacturing industry, little space is afforded for examination of the impact of extreme events on what was still a primarily agricultural society even in the central Lowlands. The focus is instead on the symbiosis of urban and rural organisms, with agriculture reconfiguring to meet the growing demand for raw materials, food and labour that came from the rapidly growing industrial centres.[2] Globalisation and its impacts, principally in terms of competition in the supply of staple commodities like grain or wool, might be noted as contributors to trends within this agricultural reconfiguration, with their impact expressed in social and cultural terms. Poor harvests and the price inflation they caused also might be noted, but not the environmental factors that delivered the crop failures and resultant subsistence crises at home.[3]

Yet, private correspondence from the 1790s and early 1800s reveals continuing deep, public anxieties concerning the precariousness of food and fodder supplies, even as the immediate impact of the 1783 Laki eruption and its aftermath faded in the collective popular consciousness; there was little obvious sign for people that better times were to hand. Cold winters and poor or failed harvests, as contemporary diaries reveal,[4] had in fact persisted through the 1780s and disturbed weather patterns continued into the 1790s as the sun moved into a new phase of low sunspot activity in the so-called Dalton Minimum.[5] As with earlier minima, like the Maunder, there was a general cooling globally, which in Scotland brought recurrent poor summer and autumn conditions threatening regional dearths. Fear of more general famine akin to the 1690s kept a firm grip on the popular imagination. A spike in emigration in the late 1780s and earlier 1790s from parishes in the north and west of the country and in the Hebrides, noted in submissions made to the *Statistical Account of Scotland* in 1792–7, should be understood in the context of poor harvests and subsistence crises.[6] Such

---

2   See, for example, the excellent summary of these processes in I.G.C. Hutchison, *Industry, Reform and Empire: Scotland 1790–1880* (Edinburgh, 2020), pp. 7–16.
3   See Lenman, *Enlightenment and Change*, p. 166 for the experience of 1795.
4   A key source for this period is the daily weather records kept by an Aberdeenshire laird's wife, Janet Burnet, which covers forty years in the second half of the eighteenth century, including the period of the Laki eruption and the extreme winters of the early 1790s: M. Pearson (ed.), *More Frost and Snow: The Diary of Janet Burnet 1758–1795*. Sources in Local History 2 (Edinburgh, 1994).
5   S. Wagner and E. Zorita, 'The influence of volcanic, solar and CO_2 forcing on the temperatures in the Dalton Minimum (1790–1830): a model study', *Climate Dynamics*, 25 (2005), pp. 205–218.
6   Figures suggest an average of 10–15 per cent of the adult population of some parishes migrated, with a rising trend from 1788. See, for example, Rev. William Bethune, 'Parish of Duirinish, Inverness-shire (1792)', *Statistical Account*, vol. 4, pp. 132–133; Rev Roderick McLeod, 'Parish of Bracadale, Inverness-shire (1792)', *Statistical Account*, vol. 3, p. 247; Rev. John MacQueen, 'Parish of Applecross, County of Ross and Cromarty (1792)', *Statistical Account*, vol. 3, pp. 376–377; Rev. George Munro, 'Parish of South Uist, Inverness-shire (1794)', *Statistical Account*, vol. 13, pp. 297–298; Rev. Colin Maciver, 'Parish of Glenelg, Ross-shire (1795)', *Statistical Account*, vol. 16, p. 267; Rev. Martin MacPherson, 'Parish of Sleat, Inverness-shire (1795)', *Statistical Account*, vol. 16, p. 537; Rev. Donald McLean, 'Parish of Small Isles, Inverness-shire (1796)', *Statistical Account*, vol. 17, p. 281.

Storm clouds over upper Strathearn, Perth and Kinross. The waves of rain-bearing clouds that moved in from the west in the saturated summers of the Dalton Minimum added to the misery of the bitterly cold winters. Peat cut in spring to provide fuel for the following winter was rarely dried adequately to burn, forcing those for whom coal was not a viable alternative into desperate measures to survive.

accounts often highlight 1791 as the worst for dearth of grain since 1782.[7] Increased volcanism in this same period, with six major eruptions in excess of Category 5 on the Volcanic Explosivity Index (VEI) (one Category 6 and one Category 7) between 1795 and 1835 added the further cooling agency of aerosol particulates to the mix, the worst of these being the 1815 Tambora eruption in Indonesia, which produced the so-called Year Without a Summer in 1816.[8] As the responses to such threats in the bad years of the 1760s and 1780s and especially following Laki showed, however, subsistence crises could be averted now by prompt, targeted interventions by local authorities. Increasingly effective government could move supplies and intervene to prevent some local crises from becoming disasters such as had scarred the seventeenth and eighteenth centuries. Beneath this increasingly effective safety-blanket, the much-anticipated catastrophe never fully materialised other than in regions far from the awareness or conscience of central government and the growing urban elites. But it is important to remind ourselves that it was not only arable production that was at risk from poor weather; production of other staple commodities was affected equally badly or worse. Fuel supply, for example, remained precarious for the thousands to whom coal was inaccessible or unaffordable, as negative summer NAO brought low temperatures and wave after wave of rain-bearing clouds in from the Atlantic over districts where traditional fuel resources were rendered almost useless; conditions in summer 1790 were so wet as to bring a fuel crisis to the Hebrides as the peat cut in May remained saturated and unusable in late autumn.[9] Such reports offer us a momentary insight into the experience of ordinary people in 'unimproved' districts remote from the sources of coal. There, weather that was increasingly viewed by the mainly urban, 'polite', coal-consuming and wealthier social strata as a personal inconvenience presented a very real threat to community subsistence.

And it was the persistence of poor conditions throughout these six decades that was most challenging to would-be Improvers and traditional commu-

---

7 See, for example, Rev. Roderick Morison, 'Parish of Kintail, County of Ross and Cromarty (1793)', *Statistical Account*, vol. 6, p. 252.
8 G. Wood, *Tambora: The Eruption that Changed the World* (Princeton, 2014).
9 Dawson, *So Foul and Fair a Day*, p. 148.

nities alike. Rain saturated the fields, flattened crops and brought floods that swept away new embankments and bridges; snow smothered sheep flocks or buried winter fodder too deep for either livestock or game animals to reach; frost delayed ploughing or damaged winter-sown grain; and drought stunted crop growth, parched pasture and hit farmers who had to make inroads on winter fodder supplies to feed their animals before even the autumn had passed. These were age-old problems for those who worked the land. Agricultural innovation was doing much to diminish their impacts and reinforce resilience, through the introduction of new types of fodder crops that stored better over winter and new forms of food crops that were more resilient in adverse conditions, but the relentlessness of the impacts was unprecedented since the 'Seven Ill Years' of the 1690s. For even the most optimistic of Improvers, poor weather often placed a brake on the rate of improvement in districts where the greatest physical and environmental challenges existed, like the cold and stony interior of Buchan or the rain-drenched mosses of Galloway.

For the gloomier recorders of events, the prognostications had looked grim from the outset of the period, for the windswept spring and the all-pervading wet they noted in summer 1790 continued through the winter into early 1791 as a series of storms battered western districts and the southern Highlands, causing spate floods that washed away bridges.[10] But amid the general air of grinding climatic challenge it is clear, nevertheless, that some regions and communities experienced what might be thought of as positive changes. Winter NAO, for example, remained generally low and brought lower rainfall to north-western mainland areas than had been experienced in the first half of the eighteenth century,[11] which serves to underscore the strongly variable local experience of this disturbed period. Low winter NAO, however, generally also meant lower temperatures, as colder air was drawn south and west from the Arctic, Scandinavia and Siberia, so any gain from drier conditions was offset by longer periods of intense cold. Similarly, the extension of cold Arctic waters southwards towards the British Isles, an element of the wider atmospheric and oceanic disturbance that displaced the Atlantic storm-track that delivered the wildest weather to Scotland, brought a fishing bonanza as the herring returned in numbers to the seas around Shetland, but the stormy conditions also extracted a heavy toll in fishermen's lives. It was adaptation to these circumstances and the application of new practices and technologies that allowed for relative agricultural success that became the hallmark of the age, in even the most challenging of conditions.

More ice?

So what were the prevailing conditions and how did they affect communities around Scotland? The following narrative, founded upon the progressively expanding records of the period, is richer and more

---

10 Pearson, *Diary of Janet Burnet*, pp. 97–101 sets out the progression from the windy but generally dry and mild early months of the year but turning cold and wet from late March. Early June was described as 'cold as winter' and the remainder of the year declined into long spells of cold, wet and windy weather. See also Adam Bald, 'Journal of Travels and Commonplace Book, 1790–99', in A.J. Durie (ed.), *Travels in Scotland 1788–1881. A Selection from Contemporary Tourist Journals*, Scottish History Society (Woodbridge, 2012), pp. 41–99 at 45–46, for the dry conditions in early April 1790 that brought a dust-storm in a 'vernal hurricane', giving way to wind-driven hail showers. Spates and their damaging impacts around the confluence of the Tay and Lyon are detailed in NRS GD112/74/232 (7).

11 Proctor et al., 'A thousand year speleothem proxy record', Fig. 4; A. Baker, J.C. Hellstrom, B.F.G. Kelly, G. Mariethoz and V. Trouet, 'A composite annual-resolution stalagmite record of North Atlantic climate over the last three millennia', *Scientific Reports*, 5 (2015), Fig, 4: DOI: 10.1038/srep10307.

Winter in the mountains around Crianlarich, Glen Dochart, Stirling. Early snowfall, severe frosts and biting winds that sapped the energy levels of humans and livestock alike brought the Breadalbane tenants in the valley to the limits of their economic and physical resilience in the 1790s.

detailed than that offered in previous discussions of climate and weather in these volumes, but it provides us with a complex and subtly textured canvas onto which our explorations of environmental themes can be projected. The abundant data for this period permits us to see trends not just on an annual basis but, at some times, on a daily and almost hourly level. This gives us some measure against which to interpret the data of earlier periods, for it gives us the ability to see seasons and their character and to understand how the seasonal variation affected different environmental sectors. We also see exactly how those variations affected the human population in different parts of Scotland and gain a sense of how the continuation of a subsistence existence affected many people in rural and coastal Scotland. In broad-brush terms, for example, we can see that although the summers were often far from optimal for crop growth, livestock husbandry or deep-sea fishing, it was the winters that drew the most contemporary comment. Biting cold through lack of fuel to heat homes was added to gripping hunger through dearth of affordable staple foodstuffs to fill bellies. Sensational accounts of metres of snow piled in the capital's streets or mail coaches overwhelmed on moorland roads in the frequent and long-lasting blizzards provide memorably graphic images of the worst episodes. But such graphic examples are not what provides the sense of what extreme winters meant in terms of the lived experience for ordinary people in either urban or rural districts. Even when we do have details of broader experience, however, such as the reports from the factors on the Breadalbane estates on the impact of the extraordinary winter weather on some tenants, we need to remember that these are still most often reports of extreme and exceptional cases. It is the more general reports, such as those which spoke of the effects across broader districts, like Glen Dochart, where winter 1791/2 saw snow starting unusually early in the season, sheep flocks having to be moved as a precaution from the upland districts to low-lying pastures, cattle that would normally have found grazing outdoors at this time requiring to have feed brought to them, and supplies of fodder running dangerously low, that provide a better sense of the extent of impacts and their effects at a community-wide level.[12] This combination of blows detailed in the Glen Dochart account challenged the resilience of even well-organised and otherwise successful farms, undermining their viability as the flocks and herds that they had built up were decimated or left undernourished, carefully managed grassland was exposed to over-grazing, and the financial reserves of the farmers were drained as they were forced to buy in feed for their hungry beasts. One bad year could be managed with support from estate owners, but successive bad years threatened the investment that was needed to continue a programme of improvement. And bad years followed one another in quick succession in this last phase of the LIA.

Modern climatological and historical meteorological research has provided modelled reconstructions of the oceanic and atmospheric conditions that injected the dynamics into the weather patterns behind the headline events.[13] Ocean surface temperatures are key contributors to wind circulation, wind strength and air temperature. At times, cold Arctic waters penetrated further south, as happened during a period of extensive and seasonally long-lasting sea-ice off Iceland in the late eighteenth and early nineteenth centuries. During this episode, the waters around Shetland were several degrees colder than at present,[14] signalling a slackening in the AMOC circulation, and the southward displacement of the thermal gradient zone, where the colder northern air met the warmer air over the mid-Atlantic, potentially produced the atmospheric dynamism which generated storms of greater intensity than in warmer periods. Storms intensified by the steep thermal gradient were a characteristic of the westerlies that were again a feature of winter 1792/3, when a series of deep low-pressure cells moving rapidly eastwards from the ocean speaks of an especially active jet stream lying low across the North Atlantic. Some of the worst storms occurred in December 1792, when

---

12 NRS GD112/14/13/7 (4).
13 The work of Hubert Lamb brought significant advances in understanding of historic climate change and its impacts at the end of the 'little ice age'. Several of the essays gathered in his edited volume, *Weather, Climate and Human Affairs* (London, 1988; reprinted 2012), set out the broad principles, while his *Climate, History and the Modern World,* 2nd edition (London, 1995) remains an indispensable text. His reconstructions of the conditions which generated the major North Atlantic and North Sea storm events, and the tracking of their movement, from the early sixteenth century onwards (see Lamb, *Historic Storms*) offer some kind of measurable sense of the human consequences of these recorded storms. See also: F. Lapointe, R.S. Bradley, P. Francus, N.L. Balascio, M.B. Abbott, J.S. Stoner, G. St-Onge, A. De Coninck and T. Labarre, 'Annually resolved Atlantic sea surface temperature variability over the past 2,900 yrs', *Proceedings of the National Academy of Sciences*, 117:44 (2020), pp. 27171–27178; V.C. Slonosky, P.D. Jones and T.D. Davies, 'Atmospheric circulation and surface temperature in Europe from the 18th century to 1995', *International Journal of Climatology*, 21 (2001), pp. 63–75.
14 The period from the 1780s to 1840s saw ice lasting on average for 17 weeks in a year, but the last decade of the eighteenth century saw ice present for around six months a year: Lamb, *Weather, Climate and Human Affairs*, pp. 153–154 and Fig. 9.9.

winds possibly in excess of 170mph at sea and registering Force 11 to 12 (around 115mph) over land first swept the Faroe and Shetland Islands before blasting the northern parts of Scotland; 'that night the wind rose to such a hight (sic) that few people remembers to have heard any thing like it', observed Janet Burnet from Aberdeenshire of the storm of 10–12 December.[15] Barely a week later, a second storm began to develop in the southern North Sea, which culminated on 23 December in a deep low-pressure cell that had moved from east of Iceland and down the North Sea, drawing Arctic northerlies with it over the northern and eastern British Isles.[16]

Equally bad weather characterised the winter of 1793/4. The worst weather occurred at the northern edge of another westerly system in late January 1794, which delivered mild and wet conditions further south in England and mainland Europe. Over the Scottish Southern Uplands, however, it produced a storm that brought widespread destruction in an alternating and quickly changing wave of snow, rain and frost accompanied by floods. In Eskdale, where this event was known as the 'Gonial Blast' (from the Scots term for a sheep found dead whose carcase was still considered edible), the remains of over 1,800 sheep were found washed up on the sandbanks at the river mouth on the Solway and a further 4,000 were found to have perished in the exposed uplands of Eskdalemuir.[17] This experience was almost repeated in winter 1794/5, when cold northerlies in January and February 1795 dumped deep snow and brought a lasting, penetrating cold to much of Scotland. Conditions were sufficiently poor to disrupt travel across central districts and force the council in Edinburgh to take emergency measures to ensure access to the city's coal depots.[18] In Aberdeen, Janet Burnet noted in her diary on 11 January, 'No Coals to be got. All the Ships gon (sic) for Coals froze in'; no vessels could manage to reach the city's port until 21 January.[19] Even upland districts, where there was perhaps some expectation and experience of extreme winter conditions, were affected badly by the depth of snow and intensity of cold. In the dire winter of 1797/8, where the cumulative effects of volcanic forcing from the 1795 eruption of Mount Westdahl in the Alaskan Aleutian Islands might have been at work, tenant farmers in Glen Lochay in west Perthshire reported the loss of most of their sheep in blizzards that swept the district.[20] Nevertheless, despite the dreadful winter weather, the summers were rarely so poor as to threaten failure of the spring-sown crops. It was only in the cold and stormy spring, summer and autumn of 1799 – the fourth-coldest year in Scotland since c.1200 and representing the nadir of the Dalton Minimum[21] – that the fears of widespread harvest failure came closest to realisation.[22] In that year, crops in lowland districts in the Lothians were unfit for harvesting until November and in upland areas they were left until late in December.[23]

---

15 Pearson, *Diary of Janet Burnet*, p. 109; Lamb, *Historic Storms*, pp. 101, 105.
16 Lamb, *Historic Storms*, pp. 105–110.
17 J. Hyslop and R. Hyslop, *Langholm As It Was: A History of Langholm and Eskdale from the Earliest Times* (Sunderland, 1912), p. 850; Dawson, *So Foul and Fair a Day*, p. 150.
18 NRS GD112/52/43 (21), GD112/74/70 (6); Dawson, *So Foul and Fair a Day*, pp. 150–151.
19 Pearson, *Diary of Janet Burnet*, p. 118.
20 NRS GD112/11/6/3/20.
21 Rydval et al., 'Reconstructing 800 years of summer temperatures in Scotland', p. 2960 and Table 2.
22 In the Moorfoot Hills on the Midlothian/Peeblesshire border, drifting snow and severe frosts with a biting easterly wind persisted into early April 1799: NRS GD113/5/448 (129).
23 Dawson, *So Foul and Fair a Day*, p. 151.

William Daniell, Aberdeen Harbour from the South. During the bitter cold of January 1795, even the brackish water of the tidal reach of the River Dee, on which the city's harbour was located, froze and trapped large sea-going vessels in the port.

Snow by the bushel and rain by the gallon

When the poor weather of 1799 continued into the new century, the stresses within the basic structures of estate organisation and rural life began to show. Emigration to North America increased rapidly as the pressures mounted from year-on-year crop failures and the struggle to subsist on small landholdings at a time of poor yields. Indeed, by 1803 the outflow of emigrants from the northern Highlands and Western Isles was such that the greater regional landowners called on the British government to step in to stem the tide. The result was the 1803 Passenger Vessels Act, which imposed a massive duty on the costs of transporting emigrants.[24] Even those who entertained no thoughts of emigration, however, saw their annual routines dislocated by the weather. For example, what should have been the straightforward business of rent agreement and lease renewal with tenants was affected, with reports from the Breadalbane estate of conditions being so bad in early February 1800 that the factor was unable to travel the 6km from Kenmore across the hill to Glen Quaich to renew crofters' leases there; he was able to visit the earl's tenants only in the lower-lying district around Loch Tay, and that with difficulty owing to the depth of snow.[25] This situation not only affected estate incomes but also injected uncertainty into the minds of tenants who were not yet benefiting from longer-term Improving leases. Another consequence was that rents on some badly affected holdings remained lower than projected, as landlords were forced to

---

24 Hutchison, *Industry, Reform and Empire*, p. 24. The duke of Argyll lamented the restraining impact of the act on emigration from overpopulated portions of his property: Cregeen, *Argyll Estate Instructions*, pp. xxxiii, 73, 201.
25 NRS GD112/16/4/4 (5). The extreme frosts of May 1801 also proved a major setback to Breadalbane's tree-planting schemes, with the pine and larch cones that were harvested for their seed being lost in the bitter weather: NRS GD112/11/7/5/53.

Ben Lawers and Loch Tay from Creagan na Beinn, Perth and Kinross. Deep snow across the Breadalbane estate around Loch Tay in early 1800 had significant economic repercussions for both estate finances and the security of the earl's tenants.

recognise that even their more progressive tenants were incapable of bearing the phased increases that had been part of the Improving leases on which they had entered their tenancies. In the west, however, and especially in coastal and island districts, where incomes were supplemented significantly by the profits of kelp processing (see below, pp. 353–358) and fishing, landlords such as the duke of Argyll were able to continue to push rents up. These estate owners benefited from the diversity of opportunity and, to an extent, from the comparatively milder oceanic climate of the Atlantic-facing coast, but also from their more benign model of estate management, which allowed tenants to participate in the prosperity that Improving methods brought.[26]

A triumph for those landlords who feared the loss of their labouring, profit-generating population to emigration, but opposed by those who recognised the ills of over-population for land of limited productive capacity, the Passenger Vessels Act consigned many of the most vulnerable Scots in the most agriculturally marginal districts of Scotland to unremitting hardship as the first decade of the nineteenth century ground on. Consistently dire weather and wider instability in atmospheric and marine systems formed a backdrop to their hardship; even the fisheries largely failed in 1802 and 1803, which suggests a significant – if temporary – change in surface temperatures and oceanic circulation. Shetland, where the fishing-crofters were already hugely vulnerable in a system which obliged them to fish as a condition of their tenures and deliver the bulk of their catch to the merchant-lairds who supplied them in return with food and equipment, was brought to its knees when the fishery failed; only a £10,000 grant from the government averted a worse crisis.[27] Fishing

26 See, for example, Cregeen, *Argyll Estate Instructions*, pp. xxix, 48–49, 50 (item 2).
27 B. Smith, 'Shetland Archives and sources of Shetland history', *History Workshop*, 4 (1977), pp. 203–214 at 205–207.

communities elsewhere around the country perhaps fared slightly better economically due to different tenancy arrangements, but they suffered as grievously from losses of boats and crews in the increasingly unpredictable seas. One particularly catastrophic event, a south-westerly gale, swept across northern England and Scotland on Christmas Day 1806, with the greatest impact in the Moray Firth district and coastal areas as far north as Orkney. In addition to heavy damage to property and woodland on shore, it brought losses of fishing-boats with all hands from Portessie, Stotfield, Burghead, Avoch and smaller communities up the coast in Caithness and the islands, with the vessels being driven away from land and eventually swamped with no hope of rescue.[28] At Stotfield, the loss of its three boats wiped out the entire adult male population of the fishertoun, underscoring the devastating consequences of such sinkings for already fragile communities.

Rain again soaked Scotland throughout 1807, damaging both grain and hay crops and delivering a late and very poor harvest and widespread livestock deaths. Such conditions raised the prospect of shortages of both victuals and fodder into the following year, which made some tenants again look to emigrate to escape the unremitting direness of their lot.[29] The year ended with worse yet, for winter snows and storms came early. In the Moorfoot Hills between Midlothian and Peeblesshire, snowdrifts were overtopping hedges, blocking the turnpike road and causing anxieties about the availability of fodder as early as 24 November 1807.[30] With animal feed already in short supply locally after the poor hay crop that summer, the snow prevented carts from bringing any supplement from better-provided areas. Some relief seemed to be in hand when 1808 started mild and the summer was warm, but the rains and high winds returned in autumn in time for the harvest. Despite that bad turn, in some parts of the country, yields were remarkably good and averted another impending crisis, which was fortunate given the deterioration into a dreadful winter with the return of intense cold and snowstorms through December and January.[31]

Bad though the previous decade had been, the 1810s were to prove worse yet and are distinguished as the coldest on record. Although the deepest trough of the Dalton Minimum was past, solar activity remained low for a further twenty years, contributing to cooler conditions globally. The major culprit in the cooling again appears to have been volcanic forcing, with dust palls of aerosol particulates from first a VEI Category 6 eruption of an as yet unknown volcano in 1808 or 1809 and then the 1815 Tambora VEI Category 7 eruption – one of the greatest in human history – circulating the globe and suppressing temperatures by up to 0.7 °C.[32] That kind of drop in mean annual temperatures was sufficient to trigger a wave of crop failures around the planet in the following year, 1816, the Year Without a Summer. They came on the back of an already poor run of years from 1810 onwards, with 1812 especially bad across the country throughout the year, starting with a late and bitterly cold and snowy spring and ending with persistent rain that drenched the already late harvest.[33] Winter 1813/14 was dreadful, with bitterly cold Arctic air flowing down the eastern side of the country. January saw the start of nearly eight weeks

---

28 Reported in *The Aberdeen Journal*, 31/12/1806, 7/1/1807, 14/1/1807, 28/1/1807.
29 NRS GD23/6/438, GD87/2/22, GD112/11/8/3/21, GD112/16/4/6 (49, 50, 53).
30 NRS GD113/5/457 (126).
31 NRS GD112/74/19 (4); Dawson, *So Foul and Fair a Day*, p. 155.
32 South Dakota State University, 'Undocumented volcano contributed to extremely cold decade from 1810–1819', *ScienceDaily*, 7 December 2009. www.sciencedaily.com/releases/2009/12/091205105844.htm.
33 Dawson, *So Foul and Fair a Day*, p. 156.

Flood levels carved into the south pier of Smeaton's Bridge, Perth. The highest level of the 1814 flood is the highest flood level recorded on the stonework, significantly overtopping the level of the devastating 1993 event that flooded large parts of the city and surrounding districts.

of sub-zero daytime temperatures with midday lows of nearly -13°C reported. Even the new Forth and Clyde canal froze, halting the flow of goods and food supplies, the Tay iced over below Perth as far as Newburgh, and ice formed on the Forth above the Queensferry Narrows.[34] At Perth, when a thaw started around 10 February, ice-floes on the Tay were dammed by the piers of the new bridge built in 1771, and by 12 February the ice dam had caused a flood over 7 metres above normal river levels. It spread across the low-lying land several kilometres upstream from the town and within the burgh left only an island occupied by the medieval parish church above the floodwaters.[35] Throughout the accounts, the primary impacts noted were on transport and the movement of goods, especially bulk foodstuffs and fuel. Almost as a footnote, it was observed that the mortality rate, especially among the poor, the aged

34 *The Scots Magazine and Edinburgh Literary Miscellany* (February 1814), pp. 149, 151.
35 D. Bowler, R. Coleman, D. Perry and N. Robertson, *Perth: The Archaeology and Development of a Scottish Burgh*. Tayside and Fife Archaeological Committee Monograph 3 (Perth, 2004), p. 12 and Illus. 4.3.

and the infirm, had almost doubled in comparison to deaths in the same period the previous year. This doubling of the mortality rate reflected the reality for the most marginalised members of Scottish society of weather bemoaned by the better-placed as 'inconvenient' or 'melancholy'.

Although the mortalities of 1813/14 were horrifying enough, far worse lay ahead. Even before the full impact of the Tambora eruption made itself evident, winter 1815/16 had started as poorly as that of 1813/14, with strong winds, heavy snow and bitter cold first affecting most of the south-west from Glasgow to the Solway from early December 1815 and then spreading eastwards. The *Scots Magazine* for February 1816 reported multiple deaths in the blizzards. When a brief thaw followed the December snowfall late in the month and into early January 1816, the resulting meltwater spate on the Clyde carried away three bridges near Hamilton and flooded the lower-lying parts of Glasgow; many other rivers that flowed out of the Southern Uplands overcarried their banks in multiple locations.[36] Reading the contemporary 'melancholy' accounts of the widespread distress caused by the extreme conditions, the all-pervading sense is one of fatigue with the constant struggle that months of ill-weather entailed. That bitter start to 1816, however, was just the beginning of a year when temperatures remained suppressed, strong winds and heavy rain lashed the country and, with snow on the northern hills by September, the prospect of a poor harvest further dampened already low spirits.[37] Against the backdrop of the generally poor conditions that had prevailed since the 1780s, however, the additional impacts in 1816 and continuing into 1817 might have been felt less keenly than in more southerly districts. Indeed, contemporary newspaper and magazine accounts appear more interested in reporting what was happening elsewhere than in Scotland, perhaps because their readers were now so inured to the awfulness at home.

Keeping a note of the weather

We are fortunate that widespread literacy and a fashion for keeping personal journal-style diaries has left for us a rich haul of data concerning weather events and their impacts to fill the gaps created by declining reportage in more traditional outlets. Such diaries are especially valuable for their local reporting of conditions during this last, bitter phase of the 'little ice age'. One particular value of these records is that they are not the self-conscious notes of political or religious leaders but are often the public observations and private thoughts of 'normal' people – farmers, innkeepers, merchants – who were noting events that had the most immediate impact on their daily lives, as well as reflections on the great affairs of the time. It is through such accounts that we capture the lived experience of less-well-off rural-dwellers, away from the large towns and cities and the professional and landowning classes that tend to dominate our records. Among published diaries is a ten-year section from the notes of Adam Mackie, a farmer, innkeeper and merchant at Fyvie in Aberdeenshire, spanning the period 1818–28. His account takes us through the decade following the dissipation of the dust-cloud from the Tambora eruption.[38] Mackie's weather observations are often subtle and generally alert us to the way in which even slight changes in prevailing conditions could affect important aspects of rural life and the farming economy. In August 1819, for example, amid his observations on the unusual heat and prolonged drought, which he understood as threatening the quality and volume of the grain yield, he

---

36 *The Scots Magazine and Edinburgh Literary Miscellany* (February 1816), pp. 73–76.
37 Dawson, *So Foul and Fair a Day*, p. 157.
38 Stevenson, *Diary of a Canny Man*.

Fyvie, Aberdeenshire, looking across the Ythan Bridge towards the former farm, inn and shop at Lewes of Fyvie, where Adam Mackie (1788–1850), farmer, merchant and innkeeper, kept a diary whose accounts provide a detailed record of the impact of weather – good and bad – on the surrounding district across the decade after 1818.

also commented on the effects of the heat on his dairying activities, in that it kept the butter soft and in need of salting to prevent it from going rancid.[39] But there was also a positive in these conditions, as the dry weather of August 1819 continued to the end of the month, enabling the harvest to be brought in relatively early, if not with so high a yield as might have been wished.

That successful harvest was fortunate, for winter 1819/20 comes across through Mackie's writing as a time of daily challenges in north-eastern Scotland. North-easterly gales, culminating in a storm from 9 to 13 October that lashed the east coast, presaged a poor winter with dire conditions, especially across four weeks between late December and the third week in January. By 28 December, snow was lying between 30 and 40cm deep, carried in on northerly winds, and further blizzards lasting over four days from 29 December to 1 January brought the general depth to 60cm and drifts three times that.[40] Sixteenth- and seventeenth-century accounts speak of wine and ale sold by weight rather than volume because they had frozen; on 30 December, Mackie's ink froze in the pot on his desk and on 18 January, despite the

39 Ibid., p. 116.
40 Ibid., p. 117.

fires in his inn, the beer froze at dinner time.[41] This season of bitter cold extended across the whole country, with reports from around 10 January 1820 of daytime temperatures as low as -14°C in the west of Scotland and ice on the Clyde thick enough to bear the weight of horses with carts.[42] Despite the dreadful start to the year, however, summer and autumn 1820 were unremarkable and, in the north-east at least, winter 1820/21 was notable for its mildness. Unfortunately, the following year brought a return to instability, with late winter 1821/22 seeing a persistent drought and by early March 1822 Mackie noted that drifts of sand and dust were being blown by the strengthening winds. Like 1820 and 1821, however, the poor starts to the year did not continue through the following months and, in Mackie's neighbourhood, the grain harvest was acceptable.

Any hope that the cold winters of the first two decades of the century had passed, however, was dashed by the return to severity in January 1823, with what in parts of the country was referred to as the 'Long Storm'. The weather deteriorated around 12 January across most of Scotland, with Mackie recording for central Aberdeenshire that it started snowing on 14 January, and a succession of storms interspersed with periods of intense frost then lasted through to the third week in February.[43] Blizzards driven by easterly gales lashed most of the east side of the country, causing severe disruption to both road and sea transport and preventing movement of food, fodder and fuel supplies into the areas worst affected. Records from elsewhere in eastern Scotland indicate that the extreme conditions were widespread, with major disruption to local and regional communications as the roads were cleared and quickly re-blocked by drifting snow. With stored supplies unobtainable, fodder ran low for livestock that had to be kept indoors. Winter grazing in any case was inaccessible beneath wind-packed and frost-hardened drifts for those animals wintered outside.[44] Mackie at first commented that the snow was less deep than what he had seen in 1820 but the cold was at least as severe, with beer and milk freezing inside his house.[45] When the storm reached its peak on 7 February, the result was drifts in excess of 6 metres deep and generally some 2 metres around buildings and yards where the walls and fences trapped it.[46] Nearly two weeks later, drifts still approaching 2 metres deep persisted around Fyvie.[47] Reports more widely from across north-eastern Scotland confirm that Mackie presented a sober, unsensationalised account of the event, revealing the scale of disruption to normal social functioning and the cost in human and livestock lives that arose from the severity of the weather and intensity of the cold. In this era before compulsory registration of deaths, we can only guess outside the large urban centres at the level of excess mortality that occurred during this episode. Against the backdrop of general shortage of fuel in rural districts – Mackie was first able to get access to peat on 10 February, a week after the start of the heaviest snowfall[48] – it is likely that the death-toll from the cold among the poorest and weakest was high.

The year that followed was one of the coldest in

---

41 Ibid., pp. 117–118.
42 Ibid., p. 118.
43 Ibid., pp. 119–123.
44 See, for example, NRS GD1/58/2/23 for expenses clearing roads in Kinross-shire between 28 January and 15 February and GD113/5/483 (3) for fodder in Peeblesshire. For the impact in north-east Fife, see Dawson, *So Foul and Fair and Day*, pp. 159–160.
45 Stevenson, *Diary of A Canny Man*, p. 119.
46 Ibid., p. 122.
47 Ibid., p. 123.
48 Ibid., p. 122.

living memory. In some districts, it was also exceptionally wet throughout the summer. Prolonged rainfall through June, July and August delayed ripening of the grain and threatened the quality of the harvest across the north-east.[49] It was not that there was no grain but more a case of it being too wet to bring in successfully. Much of Mackie's crop was damaged, and in late September, before his men started on harvesting the standing corn with scythes, they resorted to using shears to cut the rain-flattened grain before it started to rot.[50] The onset of milder conditions in late autumn saved that harvest, although the earlier rains had reduced the quality and quantity of the yield overall. Winter 1823/4 appears to have been exceptionally mild down the eastern side of the country but still interspersed with brief episodes of extreme weather. There are reports of midsummer-like conditions around Edinburgh bringing fruit trees into blossom at the beginning of February, while three days previously the depth of snow had closed the road south over Soutra.[51] It was in such conditions that workmen clearing snow at Taymouth Castle were given whisky to fortify them against the cold.[52]

With this pattern of severe winters and poor summers seemingly established through the earlier 1820s, deviation from the new norm was surprising and perplexing. Whereas previous years had been notable for cold and damp conditions, 1826 saw excessive heat and dryness. The year had not started well, with intense cold in January probably the most severe since the 1810s,[53] but in late spring it set fair and temperatures soared while rainfall plummeted.

In a reversal of previous conditions, summer 1826 saw records of a prolonged and severe drought across a zone from Bute and Galloway in the west, through the Highlands into eastern Sutherland, to Aberdeenshire in the north-east and most of Highland Perthshire.[54] River-flow was badly affected, seasonal spate-fed streams in upland districts dried up, and natural springs were severely depleted across this entire region to a degree that affected grass growth. Tenants on the Breadalbane estate in Glen Lyon, for example, were petitioning in September for access to what should have been winter pasture for their cattle as the summer pastures were so wasted by the dry conditions, while in June in Aberdeenshire Adam Mackie noted that the hay crop was around a third to a half of what he was used to, 'it is so terribly burnt with the drought'.[55] For him, however, it was the effects on the arable crop that were the worst, for by the end of June his cereal crops were barely above 10cm high, stunted and withered, while on the stonier, higher ground they were completely blasted by the heat.[56] Despite a few all-too-brief thundery downpours, the dryness was such that in upland districts wildfires swept through the tinder-dry whins and moss, possibly devastating some 500 square kilometres of the Cairngorms between upper Strathdon and the headwaters of the Dee.[57] Mackie, too, provides some memorable illustrations of the summer heat's impact on other farming activities, including the temperatures being once again so high that butter would scarcely set.[58] But the positive conditions ended spectacularly with a reversion to

---

49 Ibid., pp. 123–124.
50 Ibid., p. 86.
51 NRS GD40/9/302.
52 NRS GD112/74/486 (9).
53 Stevenson, *Diary of a Canny Man*, p. 124.
54 See Chronology of British Hydrological Events at https://cbhe.hydrology.org.uk/ for 1826.
55 NRS GD112/11/9/2/43; Stevenson, *Diary of a Canny Man*, p. 124.
56 Stevenson, *Diary of a Canny Man*, p. 124.
57 Ibid., pp. 124–125.
58 Ibid., p. 124.

The Fannaichs, Torridon and Fisherfield mountains from Beinn Dearg, Highland. Searing heat and drought in summer 1826 reversed the trends towards cold and wet of previous years, but wasted the summer pastures in the uplands to such an extent that tenants sought landlord permission to drive their malnourished livestock to the lowland winter pastures.

severe weather; north-westerly gales started in late November, driving in rain that quickly turned to sleet and snow. Storms from around 22 November were devastating in the northern Highlands, with many lives lost in the unexpected snowfall, and the winds wrought havoc on estate plantations throughout Aberdeenshire and along the Moray coast, where there was also much loss of life on land and sea.[59] This poor weather in the north continued into the early months of the New Year, and only began to abate after three months.[60] This respite for northern districts, however, did not mean that winter had ended everywhere in Scotland; March 1827 brought more snow in the west, with Ayrshire being badly affected.[61] Winter had lasted for nearly five months.

Great spates and epic storms

The climatic volatility that developed in the mid 1820s was characterised by violent and intense summer as well as winter storms. The most extreme of these summer events struck the north-east in 1829 in an episode referred to as the 'Great Flood' or 'Great Spate' of 2–4 August, with a second event affecting parts of Inverness-shire at the end of the

---

59 NRS GD44/51/169/17; Dawson, *So Foul and Fair a Day*, p. 160; Stevenson, *Diary of a Canny Man*, pp. 125–126.
60 Dawson, *So Foul and Fair a Day*, p. 160.
61 NRS GD21/452.

month.[62] Sir Thomas Dick Lauder's exhaustive account of the floods explored their impact across the catchments of the Spey and its major tributaries (Avon, Dulnain, Nethy), the Findhorn, the Lossie, the Deveron and its tributaries, the Don and the Dee, and observed that the same event affected the rivers Bervie and the North and South Esk, south of the Mounth. He narrates a truly horrific catalogue of death and devastation across a region of nearly 13,000 square kilometres. The deluge came in on a north-easterly gale, and the spate that followed, as water shed rapidly off the northern side of the Cairngorm mountains and eastern side of the Monadhliath, surged down the river valleys leading towards the Moray Firth and stripped crops and soil from the fields, uprooted plantations, destroyed bridges, fords and mills, inundated towns and farms, and drowned thousands of sheep and cattle, as well as untold numbers of rabbits, hares and smaller mammals, and an unknown number of people.[63] Although the

62 T. Dick Lauder, *An Account of the Great Floods of August 1829, in the Province of Moray and Adjoining Districts*, 3rd edition (Elgin, 1873); D. Nairne, *Memorable Floods in the Highlands During the Nineteenth Century with some Accounts of The Great Frost of 1895* (Inverness, 1895); https://cbhe.hydrology.org.uk/ for 1829; NRS GD23/4/270.

63 Lamb, *Historic Storms*, pp. 124–126 gives a broad account of the 3–4 August and 27 August events, identifying the more northerly focus of the latter. For the damage to the Spey Bridge at Fochabers and proposals for its replacement, see NRS GD44/42/4/37/4 (12–17, 19).

Sir Edwin Henry Landseer's *Flood in the Highlands*, painted a generation after the events it depicts, presents an imagined scene of the distress and destruction wrought by the Great Spate of 1829 as it passed through Moray. (Aberdeen Art Gallery & Museums)

impact of the rainstorm lessened as it moved south of the Cairngorms, the storm still inflicted severe damage across Fife and around the Firth of Forth, although here the loss of life was limited.[64] As with more recent spate events in the Tay catchment, the greatest impact of the 1829 floods in Moray was on the Improved fields around the lower reaches of the Findhorn and the Spey, where the tenants of the earl of Moray, duke of Gordon and Cummings of Altyre had been active for many decades. There, where the topsoil was not stripped down to the fluvial gravels over which they lay, the fields were submerged beneath a deep layer of gravel carried down from the upper reaches of the rivers, wiping out decades of improvement in a few hours.

After the hammer-blow of the Great Spate, a cold, stormy autumn and severe winter brought 1829 to a close,[65] with a cold easterly airflow prevailing into the early months of 1830, and marking a return to the trend of the first quarter of the century. This cycle of severe winters and poor summers affecting broad swathes of country from the Northern Isles and Highlands to the Borders continued into the mid 1830s.[66] Although not classed as one of the 'historic' storms

---

64 Dawson, *So Foul and Fair a Day*, p. 161.
65 Lamb, *Historic Storms*, pp. 126–127.
66 See, for example, NRS GD1/582/15; GD46/15/41; GD112/16/2/1 (3) for disruption to road transport and to snow impacts in Edinburgh. For a detailed discussion of the snow events, see M.G. Pearson, 'Snowstorms in Scotland: 1831–1861', *Weather*, 33 (1978), pp. 392–399.

The Great Spate in Moray, 2–4 August 1829

Sir Thomas Dick Lauder's account of the almost unimaginable impact of the immense downpour that fell over the Monadhliath and northern Cairngorms in late summer 1829 was based on written and oral testimony from those who had witnessed and directly experienced its impacts. The following description of the scene in the broad floodplain of the River Findhorn between Forres, Moy and the Moray Firth on the morning of 4 August well illustrates the scale of devastation and the cost in lives and to livelihoods that the deluge had brought.

> At day-break Dr Brands [from Forres] hurried down to the offices [the farm-court and stables of Moy House] and ascended the tower to look out from the top. The wide waste of waters was only bounded by the rising ground about Forres, skirting the flooded plain to the south – by those about Dalvey to the west – whilst, towards the north and east, the watery world swept off uninterruptedly into the expanding Firth and the German Ocean [North Sea]. He looked anxiously for the houses of Stripeside. They were still standing; but the powerful and agitated stream that rolled around them, and between them and the offices, seemed to threaten their speedy destruction. The embankments appeared to have everywhere given way; and the water that covered the fields, lately so beautiful with yellow wheat, green turnips and other crops, rushed with so great impetuosity in certain directions as to form numerous currents setting furiously through the quieter parts of the inundation, and elevated several feet above it. As far as the eye could reach, the brownish-yellow moving mass of water was covered with trees and wreck of every description, whirled along with a force that shivered many of them against unseen obstacles. Even in the immediate vicinity of the offices, the Doctor saw one of those streams, created by a hollow in a turnip field, root up two large spreading elms, and sweep them into the abyss. There was a sublimity in the mighty power and deafening roar of the waters, heightened by the livid hue of the clouds, the sheeting rain, the howling of the wind, the lowing of the cattle and the screaming and wailing of the assembled people, that rivetted Dr Brands for some time to this elevated spot
>
> (T. Dick Lauder, *An Account of the Great Floods of August 1829, in the Province of Moray and Adjoining Districts*, 3rd edition (Elgin, 1873), pp. 90–91)

The losses Dr Brand witnessed were not just of the potential harvest of that year, but were the fruits of decades of investment that had been made on the estates in the main river valleys flowing northwards from the mountainous interior. From woodland plantations snapped off or uprooted by the force of wind and water, to the fields stripped of manure-enriched plough-soil and improved grassland that had been built up over many generations, the livestock developed through careful breeding programmes, to the built infrastructure of houses, barns, mills and field dykes, this was not a catastrophe from which the affected districts and their people would recover in a single season.

by Hubert Lamb, one summer tempest which swept across the North Atlantic and Shetland Isles in mid July 1832 inflicted devastating losses on the fisher community of the Northern Isles, from which the islands took decades to recover.[67] Much further south, in Angus, the diary of James Fyffe, a small tenant farmer and even smaller-scale cattle dealer, provides a window, similar to that of Adam Mackie in the previous decade, on events from September 1836 through to February 1840. His journal illustrates weather effects on harvests, fodder availability and cattle-droving, as well as its general influence on daily

---

67 For an eyewitness account, see E. Charlton, *Travels in Shetland 1832–52*, ed. W. Charlton (Lerwick, 2007), pp. 11–19.

life and travel.[68] Fyffe's record is alive with regular observations on the shifting patterns of weather as it affected his grain harvest, starting with a series of notes on the autumn 1836 harvest, which he commenced to cut in the third week of September and did not complete until the fourth week of October because of prolonged episodes of rainfall and the need to avoid the risks and additional expenses of bringing in a wet crop. Accounts like this provide insight on the sheer physical labour and psychological pressure involved when weather at harvest periods was poor, with Fyffe agonising over potential losses and, in this pre-mechanised era, working with his men until past midnight on the few good days to reap by hand and bind the cut grain to ensure that the crop was secured.[69] He was lucky in getting his harvest in, for across Scotland the wet summer and autumn saw widespread failure of the barley and oat crops, while in the Highlands and Islands the potato crop also failed. To make matters worse, the persistently wet conditions had again meant that peat could not be dried, leading to widespread fuel shortages for those least capable of paying for more expensive coal. Although Fyffe commented that 'the poor are saying that the meal will be chepe (sic) again', the generally poor harvest after a miserable year of weather kept prices high; winter thus brought a time of dearth and price inflation for basic foodstuffs and fuel in which the poor suffered hardest.[70]

While arable farming was clearly a chancy business in the 1830s, cattle-droving caused Fyffe no fewer problems, especially during that severe winter of 1836/7, when hard frosts and snow started on 28 October and lasted into the following April.[71] Fodder was in short supply and, although a brief window of milder weather saw some grass growth in mid February,[72] he was often obliged to buy or rent additional grazing for his small herd. His difficulties highlight the impact of the prevailing cold on grass growth, underscoring the vulnerability of small farmers who lacked access to adequate foggage or better-quality winter pasture and who were unable to grow sufficient fodder crops – like that most revolutionising of late eighteenth-century introductions, swede turnips[73] – on their own ground. As Fyffe found, even buying reserves of potatoes for winter was an insecure option, as in January 1840, when his entire crop, which had been left in the ground until needed, was lost to frost.[74] A gap of nearly two years in his diary after 4 February 1838, when he recorded 'much snow on the ground so much so that I never saw as much that I mind of',[75] unfortunately spans the worst period of the dreadful winter of 1837/8 and the equally bad winter of 1838/9.[76] We know from other sources that that first season brought another extended period of intense cold and snow across north-eastern districts, lasting from late December into February, with what were described as 'killing winds', but the cold continued into late spring and snow still lay in Lowland districts into May.[77] For men involved in the cattle

---

68 J. Dundas and D.G. Orr (eds), 'Quite Happy'. *The Diary of James Fyffe, Cattle Dealer, 1836–1840* (Dundee, 2016).
69 Ibid., pp. 29–30, entries for 22, 23, 24, 28, 29, 30 September and 1–25 October 1836. Only ten days from the 25th in October were sufficiently dry for any harvesting to be undertaken.
70 Dundas and Orr, *Diary of James Fyffe*, p. 30; Dawson, *So Foul and Fair a Day*, pp. 162–163.
71 Dundas and Orr, *Diary of James Fyffe*, pp. 30–63.
72 Ibid., pp. 26, 27, 29, 55, 60.
73 Smout, *Exploring Environmental History*, pp. 141–142.
74 Dundas and Orr, *Diary of James Fyffe*, p. 71.
75 Ibid., p. 68.
76 This first period in 1837/8 seems to have seen a major storm affecting Orkney, when the beach at Otterswick in Sanday was scoured to expose an early Holocene drowned peat and woodland layer: Lamb, *Historic Storms*, p. 129.
77 NRS GD46/15/90 (4); Dawson, *So Foul and Fair a Day*, p. 163.

trade, this delayed start to the growing season was catastrophic. Disturbed weather continued throughout the summer and early autumn of 1838, climaxing on 7 September with the major North Sea storm made famous by the wreck of the SS *Forfarshire* on the Farne Islands and the rescue heroics of Grace Darling and her father.[78] The deep Atlantic low-pressure cell that had generated the storm as it moved across the British Isles was the first of several that winter, the worst of which developed on 6–7 January 1839 and inflicted immense damage across the whole of the Atlantic-facing west of these islands but also in exposed parts of north-east Scotland.[79]

Highland cow, Glen Clova, Angus. For drovers and graziers as much as those dependent on grain or root crops, the unpredictability of the weather through the 1820s and 1830s brought little respite.

Bringing in blight

No alleviation came in the new decade. Indeed, the 1840s started with a continuation of the same patterns of low winter NAO, bringing poor to severe winters with episodes of intense cold, and low summer NAO which brought generally wet, cold summers. The intense storms of the 1830s also continued, inflicting further heavy losses on the infrastructure of the Shetland herring fishery from which it took over a decade to recover. The crises that successive years of poor weather had always threatened but whose impacts had never extended beyond the local since the 1790s, however, finally erupted under the unremitting pressure of drenching rain in the mid 1840s. Word came from Ireland in 1845 of the apparent failure of the potato crop on which most of the small tenants in the north and west of that island especially depended. This was *Phytophthora infestans* at work, the micro-organism thriving in the cool and wet conditions. Before the end of the year, it had spread into Scotland to affect crops in Islay and adjacent parts of mainland Argyll, and from then until 1848 the blight ravaged potato crops throughout Scotland, triggering the worst famine in Highland districts in recorded history. The impact of that famine is discussed in Chapter 11; here I concentrate on the weather conditions that exacerbated the spread of the blight and which compounded the severest impacts of the dearth in districts least well-equipped to cope.

As in Ireland, the weather in much of Highland and Hebridean Scotland throughout summer 1845 had been unrelentingly wet. By August, there were gloomy predictions about the likely extremity of the rain's impact on crops.[80] This gloom was borne out for western and northern districts but especially for the south-west Highlands and Islands, where the blight struck first. Winter 1845/6 brought early warning signs of a gathering subsistence crisis but, in weather terms, conditions were perhaps less severe than previous years. Nevertheless, while the north and west saw a mild late winter and spring, more

---

78 Lamb, *Historic Storms*, pp. 129–131.
79 Ibid., pp. 131–133. For the impact in Caithness, see Dawson, *So Foul and Fair a Day*, p. 164.
80 Dawson, *So Foul and Fair a Day*, p. 165.

easterly parts of the country suffered the same late-lingering blasts from the Arctic as had scarred most of the preceding half century. In Teviotdale, for example, 1846 brought snow and hard frosts in mid March, which blighted the early fruit blossoms, and the lingering cold delayed grass growth on low-lying pastures.[81] A general improvement seemed to commence in late spring and, across much of the Highlands and Western Isles, June saw perhaps the highest temperatures experienced in recent years. Then, however, the benign conditions ended and incessant rain through July saturated the ground, while temperatures fell sharply; by late summer, the blight was spreading rapidly through the western Highlands and Islands.[82] Autumn 1846 continued wet and stormy, lashed by the tail-end of a tropical hurricane in late October,[83] growing colder early in the season and leading into another dire winter of early and persistent snowfall made worse by a shortage of dry fuel; the late-summer rains had wettened the peat cut for the following year's use. But the dire conditions in the west had not been replicated elsewhere in the country, and in districts from Caithness to Easter Ross, the Moray coastlands and Aberdeenshire, the grain crops had seen more success while the potatoes had failed. Here, however, encouraged until their repeal in 1846 by the price-fixing, tariff and anti-import regime of the so-called Corn Laws that still enabled them to sell even their poor-quality grain in the urban markets of southern Britain, landlords had long been exporting the yield from their estates to these high-demand centres. It was an established trade and the big northern producers continued to export grain southwards even after the repeal of the Corn Laws, despite the evident hunger and distress locally that the potato failure was causing. Their actions helped to drive up the price of the only available alternative bulk foodstuff, oatmeal, deepening the crisis for people who were already starving through failure of their own subsistence food-crop. The result was a wave of riots aiming to prevent the shipment out of grain, which were suppressed by the deployment of the military.[84] Across Europe, similar subsistence crises and repressive acts triggered a wave of revolutions that climaxed in 1848. Even in Britain there was an air of alarm in the ruling regime, which ultimately led to the repeal of the laws that had kept grain prices artificially inflated.[85]

In many parts of the country, while still peppered by violent storms like that which lashed the north of Ireland and southern Hebrides in late April,[86] spring 1847 brought the start of a period of milder conditions that again delivered a good grain harvest further east and south. The success here at least meant that there was surplus grain that those landowners in the blight-stricken areas who provided their tenants with food- or money-relief could draw upon as the distress deepened. In the north and west,

---

81 NRS GD1/938/4.

82 For detailed discussion of the evolution of the crisis and the associated weather deterioration, see T.M. Devine, *The Great Highland Famine: Hunger, Emigration and the Scottish Highlands in the Nineteenth Century* (Edinburgh, 1988 and subsequent editions).

83 Lamb, *Historic Storms*, p. 133. This event brought poor conditions to the west of Ireland but, over Scotland, it was strong gales and drenching rain rather than more extreme conditions.

84 J. Hunter, *Insurrection: Scotland's Famine Winter* (Edinburgh, 2019). For an example of riots in Wick in February 1847, see Caithness Archives Centre https://www.highlifehighland.com/nucleus-nuclear-caithness-archives/wick-grain-riots-1847/ citing the Wick Parochial Board minutes (CC/7/10/1) of 2 March 1847. Oatmeal prices contributed to unrest that resulted in a riot in Port Gordon on the Moray coast, a property of the duke of Richmond and Gordon, one of the leading figures in the opposition to Corn Law repeal: NRS GD44/44/23.

85 For discussion of the effect of these 'Corn Laws', see below, pp. 211–312.

86 Dawson, *So Foul and Fair a Day*, p. 166.

however, the weather broke before the harvest and the hoped-for oat and barley yield was almost eliminated in the weeks of rain and gales that battered the region. The 1847 harvest there was, literally, a wash-out and the dampness encouraged the further spread of blight to all but destroy the already pitifully small potato crop. A repeat of these conditions saw 1848 mark the absolute nadir, for the only thing that thrived in the damp was *Phytophthora infestans* and the potato crop failed almost completely; in Harris, for example, the yield stood at a sixth of what it had been before 1846.[87] The region's grain crops brought no relief, for the poor summer saw much of the oats and barley across the Highland districts remaining green even in October, and hopes for a late respite were dashed when the first snowfall occurred in October and flattened the unripe crop. As the decade drew to a close, the distress in much of West Highland and Hebridean Scotland had risen to a level where the only solution for many was to seek new lives in Lowland cities or to emigrate abroad.

· · ·

Few people at the end of the 1840s would have predicted confidently that the extreme weather of the preceding six decades was drawing to a close. As the new decade dawned, adverse weather conditions continued and there was no immediate sense that a corner had been turned. The impact of the potato blight in the Highlands also receded slowly, leaving the rural regime throughout much of the north and west of the country in a parlous state and fuelling a continuing outflow of people, both as emigrants to the North American and newer Australasian colonies and internal economic migrants to the Lowland industrial centres. Resilience in the face of climatic change and environmental shock had finally broken down, with the degree of collapse worsened by the overdependence on a single crop-species; this overdependence had increased with the decline of other income sources that had sustained communities and estate networks (kelp and cattle-droving especially), while population levels had continued to climb to levels unsustainable on traditional agricultural systems. What is remarkable, however, is that while famine fears ran high throughout more southerly and easterly districts, especially during the worst years of the 1810s and 1820s, there were shortages of staples rather than total failures of supply. Let's not put too shiny a gloss on that; for the poor and socially marginal, shortages brought price inflation that put adequate food beyond their financial reach, especially when their need for fuel was also squeezed upwards by the extreme cold and wet of the period. Many suffered and many died, often from the diseases connected with poverty and malnourishment rather than directly through starvation. But there was no repeat of the 1690s, despite the gloomy prognostications of contemporary commentators. Although the high tide of Improvement was yet to come, agriculture in large areas of south-eastern Scotland especially was already more efficient and more resilient in the face of extreme weather. Improved cultivation methods using better agricultural equipment, more effective drainage, better soil-conditioning techniques coupled with more efficient harvesting, processing and storage methods, and production of fodder crops for livestock, increased not just yield but also the viability of farms and the survivability of adverse conditions. Harvests rarely failed entirely on a national level, and improved transportation systems meant that bulk goods could be transported to where the demand lay. But such changes came with costs as well as benefits in human and environmental terms, and long before 1850 there were some who decried or lamented the physical changes that had been wrought on the face of the land and the devastation of the communities that had long subsisted upon it.

87 Hutchison, *Industry, Reform and Empire*, p. 32.

CHAPTER 11

Improvement Applied: Agricultural Transformation 1790–1850

Ill fa' the sheep, a grey-faced nation,
That swept our hills with desolation!

Who made the bonnie green glens clear,
And acres scarce, and houses dear;

The grey-faced sheep, who worked our woe,
Where men no more may reap or sow,

And made us leave for their grey pens
Our bonnie braes and grassy glens,

Where we were reared, and gladly grew,
And lived to kin and country true;

Who bared the houses to the wind,
Where hearths were warm, and hearts were kind.

And spread the braes with wreck and ruin,
The grey-faced sheep for our undoing.[1]

As we saw in Chapter 9, 'Improvement' of agriculture had been underway in parts of south-eastern Scotland for well over a century before 1790. Nevertheless, much of the country was still being farmed in the decades after 1800 in accordance with practices that had become traditional two hundred years earlier. As the most positive reports submitted to Sir John Sinclair's *Statistical Account of Scotland* cannot quite conceal, even on the more innovative estates there was generally a mix of 'improved' and 'unimproved' cultivation, where processes of consolidation into single-tenancy farms with tenants on long-term Improving leases were proceeding unevenly or piecemeal as opportunities and resources permitted. Elsewhere, over-ambitious schemes for rapid development had stalled or failed in the face of tenant opposition or the lairds' financial exhaustion, leaving schemes half-finished. Most commonly, however, the

---

[1] Extract from Macintyre, 'A Song of Foxes', trans. Blackie, in Dunlop and Kamm, *Scottish Collection of Verse*, p. 83.

efforts to improve were concentrated on an ambitious laird's best land, where the biggest and fastest gains could be achieved. Thus, blocks of improved cultivation and pasture were appearing as islands amid the sea of unimproved land, from which, as in earlier centuries, the work of Improvement pushed gradually outwards. But the tide of change was spreading, albeit at a pace that varied. It was not only the pace but even the fundamental nature of the changes that varied, thanks to internal factors such as emergent patterns of regional agricultural specialisation within these islands and even within the regions themselves, and there were also external factors such as the growing global trade and market network of Britain's overseas empire, with the growth and recession cycles that it brought. And its spread was inexorable. What must be stressed, however, is the slow pace of the transformation; estates that had begun the process of Improvement in the first quarter of the eighteenth century were still implementing major new developments in the second quarter of the nineteenth, all the time responding to changing climatic, environmental and economic conditions.

Political, economic and cultural drivers

As we have seen many times in these volumes already, climate was a contributor to, rather than the determinant of, changes in Scottish agricultural expansion. Just as important were human responses to weather trends and longer-term climatic change; individual or collective decisions, often culturally or politically constrained, might have lasting socio-economic consequences. In the later eighteenth and nineteenth centuries, as what we would now recognise as government policies become clearer, we are also able to identify how political decisions had ramifications with often profound environmental impacts, as much as did the social, economic and cultural theories that underpinned Improvement philosophy. One important example of political drivers is the Corn Laws, which we encountered towards the end of Chapter 10, the legislation that from 1815 to 1846 continued and confirmed protectionist measures that had first been introduced in 1773 and tightened in 1791 and 1804. These were designed to prevent the export of grain and the import of cheap foreign food staples – principally all types of cereal – until the price of British-grown grain exceeded an unrealistically high threshold of 80 shillings per quarter. At a later date they allowed imports but at such high levels of duty that they were rendered unaffordable.[2] The most obvious consequence of these laws was the artificial elevation of the price of bread and other cereal-based products, driving up the cost of living at a time of stagnating incomes. This fixing of grain prices thus became the principal subject of agitation for Corn Law repeal through the 1820s and especially in the later 1830s and early 1840s, when the political lobby against them became better organised and more articulate.[3] The beneficiaries of the laws were principally landowners with arable interests, whose often inferior-quality grain was guaranteed a domestic market at inflated prices. In Scotland, as elsewhere in the British Isles, one effect of the bar on the import of cheap, good-quality foreign grain was the expansion of arable production onto poorer-quality land and the continuation of cultivation in zones that were at best marginal for cereal production during the cold, damp conditions that prevailed for the thirty

---

2   S. Fairlie, 'The nineteenth-century Corn Law reconsidered', *Economic History Review*, 18:3 (1965), pp. 562–575.
3   J.G. Williamson, 'The impact of the Corn Laws just prior to repeal', *Explorations in Economic History*, 27:2 (1990), pp. 123–156. The opposition coalesced in the Anti-Corn Law League, a body dominated by the interests of the urban and factory-owning middle classes who saw the artificial inflation of the cost of basic foodstuffs as detrimental to economic growth as it reduced the levels of disposable income of the vast majority of the population.

Modern wheatfield, Findogask, Perth and Kinross. The high price guaranteed for cereals under the Corn Law regime encouraged cultivators to continue to sow grain crops in ground that was ill-suited for arable, in the knowledge that they had a guaranteed market for their poor-quality yields.

years down to Corn Law repeal in 1846. Where the crop was not being grown for subsistence purposes, even a marginally positive yield of inferior oats meant profit for the cultivator.

There is no way that we will ever know how much land was kept in cultivation or broken into use as arable simply to capitalise on the financial opportunity of the Corn Laws. Clearly, there was growing demand for cereals as population levels at last began to climb rapidly from the epidemic-suppressed levels of the late medieval and early post-medieval periods, and that demand stimulated an expansion in the hectarage under arable cultivation. In many districts, especially but not only in the northern and western Highlands and Islands, where something closer to a subsistence agricultural regime persisted, cereal cultivation was maintained as a supplement alongside the more general cultivation of potatoes. These, as outlined in Chapter 10, had become established by the early 1800s as the chief crop in many Highland and Hebridean districts, as rising population levels and the consequent shortage of cultivable land meant that families became increasingly dependent on that high-yield/low-area food-crop. All types of grain, however, were being grown as a cash crop, through which some landlords still required rents to be paid. On large, tenanted farms on Improving leases and on the Mains or home farms of large estates, including those of northern landowners like the duke of Gordon in Moray, the Mackenzies of

Seaforth in Easter Ross, or the Sinclairs of Freswick in Caithness, cereal cultivation was the dominant mode of exploitation.[4] Much of this grain crop was for human consumption, but we should not overlook or underestimate the volume of oats that was being produced for sale as horse fodder; horsepower was still the primary means of traction for ploughs, carriages and carts, as well as for personal long-distance travel (other than on foot).[5]

Draining, embanking and clearing

We shall return later to the impact of the inroads on mossland and wetland drainage on fuel supply (see below, pp. 343–344, 351–353), but here we shall consider reclamation of such areas for agricultural purposes. It was a process well underway in Menteith from the 1770s, exemplified by Lord Kames's efforts on the mosses on his estate of Blair Drummond, but the pace of drainage and clearance accelerated in the decades after 1790.[6] A striking illustration of this trend was the proposal tabled in 1795 by Thomas Gordon for the drainage of Spynie Loch, the largest remaining expanse of water and wetland in the Laigh of Moray, including some of the last reserved areas of raised peat moss being exploited for fuel.[7] Gordon was convinced that this operation would expose rich sediment in the former loch-bed, from which he and the other landowners with properties round the basin would benefit. When the main part of the loch – covering an area nearly 5km east–west by 1.5km north–south – was finally drained in the years after 1807, however, its bed was found to be heavily mineralised, intractable and almost barren, rendering it useless at first for agriculture.[8] Its breaking into profitable cultivation followed substantial investment in the mid nineteenth century and, although it remains an important area of cereal cultivation at present, it is a fragile zone whose soils are highly susceptible to wind erosion. Until that investment in soil-improvement happened, however, here was an example of over-optimistic prediction of the agricultural potential of the to-be-reclaimed area leading to financial loss for investors. More to the point, however, it led to loss of a significant habitat range in the wetland zone, including the watery expanses that had been home to overwintering wildfowl and other wetland and aquatic species that had been hunted and caught by local people for millennia. In terms of materials loss, its drainage to the small expanse that remains today all but destroyed the birch woods and the willow and alder carr with which it had been fringed, removing a resource that had been managed for wood-needs for wattling, fencing, building materials and fire kindling since at least the later Middle Ages.

---

4 Between 1811 and 1813 on the Gordon Castle home farm, the seed purchases were exclusively of wheat, barley and oats: NRS GD44/51/361/3. In the 1830s, the reports of the Mackenzie factor at Brahan show that sale of all grain from the estate constituted one of the principal sources of income for his employers: NRS 46/1/391. The extent of barley and oat cultivation in north-eastern Caithness in the 1810s is illustrated by the accounts of the Sinclairs of Freswick: NRS GD136/855.

5 Devine, *Transformation of Rural Scotland*, p. 37.

6 Watson and Dixon, *History of Scotland's Landscapes*, pp. 119, 122. The extent of the clearance of this moss and the proliferation of the smallholdings cleared by the 'moss-tenant' farmers can be seen on the 'Map of Blairdrummond Moss showing progress of clearing in 1813', NLS https://maps.nls.uk/estates/rec/7861.

7 NRS GD44/41/65.

8 Rev. Alexander Simpson, 'Parish of New Spynie', *New Statistical Account*, vol. 13 (Edinburgh, 1845), pp. 95–101 at 96. Drainage commencing in 1807, see NRS GD248/707/6/1. For the costs of drainage, see NRS E403/40. An undated letter in the Gordon papers includes a proposal for a 'Persian Wheel' mechanism for draining the loch: NRS GD44/43/334 (12).

Spynie Canal/Great Cut at Gilston Bridge, Moray. The channel forms the central artery of the scheme initiated in 1795 and completed after 1807, which drained the pools and wetland of the Laigh of Moray to create new land for cultivation.

Gordon and the other Improvers, who had no personal investment in the management of such resources, were unaffected by this loss but it impacted heavily on the tenants of the still largely unimproved farm properties that fringed the former loch, who had been accustomed to augment their meagre incomes from its resources.

Drainage at Spynie was purely speculative but elsewhere operations were driven by the desire to stabilise access to a land resource that was already exploited for one agricultural practice for its conversion to use by another. Increased rainfall in the early decades of the century made effective drainage and embanking essential for the success of such land improvement or use-change, especially in zones subject to seasonal flooding or inundation by the sea, or otherwise prone to waterlogging. Estates that had pioneered embanking and realignment of watercourses earlier in the eighteenth century, like the work directed by Sir James Grant of Monymusk on his properties flanking the River Don, maintained and expanded that effort in the early nineteenth century as their tenants sought further crop-growing or pasture capacity for their farms. Across Scotland, the physical legacy of this enormous effort can still be seen in the kilometres of earthwork embankment that line major watercourses where they flow through low-lying ground and their natural floodplains, and in the straightened channels that were designed to speed the transit of water past the land wanted for agriculture. The negative environmental consequences of these straightening and embanking operations are with us still, bringing increased flood risks on lower reaches of rivers where the channelled water is finally able to find places where it can overtop the banks and spread over the adjacent land. What the Improvers were converting through these operations was usually seasonal water-meadow or wet haughland that generally supported poorer-quality pasture. Even relatively small estates, like that of the Mackintoshes of Mackintosh at Moy in Inverness-shire, where large areas of wetland and meadow surrounded Loch Moy and extended from there down the Funtack Burn to the River Findhorn, were actively enclosing, draining and embanking new fields to improve pasture quality and expand the cultivable ground.[9] As at Spynie Loch, however, this 'improvement' came at the cost of loss of other managed resources, especially the loch-side hazel, willow and alder. But, in common with other forms of natural resource that were seen as the physical embodiments of retrograde, culturally conservative and inherently wasteful practices – such as peat for fuel – these often were sacrificed in the name of progress.

Another focus of activity by the Improvers was stabilisation of property on the interface between land and water, which illustrates both their determination to master nature through the appliance of reason and science and their deep conviction in the untapped fertility of the alluvium in such areas. From the lands flanking the Beauly and Cromarty Firths in the north, through the Laigh of Moray, the district around Montrose Basin, to the carses of the Tay and Forth, landowners pressed on with embanking and drainage activities. At Tarradale in Easter Ross, for example, at the head of the Beauly Firth, the Baillies of Dochfour were actively embanking the coastal wetland and preparing it for lease as farms.[10] Nearby, efforts in the 1830s to increase the value of agricultural tenancies on the Seaforth estate close to its centre at Brahan in the lower reaches of Strathconon included significant outlay on the construction of embankments along the Conon to protect new cultivation on the river's southern floodplain. There were also plans for the enclosure by embankment of the Bog of Arcan west of the confluence of the Orrin with the Conon, in a district that was subject to regular seasonal inundation.[11] There, the drainage of the bog was intended to increase the size of the Seaforth demesne farm at Mains of Arcan and augment its extent of cultivable ground.

At the opposite end of the country in the Bowmont Valley in the northern Cheviots, the patterns of land-use evident from the later fourteenth and fifteenth centuries (discussed in the volume 'Scotland AD 400–1400' and Chapter 1 above) began to change. Unsurprisingly in view of the generally prevailing wet and cold conditions of this period, the upland district around Swindon Hill saw an episode of waterlogging between about 1775 and 1850. This episode almost replicated the saturation experienced in the seventeenth century, when some of the peat basins became ponds of standing water.[12] Soil erosion and channel movement of the watercourses in the valley also accelerated during these periods of greatest climatic stress, although these processes were more or less continuous since the waning of the MCA in the later thirteenth century and were generally confined to the higher and steeper ground rather than the lower, most regularly cultivated areas. In the latter, soil-conservation techniques perhaps protected the field systems to some degree.[13] There are suggestions in the pollen record of a transition

---

9 NRS GD176/1288.
10 NRS GD23/6/564.
11 NRS GD46/1/366, GD46/1/391.
12 Tipping, *Bowmont*, p. 197.
13 Ibid., p. 198.

towards specialist pastoral production on some of the coldest and dampest upland areas above the valley from the mid seventeenth century. But cereal crops continued to be cultivated at high altitude in some parts of this district into the nineteenth century,[14] with local farmers perhaps encouraged by the high prices for all grain that resulted from the artificial stimulus of the Corn Laws until their abolition in 1846. Lower in the valley, cereals including oat, wheat and rye were being grown certainly until the 1830s, and in some parts down to c.1900. Cultivation's 'retreat from the margins' that has long been argued for as a consequence of the 'little ice age' for these upland areas of south-eastern Scotland seems, on this evidence, to have come much later than the supposed nadir of that climate episode in the late seventeenth century. Richard Tipping has proposed that the end came more through economic decisions and the socio-economic consequences of the agricultural reorganisation of Improvement than as a response to climate trends.[15] And yet, the fading out of cereal cultivation in parts of the district in the second quarter of the nineteenth century, a period of intense and long-lasting winters, poor summers and accordingly bad harvests, might indicate that a degree of environmental determinism also contributed to the decision to focus on alternative agricultural opportunities.

Soils improved

As a counter to such determinism, however, we must reflect on the increasing ability of farmers in this era to modify soil structure, its chemistry and water content in ways that made previously unprofitable land capable of bearing crops and delivering yields that represented a profit on the investment of labour and materials. Better drainage was one of the most effective methods of improving land quality and expanding the cultivable zone, but control of watering – to increase irrigation of an area of ground – was also introduced to Scotland from the late 1790s, especially for the creation of water-meadows that provided high-quality grazing and fodder throughout the year. Floodplains and haughland that had largely been avoided for intensive use historically because of the seasonal inundations they experienced, unless defended by embankments and converted to arable, were brought into a managed irrigation scheme, where water was delivered through sluice-controlled channels to keep the ground moist and supplied with water-borne nutrients.[16] First introduced by English designers on the Buccleuch estates shortly before 1800, they were quickly to be found distributed from western Galloway to Easter Ross, but with their greatest concentrations in Peeblesshire, Lanarkshire and Midlothian, before fading from use in the 1850s as better fodder-crop alternatives became more widely available. Although a short-lived phenomenon, they had an immediate impact and lasting legacy, affecting river hydrology and erosion patterns, soil conditions and riparian ecologies, especially through the loss of the scrubby carrs that had dominated the watery ground prior to its conversion to water-meadow. Such interventions, however, although profound in their local impact were minor in comparison to the effects of direct intervention in the fundamental characteristics of fields and soil.

Soil modification was, as we have already seen, a process as old as arable cultivation itself and, contrary to the dismissive views of Improving zealots, had

---

14 Ibid., pp. 199, 201–203.
15 Ibid., p. 199.
16 Watson and Dixon, *History of Scotland's Landscapes*, p. 122. For case studies in Perthshire, see I. Fraser, 'Three Perthshire water meadows: Strathallan, Glendevon and Bertha', *Tayside and Fife Archaeological Journal*, 7 (2001), pp. 133–144.

Rig-and-furrow cultivation, Lammermuir Hills, Scottish Borders. This block of cultivation, buried amid heather and cut across by erosion gullies, is a good example of the small patches of higher-altitude cultivation of medieval and post-medieval date that are scattered across the uplands of the eastern Borders.

resulted in the creation of reservoirs of built-in enrichment in expanses of anthrosols around many Scottish towns and villages.[17] Organic, manure-based fertilisation was already being augmented or replaced by mineral or chemical additives, especially of lime, in the later sixteenth and seventeenth centuries, but the later eighteenth century brought a step-change in practices. Improvement literature from the early 1700s onwards, however, reveals an obsession with not just quantity of fertiliser but also the quality of manure and how best to accumulate, store and apply it. When combined with enhancement of soil drainage, principally through trenching and construction of field drains, straightening of watercourses and embanking, application of organic and chemical fertilisers quickened the pace of large-scale ecological change across much of Lowland Scotland and many parts of the Highlands. It was the fundamental consistency and content of the soil that was being modified, changing the biodiversity of plant, fungus and invertebrate communities that were supported by and upon it.

Liming, where crushed and ground limestone or shells is applied to acid soils for the naturally occurring calcium carbonate and magnesium carbonate within that material to increase the alkalinity and raise pH levels, was the most popular method of soil-improvement. As discussed in Chapter 9, it was well

17 Oram, 'Waste management and peri-urban architecture'.

Charlestown Limekilns, Charlestown, Fife. The largest production centre in Scotland for agricultural and building lime at the start of the nineteenth century, Charlestown made use of the new canal system through central Scotland to transport bulk supplies to Improving clients.

established in Lowland districts as an agricultural method by the third quarter of the eighteenth century. Many estates were operating small-scale lime quarries and kilns where outcrops occurred on their land and where fuel could be obtained easily. In the 1760s, the scale of operation had begun to change as demand increased, and landowners such as the 5th earl of Elgin were made aware of the happy juxtaposition of beds of high-grade limestone and coal-seams close to convenient coastal sites where the processed lime – slaked for building work and unslaked for agricultural use – could be shipped to market. It was his second son, the 7th earl (of the Elgin Marbles infamy), however, who ramped up production from the complex of kilns, initially nine and increased in 1792 to fourteen, to become the largest lime-processing site in Scotland, which his father had planned at Charlestown on the Firth of Forth west of Rosyth.[18] By the 1790s, around 1,300 cargos per annum of processed lime were being shipped out through its purpose-built harbour. Proximity to the Grangemouth basin of the Forth and Clyde canal, which opened for traffic in 1790, meant that the Charlestown lime could reach western Scot-

18 B. Walker and G. Ritchie, *Exploring Scotland's Heritage: Fife, Perthshire and Angus,* 2nd edition (Edinburgh, 1995), p. 57 no. 6.

tish markets easily, as well as being shipped coastwise throughout eastern districts.

The fall in production costs after the abolition in 1793 of the high duty that was levied on shipments of Forth Valley coal north of Red Head in Angus made lime-burning in north-east and northern Scotland more economically viable, as northern proprietors had argued. Leading proponents of the abolition had included the Scotts of Dunninald, long active in eastern Angus as Improvers, whose property lay just 5km north of the Red Head boundary.[19] Lime had been quarried since the 1690s at the southernmost point of their land, at the Boddin headland at the north end of Lunan Bay, and the family had built the first kiln there, within sight of Red Head, in c.1750.[20] Access to cheaper fuel meant that production levels could be increased and, for the remainder of its period of operation down to the 1830s, it became one of the principal suppliers of both building and agricultural lime in eastern Angus and the Mearns. Another beneficiary of the abolition of coal duties was the Gordon estate, where exploitation of the limestone at Ardonald near Cairnie south-east of Keith started in 1796, using coal carted the 28km from the duke of Gordon's harbour at Tugnet at the mouth of the Spey.[21] By the 1800s, the Ardonald quarry and kiln were supplying lime for ducal tenants in the district between the River Deveron and Elgin, especially where new fields were being developed on the sandy districts either side of the Spey, where calcium carbonate leached freely and needed regular replenishment.[22] On the Breadalbane estate, kilns operated at multiple locations spread from Lochearnside, through Ardeonaig and Ardtalnaig on the south side of Loch Tay to Tomphubil above the Braes of Foss. Much of their output went into the extensive construction being undertaken by the estate across its landholding, on farms and farm-buildings, bridges and kirk manses as well as the earl's new mansion at Taymouth, but it was also supplied at good prices to tenants on Improving leases, who were bound to drain and enrich sour and acidic wetland and to plough and lime 'barren' ground.[23] Alongside such large-scale operations were multiple smaller, single kilns spread across the country, often supplying only the needs of a small estate in its building and agricultural lime and usually operating seasonally or intermittently. Where no limestone existed locally, the increasing availability of Charlestown (and other) lime at affordable prices meant that tenants on Improving leases could obtain supplies on an individual basis, if necessary.

Liming and, to a lesser extent by the second quarter of the nineteenth century, marling (using the clay/lime sediment from some lowland loch-beds), was principally used, however, to reduce the acidity of soil. Manuring was essential to increase its general fertility and improve the volume of organic material in its structure. Again, there tends to be a presentation of manuring practices as the invention of the Improvers, but it was their techniques for its better conservation and application that were new rather

---

19 In August 1793, David Scott MP wrote to Henry Dundas to thank him for his efforts in securing the repeal of the levy and advising him that the burgh of Montrose had given him free-burgess status in recognition of his efforts: NRS GD51/5/198/1–3.
20 HES Canmore, Boddin Point, limekilns: https://canmore.org.uk/site/36303/boddin-point-limekilns.
21 NRS GD44/39/30: the Gordon estate obtained coal from Midlothian. The estate papers show how the managers of the Ardonald limekiln continued to gather intelligence on methods and levels of production at other kilns on the east coast.
22 See, for example, NRS GD44/51/509/2 (14–16), for the supply of lime shells delivered to George Jamieson for the farm of Burnside of Tynet, 3km east-north-east of Gordon Castle, from the lime works of Ardonald.
23 For example, John Malloch in 1798 on the farm of Tomore: NRS GD112/11/6/4/95. For the quarry at Dalbiath or Dalaveich on Loch Earn, see GD112/11/7/7/10, GD112/74/145 (37) and GD112/74/401, and at Ardeonaig, see GD112/11/2/5/84.

than the principle itself.[24] Dung from different animals was seen as having a hierarchy of value, with the early eighteenth-century Improver Robert Maxwell extolling sheep manure above all others and recommending how the floors of the sheep cots in which the flocks were housed should be formed from layers of sand, clay, lime and organic bedding to retain as much dung and urine as possible. Management of all stabled and byre animals became an artform, and debate over dietary composition as a means of improving the quality of the resultant manure filled the pages of farmers' magazines and digests. Farm-buildings, larger and sturdier by using the lime for mortar that was produced alongside agricultural lime, were themselves developed experimentally to improve methods of capturing and containing solid and liquid manure. By the 1830s, the latter was led by floor conduit and gathered in tanks for spraying as a top-dressing on grass fields. In the arable districts of eastern Scotland, farmers adopted the overwintering of cattle in enclosed 'courts', where they were fed on swede turnips and other harvested fodder and where straw gathered during the grain harvest was used to provide bedding and to soak up the more liquid faeces these animals produce. In the spring, when the livestock was sold on, the courts – filled to a depth of almost 60cm at times with the faeces/straw mix – were then dug out and the winter's product added to the available fertiliser for the spring-sown crop.[25] Human excrement and farm household waste added to the volume and content of agricultural middens, but the greatest sources of such materials for farmers came from urban centres.[26] Shambles and fish-gutting waste, street-sweepings and stable muckings – a hugely valuable and voluminous source of manure in that age of the horse – added to the contents of privies and ash pits and were sold at public roup, the lucky buyers receiving tons of material for spreading on the fields.

Some urban sewage was already being discharged directly into streams and, from the 1770s, one of these at Kirriemuir was being used to deliver nutrients to 35 acres of water-meadow at Logie downstream of the town, to improve the grass quality.[27] The most notorious of all these experiments with sewage-fed meadows was at Craigentinny, now in the eastern suburbs of Edinburgh, where the 'Foul Burn' that drained the low ground to either side of the Old Town watered grassland leased to cattle graziers.[28] The stench from these meadows was long regarded as a miasma that carried disease back to the city, in the days before bacteria and viruses were understood, and contributed to the agitation that eventually ended sewage fertilisation at Craigentinny in the last quarter of the nineteenth century. It was, however, recognition of the negative impact of heavy metal contamination through their concentration in soil enriched with human waste that eventually led to the abandonment of this practice throughout Britain.[29]

Combined, liming and manuring not only brought an increase in yield levels across areas where arable agriculture had been established for centuries but they also enabled intensification of cultivation

---

24 Shaw, 'Manuring and fertilising the Lowlands', p. 112.
25 For discussion, see ibid., pp. 112–114.
26 Oram, 'Waste management and peri-urban agriculture'.
27 Shaw, 'Manuring and fertilising the Lowlands', p. 114; Rev. Mr. Thomas Ogilvy, 'Kirriemuir, County of Forfar', *Statistical Account*, vol. 12 (Edinburgh, 1794), p. 192. The Logie meadows were drained in the early 1800s when they were found to be sitting on top of a rich marl deposit.
28 P.J. Smith, 'The foul burns of Edinburgh: public health attitudes and environmental change', *Scottish Geographical Journal*, 91 (1975), pp. 25–37.
29 J.A. Catt, 'Long-term consequences of using artificial and organic fertilisers: the Rothamsted Experiments', in S. Foster and T.C. Smout (eds), *The History of Soils and Field Systems* (Aberdeen, 1994), pp. 119–134 at 130, 131.

Greenock, Renfrewshire, as illustrated by William Daniell. The Clyde fishing port was famed for the quality of its waste, which mixed fish-processing refuse with domestic refuse, and which was regarded by nearby agriculturalists as exceedingly effective fertiliser.

on former outfield zones and made practicable cultivation of areas previously deemed too 'barren' for productive crop-growing. The spread of these new ideas for soil-treatment was uneven, with some traditional practices prevailing in some areas where alternative fertilising means were available. Thus, in coastal districts from Berwickshire to Orkney and Ayrshire to the Western Isles, the use of storm-beached seaweed or sea-ware continued,[30] in some places well into the second half of the twentieth century. As rich, or richer, than farmyard manures in nitrogen and potassium, but deficient in phosphates, it brought a boost in the year after application, especially for grain crops, but needed to be regularly replenished, or supplemented with other, phosphate-rich, materials. Before the introduction of South American and southern African guano in bulk from the 1840s, sea-ware was perhaps the most efficacious fertiliser available in Scotland, albeit with a short effective life. It was, however, the importation of that often forgotten commodity plundered from the global south in the age of colonial empires, guano, that led to a decline in the use of many other forms of manure. Guano's concentrated levels of nitrogen, potassium and phosphates and availability in powdered form rendered it infinitely preferable to the bulkier and semi-liquid farmyard varieties. It has been argued that the transference of nutrients from

30 Shaw, 'Manuring and fertilising the Lowlands', p. 116.

source to consumer represented by the guano trade was equivalent to the precious metal plunder of the Spanish and Portuguese empires of the sixteenth and seventeenth centuries, or the African slave trade of the eighteenth and nineteenth, with access to the commodity giving an immense boost to the agricultural systems of the global north at the moment when their traditional agricultural production was hard against its Malthusian limit. The jump in productivity delivered by guano, both at home in Britain and in the agricultural systems within its empire, breached that barrier and opened the path to the full-blown industrialisation and urbanisation of the later nineteenth century. With steamships to carry the bagged guano from Peru and Namibia to British ports and a mature rail network to deliver it to depots in almost every corner of the country, its impact on Scottish farming was immediate and immense.[31] Although we now know that an unwanted companion of imports of Peruvian guano was the *Phytophthora infestans* potato blight from South America, which brought untold death and suffering to Europe and especially to Ireland and Highland Scotland in the 1840s, it was the nitrogen, potassium and phosphate boost to grain production that helped to prevent that disaster from having far wider and greater impacts.[32]

Soil enrichment and modification were major agents of environmental change in the first half of the nineteenth century, producing very visible results in terms of the health and quality of grown crops and grass, and the yields they delivered. They also transformed the colour and texture of the land as plant ecologies responded to the changed soil chemistries or were changed by introduction of new sown species. The vivid greens of artificially enriched grassland in upland districts is still strikingly visible in the landscape today. These were among the many transformations that Walter Scott lamented from his historical perspective and the Romantics bewailed from their idealised rural social viewpoint. But the shape of the land was also changing fast as the new methods spread, organisational structures were reconfigured, and new agricultural technologies introduced. From our modern vantage point, it is too easy to underestimate the significance of the changes in agricultural equipment, with first the adoption of lighter swing ploughs following Joseph Foljambe's 1730 patented 'Rotherham' form, and especially from 1763 with John Small from Berwickshire's cast-iron 'Scots Plough', which was the first capable of mass production.[33] These new ploughs were lighter, requiring fewer draught animals than the heavy ploughs that had been used from the eleventh to eighteenth centuries, whose long ox-teams had required room to turn. This turning room was responsible for creating extended 'headlands' at either end of the sinuous, reverse-S ridges, which were themselves a necessary consequence of the wide turning arc.[34] So too the new ploughs changed the stock balances on farms, as numbers of oxen declined, and horses, the preferred source of traction for the new ploughs, increased. But it was the shape and nature of the ploughing that changed most dramatically, with the unenclosed broad rig-and-furrow form created by the old-style plough and ox-

---

31 For the global impact of this 'wonder fertiliser' see G.T. Cushman, *Guano and the Opening of the Pacific World: A Global Ecological History* (Cambridge, 2013).
32 A.C. Saville, M.D. Martin and J.B. Ristaino, 'Historic late blight outbreaks caused by a widespread dominant lineage of *Phytophthora infestans* (Mont.) de Bary', *PLoS One*, 11 (2) (2016): doi:10.1371/journal.pone.0168381.
33 National Museums of Scotland, *The Story of the Plough* https://www.nms.ac.uk/explore-our-collections/stories/science-and-technology/ploughs.
34 P. Dixon, 'Field systems, rigs and other cultivation remains in Scotland: the field evidence', in S. Foster and T.C. Smout (eds), *The History of Soils and Field Systems* (Aberdeen, 1994), pp. 26–52 at 37–38.

Menstrie Glen, Clackmannanshire. The upland glen on the south side of the Ochil Hills was a scene of intensive Improvement activity at the turn of the eighteenth and nineteenth centuries; its legacy can be seen in the extensive areas of cultivation rigs that cover the south- and west-facing slopes.

teams, whose presence can be seen almost universally across Scotland outside the Highlands, Hebrides and parts of Galloway[35] in the mid-eighteenth-century maps of General William Roy. In their place came narrow, straight rigs with smaller headlands on which the horse-teams and smaller plough turned. These prevailed until the widespread introduction of underground ceramic field drains in the later nineteenth century, which brought a further change in ploughing technique.[36] Armed with this new technology, would-be Improvers turned their attention to land that had always lain at the environmental and technological margins, pushing back into upland districts and creating the distinctive surface forms that survive as relict landscapes in places like Menstrie Glen above the steep escarpment of the Ochil Hills in Clackmannanshire.[37] The legacy of this change, however, is not just in the corduroy of old plough ridges, for the local biota changed as drainage patterns changed, soil was aerated and became uncompacted, something that is often immediately visible in the lines of rushes that fill the wetter furrows, while the ridges are crowded with plants that prefer drier conditions. From the pattern of farming settlement to the shape of fields,

---

35 The narrow curving rigs identified in Galloway and parts of Argyll have not been dated but are certainly pre-Improvement: ibid., pp. 39, 41.
36 Ibid., p. 41.
37 RCAHMS, *'Well shelterd and watered': Menstrie Glen, a farming landscape near Stirling* (Edinburgh: RCAHMS, 2001).

soil conditions and crop types, to the non-cultivated biota upon and within the ground, the transformation was radical and delivered a new prescription that many deplored. For that disaffected community, despite the new techniques' manifest economic advantages for a fortunate few, they marked the first wave of a profound social and economic dislocation that forever tipped the balance between rural and urban populations.

Dislocation, dispossession and resettlement

We have for long been too glib in passing over the negatives when we talk of 'agricultural Improvement' in the later eighteenth and nineteenth centuries. Improvement did mean the end of old-established rural regimes, principally the demise of the multiple-tenant fermtoun of the Lowlands and Highland fringe and its replacement with the rise of the single-tenancy farm. Consolidation into single tenancies usually meant social degradation for those displaced and incapable of securing a tenancy elsewhere, as they declined into the ranks of the landless rural labourers or became economic migrants seeking employment in the new industrial centres or, perhaps, emigrated. We are probably all familiar with the conventional perception of the impact of Improvement in the Highlands and Hebrides, where the Clearances were long presented as perpetrated solely in the name of reason and profit – or greed – and implemented with a brutality and ruthlessness that could shock and outrage generations later. It was not just Clearance that yielded the legacy of depopulation, ecological transformation and an enduring cultural cleavage that maintains a powerful hold in the popular consciousness. That dominant discourse has meant that we have only recently regained awareness of the impact of similar processes at work in the Lowlands from far earlier and lasting far longer, where far greater numbers were affected, but where the steady removal of people from the land occurred with little public comment or criticism and has consequently received far less popular attention.[38] Here, however, unlike in most of the Highlands, there were nearby alternatives that could absorb dispossessed tenant populations, either in alternative rural employment or through migration into the rapidly expanding manufacturing towns. There were also counterflows, with multiple new single-tenancy holdings being carved from single larger and older properties. This was the era, too, of the first development of planned villages by estates, often to house dispossessed tenants who were still needed as labourers, but also to form nuclei for small-scale rural industrial developments. In the Machars of Galloway, for example, retired army officer James McHaffie purchased the Torhousemuir estate in 1827, enclosed it with a ring dyke and subdivided it into small farms and crofts rentable at £10 and £5 respectively as holdings on Improving leases, with his own mansion at its heart, at a time when other landowners were removing such clusters of small and relatively unproductive farms.[39] Elsewhere, it was not simply a matter of replacement of a community with a single farm, for the transformation

---

38 Devine, *Transformation of Rural Scotland*, especially chapters 7 and 8, provides an overview of the pre-1815 period, but the subject is covered in most detail in T.M. Devine, *The Scottish Clearances: A History of the Dispossessed 1660–1900* (London, 2018). For an accessible overview, see Aitchison and Cassell, *The Lowland Clearances*.
39 R.D. Oram, *Torhousemuir Historical Account:* Report on Programme of Historical Research undertaken by Dr Richard Oram/Retrospect Historical Services on behalf of Scottish Natural Heritage, February 1995. McHaffie's motivation is unknown, but the arrangement of the smallholdings has a distinctly Romantic and paternalistic air to it. Many of the crofts, especially in the wet and stony moorland portion at the northern end of the property, were economically unviable and failed before 1850, being merged into some of the more successful larger holdings.

Field clearance, Hawksnest, Ale Water, Scottish Borders. The valley of the Ale Water had been the focus of agricultural activity since the later twelfth century, mainly for large-scale cattle and sheep grazing, with limited agriculture around the towers of the minor lairds, but in the early 1800s an intensive episode of agricultural expansion saw stone-clearance to carve new arable fields from the upland pastures.

extended to a realignment of boundaries and reallocation of the resources available on the land, from access to fuel reserves, woodland and quarries for building stone, to allocation of summer grazing. And these boundaries were real and substantial, with new systems of stone dykes creating new field enclosures that zoned the landscape into new shapes and patterns of use. As the records of the Breadalbane estate alone serve to illustrate, dyke-building constituted one of the biggest operations underway from the 1790s onwards, carving new lines through old properties, cutting unimproved farms off from areas of traditional grazing in upland districts that had been associated with them for generations, providing enclosed plantation ground for new trees, and dividing the lowland zone into blocks of intensive cultivation, woodland and managed grassland.[40] Such divisions between properties were more rigid and physically prominent than before, giving hard edges to zones of different land-use and management.

40 See, for example, in a sample for the 1790s alone NRS GD112/1/786, GD112/1/787, GD112/1/788, GD112/1/799, GD112/11/2/5/38, GD112/11/2/5/59, GD112/11/3/2/64, GD112/11/4/2/47, GD112/11/5/2/117, for estimates of building costs, assessment of work undertaken, complaints and petitions over new boundary lines, petitions concerning maintenance of the dykes and for access through them.

Contrasts between areas of 'improvement' and traditional exploitation, often reflecting the financial capacity of different owners to invest in new methods and equipment, could thus become starker still. The stone and turf march-dyke that separated the financially sound duke of Argyll's land in eastern Coll from the impoverished Maclean estate in the rest of the island in the 1790s is an excellent example of the cleavage in landscape character and underlying ecological pattern that such divisions established.[41]

Dyke-building, however, was more than a matter of better enclosure to protect growing crops and trees from unsupervised grazing animals; it was also a primary mechanism for ground-clearance and improvement of soil conditions to make it easier to plough and harrow. Stone clearing from cultivated ground is as old as the practice of agriculture and has left a legacy of clearance cairns (where the stones were simply heaped) or spreads (where the stone was dumped over the edge of a steep, uncultivable slope) that developed over nearly five millennia, but this work was now being undertaken systematically and on a scale that changed the character of great tracts of land.[42] The new field enclosures consumed much of the stone cleared from the ground and became referred to as 'consumption dykes'. It was after 1850 that the greatest of such dykes were built, mainly in the north-east where the Improvement of the interior of Buchan especially was underway,[43] but the substantial stone walls that elsewhere carve up much of Scotland's modern agricultural landscape were largely the product of the first half of the nineteenth century. Often overlooked or unremarked upon, these structures are among the greatest visual indicators of the scale and extent of change being wrought within just a few decades. From the outset, they could form a nigh-impermeable edge between zones of different land-use within and between farms and between arable land and pasture. Scotland's environment was being reconfigured to fit a new prescription.

Droving and grazing

Reconfiguration was not a smooth or even process. It offered and withdrew opportunity, saw brief efflorescence and rapid demise within different agricultural sectors across the six decades explored here. All left some mark on the land, and some contributed to environmental change on a grand scale. Some of the greatest changes occurred in the livestock sector, especially within upland districts, shifting balances from cattle to sheep and from small flocks and herds in their tens held by individual small tenant farmers to large undertakings of hundreds, if not thousands of animals controlled by capitalist adventurers. The post-Union growth of the domestic cattle trade and its increasing onward flow to England, discussed in Chapter 4, had accelerated in the middle of the century and brought a shift in land management along the routes down which the cattle were driven to the markets of central Scotland. Large parks or stances, like those developed at regular intervals through the Breadalbane lands in Glenorchy and Glendochart, on the Campbell of Argyll and Campbell of Knockbuy's Lochfyneside properties, or those at Ingliston west of Edinburgh (discussed in Chapter

---

41 Cregeen, *Argyll Estate Instructions*, pp. 19, 20.
42 Impressive spreads of clearance material of early nineteenth-century date can be seen at Ballinlagg near Grantown-on-Spey, where the Grant estate was developing improved farms from the 1760s and with renewed intensity from *c*.1802. See NRS RHP8982 and RHP98239.
43 Some of the finest surviving are to be seen near Kingswells, west of Aberdeen, where they were built from *c*.1854 on the instruction of the landowner to 'consume' the stones from the unimproved areas of his estate: HES Canmore, https://canmore.org.uk/site/19313/kingswells-consumption-dykes.

Planning the next leap forward

Improvement's progress was still uneven around the country by 1850. It was not a question of advanced and developed Lowlands versus retrograde and underdeveloped Highlands, for landlords throughout the country had introduced Improving leases generally by the 1830s. The result was a patchwork of areas of established Improved agriculture and grassland spread from Wigtownshire to Orkney alongside tracts of land viewed as capable of development but where finance and technological limitations had deterred any investment of labour and resources. A brief survey of the parish reports in the *New Statistical Account* of the 1840s, undertaken to provide data on the progress made since Sir John Sinclair's 'old' *Statistical Account* of the 1790s, reveals this picture of unevenness. What they also present is a record of work in progress and of planned activity that would accelerate in the coming decades. It was not climatic amelioration that fuelled that expansion, for the 1840s – and, indeed, the 1850s – saw some of the poorest weather conditions since the 1810s. Instead, it was greater access to the means to effect change. Access to capital grew as families participating in Britain's burgeoning empire funnelled their wealth homewards, and access to credit likewise grew as the banking network expanded and sought ways to give investors returns. Materials could be moved faster due to the extension of a road system that was capable of bearing heavy, wheeled traffic and the spread of the railway network. And bubbling away behind this was further experimentation with new crops, new tree types and selectively bred livestock who were better suited to the more intractable areas of the country that were being eyed for the next phase of Improvement. It was to these that the efforts were to turn with a vengeance in the decades after 1850, as tenants expanded their cultivated ground or improved grassland into the as yet unimproved areas of their holding. In this, they were aided by the increasing availability of lime and guano to enrich even the most hostile of soils. In the coming decades, the introduced fertilisers would bring the transformation of Scotland's landscape to what can be viewed as both a high tide of Improvement and the lowest ebb of environmental degradation.

One area where the next leap forward was poised to begin was Buchan in north-eastern Aberdeenshire, a district then notorious for its restricted patches of good and improved ground interspersed through areas of poor, rocky soils, lowland mosses and wetland hollows, exposed to the worst that the prevailing northerly or easterly winds could bring. Here, Improvement had been an early arrival on estates like those of the Gordon earls of Aberdeen at Haddo or on the plantation-funded development of the Russells at Aden, but in most places the investment had concentrated on the already worked land that was clearly capable of improvement. The description of the parish of New Deer in the *New Statistical Account* captures this situation at the moment when change was about to take off:

> The soil in general, with few exceptions, is light and shallow. It would answer well for agriculture if it were not for the climate and the subsoil. A great proportion of the parish rests on a hard rock pan of from 6 inches to 2 feet thick, which prevents the surface water from sinking into the earth, and keeps the soil wet till the sun evaporates the moisture. This pan prevents trees from thriving and coming to any size. Attempts have been made, and in some places successfully, to break up this pan, and give the surface water a passage into the earth.
>
> In some parts the subsoil is moss on coarse clay, in others it is mixed with a coarse granite. Lime of indifferent quality is found in the land of Barrack. The farmers quarry it for themselves, and burn it either for building or putting on the land.
>
> There is plenty of moss in the parish, though it is wearing away apace, either through improvement, or by consumption as fuel. The part of it that has been cultivated, produces excellent crops when mixed with shell sand from the sea side. The country appears to have been once covered in wood, from the remains of trees that are dug out of the mosses, though it is somewhat remarkable that none of those trees are fir
>
> (Rev. James Welsh, 'New Deer, County of Aberdeen', *New Statistical Account of Scotland*, Volume 12 (1845), p. 176).

Here we are presented with a record of improvement underway, with farmers who were already exponents of Improvement practices on the more tractable land labouring to break through mineralised subsoils to enhance drainage, quarrying their own lime to burn in their own kilns and use on their fields or to build new farm-steadings. We hear of wetland being drained and mosses stripped for fuel or converted into cultivable ground through the application of shell sand brought in from the nearby coast on the new roads. The Rev. Welsh's account makes it sound straightforward, but what he was recording was the opening stage of one of the most extensive and rapid episodes of anthropogenic environmental change in any region of Scotland, which transformed the regional ecology and created the managed landscape of modern Buchan.

9), where the animals were rested and fattened before being driven into the city, proliferated from the 1750s on the drove routes through Argyll, western Perthshire and Dunbartonshire.[44] The creation both of this droving infrastructure and of farms oriented towards large-scale cattle management was not victimless; it has been observed that population decline in the southern Highlands after 1755 was linked directly with the creation of such parks, stances and grazing farms and the eviction of the tenants who had previously farmed that land communally.[45] The result was often an abrupt end to arable cultivation within extensive blocks of land, intensified grazing and browsing pressures over wider areas, construction of double-dyked loans, drove-roads and enclosures that cut across traditional areas of common resource including peat and pasture, the seeding of newly enclosed pasture ground with new grass varieties; all of these contributed to wider ecological changes on a local level.[46]

Various estimates have been made of the scale of the drove trade, with most agreeing that something in the order of 100,000 cattle per annum were driven south into England by 1800, with many thousands more being disposed of for consumption in Scottish towns.[47] If the one-fifth estimate of numbers for cattle being sold annually in the early 1800s is correct, it points to a national herd in excess of half a million head – and rising – at that time.[48] The sources within the country from which they were obtained, however, changed from the eighteenth into the nineteenth centuries, suggesting that regional specialisation was stronger in some areas than others. Thus, although the Hebrides, which had dominated the drove trade in the mid eighteenth century, remained an important source of cattle into the second quarter of the nineteenth century, the numbers being moved from there had peaked by the 1770s and were already in decline by the 1790s, remaining relatively stagnant through the 1800s.[49] Some tacksmen in the region had moved down the path of developing their property for exclusive cattle-grazing, such as undertaken by Duncan Campbell at Aros on Mull,[50] but more generally in the Highlands the continuance of traditional division of land and its resources between multiple small tenants acted against the development of such consolidated farms

---

44 See Haldane, *Drove Roads*, chapters 4–7 for the stances and routes in and from the main cattle-producing districts. Cregeen, *Argyll Estate Instructions*, pp. xi–xii, xix–xxi. One major beneficiary of the rise of droving from Argyll and Perthshire along the routes that converged at the southern end of Loch Lomond was the duke of Montrose, who in the 1720s and 1730s had developed extensive parks on his property around Buchanan. See, for example, NRS GD220/5/882 (14) for the leasing of the Buchanan parks.

45 E. Richards, *The Highland Clearances*, new edition (Edinburgh, 2018), pp. 88–90.

46 Just one of many local examples of entrepreneurial responses to the opportunities for profit through cattle-grazing is the 'park of Laguisan' on the Breadalbane property at Moness, south of Aberfeldy, where the tenant promised to either 'dung and manure it' for arable or 'sow it with rye grass and clover' to improve its grazing value: NRS GD112/11/3/2/61. The scale of the infrastructure development to support the drove trade can be seen clearly in the great road engineer Thomas Telford's 1811 report and construction estimates for a planned route across Rannoch for bringing the herds from the Hebrides and north-west Highlands into Perthshire: Haldane, *Drove Roads*, Appendix B, pp. 227–234.

47 Haldane, *Drove Roads*, p. 205.

48 See, for example, Sir James Macdonald, *General View of the Agriculture of the Hebrides or Western Isles of Scotland* (Edinburgh, 1811), p. 422.

49 Haldane, *Drove Roads*, pp. 205–206.

50 Cregeen, *Argyll Estate Instructions*, p. xix.

Highlandman Loan, Crieff, Perth and Kinross. One of the most famous of the surviving cattle drove routes on the route from Strathbraan via the Sma' Glen to Dalpatrick Ford over the River Earn and on towards the great market at Falkirk, the Loan still runs a straight route between dykes, hedges and embankments through Improved land. This section near Monzie is bounded by stone dykes where the drove-road dropped from the open country around Foulford to the intensely improved land of the Campbells of Monzie and the Maxtone-Grahams at Cultoquhey.

and kept herd numbers small. For the drovers and graziers who were the middlemen who brought the cattle from source to market, there was a disincentive in dealing with multiple small providers, especially when Improvement practices elsewhere had created larger enterprises.

The principal zones of cattle specialisation by the early 1800s were Aberdeenshire, Angus, Kincardineshire and Moray, although farmers there often sourced their cattle in the northern and western Highlands and Islands and brought them to fatten in the North-East before selling them on.[51] This north-eastern drove trade reached a peak in the later 1830s but began to decline swiftly thereafter as a variety of factors undermined the financial basis on which droving was founded. Importantly, there were now alter-

---

51 Haldane, *Drove Roads*, pp. 206–207. James Fyffe at Kirriemuir is an example of one such Angus dealer who obtained his herds from multiple sources, including Highland Aberdeenshire and north Perthshire, which he fattened in Strathmore before driving south to Edinburgh: Dundas and Orr, *Diary of James Fyffe*, pp. 23–24, 26–28, 30–37, 39–40, 54–56, 58–67, 69–75.

native means of moving large numbers of animals south faster and more directly than on the drove-roads, and farmers consciously switched from general stock breeding to increasingly specialised breeding of beef cattle for stock fattening and sale through regional markets that were tied into the new steamship and rail networks.[52] Adding to the pressure of that reorientation, whereas in the 1790s and earlier 1800s most of the grazing along the routes had been freely available, including at the overnight stances, by the second quarter of the nineteenth century the quality of grassland had been modified significantly by landowners and tenants along the traditional routes, at a cost to them for their efforts. Proprietors were therefore now charging for pasture on the improved grasses that had once been free and were seeking to dictate where night-time stances could be located.[53] Physical constraints in a landscape that had once been largely open but was now criss-crossed by stone dykes, and where roads now lay between continuous boundary walls, banks or hedges, also reduced access to the wayside grazing on the still rough ground through which the drove routes passed. Furthermore, as landowners looked to exploit their estates more rigorously to capitalise on the new market opportunities, droving came into conflict with alternative land-use, principally at first in some parts of the northern and western Highlands from the development of large-scale sheep-farming ventures, but from the late 1830s by the almost ubiquitous trend towards the creation of deer forests and grouse moors from which domestic livestock were largely excluded.

Droving's rise and fall was both cause and symptom of multiple changes that affected mainly but not exclusively the upland districts of Scotland from the Hebrides and northern Highlands to Galloway and the Southern Uplands. Political as well as socio-economic change stimulated the development of the trade. Its decline was as much a consequence of technological advances as other socio-economic factors and market-driven reorientation, but the impulses and waves within it were largely generated by the philosophy and practice of Improvement. Slow adoption of Improvement practices in the Hebrides and much of the northern and western Highlands limited the potential of these districts to develop their cattle-breeding practices beyond the ceiling that had been reached in the mid eighteenth century, while their adoption in the North-East created the conditions for sustained growth. Reliance on traditional grazing practice kept carrying capacities low in some districts, while policies of enclosure and improvement of grazing in others, coupled with the increasing cultivation of new fodder crops like swede turnips, pushed up grazing levels and herd sizes, and changed the seasonality of cattle husbandry. Although broad swathes of grazing had gained some benefit from the enrichment enabled through the dung of the passing herds, especially around the traditional night-time stances, its unfocused impact was minimal when compared to the concentrated investment of the large estates and their Improving tenants in increasing pasture quality. Although much has been made of the change in nutrient flow that resulted first from the containment of the driven cattle and then from their sharply falling numbers, especially across land that was otherwise used for seasonal grazing, it was essentially only at the stances that an appreciable change to local ecologies would have been visible or significant. Major environmental change came elsewhere, as farmers modified their grassland and changed land-use to maximise its fodder potential and expand its carrying capacity, which created new regional centres of livestock breeding and fattening in the 1800s. As those areas developed, principally in those north-eastern districts identified above, it was in certain more traditional source areas in the Highlands, where funda-

---

52 Haldane, *Drove Roads*, pp. 217–220.
53 Ibid., pp. 207, 210–213.

mental shifts occurred in livestock balances from the 1790s onwards, that some of the most profound environmental impacts are believed to have occurred.

Towards an 'empire of sheep'

Writing in 1954, the ecologist Frank Fraser Darling was convinced that in the Highlands one of the greatest engines of that reconfiguration was set in motion by 'the migration northwards of the Southern Upland flockmasters with their flocks of Black-faced sheep'.[54] Decrying what he saw as their ill-informed overstocking of land where cattle had previously been dominant, pushing sheep numbers to levels sustainable in the Cheviots but far beyond the carrying capacity of the grazing resource of more northerly and westerly districts, he saw their arrival as marking a point where Improvement crossed an environmental boundary and wrought far greater harm than good. Fraser Darling's denunciation is usually seen as the start of the long and continuing negative presentation of sheep-farming, especially in the upland zones of Britain, which reached new heights in the 2010s with George Monbiot's now ubiquitous descriptions of hill-grazed sheep as 'woolly maggots' and a 'white plague'.[55] Yet, as early as the 1790s, Sir John Sinclair was already cautioning against large-scale introduction of sheep into the northern and western Highlands, suggesting that flocks of a few hundred were advisable rather than the thousands that came to be grazed across the region. For Fraser Darling, the mass introduction of sheep catastrophically multiplied the effects of the deforestation that he believed had started to accelerate in the sixteenth century and which reached a peak in the eighteenth; tree-loss and sheep, he opined, went hand-in-hand to complete the devastation of the Highlands' ecosystem, a view that is still prevalent in some of the more ill-informed debate surrounding ecologically restorative proposals in upland districts. The trees and the seasonal use of the woods by cattle-herders for wood pasture had, in his view, created a rhythmic cycle of woodland regeneration, soil amelioration, herbage production and cattle nutrition. That cycle, however, was overturned by 'the coming of the sheep', which 'caused a revolution in natural history quite as certainly as it altered the lives of the people of the Highlands, and the subsequent history of the people has been quite definitely affected by that revolution in natural history and the new eco-systems it brought about'.[56] Deforestation, he argued, was the first stage in the process, the second was depopulation through the process that has become known as the Highland Clearances. With the removal of people went the removal of their cattle, abruptly changing livestock ratios, introducing new grazing practices and pressures, and ending the last vestiges of Fraser Darling's old seasonal rhythm.

Fraser Darling's uncritical acceptance of the nineteenth-century mythology of exclusively anthropogenic, sheep-farming-led woodland clearance – and the enduring legacy of his broadcasting of that perception which was cemented through his well-deserved scientific reputation – has created a now-traditional schema of linear decline from early surplus, through precarious medieval balance, to early modern devastation caused chiefly by avaricious capitalist landlords and their sheep flocks. Even in districts like the central Cairngorm massif, where large-scale sheep-farming was never introduced and the transition was directly from subsistence farming to deer forest, the environmental degradation of the montane zone has been attributed to over-grazing

---

54   Fraser Darling, *West Highland Survey*, p. 167.
55   G. Monbiot, 'Sheepwrecked', *The Spectator*, 30 May 2013; G. Monbiot, 'English countryside "shagged" by the "white plague" of sheep', *The Week*, 20 October 2015.
56   Fraser Darling, *West Highland Survey*, pp. 167–168.

The 'woolly maggots' are widely regarded as the scourge that almost singlehandedly reduced the Highlands to a desert of ecologically devastated, over-grazed grassland.

by sheep. As explored earlier, much of what Fraser Darling accepted as historical 'fact' concerning the form and fate of the Highland woods through the seventeenth and eighteenth centuries can be shown to be the product of nineteenth-century polemic or rose-tinted Romanticism. The removal of the pegs provided by the narrative of greedy ironmasters, tanbarkers and impecunious lairds leaves Fraser Darling's broad historical thesis adrift and unsustainable. Those elements charting the supposed ecological consequences of the Clearances, however, are more soundly based, where there was a substitution of the old, cattle-based subsistence economy with the new, commercialised and sheep-based systems. The first of these consequences might be seen in the effect of substitution of cattle with sheep on previously afforested land and on land that was still wooded (he assumed that *all* surviving woodland in the Highlands was unenclosed and consequently grazed through). Here he saw the principal impact as coming from the different grazing styles of cattle and sheep, with the former tearing off vegetation with their tongues and the latter nibbling and pulling with their teeth, coupled with the multiplier factor of one bovine being replaced with five to seven sheep. It was his contention that this change in grazing would have led to a rapid transformation of the herbage floor and a sharp decline in any opportunity for woodland regeneration, as tree-shoots would have been quickly nibbled down.

But there was a second factor in Fraser Darling's calculation: sheep are highly selective grazers. For him, their selectivity meant the development of areas where unproductive grasses and other non-grazing plant species flourished, for which the only remedy for the sheep-farmers was fire, which destroyed all growing species, impoverished flora diversity and accelerated soil erosion or transformation. In his

Sheep stell or fank, An Caorann Beag, Kintail, Highland. The circle of the stone fank lies on a hillside stripped bare of scrub woodland through two centuries of unrelenting grazing, first by sheep, and more recently by rapidly multiplying numbers of red deer.

words, 'the combination of fire and tooth is invincible'.[57] From this position, Fraser Darling went on to assert that 'the continuance of heavy sheep-grazing and burning for 200 years on ground previously growing woodland and savannah and grazed predominantly and only seasonally by cattle, has changed the "soil" of much of the West Highland hills from a friable mould with some mineral particles to a tough, dense, rubbery peat'.[58] In his view, the Clearances meant not only the destruction of traditional Highland society and culture but also, through its impact on the soil that was the foundation of all life, human and non-human, of the region's wider environment.

High on emotive polemic but low on historicity, Fraser Darling's thesis, and the more recent and better-evidenced environmental critiques of George Monbiot, inspire followers among new generations of often armchair ecologists seeking to right relatively recent historical-ecological – and social or cultural – wrongs. It was not, however, the only prescription and in the decades between Darling's *West Highland Survey* and Monbiot's 'Sheepwrecked' the debate over the environmental or ecological effect of clearance for sheep took many turns. Perhaps the principal of these was over whether the flocks directly caused widespread environmental degradation or instead exacerbated processes that were already long established and causing progressively more negative conditions. Some arguments were semi-apologetic, focused on the inhospitable nature of much of the cleared land and its incapability of supporting any form of economic agricultural regime other than sheep-

57 Ibid., p. 170.
58 Ibid., p. 171.

farming,[59] effectively offering an 'if not this, then what' challenge. By the 1990s, studies of areas with reduced sheep numbers were suggesting that there were limited regenerative effects on the plant ecology and that deer-grazing and rabbit and small rodent activity possibly had an impact as negative as or greater than that of sheep.[60] Subsequent research has tended to confirm the negative impact of sheep grazing, but the absence of a 'bounce-back' when only sheep were removed from the equation pointed to a broader suite of contributory factors in the process of ecological deterioration than simply the introduction of the flocks onto land. More recently, long-term research findings have confirmed the active role of sheep in the process but also have stressed the complex interplay of climatic, wider environmental and anthropogenic agencies at work in the environmental degradation of the Highlands and the fallacy of quick-fix solutions. The strength and enduring appeal of the Fraser Darling thesis, however, is in just that simplistic cause-and-effect nature and its binary opposition of a 'good' (the forest and cattle model) with an 'evil' (sheep and rapacious landlords). That appeal is coupled with the incontrovertible facts that sheep's grazing manner and the impact of muirburn can be as deleterious as Fraser Darling claimed and that fieldwork and government-funded research since the 1970s have confirmed the strongly negative consequences of overloading the land with sheep.[61] But its weakness is also that simplicity and the presentation, first, of sheep as the prime cause of the current ecological state of much of the Scottish Highlands (and Southern Uplands) and, second, of the impact of sheep and muirburn in the Scottish Highlands as a 200-year-long continuum built on the back of a millennium of progressive anthropogenic woodland clearance. Nothing in the environment, however, is ever that simple.

The human history of the Highland Clearances has been long and well researched. Within that research, there is broad acceptance of a base narrative of a progressive spread of southern Scottish sheep-farming practices northwards, starting in the late seventeenth century but accelerating from the middle of the eighteenth.[62] It is also recognised that clearance arose for different reasons in different parts of the country, not all associated with the financial lure of what has been termed 'the empire of sheep'.[63] What Fraser Darling had in mind when he articulated his generalised case was principally the progressive programme of eviction and creation of large sheep farms across the Sutherland estate from Assynt in the west, through Lairg to Kildonan in the east, and north to Strathnaver, starting in 1807, climaxing between 1814 and 1819, and effectively completed by 1820.[64] As an exercise in applied capitalism at its

---

59 M.L. Ryder, 'Sheep and the Clearances in the Scottish Highlands: A Biologist's View', *Agricultural History Review*, 16:2 (1968), pp. 155–158.

60 D. Hope, N. Picozzi, D.C. Catt and R. Moss, 'Effects of reducing sheep grazing in the Scottish Highlands', *Journal of Range Management*, 49:4 (1996), pp. 301–310.

61 R.H. Marrs, H. Lee, S. Blackbird, L. Connor, S.E. Girdwood, M. O'Connor, S.M. Smart, R.J. Rose, J. O'Reilly and R.C. Chiverrell, 'Release from sheep-grazing appears to put some heart back into upland vegetation: a comparison of nutritional properties of plant species in long-term grazing experiments', *Annals of Applied Biology*, 77:1 (2020), pp. 152–162.

62 For histories of the Clearances, see Richards, *The Highland Clearances*; J. Hunter, *Set Adrift Upon the World: The Sutherland Clearances* (Edinburgh, 2015); Devine, *The Scottish Clearances*.

63 Richards, *The Highland Clearances*, pp. 87–91; for the term, see p. 100.

64 Ibid., chapters 8–11. For a summary of the Strathnaver clearances, see M. Bangor-Jones, 'From clanship to crofting: landownership, economy and the Church in the province of Strathnaver', in J.R. Baldwin (ed.), *The Province of Strathnaver* (Edinburgh, 2000), pp. 35–99 at 72–74.

Shearing pens and grassland near Shesgnan, headwaters of the River Spey, Highland. This was once the common grazing of a *dabhach* whose main settlement lay further east and lower down the valley. The land supported birch wood pasture for cattle, grass, small patches of cultivation and forage ground for other natural resources. However, in the early 1880s the creation of single-tenancy holdings here introduced large-scale sheep farms and a sheep-run across this land.

most cynical and degrading, it had no precedent in Scotland and was unsurpassed in the scale and spread of its impact through the almost total depopulation of whole valley systems that had sustained mixed agricultural communities for millennia. Leaving aside arguments about the economic viability, environmental sustainability or cultural vitality of the pre-Clearance Highland rural society that was jettisoned in pursuit of the high-rental income and wool-profits that the new regime delivered, it must be recognised that in the cleared areas there was an abrupt ending of long-established agricultural practices, both arable and pastoral, of woodland management, and of peat extraction. Ecological balances would not have changed immediately, but once the sheep flocks had been established there was a progressive and accelerating shift away from the position that had prevailed in recent centuries. The suddenness of the change has been identified in places like western Glen Affric through the disappearance of cereal pollen and a decline in the pollen of weeds of cultivation.[65] But here we need also to apply a corrective, as further shifts in the pollen record indicate that the sheep-dominated grazing regime was itself most often transitory; in many parts of the Highlands the trend towards large-scale sheep-farming peaked before 1830, plateaued through the 1840s, was in decline by the 1860s and in full retreat by the 1870s, as global wool markets shifted and an alternative source of high income from shooting-rents produced a landslide towards the creation instead of deer forest.[66] Methods of sheep-husbandry might have introduced

---

65 See, for example, A.L. Davies, 'Upland agriculture and environmental risk: a new model of upland land-use based on high spatial-resolution palynological data from West Affric, NW Scotland', *Journal of Archaeological Science*, 34 (2007), pp. 2053–2063 at 2057, 2059–2060.
66 Richards, *The Highland Clearances*, pp. 369–371.

March-dyke between Connachan and Achnafree, Glen Almond, Perth and Kinross. The early nineteenth-century dyke imposed a barrier that cuts north–south across the land from valley bottom to hilltop. The different management regimes of adjoining landowners soon created significant ecological differences that are still evident in the vegetation on either side of the boundary.

new environmental stresses into the north and west Highlands, but their direct impact was neither as long nor as ubiquitous as Fraser Darling proposed.

• • •

In summary, as themes explored in this chapter have served to illustrate, there was almost no part of the country into which some dimension of agricultural Improvement did not reach. After a slow but gradually accelerating start through the eighteenth century, the early nineteenth century brought a sharp quickening of the pace as Improvers gained access to the means to realise their ambitions. Political, social and economic conditions enabled and encouraged landowners to pursue the radical reconfiguration of their estates, applying the theories of Enlightenment and Improvement to align them better with the new markets for their products that were emerging at home or to satisfy the demand for materials of the expanding imperial apparatus abroad. From the physical reorganisation of the landscape through consolidation of tenancies and reduction in the size of the population subsisting directly on the land, enclosure and realignment of boundaries, introduction of new crops and cropping methods, of grasses and fodder production, to the modification of the soil itself through draining, ditching, liming and manuring, this period witnessed a fundamental transformation of the Scottish countryside. Perhaps the biggest change was through the almost total eradication by 1850 of the multiple-tenancy fermtouns that had once formed the basic component of Lowland and sub-Highland rural social organisms, and the redistribution or redundancy of the mechanisms for natural resources allocation that had sustained them for centuries. This was the era when most of the remaining commonties and the shared zones of common grazing or communal fuel supply were swept away, with new boundaries imposed on the land, dividing it into the zones which still constitute the basic building blocks of our modern rural landscapes.

The legacies of that episode of transformation remain today as some of the most visible features of the modern countryside of both Highland and Lowland Scotland. Walls constructed to signal where one person's property ended and another's began, and to keep their livestock from encroaching on their neighbour's land, quickly became dividing lines between management regimes that progressively accentuated the differences in environmental impact of the strategies pursued by adjoining landowners. Whether maintained or not, the drystone dykes and estate fences that compartmentalised the land enabled the creation of the fragmented patchwork of mini-ecosystems that characterise the landscape around us. It was not, however, purely that anthropogenic redrawing of land divisions that created the environmental character of what we see today. It was, rather, just one factor that has sharpened wider changes that were occurring against a backdrop influenced on the one hand by socio-economic and political change remote from the land affected and on the other by the far greater environmental stresses and opportunities of the last phase of the 'little ice age'. Together, the interplay of all these forces, coupled with an Improving mindset that saw all land as having the potential to be brought into what was then regarded as productive use, laid down the patterns, colours and textures that we can still see today. On the one hand, those visible markers of anthropogenic change are the clearest evidence for the cultural nature of the landscape of Scotland, where even the highest upland zones were being partitioned, enclosed and transformed and are powerful symbols of human endeavour that form an irreplaceable component of our cultural heritage. On the other, however, those same patterns, colours and textures are evidence of the profoundly negative impact of that same endeavour on our wider environment, memorialising how we brought about and maintained a degree of ecological change the magnitude of which is incomprehensible to the majority of us today, and with whose consequences we still live and contend.

CHAPTER 12

Fuelling Scarcity, Gathering Abundance and Netting Loss: Peat, Coal, Kelp and Fish

———

> The fish which frequent the coast are herrings, ling, cod, mackerel, haddock, flounders, sye and cuddies. Herrings and ling are exported; and when sold in the country, a barrel of salt herrings brings about 16s. and ling L13 the ton. The parish exports about 20 tons of ling yearly at an average. There are likewise some seals, otters; and whales of a large size are seen in the Channel, between this and the Long Island. Every species of sea weed is reckoned most excellent manure, though it is thought to burn and waste the soil, when it is not mixed with earth, and it is reckoned more profitable to convert it into kelp, than into manure. There are some kelp shores, and the parish makes about an 100 tons annually.[1]

In common with the orthodox historiography of the eighteenth century, our traditional histories of Scotland between the 1790s and the mid nineteenth century are dominated heavily by political, social and economic narratives. From a rural perspective, the central storyline is usually the slow unfolding of the Highland Clearances through successive waves of forced removal and displacement of population in favour of the profits first from large-scale sheep-farming and subsequently from creation of deer forest for elite sporting rentals.[2] Despite the strong emotions that still circulate around those masterclasses in the inhumanity of Improvement principles, applied without thought of the human or environmental consequences, in the general historical narratives Clearance is a symptom of the wider socio-economic upheaval made in response to industrial expansion that met growing consumer demand. Consequently, the primary consideration in such work revolves around the trends towards accelerating growth of urbanisation and industrialisation, which brought rising demand for labour, fuel and raw materials.

There is no question that these two phenomena were central elements throughout the nineteenth century in the making of modern Scottish culture and identity. They have been instrumental in framing

---

1 Extract from Rev. Mr William Bethune, 'Duirinish, County of Inverness', *Statistical Account*, vol. 4 (Edinburgh, 1792), p. 131.
2 The grand narrative here is dominated by the work of Eric Richards, which spans the phenomenon from the later eighteenth to mid nineteenth centuries. See Richards, *The Highland Clearances*.

many of the most widely held popular perceptions in our contemporary, mainly urbanised society of the environment around us. But the emphasis in the discourse perhaps needs to be recalibrated to reflect that urbanisation and the drift from subsistence agriculture and rural labouring employment into waged town-dwelling factory-workers was still not the majority experience for Scotland's people by 1850. Yes, some towns were witnessing remarkable growth; Glasgow's population increased fourfold between 1801 and 1851 to reach around 329,000 in the second UK census. That figure was almost double that for Edinburgh, which had been for centuries before that Scotland's largest burgh, and five times the size of the second-largest industrial centre, Dundee. From a national population of some 2.9 million, however, some two-thirds of Scotland's people were still living in rural districts or towns with populations less than 10,000, spread through mainland Scotland from Galloway and the Scottish Borders, the Central Belt and round the eastern margins of the Highlands to Easter Ross. The explosion of urban growth, which saw one in three people located in just the four largest centres of Glasgow, Edinburgh, Dundee and Aberdeen by 1901, occurred in the half century after 1851. Outside those large cities, revolutionary transformation was still in progress as the 'little ice age' drew to a close and it still continues today, despite a tendency to present Improvement in Scotland as a product that was largely finished before the end of the nineteenth century. When viewed in those terms, these dominant narratives of urbanisation and industrialisation are seen to be 'national' only in that the themes they explore affected the entirety of Scotland's land and people through their hunger for manpower and materials. They are, however, markedly skewed by the strengthening emphasis on the experience of the central Lowlands – especially the west-central area – and the larger conurbations, as being representative of some nationwide norm. More recent historiography has highlighted the false dichotomies, and especially the narratives of decline, focused on the Highland and wider rural experience that have been founded upon that construction, emphasising instead the diversity of societies – urban and rural – that prevailed into the later nineteenth century.

Among the many consequences of the urban/industrial skewing and the determinedly Whiggish narrative of continuous social, economic and cultural improvement is the way aspects of daily interaction with the environment of the majority of Scotland's people have dropped in and out of view. This chapter, with its somewhat eclectic content, is intended to bring into focus for this final phase of the 'little ice age' three important topics that are largely neglected in the mainstream historiography of this period: the myth of the great fuel transition of the era; the brief efflorescence of the kelp-processing industry; and the faltering state of the sea-fisheries before their explosive growth in the late nineteenth century. A chief dimension of these topics is the exploitative practices that had dominated the human struggle to subsist in marginal if not hostile environments for centuries. Alongside them, however, are the short-term, opportunistic responses to market demand for certain commodities that had emerged, which offered a means to generate new income streams from sources that were traditionally exploited for other needs. Despite their novelty, only the economic significance or failure of those alternative strategies have received much attention, while their environmental consequences and social costs have barely been noted. Some of that fading from view happened because of the disdain for traditional practices that were considered culturally inferior, and testifies to the lack of interest of those who composed our sources of contemporary evidence in anything other than the financial importance of the material object. Most commentators were, by the second quarter of the nineteenth century, living so far removed from the production of their dietary staples or from the expenditure of labour to secure their annual thermal and lighting energy needs that provision of these essential commodities no longer impinged on their conscious awareness. Supplied by merchants who delivered orders to their doorstep and personally affected only by cost and seasonal availability, they had no direct involvement in the production or processing of what they consumed. Flour came ready processed, meat already butchered, and fuel – almost ubiquitously for those modern consumers, coal – came in sacks from depots that were themselves remote from the mines where it was won. For them, this was the norm; the continuation of age-old practices was an object of scorn and derision, to be denigrated as evidence for cultural inferiority and paraded as proof of moral failure.

From the perspective of the disparaged many who were still dependent on their personal labour to secure these necessary articles for their own and others' benefit, the impact of the Improvement being urged on them was as often negative as it was positive. The enclosing and division of commonties, reordering of the landscape to create the new single-tenancy farms, and conversion of mossland and moor to arable and improved pasture all reduced access to traditional resources. Meanwhile adverse weather was adding to the difficulties of securing adequate provision of fuel and fodder. The result was often intensification of pressure on places where environmental stress was already high, with increased demand from a greater number of users for resources that were already diminishing. New conflicts emerged where there were competing or alternative uses for the same resource. New dependencies were created where the patterns of availability of certain resources changed, such as the migration track of certain fish species, or closure of a foreign supply source creating a short-lived opening for an alternative available within Scotland, or landlord changes to tenancies by removing traditional access rights to once common resources. Through the first half of the nineteenth century, largely masked by the grand narrative of progress, economic development and, of course, urbanisation and industrialisation, the impacts, sometimes subtle, sometimes far-reaching, of these complex processes on land, water and air throughout Scotland were proliferating, ramifying and deepening.

Fuel: exhausting the inexhaustible

Perhaps the greatest concern of most ordinary Scots in the final decades of the 'little ice age' is a topic that haunts us still: the quest for cheap, abundant and available fuels to provide thermal energy. In the last decade of the eighteenth century, there was real concern at many levels of society about the possibility of exhaustion of all types of fuel, but principally peat and coal. There was animated discussion of mechanisms and methods to avoid that looming catastrophe, from the lunatic to the visionary.[3] Some of the concern was generated by the cross-over of scientific theory with theology and a 'shallow time' view of the world's age and the essentially finite nature of a divinely created resource that only needed to sustain humanity for a fixed period. But much of it was a combination of observation, of an increasing understanding of the geological structures of coal-bearing strata and their limited distribution within Scotland, and of arithmetical calculation of suspected accessible volumes of fuel and their rate of consumption. Until recently, however, the fact that these fears were – fortunately – never realised ensured that this paroxysm of fear never entered the mainstream historiography; abundant supplies of coal were in fact available in Scotland and more widely in Britain to power the Industrial Revolution through the nineteenth century. If we were to accept the conventional historical narrative, coal was simply

---

3   For detailed exploration of that sense of crisis and responses to it, see Albritton Jonsson, *Enlightenment's Frontier*, chapter 7.

the ubiquitous and most abundant fuel of the nineteenth century, quite literally powering the engines of industrialisation and feeding the domestic fires of heavily urbanised Scotland. But it was, for the vast majority of the population in the early 1800s, still just one component in a diverse mix of fuel types that had been used for centuries. Alongside poor, low-calorific fuels like dried seaweed or animal dung used by those on the socio-economic margins in fuel-impoverished regions, wood and above all peat were still, and were to remain for decades yet, more affordable and quite simply more readily available than coal. That being said, that fear of exhaustion of peat – or rather, exhaustion of the accessible, workable sources – which had stalked Scottish rural communities since the later fifteenth century, maintained a powerful hold in the consciousnesses of peat-dependent Scots. Exhausting the supposedly inexhaustible supplies of peat and coal was a possibility that exercised the minds of policy-makers, landowners, intellectuals and consumers well into the nineteenth century.

For all its versatility as a material, other than when used in the form of charcoal, wood had almost disappeared as a primary fuel source in most parts of the country long before the eighteenth century, even in places where it remained relatively abundant. Still used in displays of conspicuous consumption with the burning of logs or billets by the wealthy, or in desperation with the windfall and deadwood gleanings in the fires of the poor who lacked access rights to peat, across most of Scotland firewood remained a fuel of polar opposites on the social spectrum. It is wrong, however, to say that wood had ceased to feature in the thermal energy needs of the many. In addition to pine torches and resinous spills used as tapers, smallwood, driftwood, carpentry trimmings and waste wood salvaged from old furniture, broken utensils and equipment or building parts provided kindling for coal and peat fires, even in areas where woodland of any kind was scarce. If it could not usefully be recycled, it could be burned. Some crafts and specialist industries also still consumed it as their principal fuel. Wood was certainly still used in some food preparation, but chiefly for its smoke rather than its thermal properties. Smoked fish and meat processing, however, were scarcely major consumers and could use trimmings and gleanings rather than cut billets or faggots. When the great planting landlords of the early 1800s looked at their woodland, among its many envisioned future uses consumption as firewood probably never entered their minds, despite the apparently ever-present sense of supply crisis for fuel in many parts of the country. Yet, as we have already seen throughout our exploration of weather from the post-1790 period onwards, the problem of securing adequate supplies of fuel ran as a continuous strand. It was often cold and it was regularly wet – in June 1790, Janet Burnet recorded only two days at Kemnay in central Aberdeenshire when it was sufficiently warm to make it unnecessary to light a daytime fire in the house[4] – which pushed up demand for both domestic fuel and fuel for corn-drying.[5] As had been experienced by earlier generations during the depths of the 'Seven Ill Years' or the Spörer Minimum before that, such persistent wetness badly affected the ability to secure a suitable and abundant supply of what was still the most common high-calorific source of thermal energy, peat.

Although coal use had been increasing in central and eastern lowland districts of Scotland since the early sixteenth century, it is important to emphasise that in 1790 peat was still the primary, if not only, fuel burned by the vast majority of Scots who lived outside those areas and by a great many who lived

---

4 Pearson, *Diary of Janet Burnet*, p. 99.
5 The labour cost of obtaining peat for corn-drying was a major expense on the Mackenzie of Seaforth estate at Brahan in Easter Ross in 1790: NRS GD46/1/418 (11).

Dead oak, Caerlaverock Woods, Dumfries and Galloway. Dead and windfall wood, like the branches lying beneath this long-dead tree, were essential sources of fuel for those on the economic margins and would have been foraged by locals under the close supervision of the proprietor's wood-officers.

within them.[6] Indeed, even within those areas where coal was available, peat continued to be consumed in bulk by those who either could not afford the alternative or lacked the facilities to burn it. As the partly infilled and reduced fireplaces visible in many historic ruins attest, different forms of hearth – with or without grates – were needed to burn different types of fuel and required adaptation if there was any change in the main material being burned. Adopting or adapting technology to burn alternative fuels was not straightforward and, for those on the economic margins, unaffordable. Consequently, even in major urban centres where there was an established coal trade dating from the later medieval period, there remained buoyant demand for peat, which created a market that rural populations in the burghs' hinterlands were swift to exploit. Traditionally cut in late spring and left to dry on the moss in the summer months, peats were gathered, transported and stacked in the autumn for use over the following twelve months. With around 200 'loads' or 'leads' of peat estimated as the annual requirement for a single hearth, we can calculate the sheer bulk of the material that was needed for a one-fireplace house; the volume consumed in larger establishments or whole communities represented hundreds of cubic metres removed from the mosses per annum. But if it rained and rained persistently across what should have been the drying season, even the seemingly inexhaustible supplies of peat with which some parts of the country were blessed became of little value as it was too wet to burn. Sometimes it was so wet that the cut blocks disintegrated where they had been laid to dry, and what was salvageable required to be brought into a warm, dry place to be air-dried sufficiently to be usable. The folk narratives of the early nineteenth century are replete with tales of the inhabitants of peat-dependent regions resorting to draw lots as to which of their houses would be dismantled to provide combustible materials for their fellows to burn during some of the worst years in which inadequate amounts of peat had been secured.

But peat was not abundant everywhere by 1790; millennia of peat consumption had seen vast areas of peat moss already stripped by the later medieval period and conflict over access to and control of peat sources stimulated by insecurity of supply had become widespread since the late fifteenth century (see above, pp. 44–51).[7] Although at the end of the twentieth century it was reckoned that as much as 1 million hectares of peatland remained in Scotland, only a tiny percentage of that amount was anywhere near 'pristine';[8] the majority was significantly altered by centuries of human inroads for fuel, drainage and clearance for conversion to farmland, and more recently by widespread planting with forestry. Often, centuries-worth of peat growth had been stripped, leaving just the stony 'foot' of the moss at the interface with the podsols or clays below. Expanses of 'scalped' or 'flayed' ground, where all consumable materials have been removed and just very modern moss-regeneration has occurred, can still be seen in all parts of Scotland. As we have already explored in previous chapters, in some areas of the country where population densities were highest and demand for fuel greatest, most of the convenient peat sources had been exhausted long before 1790. References to former peateries in places that had been under

---

6 See discussion in T.C. Smout, 'Bogs and people in Scotland since 1600', in T.C. Smout (ed.), *Exploring Environmental History: Selected Essays* (Edinburgh, 2009), pp. 99–113 and T.C. Smout, 'Energy rich, energy poor: Scotland, Ireland and Iceland, 1600–1800', in T.C. Smout (ed.), *Exploring Environmental History*, pp. 113–133 at 124–127. The general situation is set out in Oram, 'Arrested development?' and explored in depth in Oram, 'Social inequality in the supply and use of fuel'.

7 Explored in detail for Moray in Oram, 'Abondance inépuisable?'.

8 Smout, 'Bogs and people', p. 99.

Flanders Moss from near Ruskie, Stirling. The flat fields in the valley bottom once lay several metres beneath a raised bog that stretched from the hills around Buchlyvie and Gartmore east to Stirling, which was cleared for agriculture in phases from the mid eighteenth century.

cultivation since the later Middle Ages and observations on moss 'improvement' in the Statistical Accounts provide us with some sense of the scale of the depletion of mosses and wetlands already before 1600 and the acceleration of the depletion in the nineteenth century.[9] The rate of loss had been accelerating through the seventeenth and eighteenth centuries as the population level recovered after its long stagnation after the Black Death and as the cooling climate brought increased demand for sources of thermal energy. This is possibly reflected in the mounting evidence for efforts to better manage the remaining common peat resources accessible to townsfolk especially (see Chapters 4 and 8). It is likely that there would have been many more willing converts from peat use to other fuels, coal in particular, had that alternative been readily available, but high pricing was worsened in the eighteenth century by the hefty government-imposed duty on 'coastwise' movement of coal north from the main coalfields around the Firth of Forth that we have already encountered.[10] Even in areas not subject to that duty and in relatively close proximity to mines, however, short-range transportation costs – as numerous respondents to the Statistical Account observed – rendered coal unaffordable by most people in rural districts.[11] Coal, thus, remained in the 1790s a luxury that few could afford to use in bulk, if at all, and the further one lived from the Scottish mining centres the more likely it was that you remained reliant on

---

9 For example, the land on the north bank of the River Isla from the neighbourhood of Bendochy and extending to Blairgowrie had been moss exploited for fuel in the twelfth and thirteenth centuries but was completely stripped and laid in fields by the early 1800s.
10 Duckham, *Scottish Coal Industry*, vol. 1, p. 37. As Duckham observed, it was cheaper for ports from Montrose north to import coal from Northumberland than from the Forth, but even that was still prohibitively expensive.
11 This is very clear in the accounts for inland parishes bordering Edinburgh to the west, where want of access to affordable and abundant coal was lamented by the respondents: Rev. Mr James Oliver, 'Parish of Corstorphine', *Statistical Account*, vol. 14, p. 456; Dr William Nisbet, Physician in Edinburgh, 'Parish of Currie', *Statistical Account*, vol. 5, pp. 324–325.

peat or the gleanings of windfall and deadwood in estate woodland.

Access to fuel was thus one of the greatest constraints on urban development at scale throughout the pre-modern era in Scotland and played a significant part in the relative success or failure of existing urban centres in the post-medieval period.[12] To be clear, however, we should reiterate that even the largest and most successful Scottish towns at this time – Aberdeen, Dundee, Edinburgh and Glasgow – still consumed a mix of fuel types into the second quarter of the nineteenth century; even the smoky haze of long-established coal-burning 'Auld Reekie' (Edinburgh) included peat-reek from consumption of peat carted into the town from mosses in the Pentland and Moorfoot Hills and the muirs west of the Almond in West Lothian. Away from the Lothian and Fife coalfields, in districts where until 1793 coal duty inflated its prices, Aberdeen spent up to £4,000 per annum – an enormous sum for the time – on peat brought in from its rural hinterland.[13] Demand, it is clear, remained high, even in towns like Stirling and rural neighbourhoods that were close to mining centres. It was sufficiently high that in 1805 a proposal was put to George Drummond of Blair Drummond, who was continuing the clearance of the mosses on his lands that had been started by his father-in-law Lord Kames in the 1770s, to process and sell the peat that his tenants were stripping away rather than simply flushing it through the sluices into the Forth.[14] The proposer, Sir John Dalrymple, understood that there was still huge demand for the traditional fuel type, for he recognised that adoption of coal carried other costs than simply that of the fuel itself, requiring changes to hearth and cooking apparatus, and that many rural-dwellers and townsfolk at the lower end of the social hierarchy could ill afford to introduce these additional changes involved in adoption of coal to their homes. Such economic marginality created a dependency that Dalrymple looked to exploit through access to an abundant source of peat that was otherwise being cut simply to be discarded.

Such a dependence on an insecure, imported commodity, as Dalrymple saw, had broader implications. Looking at rates of urbanisation in Scotland before c.1700, it is clear that the largest burgh with the easiest and most secure access to coal at that time – Edinburgh – experienced the most rapid growth, while most communities, other than the already well-developed seaport of Dundee, which was importing Forth or Northumberland coal in the Middle Ages, remained dependent on peat and grew only marginally beyond their medieval limits. Demand for coal, however, was there and mine-owners and merchants were determined to maximise opportunities to meet that demand, often claiming that Scotland's coal resources were inexhaustible. As the Alloa-based mine engineer Robert Bald put it in 1808, however, 'Even if the Grampian mountains were composed of coal, we would ultimately bring down their proud and cloud-capped summits, and make them level with the vales', for he saw that no resource was inexhaustible, especially with the rate of expansion of Scottish demand for fuel.[15] Confident in the skill of the mining engineers to extract the fuel regardless of geological or geographical challenges and certain of the market that was hungry for coal, although he was trying to highlight the finite nature of the coal reserve, Bald's words exposed a mentality among some Improvers, who saw all of nature as a resource that could be bent to the will of the Improvers and new industrialists. Clearly, with the

---

12 Oram, 'Environmental change, resource conflicts and social change'.
13 Smout, *Exploring Environmental History*, p. 103.
14 NRS GD24/1/802 (4).
15 R. Bald, *A General View of the Coal Trade of Scotland, Chiefly that of the River Forth and Mid Lothian* (Edinburgh, 1808), pp. 97–98.

Detail from John Slezer's late seventeenth-century view of Arbroath from the south-west, showing two men carrying packs and a third leading a horse-drawn sledge with wooden slatted sides. The middle figure has a creel on his back. Creels and sledges continued to be the main means of transporting peat from rural districts to urban markets into the nineteenth century. (National Library of Scotland)

constraints of contemporary transportation networks and artificially inflated prices, however, those towns that grew had already unrestricted access to energy sources and raw materials and were integrated into already established distribution networks; those that did not languished or declined.[16] Indeed, the limitations of the transportation networks beyond the Central Belt as late as the mid 1840s gave a built-in advantage to those communities closest to the coalfields,[17] entrenching their population dominance and the markets for goods and energy within them and the reservoirs of consumers for both. The percentage of Scots living in towns of more than 10,000 people almost doubled between 1700 and 1750 from 5.3 to 9.2 per cent and then almost doubled again to 17.3 per cent by 1800, with the vast majority of the new urban-dwellers located in the emerging manufacturing centres spread across the central Lowlands; by 1850 most of the 32 per cent of Scots who lived in large towns and cities were found in this zone.[18] It is no surprise that this growth occurred in the region where the coal seams were most accessible, providing

---

16 Oram, 'Social inequality in the supply and use of fuel', p. 213.
17 Smout, *Exploring Environmental History*, p. 103. The comment on the limitations of the rail network in 1843 needs qualification. The Dundee–Newtyle railway, for example, had opened in 1831 and by 1838 linked the port of Dundee with communities spread from Coupar Angus to Glamis in Strathmore, providing a means of transporting bulk agricultural produce out from the rural districts and commodities like coal to the rural railheads. A similar means of bulk carriage of goods to and from Aberdeen after 1796 was provided by the Aberdeen–Inverurie canal, although its regular icing in winter meant that it was less useful as a transport means for bulk fuel supplies.
18 J. de Vries, *European Urbanisation, 1500–1850*, 1st edition (London, 1984), pp. 39–48.

Wood gleaners *c*.1870. These two women are carrying bundles of offcut branches discarded by the workers in the wood yard behind them when the plantation on the hill behind was felled. Many on the socio-economic margins – rural and urban – supplemented their personal fuel stores and their incomes from selling on surpluses through foraging for waste or unwanted offcuts, windfall and deadwood.

a dependable and locally obtainable source of thermal energy for people and, once steam power became widespread in manufacturing processes, for industry.

Much of this new urban population was comprised of economic migrants from rural districts who had transitioned from usually landless and often seasonally employed labourers into full-time waged workers in the new urban-based industries. A key dimension to this transition from countryman to townsman was that families who previously had obtained their annual fuel needs through expenditure of their own time and labour became consumers of fuel purchased with part of their wage-earnings from the commercial suppliers within the towns. These suppliers were the successors to those speculators in whose hands formerly common peat resources had been essentially privatised in the seventeenth and early eighteenth centuries, and who, once the lifting of coal duty had helped to stimulate demand for that commodity, diversified where possible into a trade in alternative fuel types. Dealers continued to trade in peat, but their most important and lucrative customers were among the growing number of coal users. The most prominent in that new community were the prosperous urban professional and merchant classes. They were usually members of the burgess electors and councillors, the same group who controlled and regulated the licensed supply of peat from the largely privatised former common burgh mosses. Driven as much by fashion as by the socio-economic theories of Improvement and Modernity, in which coal was a symbol of progress and cultural sophistication and peat was emblematic of primitivism and retrograde cultural conservatism, this class embraced the new fuel and lost interest in the conservation and regulation of the old.[19] One indirect

---

19 Ibid., pp. 216–220.

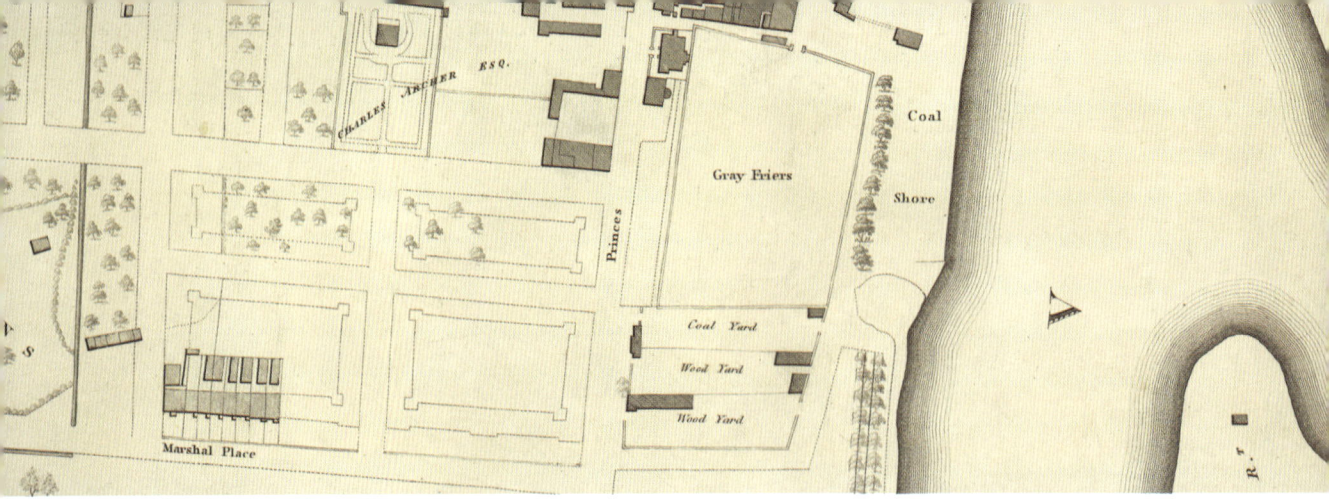

Top. Detail from Robert Reid's 1809 plan of Perth, showing the Coal Shore and the coal and wood yards south of the town.
(National Library of Scotland)

Above. Pow Valley, Perth and Kinross, looking east towards Perth over the location of Perth's former common moss towards Blackruthven and the burgh beyond. As the availability of coal shipped into the burgh's Coal Shore increased in the 1800s, this former landscape of conflict lost its importance to the burgh councillors, who disposed of it for cultivation.

consequence of that fashion-consciousness was that from the 1780s, with their thermal energy needs secure through alternative sources, the privatised former common peat mosses that they had overseen ceased to be a personal concern and many were rapidly sold by burgh councils or leased for private agricultural uses.[20] This trend accelerated the loss of remaining peatland in many lowland districts close to the towns, as it was drained, burned or stripped to create new fields,[21] and thus further stimulated the demand for alternative fuel types, i.e. coal, as the availability of peat diminished.

We should not overstate the importance of the lifting of coal duty in accelerating the spread of coal use outside urban centres beyond the well-networked central districts of the country. Bulk transport by rail developed slowly only after 1831, and shipment by sea only took off around the same time as steamships

---

20 Ibid., p. 220.
21 This approach to clearing peat to create arable had been already established in the late seventeenth century but accelerated rapidly from the 1780s: Smout, *Exploring Environmental History*, p. 101.

of large enough burden came into service. As a consequence, the cost-multiplier effect of small-scale transport of coal further removed from the pitheads kept it prohibitively expensive, and it was even dearer in winter when demand was high and the difficulties of transportation were worse. Price data and trends in consumption show that cost restricted coal use for all but the most well-off, even in large regional centres like Aberdeen, let alone smaller towns like Crieff, even though it lay only 37km from the nearest mine.[22] Even for the greater landowners in interior areas remote from the coalfields, or in northern and western districts, who perhaps burned coal in the grates in the principal apartments of their homes, peat remained the essential fuel commodity for their tenants and for the fires in the 'below stairs' parts of their mansions into the middle of the nineteenth century.[23]

A long historiographical tradition of seeing later eighteenth- and earlier nineteenth-century landowners as largely indifferent to the difficulties of their tenantry in their relentless drive for Improvement, ignores their evident awareness of the pressure on fuel from depletion and from weather-related impacts. At the very least, they understood the threat to their rental income if communities failed through want of peat (or any fuel). The Campbell of Argyll estate records, for example, show that the 5th duke of Argyll knew that peat had become increasingly difficult to obtain in parts of his property in the Inner Hebrides by the last quarter of the 1700s, a problem exacerbated to a degree by past poor management of the resource.[24] Centuries, if not millennia, of peat-cutting had seen reserves become badly depleted by the 1780s and instructions began to be issued to the managers of the worst-affected portions of the estate to identify suitable replacement sources. On Tiree, Argyll's chamberlain reported in October 1787 that the tenants in Caoles at the easternmost end of the island should be given access to peat at Cornaig on the north-east tip of neighbouring Coll, a sea-journey of nearly 20km.[25] With no available source on Tiree that was not already allocated to other communities there, Argyll's chamberlain had inspected mosses at Friesland on the south-east coast of Coll and only half the distance away compared to Cornaig, but found that 'the moss in Friesland is very much exhausted already' and recommended that what remained should be reserved for the duke's tenants at the west end of Coll.[26] Despite the efforts of the landlord to conserve the mosses and find convenient alternatives, by 1801 the shortages had become critical. Argyll imposed a prohibition on further peat-cutting on Tiree in October that year, proposing to start a regular supply of coal that his tenants could buy 'on as easy terms as possible', with the alternative

---

22 Oram, 'Social inequality in the supply and use of fuel', pp. 225–226.
23 For this differentiation in who and where coal was used, see Oram, 'Social inequality in the supply and use of fuel', p. 220. For the coal/peat mix on the Breadalbane estate in the mid 1790s, see NRS GD112/16/13/7. In 1802, the duke of Gordon's estate was providing access to new sources of peat for his household at Fochabers: NRS GD44/51/568/1.
24 This was the case reported to the duke of Argyll by his factor for Mull in respect of mosses supplying Corkamull and the island of Gometra, which had been wasted by the tenant at Laggan: Cregeen, *Argyll Estate Instructions*, p. 104. In 1787, the duke's factor was also instructed to enforce good moss management at Corkamull 'so as to do the least hurt to the mosses': p. 146.
25 Ibid., p. 9. In 1785, it was also reported that the duke's tenants on Iona found all their peat at Creich at the western end of the Ross of Mull, where they required grazing for their horses used in transporting it from the moss to the shore: ibid., p. 137.
26 Ibid. Argyll also issued instructions that peats were to be reserved for domestic use and the licensed whisky distillers of Tiree were forbidden from using them in an attempt to reduce the pressure on the remaining mosses. Coal was prescribed as the solution to the distillers' needs: ibid., pp. 19, 42.

Friesland Moss from Dun an Achaidh, Coll, Argyll and Bute. By the last quarter of the eighteenth century, this area of moss was one of the last remaining sources of fuel on the island. Identified for drainage and Improvement in the nineteenth century, it defied the effort needed and is today an almost unique survival on Coll.

being permission to find peats in the Ross of Mull, 38km to the south by sea.[27] A second example can be found on the Mackenzie of Seaforth lands on Lewis, where John Mackenzie was looking in the cold and wet late 1820s to make the local fuel resource more usable as fuel both for his tenants and for the estate's distillery and kelping operations, to avoid the cost of importing coal. He experimented in the mid 1830s with peat compressors to reduce the water content mechanically and so render the peat more readily dryable for burning at a time when the poor weather was making it virtually impossible to secure adequate supplies of dry fuel in an island largely still blanketed with moss.[28] His efforts, sadly for him, bore little fruit in that excessively cold and wet decade.

Measures like those explored by the Argyll estate may have made some difference to the experience of communities who were struggling to secure fuel in their traditional supply locations. They were, however, merely a deferral of the inevitable rather than a solution. Many of the locations offered were too remote from the settlements to be worked viably as and when the weather conditions deteriorated or they contained far less peat than the optimistic promises of 'inexhaustible abundance' suggested.[29]

27 Ibid., pp. 56, 68–69. In 1803, the chamberlain was arguing that the time and labour spent in going for peats in Mull was so great and occurred at times of the year when that time and labour would be better employed in kelp production, that an arrangement for the Tiree men in eastern Coll would be preferable: ibid., pp. 83–84.
28 NRS GD46/1/194 (3), GD46/1/539, GD46/17/76.
29 Oram, 'Abondance inépuisable?', pp. 31–32.

On the Breadalbane estate, for example, a dispute had started in 1786 between the township of Middle Lix in lower Glen Dochart and the nearby village of Killin at the mouth of the glen, over the latter's access to a moss on the former's lands. It had appeared to be resolved in 1792 when the earl gave Killin a supposedly limitless alternative, but deferred a fuel crisis only until the early 1860s, when that new 'inexhaustible' supply was exhausted.[30] Here, a combination of increasing population with rising demand for fuel and poor moss management through the extreme weather of the early decades of the nineteenth century had seen the rapid depletion of the blanket moss and its degradation through water and wind erosion.

During the worst years of the early nineteenth century, it was the wetness of the peat rather than its availability that was the principal issue, especially in the Highlands and Islands, and there was little prospect there of a developed coal-distribution network to provide an alternative. But even coal supplies were not entirely reliable in the main consumer centres, for deep snows closed off access to the suburban 'coal hills', as occurred in Edinburgh in 1795, or storms prevented the sailing of coal cargos, as occurred in Aberdeen that same year (see above, pp. 343–344, 351–353). For many of the poorest urban families, however, even in times when supplies of fuel were readily available for purchase, scavenging was an essential means of supplementing their fuel supply over and above the meagre charitable allocations of coal that were distributed during the periods of most acute need.[31] When cold, wet summers piled pressure on top of long, cold winters, often with the added stress of falling incomes as was experienced in the 1820s and 1830s, traditional scavenging for windfall and driftwood, or sea-coals on the Fife and Lothian coasts, strayed into illicit wood-cutting and theft. All across the country, from Dunbartonshire to Inverness-shire and eastern Perthshire, periods of economic depression, severe weather, or both, witnessed an explosion of thefts of wood from fences and outbuildings, or cutting of greenwood in plantations and from hedges.[32] Fuel poverty among the poorest was compounded, moreover, by the accelerating pace of clearance by draining, burning or stripping and conversion into arable or grassland in regions of rapidly spreading Improved agriculture.

A sense of the inroads made into the mosses across this era can be gained from submissions to the *New Statistical Account* in the mid 1830s and 1840s. Renfrewshire provides several instances of large-scale clearance that had occurred between the time of the compilation of the original *Statistical Account* in the early 1790s and the second account in the later 1830s and 1840s. For example, the mosses along the River Cart from Paisley to the Clyde had been removed almost entirely by this latter period, while at Neilston two-thirds of the moss had been stripped.[33] At Houston and Killallan, reference was made to largely unsuccessful planting of the moss there with conifers, and recommendation was made for its conversion instead to profitable arable through a process that involved application of whale blubber or oil to ferment the peat and create a 'black, greasy, rich mould'.[34] Experiments of that kind, encouraged

---

30 NRS GD112/11/1/4/96, GD112/11/10/14/33, GD112/14/6/7 (25).
31 Oram, 'Social inequality in the supply and use of fuel', pp. 224–225.
32 Ibid., pp. 227–228.
33 Rev. Robert McNair et al., 'Town and Parishes of Paisley', *New Statistical Account of Scotland*, vol. 7 (1845), pp. 135–306 at 156; Rev. Alexander Fleming, 'Parish of Neilston', *New Statistical Account of Scotland*, vol. 7 (1845), pp. 307–352 at 319–320 and 351–352.
34 Rev. John Monteath, 'United Parishes of Houston and Killallan', *New Statistical Account of Scotland*, vol. 7 (1845), pp. 46–56 at 55.

Making the most of what you've got

Our modern image of the production of peat for consumption as fuel is of neat, rectangular blocks cut from a vertical face and thrown onto the ground surface behind the cut to be arranged into small stacks for drying before being transported homeward for assembly into the family's peat stack. Numerous paintings and photographs, from William MacGeorge's painting of peat-cutting in the Galloway machars, to the Angus glens, Western Isles and Shetland, all show this same, basic method with a few minor regional variations. All present to us the ideal image, where peat was abundant, survived in depth and with highly compacted deep, black material towards the base, and was cut moving progressively backwards from a straight edge, and with the moss well managed and regulated. This ideal, however, was not always possible, as the situation at Elgin explored in Chapter 4 illustrated, or which occurred when a moss was nearing exhaustion, and especially during the periods of increased precipitation at the end of the seventeenth and beginning of the nineteenth century.

Although it records a late seventeenth-century example, the description from Galloway by Andrew Symson of an alternative method to producing a supply of adequate, combustible peat for fuel provides us with a glimpse of the realities which some Scots faced in trying to satisfy their thermal energy needs. He tells us that at Whithorn, a parish whose shortage of access to a secure source of high-quality peat continued to be problematical throughout the eighteenth century (see above pp. 239–240):

> because severall of them are a considerable distance from the peit moss, they have a fewell, which they call baked Peits, which they take out of a stiff black marish ground in the summer time, work with their hands, and making them like very thick round cakes, they expose them to the sun, and after they be throughly dry they yeild a hot and durable fire
>
> (Symson, 'A large description of Galloway', in Mitchell (ed.), *Macfarlane Geographical Collections*, vol. 2, p. 100).

Wooden mould used for forming peat blocks, L'Aisne Valley, France.

This use of more liquid, uncompacted peat highlights the critical shortages and supply difficulties faced in some parts of the country but demonstrates also the application of practical solutions. There, where the peat had lost cohesion, broken down into small fragments by erosion and saturation, and semi-liquified by rainfall, this process of straining, moulding and compaction to restore integrity and enable the material to be stored dry for future use, prolonged the extraction of peat from sites where the moss had been reduced to its base layers and become waterlogged. Similar methods of making combustible fuel from waterlogged peat are still practised in the wetlands of the Grand Marais near Liesse-Notre-Dame in the Aisne valley in north-eastern France, where – as illustrated here – historic peat extraction has left a landscape of water-filled hollows from whose bottoms and sides, wet or semi-liquid peat is still dug in the twentieth century, mixed to a pasty consistency, put into moulds and compressed, then left to sun-dry.

since the 1790s by prizes for the best methods awarded on an annual basis,[35] were being made all across the country, usually utterly useless but quite devastating to the mosses on which they were tried. Where such efforts failed, draining, paring, burning and ploughing were the solution, leading to the loss of hectares more peatland annually in the most populous districts. Here, however, there was no speculator like Dalrymple to suggest processing and selling the unwanted peat as fuel to those most in need, although some observers saw this as an obvious solution in the then-current conditions. There is no sense of irony in the Rev. John Sommer's observation in the *New Statistical Account* entry for his parish of Mid Calder that 'the vast quantity of coal now annually consumed by the manufacture of iron and by steam-engines, has raised the price of this necessary article of daily use so high, as to render it scarcely attainable by the poorer classes of the community. The conversion of common peat moss into a fuel has, therefore, become an object of considerable importance'.[36]

Such advocacy fell on deaf ears. Reversion to peat use in regions where coal was readily available, albeit at high price, was viewed by most landowners as a retrograde step. Furthermore, many of them and their Improving tenants had already invested heavily in drainage and clearance schemes on their land, often on peat mosses, and were unprepared to surrender that Improved land for stripping for fuel. In some districts, pressure was relieved by the bulk carriage opportunities of coastal steam packet shipping and the progressive expansion of the railway network in the late 1830s and 1840s. That expansion added significantly to the distribution network for coal and began to drive down its cost, though still mainly in those locations where there was a large market for it among the social classes who could afford to buy their fuel rather than cut it with their own labour. By the 1840s, as we have already seen in places like New Deer in Buchan (see Chapter 11, p. 327), peat mosses were being drained at an increasing rate as landowners and their tenants sought to improve land quality and diminish flood threats,[37] stabilising agricultural income and encouraging incoming tenants to aspire to improve their land and adopt modern methods and materials, including that key symbol of Modernity, coal.

Kelp

Among the range of potentially combustible alternatives materials that had been exploited for millennia in districts where wood was scarce and peat supplies inadequate, the dried masses of seaweed driven ashore on Scotland's beaches were the most widespread and significant. Hearth residues from the Northern and Western Isles reveal its near ubiquity as a fuel in those places from at least the Bronze Age and it continued to be used as a low-calorific supplement in domestic fires into the twentieth century. North Ronaldsay's seaweed-eating sheep are probably the best-known exponents of another of its principal uses, but *Statistical Account* respondents in the Western Isles especially also referenced its role as a winter fodder source for cattle.[38] Its value as a soil-enriching agent also had been understood for centuries in some areas, but it was in the 1790s that certain varieties of marine algae were found to have even more valuable properties, leading some

---

35 Albritton Jonsson, *Enlightenment's Frontier*, p. 120.
36 Rev. John Sommers, 'Parish of Mid-Calder, County of Edinburgh', *New Statistical Account of Scotland*, vol. 1 (1845), pp. 356–380 at 380.
37 For example, on his land flanking the lower Spey, the duke of Gordon in 1847 was overseeing the laying of networks of tile drains through the mosses that flooded regularly in winter and during spates: NRS GD44/43/420.
38 Rev. Mr Allan MacQueen, 'Parish of North Uist', *Statistical Account*, vol. 13, p. 301.

Sròn a Chlachain, Killin, Stirling. The supposedly 'inexhaustible' blanket of peat on the hill overlooking Killin at the western end of Loch Tay, in the centre of this view, was given to the villagers for fuel by the earl of Breadalbane. It was intended to end the conflict with the people of Lix in Glen Dochart (to its south) over the moss there. The Killin folk exhausted their new source within a few decades.

proprietors to ban their tenants from applying sea-ware as fertiliser on their fields.[39] Manufacturing industry's insatiable demand in this period for fuel and raw materials to feed its processes brought new pressures to bear on Scotland's natural resources, from woodlands to water. Large-scale production of commodities, from cured leather and textiles to glassware, required access to abundant sources of energy and ever-growing quantities of chemicals, most of which were still derived from primary processing of natural resources like oak bark and pinewood. These woodland resources, discussed in Chapter 13, had been exploited in this way for centuries, but the extension of industrial-scale exploitation to seaweed found in the tidal waters off Scotland's shores was something new and, for a brief period at the end of the eighteenth century and down to the 1830s, placed unheard-of human pressure on an important marine ecosystem. Sea-ware – the storm-dislodged and beached masses of all varieties of seaweed – as we have already seen, however, had been gathered and used for fertiliser since prehistory. This strand debris, however, contained a mix of species that did not meet the requirements of new industrial processors, who had identified specific varieties as being rich in certain highly prized chemicals. They wanted a pure crop of one species that they knew would deliver them the product they wanted, kelp. In addition to their value to farmers as sources of fertiliser for the calcareous soils of the islands, the kelp beds in Hebridean and Orcadian waters thus became of increasing economic significance to estates in the second half of the eighteenth century.

Kelp's importance lay in its rich iodine and alkali content, the latter, comprised principally of sodium carbonate, being obtained when the kelp was burned to produce soda ash. Until the last quarter of the eighteenth century, the bulk of Britain's soda ash needs were met by imports of ash from the Mediterranean barilla or opposite-leaved saltwort plant (*Salsola soda*), mainly from Spain. But domestically available kelp was known from at least the later

---

39 Rev. Mr Edward MacQueen, 'Parish of Barray', *Statistical Account*, vol. 13, pp. 330, 331.

Kelp-drying dykes, Links of Noltland, North Ronaldsay, Orkney Islands. These low drystone dykes were used to raise the gathered seaweed and aid in its turning and air-drying before it was burned.

seventeenth century to be a source of this in-demand commodity, which was used mainly in the manufacture of soap and glass. By the late 1700s, it was of growing importance in production of bleach and ammonia for Britain's rapidly expanding woollen, linen and cotton textiles industry. From a standing start in the seventeenth century, Scottish lairds had begun by exploring the viability of an export trade in cut kelp to glass manufacturers and pottery-glazing businesses in the Netherlands during the domestic economic crisis of the 1690s. Domestic consumers lagged not far behind: East Lothian glassworks in the late 1720s were seeking sources of '(sea)ware fit for being burnt and kelp' in Caithness and Easter Ross.[40] Thomas Fordyce, proprietor of the Port Seton glassworks, was seeking sources of kelp to bring to his works and process there, but a growing volume was being burned close to the source and the soda ash bagged and shipped to the end-users. Cut kelp was still being exported unprocessed from Mull in the late 1750s, but kelp-burning operations and a trade in soda ash from 'box-wrack' cut from tidal rocks were already established on the south Ayrshire coast by 1709, in the Uists by the 1730s and in Orkney by the 1740s.[41] From the 1750s, as part of the effort to increase economic activity on the estates of forfeited Jacobite landowners, the Commissioners for both Annexed and Forfeited Estates encouraged the more systematic exploitation of the coastal kelp resources, but other landowners were also moving to cash in on the growing demand for soda ash.[42] By the late 1760s, the forfeited Cromarty estate's 'sea-ware' and

---

40 NRS GD1/1003/5, GD24/3/237, GD122/3/13 (2).
41 NRS GD25/8/862, GD150/3482/58. The origin of kelp production in North Uist was attributed to a trial made for Hugh Macdonald, tacksman of Baleshare, in 1735: see Rev. Mr Allan MacQueen, 'Parish of North Uist', *Statistical Account*, vol. 13, pp. 305–306. The Rev. MacQueen commented on the 'great quantities' of seaweed washed ashore during episodes of stormy weather: p. 301.
42 See, for example, the earls of Morton on their Orkney properties and MacDonald of Clanranald on his Moidart, Arisaig and South Uist lands: Orkney Records GD150/2520, NRS GD201/2/15, 201/2/16.

'kelp shores' on the coast of Coigach and at Barrisdale were being leased, with kelp merchants based in Edinburgh and more locally at Lochbroom pursuing the leases.[43] This intensification in exploitation of a commodity which grew in abundance, but whose economic potential had never been realised by most landlords on whose property it grew, marked the start of a boom that lasted into the 1820s.[44]

From being an essentially unvalued but fortuitously available resource used for its soil-improvement qualities by agricultural tenants around the Scottish coast, kelp went to being a prized, high-value commercial asset over which control was disputed between landowners.[45] High profits could be made from a commodity that grew and replenished naturally and which could be harvested and processed cheaply by local peasant labour. Consequently, the two decades down to 1790 saw a steady ramping up of kelp processing around the Scottish coast from East Lothian to Galloway, but especially in the Hebrides and Orkney.[46] In the early 1790s, therefore, kelping was an established venture, albeit still relatively small-scale in comparison to the dominant Spanish barilla industry, whose product was well known and of a consistently high standard.[47] That changed from 1793 when barilla became subject to high import tariffs imposed at the start of the war with Revolutionary France and there was the added threat of trade embargoes and disruption through warfare. Scottish soda ash production, therefore, geared up rapidly in the 1790s to meet the rising demand from domestic industrial consumers.

The socio-economic impact of the kelp industry on island communities was profound but it also provided an often life-saving source of cash income during the worst years of crop failure and fuel shortage. It was an activity on which by c.1820 between 15 and 25 per cent of the population had become utterly dependent for employment, in cutting, drying, burning and bagging kelp and soda ash, to generate the income necessary to meet rising land rents.[48] Inroads on the kelp were as profound, with an estimated 23 tons of algae needed to produce one ton of soda ash; in 1791, 244 tons of kelp (over 5,600 tons of raw seaweed) were processed on Tiree alone, with the total rising year-on-year thereafter while the boom lasted.[49] Fortunately, the regenerative ability of the algae is such that even the intensified cutting that occurred in the quarter century down to c.1820 failed to diminish its availability, safeguarding proprietors' incomes both from rents and from the processed kelp and giving some security to the labouring workforce during the economic recession immediately after the Napoleonic Wars and during the weather-related subsistence crises. By 1810, the value of Scottish kelp processing had quintupled over its 1750 level, with

---

43   NRS E746/98, E746/126, E746/127, E746/176.
44   Hutchison, *Industry, Reform and Empire*, pp. 30–31.
45   NRS GD128/47/5, dispute over kelp rights in Loch Eynort, Skye; GD174/1411, dispute over kelp rights in south Mull.
46   The Clanranald Macdonald and Mackenzie of Seaforth estates provide excellent evidence of this intensification, with kelp operations on the former's island properties in Benbecula and South Uist and on the mainland, and the latter's Lewis and Wester Ross estates booming post-1760: (Clanranald) NRS GD201/2/25, 201/2/36, 201/2/38, 201/2/41, 201/2/42, 201/2/45, 201/2/49, 201/2/50, 201/2/51, 201/2/55, 201/2/56, 201/2/60, 201/5/1232; (Gillanders of Highfield, factors for earl of Seaforth) GD427/66, 427/87, 427/105–132, 427/136–138, 427/197, 427/233, 427/252. The duke of Argyll's factor on Tiree commented in 1794 that kelp manufacture there had only commenced 'but in a small degree' in the 1770s: Cregeen, *Argyll Estate Instructions*, p. 32.
47   For the poor quality of much Scottish processed kelp, see the duke of Argyll's comments to his bailie in Tiree in 1792: Cregeen, *Argyll Estate Instructions*, p. 25.
48   Hutchison, *Industry, Reform and Empire*, pp. 30–31.
49   Cregeen, *Argyll Estate Instructions*, p. 22.

Kelp-burning, Gribun, Mull, Argyll and Bute. William Daniell's record of the processing of the kelp on the otherwise inhospitable shore captures the impact of industrial activity in the Hebrides just as it collapsed following the kelp boom of the Napoleonic Wars.

most of that increase occurring after 1793. Recognition of the profits to be made led to speculation and, in the 1800s, to a surge in illicit production by small-scale operators, which controlling landlords moved swiftly to shut down.[50] The return of competition from Spanish barilla in the 1810s, however, burst the bubble of high profits and, although the industry continued to play a significant part in estate economies into the early 1820s, it was clear that it would no longer sustain the levels of employment and income that it had done during the boom. The real crash came in 1825 when parliament abolished the high duty charged on salt, which made the cheap production of sodium carbonate from salt, using the method devised in 1791 by the French chemist Nicolas Leblanc, commercially viable in Britain. From highs of nearly £20 per ton in the mid 1810s, by 1827 Scottish producers were receiving around £4 per ton, as the all-important soap and textile manufacturing consumers of soda ash switched to the high-quality

50 For the steadily increasing pursuit of individuals engaged in unlicensed soda ash production after 1800, see the Orkney sheriff court records: Orkney Archive, SC11/5/1800/4, 11/5/1802/93, 11/5/1804/115, 11/5/1805/5, 11/5/1805/55, 11/5/1806/119, 11/5/1806/124, 11/5/1806/130, 11/5/1806/163.

new source. Demand remained for the second major product obtained from kelp, iodine, but the late 1820s and 1830s were characterised by the steady decline of kelp processing as the market contracted and various efforts to protect producers and promote kelp ash use failed to secure government backing.[51] Although kelp processing continued to figure in estate incomes into the later nineteenth century, the bursting of the bubble in the 1820s effectively ended the most intensive phase of exploitation of this marine resource around Scotland's coastlines until its late twentieth-century use as a source of polysaccharides in pharmaceuticals.

Sea-fisheries

No review of Scotland's environmental history in this era would be complete without a brief discussion of the contemporary development of that far better-known marine resource, the fisheries in the waters that surround it.[52] As we have seen previously, there had been major episodes of intensive exploitation of the sea-fishery by Scots in the Middle Ages, especially on the east coast in the thirteenth century and the west in the fifteenth century. Through the sixteenth and seventeenth centuries, however, there was generally widespread lamentation from monarchs and their privy councillors that the wealth from Scottish waters was going largely into foreign pockets. Attempts to kick-start the fishery in the reign of James VI and again at the end of the seventeenth century through encouragement of Lowland investors had come to naught in the long term, and a combination of political rivalries that fed local hostility and, on a grander scale, a complex mesh of international politics, had helped to ensure that no operation was successful in the pre-1707 era.[53] It was only in the last quarter of the eighteenth century and the foundation in 1786 of the British Fisheries Society that plans for an indigenous fishery finally bore fruit.[54] As the published history of the Society illustrates, however, it is the social, economic and political effects of its establishment that have dominated past discussion, along with the foundation of its planned settlements at Tobermory on Mull, Lochbay near Dunvegan on Skye, Ullapool in Wester Ross, and Pultneytown across the river from Wick in Caithness. The widely varying success of these communities, from the still thriving port at Ullapool to the small, huddled cluster of houses at Lochbay, had different levels of impact on their hinterlands through the market and employment opportunities they afforded, but it is their effect on marine fish stocks with which we are concerned here.

For the old and well-established fisheries of the east coast, the eighteenth century had seen general decline as recurrent failure of the herring shoals to appear in the waters of the outer Firth of Forth and adjacent North Sea areas also saw declines in numbers of gadids – mainly cod and haddock – that fed on herring eggs and fry.[55] Ministers from the

---

51 For the impact on the Seaforth estate on Lewis, see NRS GD46/1/307, 46/13/129–135.
52 For general and very accessible histories of the cod and herring fisheries in the North Sea and North Atlantic, see M. Kurlansky, *Cod: A Biography of the Fish that Changed the World* (London, 1999) and M. Smylie, *Herring: A History of the Silver Darlings* (Stroud, 2004).
53 As proposed, for example, by Martin in 'A Brief Account of the Advantages the Isles Afford by Sea and Land, and Particularly for a Fishing Trade', in *Description of the Western Islands*, pp. 349–359.
54 For the history of the Society, see J. Dunlop, *The British Fisheries Society 1786–1893* (Edinburgh, 1978). Schemes to introduce Dutch settlers to teach the techniques of commercialised fish-curing to the people of the Hebrides were still being proposed in the 1790s: see NRS GD51/5/210.
55 Smout and Stewart, *Firth of Forth*, pp. 38, 40; M. Gray, *The Fishing Industries of Scotland 1790–1914: A Study in Regional Adaptation* (Oxford, 1978), p. 9.

coastal parishes of Fife and East Lothian who submitted reports to Sir John Sinclair for the *Statistical Account* presented a generally bleak view of the fishery's state. 'About the beginning of this century', stated Andrew Bell, minister of Crail,

> Crail was the great rendezvous for the herring fishery in the frith (sic) of Forth. Beside a great number of boats fitted out and manned by the fishermen and others belonging to the town, several hundreds assembled from different parts of the country, particularly from Angus, the Mearns and Aberdeenshire. These were supplied by the inhabitants with nets, for the use of which they received a certain proportion of what was caught. Immense quantities of herring was cured for home consumption, and for exportation. The *Drave*, as it was here called, was seldom known to fail. The fisherman expected it as certainly as the farmer did his crop. Almost all the people in the place derived their support from it, the other fisheries, and the trade and manufactures which were immediately connected with them. A sad change has now taken place; and we listen as to a fairy tale, to accounts given by old people of what they remember themselves, or have heard related by their fathers. For half a century, the fisheries here have been gradually declining. The herrings, for several years past, have neither visited the coast in any considerable quantity, nor remained long enough upon it to spawn as formerly.[56]

Bell saw no single cause for the decline, pointing to the poor weather across the preceding decades, the rise of intensive new fishing operations in north-eastern Scotland, and – of course – the inroads of the large-scale fishing busses of the English and Dutch especially. For an example of the effects of these latter, he pointed to their operations off the Norwegian and Swedish coasts, where 'millions' of herring were caught and rendered down for oil. Certainly, the large-scale Dutch and English fisheries were having an effect, but the principal cause of any decline at this time is likely to have been related to oceanic circulation and the southward spread of cold water from the seas around Iceland connected with the general climatic instability.[57] Bell's story is similarly reported in other parishes fringing the Firth, presenting a picture of total collapse of the herring and haddock fisheries and consequent economic hardship in communities that had grown to be reliant upon them. There were some attempts at diversification, with a number of the fishing ports supplying large numbers of lobsters, for example, to the London market, while other fish that had once been thought of as unsaleable rubbish, like turbot, had become fashionably in-demand.[58] None of these, however, came anywhere near the level of value – let alone number – that had been caught in that late-summer Drave.

While the outer Firth ports were in decline through the haddock and herring failures, the inner Firth parishes reported a boom. This was from the so-called 'winter herrin', a migration of fish that were genetically distinct from those of the Lammas Drave, which entered the Forth to spawn between mid October and late February. Smaller in scale to the late-summer migration, it had generally been ignored by the fishermen of the outer ports, but faced with the decline of their traditional livelihood, they and fishermen from around the country began to

---

56   Rev. Mr. Andrew Bell, 'Parish of Crail', *Statistical Account*, vol. 9 (1793), pp. 439–458 at 444–445.
57   Smout and Stewart, *Firth of Forth*, pp. 40, 43, 76.
58   Ibid., pp. 41–43.

Sunset over the Firth of Forth from Largo Bay, Fife. The waters of the outer Firth saw recurrent cycles of collapse and recovery in their traditional herring fishery through the later eighteenth and early nineteenth centuries.

descend on the zone of the firth from Bo'ness in the west to Burntisland in the east to reap this bounty. By 1798, it was reported that the fleet – possibly around 1,200 vessels of all sizes – came from fishing ports spread from Caithness to Kent and, thanks to the opening of the Forth and Clyde canal, from Liverpool, Bristol and the east coast of Ireland.[59] Such was the success of this fishery in a few short years that by 1800 it seemed as though Burntisland was set to become the base for the large-scale Scottish herring fishery that had been under discussion politically for decades.[60] But as suddenly as it had emerged it was gone and by 1805 both the 'winter herrin' shoals and the seasonal fleet had disappeared. Again, contemporary observers were at a loss to explain this sudden appearance and dissolution, seeing in it an 'act of God' that had brought bounty in a time of dearth and removed it to curb the greed and profligacy of those over-exploiting the fishery. Unlike the Lammas Drave migration, however, this smaller and genetically distinct herring migration might have succumbed to over-fishing across the decade from 1793, leading to its collapse. Taken together, the demise of the two seasonal migrations of herring upon which both other fish populations and human communities had depended provides illustrations of the complex interplay of environmental, climatic, species-specific and anthropogenic factors at work.[61] The interplay is emphasised all the more in that the shoals associated with the Lammas Drave made an equally dramatic reappearance in 1816, stimulating a rapid revival in the Forth fishery, declined again after 1822, then returned in numbers from 1836 until late in the nineteenth century.[62] Human greed might

59 Ibid., p. 44.
60 Gray, *Fishing Industries of Scotland*, p. 27.
61 Smout and Stewart, *Firth of Forth*, pp. 75–76 for discussion of the spawning patterns of herring.
62 Ibid., p. 45.

have hastened the collapse of the 'winter herrin' but it seems more likely that the protracted climatic disturbance discussed in Chapter 10 was the prime driver in the ebb and flow of the main east-coast fishery.

Changes in migratory patterns, perhaps in response to changes in oceanic circulation, water temperatures and the consequent presence or absence of the organisms on which herring feed, meant that while some traditional fisheries experienced crisis from the 1790s others enjoyed a boom. This was the case with Caithness, where the quarter century down to 1815 saw a major expansion of activity associated with what is known as the Buchan herring population in the wide waters of the Moray Firth. The Buchan shoals spawned in the waters off Shetland and the north-east mainland of Scotland in early summer, peaking in July, so came between the migrations of the 'winter herrin' and the Lammas Drave. During the seventeenth- and early eighteenth-century peak of the Lammas Drave, this north-eastern fishery had drawn little interest, but activity picked up as the eighteenth century progressed, with boats at first operating principally from the east-coast parishes of Loth, Latheron and Wick.[63] Earlier in the eighteenth century, Caithness fishermen had been supplying mainly cod to south-east Scottish merchants, but from the 1770s they began to deliver a growing volume of herring to southern buyers and processors, probably in response to the decline in the Lammas Drave. The next logical step was for these southern operators to establish bases in the region and by the 1780s they had fixed processing and curing stations on the Caithness coast. By the early 1800s, Wick had become established as the largest centre for such operations.[64] Between the mid 1790s and

---

63 Gray, *Fishing Industries of Scotland*, pp. 28–38.
64 NRS GD9/270 for details concerning establishment of a curing house at Wick in 1803.

1808, there were around 200 boats operating from the eastern Caithness havens, jumping to over 800 in 1814 and by the mid 1820s to over 1,000, at which level it was sustained into the last quarter of the century. As the largest concentrated fishery in Scotland, its value was recognised in the early 1800s when the British Fisheries Society started to develop Pultneytown at the mouth of the River Wick, completing its harbour in 1811.[65] While herring remained the mainstay of the Caithness operation, the weakened state of the Forth fishery increased demand for its white fish potential, thus spreading the fishing opportunity onto a year-round cycle. Many of the boats and crews were drawn from the Forth, where the fishery locally had effectively collapsed, enabling the maintenance of a well-equipped fleet in the Fife and Lothian ports that was able to respond to the revival of the Lammas Drave in the early 1820s and especially after 1836.[66] The level of success of the fishery and the construction of the Pultneytown facility also saw major changes in the local population dynamic, economic networks, seasonal employment and demand patterns, which introduced new stresses on the supply and support side that were to surface dramatically during the potato blight era in the 1840s. Until that time, however, the Caithness fishery presented an image of sustained success drawing on a resource of seemingly inexhaustible abundance.

Although the British Fisheries Society had been actively developing fishing operations on the north-western coast from 1788, when work started on the development of what became Ullapool, neither that location nor the other developments at Tobermory and Lochbay matched the scale of the eastern fisheries. Part of the issue in this region seems to have been the low baseline from which it started, with much of the fishing being a subsidiary activity to the mainly farming economy of the Highlands and Islands. This situation remained the case into the first half of the nineteenth century, despite some landlord efforts to force tenants into the sea-fishery to augment incomes following evictions from cleared estates to new settlements on the coasts.[67] Added to this was the shortage of supplies available locally of the essential commodities needed to sustain a large-scale fishery of this era, principally salt and barrels.[68] The result was that into the late eighteenth century the most concentrated fishing efforts were organised from Clyde estuary ports, Greenock especially; they brought their supplies with them, and operated from temporary shore bases where the catches were processed and barrelled.[69] This activity continued into the mid nineteenth century, despite the establishment of the British Fisheries Society settlements in the region, which had barely survived the economic upheavals of the early 1800s and which competed unfavourably for local backing in the face of the high profits available at that time through kelp processing.[70] Tobermory struggled as a fishing station and Lochbay was bedevilled by misfortune and bad management from the outset. By 1810 the former was already operating more as a general port

---

65 Gray, *Fishing Industries of Scotland*, p. 33; Dunlop, *British Fisheries Society*, chapter 11. For British Fisheries Society papers concerning the development of the town and harbour, see NRS GD9/275.
66 Smout and Stewart, *Firth of Forth*, pp. 80–81.
67 Gray, *Fishing Industries of Scotland*, pp. 101, 103–104.
68 The duke of Argyll's chamberlain in Mull saw opportunity to supply materials for barrel-making to Tobermory but it was mainly opportunities to provide materials or resources for the non-fisheries activities of the inhabitants that the estate moved to meet: Cregeen, *Argyll Estate Instructions*, pp. 158, 165, 172, 174, 177, 192, 193. The Breadalbane estate was supplying salt from its pans near Benderloch: NRS GD112/7/16 (5).
69 Gray, *Fishing Industries of Scotland*, p. 105.
70 Cregeen, *Argyll Estate Instructions*, pp. xxxi–xxxii.

South Heads old pier, Pultneytown, Highland. The storm-damaged and partly collapsed outer pier is a legacy of the great new fishing-port development undertaken here in the first decade of the nineteenth century.

than a centre of the fish trade and would survive on that basis, while the latter struggled on through the 1820s and was eventually sold in 1837 to a local landowner who was interested in its land-value and not its fishery.[71] The failure of these bases to take off had been exacerbated by the unpredictability of the herring migration, with years of bounty followed by runs of almost total failure in the western sea-lochs, the Inner Sound off Skye, and the Minch. It would take a significant shift in market demand, external commercial interest and local opportunity before the north-western fishery began to develop on any scale, in the 1840s; that decade saw east-coast curers establish permanent but seasonally active bases in the region, especially on the eastern coast of Lewis, encouraged by a return to more reliable annual shoal-migration patterns.[72] Here, where there was no recent tradition of intensive herring fishing, far more clearly than on the east coast, it is likely that wider environmental factors affecting the sea conditions favoured by the migratory shoals were at play, creating an unpredictability that commercial investors found discouraging.

As had happened in the Middle Ages, it was the southerly part of the west coast fishery and especially that in the Firth of Clyde and its northern sea-lochs – Loch Fyne, Loch Goil, Loch Long etc. – that saw success in this period. It is not known why Loch Fyne

71 Dunlop, *British Fisheries Society*, pp. 181–182. For the state of the settlement from 1802 to 1823, see the bundled papers in NRS GD9/166 and GD9/196.
72 Gray, *Fishing Industries of Scotland*, pp. 109–111.

Looking south into Loch Bay and Loch Dunvegan, Skye, Highland. The sea-lochs of northern Skye and the adjacent waters of the Minch had long been regarded as prime locations for development of a herring station. Despite repeated efforts, the plans for Loch Bay failed to mature, with Ullapool instead emerging as the regional centre.

was the most favoured location for herring, seeing the greatest regularity of visitation by the shoals compared with any other sea-inlet around the British Isles, but it enjoyed a reliability through the period that no other eastern or western fishery saw.[73] Benefiting also from its proximity to Glasgow and the rapidly growing west-central Scottish urban centres, where there was a market for its produce, and the ports that sent processed herring to the Caribbean where they were a mainstay of diets on the slave-worked sugar plantations, the fishery remained buoyant and profitable where others struggled. Down to the 1830s, this was principally a drift-net fishery but from then there was an increasing number of ring-net operators who, despite efforts to control their activities and curb the depressed prices and supply fluctuations that they caused, made serious inroads on the shoals. By the 1840s, there was a growing body of opinion – supported by some evidence – that the ring-netting operation was threatening the sustainability of the shoals and in 1851 this style of trawling was prohibited on Loch Fyne.[74] It was a warning for the future that few fishermen, fish-curers or investors were prepared to heed, for in the second half of the nineteenth century the sea-fisheries around Scotland were moving rapidly into an era of seemingly limitless boom, and those directing what was becoming an industry looked to use whatever methods delivered the greatest possible return.

• • •

In these three topics, we have seen the nature of some of the tensions that were still being played out in the first half of the nineteenth century between tradition and innovation and the impacts of those tensions on niche elements of Scotland's environment. Each has provided insights on the shifting nature of anthropogenic change as Improvement drew towards its high tide in the second half of the nineteenth century. The priorities of estate managers, businesses and consumers were also changing as they began to settle into new, market-oriented and demand-driven configurations. All three have also exposed important divergences from the 'norms' presented in the conventional historical narratives, but especially with regard to the continuing significance of peat and peatland in the lives of the majority

73 Ibid., pp. 118–119.
74 Ibid., pp. 120–121.

of Scots. Here, on the one hand, was what for many Scots, both urban and rural, was the only form of fuel available to them in large volumes, much of it still gathered personally by hand, while on the other it occupied ground that could either be stripped of its peat overburden to reveal the supposedly fertile subsoils or itself converted into high-quality arable land and pasture. The conflict was most often played out in terms of the waste of time and energy that was represented by the annual cycle of peat-cutting, drying, transporting and stacking, primitive activities that held back the Scottish peasantry – especially in the Highlands – from joining the march of progress to a higher social, cultural, economic and moral state. The employment of an individual's personal labour was, in the minds of Improvers, an unconscionable waste of the opportunity value of that energy expenditure, when there were – in their minds – alternatives available through the open market. The fact that the main alternative – coal – was neither affordable nor readily accessible by the bulk of the population was irrelevant. Demand would solve both issues as the market adjusted to supply the potential customers in areas remote from the production centres. But that adjustment was slow and, in many areas, peat mosses were being removed as common sources of fuel long before alternative sources for the thermal energy needs of local communities had been secured.

Hidden within these debates, we can also see the advance of one of the biggest and most rapid episodes of large-scale landscape transformation to occur in Scotland in the historic era. All across Scotland, peat had been extracted progressively to satisfy community fuel needs for centuries, removing an enormous volume of material in Lowland districts especially, where the larger urban communities had been voracious consumers of fuel. In the volume 'Scotland AD 400–1400' and in Chapters 2, 4 and 8 above, we encountered the cycles of perceived crisis in access to peat but, despite the recurrent sense of imminent failure of supply which was expressed in many communities and the conflicts that arose when climate change increased the obstacles to securing adequate supplies, peatland continued to represent one of the most common landscapes encountered in both Lowland and Highland areas. And it was that prevalence of peat that added to the pressures upon it, for those fired with Improving zeal saw it not as a convenient thermal energy source but as potential farmland ripe for conversion from wet waste into profitable productivity. Men such as Lord Kames and his son-in-law led the way on Blair Drummond and Flanders Moss, but their extensive and decades-long programmes of clearance and conversion were matched by countless smaller endeavours by landowners and farmers from Shetland to the Mull of Galloway. As Improving landlords and tenants completed their reconfiguration of the blocks of long-established cultivable ground, they began to turn their eyes to the tracts of raised bog and blanket peat that filled the undeveloped parts of their property. As the Rev. Welsh observed for New Deer in Buchan in 1845, it was the mosses – and the iron pan layer that prevented the free drainage of surface water – that were the focus of efforts at Improvement. Over the next half century, an as yet uncalculated extent of peatland disappeared through drainage, stripping, soil-modification efforts, or woodland plantation. With it went not only the millennia-old carbon sink of the peat and the cultural heritage of peat-cutting in areas where it was both supplanted by coal and obliterated as a visible, physical presence, but also entire mossland ecologies embracing plant and invertebrate species, birdlife, reptiles and small mammals. Other resources vanished with the mosses, from wildfowl that overwintered on their pools to the reeds that fringed them, which had provided food and materials for the same communities that exploited the peat for fuel. Unloved and forgotten, the destruction of the mosses passed into the positive narrative of the benefits of Improvement that we are only now beginning to re-evaluate as a costly loss that diminished us all.

CHAPTER 13

Trees: For Beauty, Effect or Profit

―――――

> [W]hen you have found the shrubbery, you must cut down
> the mightiest tree in the forest . . . with a herring![1]

'Figure to yourself a rich and beautiful valley extending itself for several miles and yielding almost spontaneously the various productions of the Earth enriched and enlightened by the streams of a large and magnificent river circled by a zone of mountains through whose fissures thousands of young trees are seen struggling for existence among the aged and lofty pines which already adorn their sides.'[2] In these terms, an anonymous English visitor to Scotland in early autumn 1817 described the vista of Strath Tay that he observed from near Balnaguard as he travelled from Aberfeldy to Dunkeld. That same year, Sir Bourchier Wrey, arriving at Dunkeld from the west, recorded with an air of wonderment the 'whole range of mountains covered with larch fir principally and many other trees of infinite variety'.[3] Wrey, whose native Cornwall was still a land of largely treeless moors, was astounded by the forest of tall conifers that stretched as far as he could see. Similar wonder can be read in the words of the anonymous traveller, who, although rather overstating the ruggedness of the country through which he was travelling, captures a moment at the high tide of the planned afforestation of parts of Highland Perthshire under the direction of the 4th duke of Atholl, the 4th earl of Breadalbane, and a number of smaller landowners. Their efforts represented continuation of policies set in place by their predecessors that extended from the early eighteenth century (see Chapter 7), who had sought to establish their properties as a sustainable source of bark and timber to meet the growing demand for these commodities as new markets opened in post-1707 Britain. But, as we have already encountered in Chapter 7, John Murray, 4th duke of Atholl later enthused that in so doing they were also creating designed landscapes of wonder and beauty to rival the forests of the world.

Setting aside the wonder and beauty, it is that designed character of what was being created that is the most significant aspect of this phenomenon.

---

1 Terry Gilliam and Terry Jones (Directors), *Monty Python and the Holy Grail* (Motion picture). EMI Films (1975), challenge of the head of the 'Knights who say "Ni"' to King Arthur.
2 'Anon: Tour to the Highlands 1817 and 1818', in Durie, *Travels in Scotland*, pp. 100–130 at 116.
3 Sir Bourchier Wrey, 7th Baronet Wrey, *Journal of a Tour to Scotland, 1817*: Edinburgh University Library Special Collections, quoted in Fowler, *Landscapes and Lives*, p. 94.

Confluence of the Tay and Tummel at Logierait, Perth and Kinross. The hills flanking the river valleys were those seen by Bourchier Wrey as cloaked in a haze of young conifers. The duke of Atholl's larches are long gone, replaced today mainly by Sitka spruce plantations.

Human design rather than the subtle influences and interplay of anthropogenic management and natural regeneration on woodland structure signalled the most profound transformation of woodland character to have occurred down to that time. Over the course of the eighteenth century, demand from contractors for the most sought-after wood types encouraged the extirpation, or at least persecution, of less commercially valuable tree species – species whose value may have been far greater to local populations, for everything from building materials to livestock fodder – in much the same manner as gamekeepers were actively exterminating the 'vermin' that reduced the quality and quantity of the desired game birds. Factors and wood-managers charged with maximising the production and extraction of oak or pine did not want woods filled with birch, holly, rowan and aspen, which reduced the overall value of an area of woodland, made access to the commercially important trees difficult for the wood-cutters, and occupied ground that could produce more valuable timber. These species came to be regarded as 'weeds' that had sprung up unwanted on land and they were believed to threaten to overwhelm slower-growing hardwoods and pines.[4] This intensified management of the native woodlands became almost scientific in the course of the nineteenth century and shaped perceptions of the relative values of different tree species in purely financial terms into the late twentieth century. Consequently, they framed professional foresters' attitudes towards once dominant species, like birch and aspen. Already by the start of the period under review in this chapter, however, the character of managed native woodland across most parts of Scotland had changed dramatically from the rich and diverse

---

4 For this trend towards less diverse woodland structures, see Smout et al., *History of the Native Woodlands*, pp. 74–75.

constituents described by sixteenth- to eighteenth-century observers.

It was in the proliferating plantations of this era that such visible transformation in woodland character would have become most evident. As travellers' descriptions of the new planting on the Atholl, Breadalbane or Grant estates suggest, single species dominated swathes of land in stands of uniform age, height, shape and colour that contrasted starkly with the varied form of the native woodlands. These were, first and foremost, crops and, like the enlarged arable fields of Improvement agriculture, every effort was taken to reduce the intrusion of undesirable 'weeds' from among the planted trees to maximise the yield volume and value. Less desirable species were never wholly eradicated but they were marginalised, subjected to regular purge clearance, and generally discouraged. Variation in colour, texture and shape in these new plantations came from selection of species by the planters and not as a result of the random, natural seeding of the semi-natural native woodlands. The 4th duke of Atholl extolled the beauty of his expansive larch woods, where only some Scots pine and other, non-native, conifers were planted to give some variety, but where lesser species were managed almost out of existence. Certainly, the seasonality of the deciduous larch, with its progression from spring green, through vibrant autumnal gold, to winter bareness, brought an ebb and flow of colour to those woods, but even with this species the monotony of type meant that beauty was very much in the eye of the beholder.

Ornament, enjoyment, enrichment and profit

Taking pleasure from woodland was hardly a new phenomenon in the early nineteenth century, for the same motivations of ornament, enjoyment and enrichment, as well as future profit, are evident in tree-planting practices from the late seventeenth century. Indeed, we can even detect an undercurrent sense of hoped-for aesthetic betterment in James V's sixteenth-century desire to have Scots plant trees 'for policy to be had within the realm', where 'policy' carried a meaning of social and cultural enhancement.[5] Rather than being a binary opposition, however, caricatured in some modern observations in a false dichotomy between Improvement and Romanticism, beauty and profitability could be two sides of the same coin, as we have already seen in respect of Breadalbane's romantic woodland around the hermitage and falls at Acharn. We need also to recognise that planting of trees on a grand scale across open country was generally undertaken in a different manner and for different purposes than the planting around the residences of the landowners. While the walled 'policies' around the mansions still could contain blocks of woodland plantation intended for future sale for timber and bark, they were increasingly being laid out for leisure and enjoyment, containing specimen trees and, especially from the 1830s under the influence of Scottish landscape gardener and theorist John Loudon, arboretums.[6] Exotic tree species, like the North American conifers introduced by explorer and seed collector David

---

5 See above, pp. 129–130 for discussion of James V's parliamentary acts concerning planting.
6 The principles of arboriculture, keeping and importance of arboretums, and discussion of the hardy tree species capable of being grown in Britain generally were set out in J.C. Loudon, *Arboretum et fruticetum Britannicum; or, the trees and shrubs of Britain, native and foreign, hardy and half-hardy, pictorially and botanically delineated, and scientifically and popularly described; with their propagation, culture, management, and uses in the arts, in useful and ornamental plantations, and in landscape gardening; preceded by a historical and geographical outline of the trees and shrubs of temperate climates*, 8 volumes (London, 1838). The evolution of Scottish landscape design is examined in detail in A.A. Tait, *The Landscape Garden in Scotland 1735–1835* (Edinburgh, 1980).

The David Douglas Fir, Scone Palace, Perth and Kinross. The original Douglas fir, grown in 1827 from seed sent back from what is now British Columbia on the Pacific Coast of Canada, is representative of the environmental legacy of empire which saw species from the farthest-flung reaches of British control sent back as specimens, trial trees for possible commercial planting, and ornaments that symbolised power and connection and beautified the gardens of the empire's masters.

Empire, trees and escapees

Among the most powerful symbols of past ecological imperialism to be seen in twenty-first-century Scotland are the pinetums that still dominate the designed landscapes around many of the mansions of Britain's nineteenth-century captains of empire. Providing both a testing ground for species that might be grown productively – commercially – for the benefit of the owner and also a social statement that displayed the connections of the family that extended to the furthest reaches of Britain's global empire, the pinetum is an expression of human power and control over nature at home and abroad. The finest of these tree collections can be seen, for example, in the grounds of Airthrey Castle (now home to the University of Stirling), at Blairadam in Perth and Kinross, at Ardkinglass in Argyll and Bute, but above all at Scone Palace near Perth. The pinetum there was largely the work of locally born David Douglas, who sent back seeds and seedlings from his travels across North America, including the example of the fir species named after him – the Douglas fir – whose seeds were part of a package of material sent to the earl of Mansfield in 1827. Around it are avenues of Wellingtonia, noble fir and western hemlock, species introduced from the western districts of Canada into which British expeditions were penetrating at that time. While we can wonder at the scale, majesty and beauty of these two-hundred-year-old trees, which represented the advance guard of a legion of 'alien' tree species that was introduced in the quest to satisfy Britain's demand for fast-growing and high-quality timber, we should reflect on the ramifications of their introduction for native species, which sank in value and demand. We should also reflect on the fact that, along with the trees, the seed- and plant-collectors were also sending back other exotics whose presence is regretted far more, including *Rhododendron ponticum* and *Heracleum mantegazzianum* (giant hogweed), whose garden 'escape' later in the nineteenth century has resulted in invasive and highly destructive occupation of Highland hillsides and Lowland riverbanks.

Douglas to the earl of Mansfield's pinetum at Scone Palace, were grown in arboretums as much for wonderment and to broadcast the reach of the garden's owners and the empire they served, as they were to trial potential commercially exploitable varieties. Successors to the tree nurseries of the earlier eighteenth-century seedsmen, these carefully nurtured collections provided seed for broader planting on the estate or to be given to friends and colleagues for their own enjoyment and benefit. The legacies of such planting from the early nineteenth century can still be seen in the magnificent specimens at Scone, Airthrey, Blairadam and Murthly.[7] But these exotic groves were tiny in hectarage compared with the often hundreds of hectares of native Scottish or British tree species in the plantations and parkland around landowners' houses.

The driving agency in seventeenth- to nineteenth-century woodland expansion is often attributed to the efforts of individuals in different periods, and Sir Walter Scott has long been seen as possibly the most influential figure in the expansion of planting across Scotland in the first quarter of the nineteenth century.[8] Even in his much-lauded but ruinously

---

7 House and Dingwall, 'A nation of planters', pp. 142–153.
8 See, for example, Fowler, *Landscapes and Lives*, chapter 9; J.F. Ogilvie, 'Sir Walter Scott and forestry', *Scottish Forestry*, 39:1 (1985), pp. 13–22; J.F. Ogilvie, 'Sir Walter Scott and Scottish forestry', *Native Woodland Discussion Group, Scottish Woodland History Conference Notes XVIII, Plantations in Scotland* (2016), pp. 40–46. Scott's role has more recently been reassessed and further elevated in S. Oliver, *Walter Scott and the Greening of Scotland: Emergent Ecologies of a Nation* (Cambridge, 2021), where the excellent ecocritical literary approach is not matched by the historical, especially environmental-historical, contextualisation. For all the more extravagant claims of Scott's centrality in Scottish forestry matters in the first half of the nineteenth century, however, he secures no mention whatever in Smout et al., *History of the Native Woodlands* as either a conserver or promoter of planting with native species.

Blairadam from Benarty, Perth and Kinross. Much of the contrived wildness of Scott's day has long gone, but the core ornamental woodland around the house survives to give a sense of the Romantic landscape of the early 1800s.

expensive improvement and planting on what became his estate at Abbotsford, however, Scott was an emulator rather than an innovator or originator, adopting and developing ideas shared with many of his contemporaries. His greater impact came from his undoubted triumphs as a networker and, through the fictive landscapes of his novels and the enthusiastic prose of his practical pamphlets and journal articles, from his ability to reach a national – and international – readership. John Fowler has pointed to Scott's membership of the so-called Blair-Adam Club that met annually from 1817 to 1831 at Blairadam in Kinross-shire, the house at the heart of an agriculturally poor upland estate at the east end of the Cleish Hills that successive generations of the Adam family had been steadily planting with trees since the 1730s, to create what was called in the fashion of the age a *terre ornée* or adorned estate.[9] William Adam's *Remarks on the Blair-Adam Estate*, written and published privately by their host at Scott's suggestion, charted the progress of plantation from bare and unproductive low-value grass and heath, through the formal woodland divided into blocks by avenues and rides popular in the mid eighteenth century, to the Romantic irregularity with eye-catching plantations that emphasised natural landforms which Scott and his fellows lauded and copied.[10] The reach of Blairadam and its influence in both landscape design

9 Fowler, *Landscapes and Lives*, pp. 101–102. For discussion of Blairadam, see Tait, *Landscape Garden*, pp. 94–101.
10 W. Adam, *Remarks on the Blair Adam Estate* (private circulation, 1834). The Adams' work on their property was matched on similarly limited scale, and with identical purposes of beautification in advance of the economic importance, in the early 1800s by the Stewart earls of Galloway on their Wigtownshire estate, where the earl commented, 'I am an Economist about buying Trees, yet I must beg you to be extravagent (sic) in beautifying about Galloway House, all my children are so partial to it that I want to make it as pretty as possible': NRS GD138/3/21.

and plantation policy was truly national, with even the infamous James Loch on the Sutherland estate citing it in 1839 as the exemplar for his planned replanting of the 2nd duke of Sutherland's woods in Dunrobin Glen.[11] Loch's notes show that visual impact, colour and texture mattered as much as, if not more than, future timber value. This perspective of a weighing of aesthetic and economic value can be seen in William Adam's words: his *terre ornée* was a means to 'combine usefulness and profit with enjoyment and ornament' while the *ferme ornée* his family had already created in the mid eighteenth century at North Merchiston, now lost under Edinburgh's suburbs, was a smaller version that combined ornamental gardens with cultivated ground.[12] It is in the mixed woodland, walled garden and arboretum, sinuous woodland rides and paths, sudden clearings to create directed views to vistas with distant landscape features like Loch Leven Castle,[13] and soft-edged plantations of Blairadam's *terre ornée* that one inspiration for Scott's planting around Abbotsford can be found.[14]

Although the voluminous record of Scott's *Sylva Abbotsfordiensis*, the notebook in which he kept memoranda from January 1819 onwards concerning his estate woodland, has been mined regularly to illustrate his unquestioned passion for tree-planting and success as an advocate of plantation policies, it also reveals him to have been a magpie for others' ideas and a not altogether successful experimenter. Acorns were planted by the bucket-load, supplied by friends and correspondents in England and Ireland, but were eaten in almost equal volume by ravenous rodents who gorged on the bounty.[15] Some ground was over-planted with beech, elm and oak, requiring many healthy young trees to be cut down to allow room for others to grow. Elsewhere, despite his supposed sensitivity to the environments that supported various native species and the historical significance of such trees, he attempted to grow the same introduced broadleaf species on perennially waterlogged and seasonally flooded land by the Tweed, replacing the riverside alder, hazel, willow and birch carr that had flourished there. In such actions, he was diminishing the mix of species that gave character to the old, semi-natural native Scottish woodland, implementing locally a trend away from species diversity to more limited composition that was part and parcel of the 'Improved' and commercialised woodland that was being developed across much of Highland Scotland.[16] Although he resisted the lure of in-demand Scots pine, which had been very much in vogue in eighteenth-century planting, in the 1820s he followed the duke of Atholl's example and planted larch seed and seedlings to replace young hardwood broadleaf trees that had suffered in the drought of 1821/2. Even for Scott, it seems, the championing of what he believed to be the historic native woodland trees of the Scottish Borders was tempered with occasional realism and recognition that some introduced species were better suited than others to the terrain and enriched the colour and texture of the landscape he was modelling. It was such pragmatic views of the relative benefits of native versus introduced species, shared by Scott and many of his contemporaries, that perhaps contributed most to the continued – and accelerating – contraction of the surviving areas of native wood-

---

11 M. Bangor-Jones, 'Plantations on the Sutherland Estates', *Native Woodland Discussion Group, Scottish Woodland History Conference Notes* XVIII, *Plantations in Scotland* (2016), pp. 34–39 at 36.
12 Tait, *Landscape Garden*, p. 101.
13 House and Dingwall, 'A nation of planters', p. 133 plate 6.4.
14 Oliver, *Walter Scott and the Greening of Scotland*, pp. 126–127.
15 Fowler, Landscape and Lives, p. 105.
16 Smout et al., *History of the Native Woodlands*, pp. 74–75.

Scott's View, looking west up the Tweed towards Melrose and Abbotsford. Much of the broadleaf planting visible in this section of the valley was the work of the Scotts of Buccleuch, the Southeys at Pavilion, and their neighbours, perhaps inspired by Sir Walter Scott's experiments around his own house.

land in Scotland, or its systematic transformation into species-poor blocks through the 'weeding' of uncommercial trees from the mix of species that once grew there, through the second half of the eighteenth and into the nineteenth century. By 1850, perhaps as much as half of the areas of ancient native woodland that had existed a century earlier had been lost.

Perhaps the clearest record of Scott's embracing of both the practicality and necessity of Improvement and the desirability of the aesthetics of Romanticism can be seen in his 1827 review essay 'On Planting Waste Lands', which offered a critique of Sir Robert Monteath's 1824 *Forester's Guide and Profitable Planter*, written from Scott's personal perspective and experience as, by then, a significant planter and adviser on planting.[17] Described as containing 'a practical treatise on planting moss, rocky, waste, and other lands, also a new, easy, and safe plan of transplanting large trees, and of valuing growing wood and trees of all descriptions', Monteath's work had been aimed unashamedly at would-be developers of large-scale commercial planting on otherwise 'valueless' land. What could be labelled a 'we can plant anywhere and everything' mentality, coupled with tendencies towards rigid-edged monotony and the usually oak monoculture that Monteath encouraged, was condemned by Scott

---

17 Sir Walter Scott, 'On Planting Waste Lands', *Quarterly Review* (October 1827).

and avoided on his own property.[18] As his *Sylva Abbotsfordiensis* notes and his review of Monteath demonstrates, he saw in Blairadam's mature woodland and his own still immature trees at Abbotsford, planted in an 'ebb and flow like a green tide',[19] organisms that conformed to the landscape rather than ignored natural shapes, and alive with trees that occupied 'precisely the place in [that] landscape where nature's own hand would have placed them'.

Scott might well have reflected on his own words when he tried to plant inappropriate trees on unsuitable ground that then required the introduction of drainage, riverbank stabilising and soil-improvement to render it usable. And, although he had a clear preference for 'native' species that aligned with his conception of the nature of the historic landscape of Scotland, the sourcing of seed for much of his planting led him to introduce varieties from England, Ireland and elsewhere in Europe that were alien to Scotland, particularly of English pedunculate oak in preference to Scottish sessile varieties, which led to marked changes in visual appearance, ecology and ultimately genetic structure of the woodland. In his philosophy and approach, moreover, there was a blend of the almost theatrically contrived betterment of nature advocated by the Rev. William Gilpin (1724–1804) in his theories of the 'picturesque', the full-blown Romanticism of his friends and correspondents William and Dorothy Wordsworth, and the 'evangelical enthusiasm' of the landscape designer Humphrey Repton.[20] Although much recent work has tended to focus on Scott's social-historical perspectives, his love of the Scottish landscape and the historical record embedded within it, and what might be regarded as his wider ecological interests,[21] he attempted – perhaps half-heartedly even in the face of his own financial difficulties – to marry picturesque, Romantic naturalism with the pragmatism of the *terre ornée* concept. For him and the countless other landowners who looked to enhance their properties with trees, economic necessity competed strongly with romantic idealism and emotional attachment and, whereas in Scott the latter outweighed the former, for others the emotional attachment to the land that steered his approach was an indulgence that could not be afforded. Scott, it must be stressed, was nevertheless no misty-eyed Rewilder before his time, looking to put back something that had been lost and to restore what was already in the early nineteenth century a land transformed by a tide of Improvement. Although perhaps 'greener' in an early twenty-first-century sense, he was unquestionably a leading figure in driving forward one of the most profound episodes of anthropogenic environ-

---

18   Monteath's enthusiastic promotion of oak is illustrated in R. Monteath, *Miscellaneous Reports of Woods and Plantations* (Dundee, 1827), where he commented that 'Oak, and nothing but oak, is the only profitable tree for coppice cuttings, and whenever such a plan is intended, nothing else should be reared.'
19   Fowler, *Landscapes and Lives*, p. 106.
20   Three of Gilpin's many works on what he termed 'the picturesque' can be seen as forming the notions that underpinned much of the landscape design at Blairadam that Scott so admired: W. Gilpin, *Observations, relative chiefly to picturesque beauty, made in the year 1776, on several parts of Great Britain; particularly the High-lands of Scotland* (London, 1789); W. Gilpin, *Remarks on forest scenery, and other woodland views (relative chiefly to picturesque beauty), illustrated by the scenes of New Forest in Hampshire* (London, 1791); W. Gilpin, *Three essays: on picturesque beauty; on picturesque travel; and on sketching landscape: to which is added a poem, On landscape painting* (London, 1792). For the Wordsworths' 1803 tour of Scotland, accompanied initially by Samuel Taylor Coleridge, see Dorothy Wordsworth, *Recollections of a Tour Made in Scotland*, ed. C.K. Walker (London and Newhaven, 1997). Coleridge's own view of that tour and his solo journey after leaving the company of the Wordsworths at Arrochar is presented in C.K. Walker (ed.), *Breaking Away: Coleridge in Scotland* (Newhaven and London, 2002). Scott's views on Repton are discussed in Tait, *Landscape Garden*, pp. 203–206.
21   Oliver, *Walter Scott and the Greening of Scotland*.

mental transfiguration in Scotland's history.

At Abbotsford, Scott's re-imagining of a wooded landscape was limited only by the boundaries of what land he could purchase and the burden of his debts. Elsewhere, he had far greater latitude as an adviser to some of the leading landowners of his day on how they could transform the country around their residences. Through the 1810s, Scott was an almost daily adviser to the 3rd duke of Buccleuch, who in 1810 had inherited the Queensberry estate centred on Drumlanrig Castle in Dumfriesshire from his childless second cousin, the 4th duke of Queensberry. Fabulously wealthy in his day, Queensberry (better known as 'Old Q') had nevertheless allegedly plundered his estates of their mature woodland with no policy of replenishment. It was for such wasting of his estates' woodlands that in 1803 he won William Words-worth's contemptuous label as 'Degenerate Douglas' in an excoriating sonnet that lamented the felling of 'a noble horde, / A brotherhood of venerable Trees' around Neidpath Castle outside Peebles.[22] Words-worth witnessed the recently felled oakwoods on the slopes above the Tweed around the castle, seeing them as an almost apocalyptic landscape left wasted by an owner who had not one drop of romanticism or aestheticism in his shrivelled soul, and was outraged by what he saw as wanton destruction. Yet, the felling of the mature Neidpath oaks was, in the eyes of the duke's estate managers, simply the realisation of an estate asset and part of a programme of felling and regeneration that had been started by its previous owner, the famous early Improver and planter the 2nd earl of Tweeddale (see Chapter 4, pp. 199, 201–202) in the 1670s. The Neidpath oaks were capable of regeneration as coppice, but elsewhere on the Queensberry estate the trees being felled were not of species given to regrowth in this manner, and a policy of investment in fresh planting was needed to replenish what was still considered to be an economic resource rather than an object of aesthetic delight. It was Scott who advised Buccleuch on the replanting of the woodland on the Queensberry properties that followed his inheritance of the estate and titles

---

22   William Wordsworth, *Sonnet Composed at* [Neidpath] *Castle* (1803):

>   Degenerate Douglas! oh, the unworthy Lord!
>   Whom mere despite of heart could so far please,
>   And love of havoc, (for with such disease
>   Fame taxes him), that he could send forth word
>   To level with the dust a noble horde,
>   A brotherhood of venerable Trees,
>   Leaving an ancient dome, and towers like these
>   Beggared and outraged! Many hearts deplored
>   The fate of those old Trees; and oft with pain
>   The traveller, at this day, will stop and gaze
>   On wrongs, which Nature scarcely seems to heed:
>   For sheltered places, bosoms, nooks, and bays,
>   And the pure mountains, and the gentle Tweed,
>   And the green silent pastures, yet remain.
>   (From Wordsworth, *Recollections*, p. 201.)

The 'venerable' oaks whose loss was lamented by Wordsworth were themselves part of a planting policy on the part of John Hay, 2nd earl of Tweeddale, whose Yester estate in East Lothian was one of the first in Scotland to see large-scale tree-planting in the third quarter of the seventeenth century. In 1803, they were a little over 130 years old.

Neidpath Castle, Peebles, Scottish Borders. The castle is set nowadays among the remnants of the conifers planted by Old Q's Wemyss successors and some regenerated oaks from the woods felled in 1803, whose cutting down so outraged William Wordsworth. The hillside behind is covered mainly by twentieth-century planting.

(but not Neidpath, which passed to the Wemyss family) and also on expansion of the woods around the 3rd duke's new house at Bowhill at the junction of the Ettrick and Yarrow valleys west of Selkirk.[23] At Bowhill, the primary intention at first was for ornamental planting, the duke's friends recommending larch and black Italian poplar as suitable species to provide eye-catching colour contrasts.[24] Much of the broadleaf woodland – mainly 'alien' pedunculate oak and beech – that fills central Nithsdale around Drumlanrig and the slopes and valley bottom of Yarrow around Bowhill originated at this time, although the designed elements in the landscape rather than Scott's random planting were implemented after his death and gave the woodland much of the character it has today.

Planting for future benefit

When Sir Walter was advising the 3rd duke of Buccleuch about landscaped planting of trees, the duke's family already had over a century of commercial woodland plantation experience behind them, and they had extensive mature woodland in various locations on their Midlothian, Roxburghshire and Dumfriesshire properties. Plantation policy, indeed, formed a significant element of the day-to-day management of the sprawling Buccleuch estate, with the woodland of Eskdale and Ewesdale in eastern Dumfriesshire a particular focus of managed thinning, extraction and sale in the first quarter of the nineteenth century.[25] The estate's factors valued the woodland highly and were no less rigorous in their efforts to halt illicit cutting of timber than they were in their actions against poachers, deploying spring-guns and mantraps as deterrents.[26] The Scotts' principal estate interests, however, continued to be focused on their income from rents, arable yields and livestock, mainly sheep on their Southern Uplands properties rather than timber and wood products. Nevertheless, they had also positioned themselves at the start of the wars with Revolutionary France in 1792 to supply materiel to the government, including the broad range of wood and wood products – charcoal for gunpowder, oak bark for tannin – from their plantations. The Scotts' effort, though, were vastly outstripped by more northerly landowners such as the dukes of Atholl and Gordon, earls of Breadalbane and Seafield, and smaller estate owners such as the Grants of Rothiemurchus. Those landowners' early and less systematic planting activities gained fresh impetus and direction in the evolving fiscal-military state of the Revolutionary and Napoleonic Wars. New demands for warships and military equipment forced up the price of timber and wood products and quickly alerted landowners to opportunities to meet future demand for timber to renew the nation's rapidly expanding navy. The scale of their ambition can be seen, for example, in the March 1802 report sent by Breadalbane's factor at Taymouth to his employer, which noted the planting of upwards of three million oaks the previous season.[27]

At the forefront of these opportunists, as first explored in Chapter 7, was the so-called 'Planting Duke', the 4th duke of Atholl, who had blanketed

---

23 Tait, *Landscape Garden*, pp. 204, 206.
24 NRS GD224/628/5 (53).
25 See, for example, NRS GD224/522/3 (66–74, 85, 86).
26 Ibid., no. 32.
27 NRS GD112/11/7/5/53. Although the estate accounts show that English pedunculate acorns were being purchased by the Breadalbane estate since the mid eighteenth century, it is clear that the earl's managers were also sowing locally gathered sessile acorns. The 1801/2 accounts include payment to boys gathering acorns and rowan berries on the estate for sowing in the nurseries: GD112/74/423 (17–23).

Loch Ordie, Guay, Perth and Kinross. Only a few scattered remnants, like this one example on the hillside above the loch, survive of the millions of larches planted across the Atholl estate. Fires, storm damage and large-scale felling in the later nineteenth century saw much of the area revert to open country, with regeneration prevented by the rising deer numbers that were encouraged for sport shooting.

his lands in Strath Tay and Strath Tummel with between 14 and 27 million larches (according to different estimates) by the time of his death in 1830.[28] Traditions of his use of cannon to scatter larch seed on the ledges of the rock faces flanking the Pass of Dunkeld might be apocryphal – he lowered boys on ropes to reach these exposed perches – but the wildly Romantic landscape he created shows that he was not interested solely in the commercial value of the tree he championed, recognising in its shape and changing, deciduous conifer colours a species that added beauty to what one traveller had once styled 'the desert and dreary ranges of the valleys of the Tay and Tummel'.[29] Already by 1803, when the Wordsworths visited Dunkeld, the duke's larch plantations – labelled 'fir-trees' by Dorothy Wordsworth – covered the hillsides of the Tay and Tummel valleys for around three miles north of the town. 'In forty or fifty years', she opined, 'these plantations will be very fine, carried from hill to hill, and not bounded by any visible artificial fence.'[30] Sir Walter Scott would have approved enthusiastically.

For a while in the 1810s, the duke's forestry activities were the subject of widespread comment and conversation, for other landowners were keen to learn more of the commercial potential in larch planting.[31] This interest should be seen in the context of the energetic public discourse and political lobbying surround-

28 Fowler, *Landscapes and Lives*, pp. 87–94.
29 Ibid., p. 94.
30 Wordsworth, *Recollections of a Tour Made in Scotland*, p. 173.
31 NRS GD124/15/1735 (50).

ing larch and the supply of shipbuilding timber to the Royal Navy explored in detail by Fredrik Albritton Jonsson,[32] and not simply as an exercise in polite exchange of Improving ideas. Atholl's enthusiasm for this species, which he favoured over slower-growing, twist-prone and knot-prone Scots pine, was slow to win converts among the government's purveyors and naval shipwrights, only making a brief breakthrough in 1816 when he supplied the materials from his estate to build the frigate HMS *Athole*.[33] By that date, he had already calculated that his family lands could supply all British shipbuilding timbers by the end of the nineteenth century and he enthused that if other estates followed his practice then Britain could become a net exporter of timber, reversing its already worrying dependence on foreign supplies. By 1816, however, the French wars were over and, while Britain still needed to maintain a large navy to protect the far-flung empire that had grown rapidly over the course of the conflict, it did not need the number of vessels that had been constructed and was already breaking up older ones. Furthermore, influential figures in government, including the key Scottish political fixer Henry Dundas, favoured alternative sources – in British North America and South Asia – and different tree species, principally imported teak, which they promoted aggressively and much more successfully in decision-making committees at Westminster. Nevertheless, Atholl pressed on with undiminished confidence that larch would soon wholly replace oak as the predominant shipbuilding timber, and in the same year as HMS *Athole*'s launch started on a ten-year programme of planting across nearly 1,300 hectares of fenced upland around Loch Ordie, north of Dunkeld, through which he constructed a network of carriageways, tracks and bridges that made management and future extraction easier. This was forward planning and commercial land management on a grand scale, creating a blanket of almost monoculture larch such as was not to be seen again until the mass planting of Sitka spruce by the mid-twentieth-century Forestry Commission. When he died in 1830, the duke had still not secured the permanent supply contract for the Royal Navy that he craved and was, happily, oblivious to the future of warship-building which was soon to render redundant both the 'larch autarky' and the alternative 'empire of teak' that had been lobbied for by Atholl and Dundas. It was not wood but metal that won the day: the future lay in iron-clad and then steel-hulled vessels, the first of which, the French *Gloire*, was launched nearly thirty years after the 4th duke's death.

Despite his lack of headway in the politicised debates of the 1810s and 1820s, Atholl's proselytising for the adoption of larch was not scorned by neighbours or rivals, who were quick to recognise that what had once been thought of as a delicate garden specimen tree was ideal for the rocky and peaty terrain of the central southern Highlands. On the neighbouring Breadalbane estate, larch was first being obtained from James Seton's nurseries at Crieff in the 1790s. These were supplying tens of thousands of seedlings annually to plant across the estate.[34] Losses of cones in the late frosts of 1801 hampered the Breadalbane estate's larch-planting policy and forced the earl to make up for the deficiency with the purchase of seedlings in London.[35] Breadalbane was also involved in the sale of larch seed and seedlings, with demand for the species soaring in the 1810s,[36] perhaps buoyed by the apparent successes

---

32 Albritton Jonsson, *Enlightenment's Frontier*, chapter 6.
33 Fowler, *Landscapes and Lives*, pp. 90–91.
34 NRS GD112/74/418 (21). Seton, however, was also supplying oak in bulk, probably pedunculate varieties: GD112/74/423 (16).
35 NRS GD112/11/7/5/53, GD112/74/145 (4).
36 NRS GD112/74/72 (15).

Priory Island and Drummond Hill, Loch Tay, from Croft-na-Caber, Perth and Kinross. Little of the Breadalbane planting that once cloaked the hillside north of the loch survived the financial collapse of the Campbells in the nineteenth century and demand for wood in the First World War, but the band of mixed deciduous and ornamental conifer trees along the lochside beneath the blanket of Sitka spruce is a remnant of the millions of trees once planted on the estate.

of the Atholl estate. Regardless of the ultimate failure to secure the Royal Navy supply contract, the Atholl larch plantations were widely recognised as a timber-growing success, and other estates, including Buccleuch across the Southern Uplands and the Mackenzies in Ross-shire, were actively planting larch in high numbers into the 1830s or gathering information on larch-planting practice.[37] Estimations of the growing timber numbers and values on the Atholl estate, circulated along with official reports on the properties of larch timber to encourage planting among other estate owners, confirm that the 4th duke's plantings were intended to contribute – without any replenishment – to the Murrays' income into the twentieth century.[38]

Among the tide of plantation and introduced species, it is possible to lose sight of native woodland and its continuing exploitation as sources of building timber, wood for barrel-staves and hoops, turning, tanbark and charcoal, as well as contributing to fuel needs, both on a commercial basis and for the needs of the local communities. Most of the largest estates

---

37 NRS GD46/1/476, GD46/15/237 (10, 11) [1837 accounts for purchase from seedsmen in London of 'best Scotch' and 'foreign' larch seed], GD46/17/50 [1819 letter reporting the purchase of Italian larch seed in Genoa]; GD224/588/7 (17).
38 The success of this publicity can be seen in the copies of this material kept by one Peeblesshire proprietor: NRS GD504/3/78, GD504/3/150.

engaged in large-scale planting also contained extensive stands of native woodland, primarily sessile oak and Scots pine, often with elm and ash, but also managed areas of fast-growing alder, aspen, birch, blackthorn, gean, hawthorn, hazel and willow, from which tenants' wood needs had been supplied, albeit under strict supervision. But not every great landowner embraced the tree-planting zeal of the era and some, instead, continued with the by-then traditional modes of exploitation of the existing areas of managed woodland, hagging, coppicing and regenerating where possible and regulating the inroads of their tenants. Where they were involved in creation of plantations, it was often through closer management of certain native tree species and specialisation of production for very precise uses, like barrel hoops, rather than planting of new, introduced timber species.

Where such specialised and commercialised production was to be undertaken, woods had to be enclosed to prevent human and animal inroads and cleared of low-value brushwood to allow the in-demand species to thrive, changing delicate ecological systems in the process.[39] Some of the greatest impacts were on the least valued wood species, such as juniper, which had been in natural decline for millennia as it was out-competed by the more aggressive birch, but which retained an extensive presence in and around the Cairngorms and Strathspey, and also in parts of the northern Highlands.[40] In 1803, Coleridge noted 'a good deal of low cowring Juniper with its fruit of various years, purple and green' as he traversed Strathdearn on his route from Inverness to Kingussie. Some decades later when plans were being drawn up for an estate road in the heart of the Coignafearn deer forest at the head of the strath, it was noted that there was still an abundance of juniper wood locally.[41] Although there are traditions of juniper wood being valued alongside peat as a 'smokeless' fuel by illicit distillers who were active into the second quarter of the nineteenth century, it is almost wholly invisible as a resource in estate records and unmentioned in the Statistical Accounts.[42] This perception of minimal utility, which also affected views of alder and aspen as expressed by respondents to Sir James Sinclair's survey, led to no efforts to conserve juniper. Consequently, expanses of juniper were displaced for Scots pine plantation on the Grant and Mackintosh estates in Strathdearn and Strathspey from the 1790s onwards. Meanwhile the introduction of increasing numbers of grazing animals – and from the 1840s onwards the encouragement of greater deer numbers for stalking – accelerated the long process of natural decline by halting regeneration through natural seed dispersal.[43] It is only in recent years, with the inclusion of juniper

---

39 A. Hampson and C. Smout, 'Trying to understand woods', in T.C. Smout (ed.), *Understanding the Historical Landscape in its Environmental Setting* (Dalkeith, 2002), pp. 87–95 at 89–91.

40 Smout, 'Highland land-use', p. 11 for the continuing presence of juniper in the woods of Strathnaver. The inferior status of juniper was embedded in Irish lawcodes from the eighth century, which grouped tree species into classes from 'nobles' to 'slaves' that reflected the economic significance of different trees. Juniper was classed as a 'slave' species, which underscores its most inferior status in Gaelic cultural terms: Smout et al., *History of the Native Woodlands*, p. 79.

41 Walker, *Breaking Away*, p. 172; NRS RHP2191.

42 Smout et al., *History of the Native Woodlands*, p. 98. Juniper berries were gathered for flavouring food and alcohol. A visitor to the Grants at Castle Grant in 1796 wrote to the factor there to request that items he had left behind, including juniper berries, be sent to him: NRS GD248/456/5.

43 It has only been since the 1960s that changes in land-use and increasing human presence in areas like the northern Cairngorms, which has disturbed deer, has seen recovery of juniper on the mountain slopes: R. Worrell and N. Mackenzie, 'The ecological impact of using the woods', in T.C. Smout (ed.), *People and Woods in Scotland* (Edinburgh, 2003), pp. 195–213 at 201.

among species being planted in upland conservation and restoration programmes, that the populations of this undervalued tree have begun to recover.

A balance between traditional and commercial exploitation of native woodland can be seen in operation on the Argyll estate in the two decades either side of 1800. There, the Atlantic oakwoods that fringed the mainland coast and the eastern side of the larger islands, such as Mull, continued to meet domestic demand at the end of the eighteenth century, but were coming under intensifying pressure from the duke of Argyll's ever-increasing tenant population. In 1786, for example, the duke's chamberlain of Mull, whose territory also covered the woodland on the shores of Loch Sunart, was reporting that folk from treeless Tiree, who had traditional access rights to the Mull and Sunart woods, were abusing those rights and, he warned, 'in a few years they will utterly destroy the woods'.[44] In response, the duke insisted that his chamberlain increase the rigour of his oversight of the tenants' activities: 'I insist that you take measures for preventing their getting a single stick without your order and your knowing what use it is for.'[45] Concern over the sustainability of supply from these heavily exploited Sunart sources for local use and to provide tanbark for sale and charcoal for the Argyll iron furnaces led to comprehensive surveys of remoter parts of the duke's possessions. In 1790, it was Moidart to the north of Sunart that was inspected, the report suggesting widely varying local conditions, from areas of limited surviving oakwood but more abundant 'blackwood' (the mix of scrub species dominated by birch), to land where 'a close, thriving stool' of valuable oak coppice remained, but overall the trees were scattered thinly and widely over the district which rendered them unviable to enclose or exploit profitably.[46] Although much of the area around Kinlochmoidart today is swathed in conifer plantation, it is intermingled with expanses of predominantly birch and oak wood, a species-impoverished successor to woodland exploited since at least the late sixth century (see the volume 'Scotland AD 400–1400') and which continued to be used by Argyll estate tenants through the nineteenth century, despite the pessimistic reports of the duke's wood-surveyors. Here, though, estate policy saw commercially less-valued but locally in-demand species, such as ash and elm and the broad range of smaller species dominated by alder, blackthorn, hawthorn and rowan, weeded out to allow the oak especially to thrive.

Another area of interest closer to the economic centres of the estate, at Ardtornish on the mainland opposite Mull, had been identified in 1789 as a suit-

---

44 Cregeen, *Argyll Estate Instructions*, p. 7. The woods were mapped in 1733, showing dense areas remained, extending along both sides of Loch Sunart east of the island of Carna almost to its head near Strontian: Alexander Bruce and Richard Cooper, 'A plan of Loch Sunart . . . become famous by the greatest national improvement this age has produc'd / survey'd &c. by Alexr. Bruce; R. Cooper sculp.' (Edinburgh, 1733) at NLS https://maps.nls.uk/view/00000545. Pretty much the same extent of wood was shown in George Langlands' 1801 survey: George Langlands, 'This map of Argyllshire taken from actual survey is . . . dedicated to . . . John Duke of Argyll . . . / by . . . George Langlands & Son . . . (Engraved by S.J. Neele)' (Campbeltown, 1801) at NLS https://maps.nls.uk/joins/7382.html and in John Thomson and William Johnson, 'Northern Part of Argyll Shire. Southern Part' (Edinburgh, 1832) at NLS https://maps.nls.uk/atlas/thomson/506.html. What these sources cannot show is how a rolling estate policy of coppice/clear-felling and regeneration might have moved across the areas mapped across the century between 1733 and 1832, but they do suggest a greater resilience than the chamberlain's fears implied.

45 Cregeen, *Argyll Estate Instructions*. The resulting record of 'Orders for Timber' from November 1786 to September 1789 reveals high demand for roof couples, with requests for 472 (pp. 12–15) but the account of wood supplied from Loch Sunart shows that only 312 were provided (pp. 17–18).

46 Ibid., pp. 160–161.

Juniper tree, Muckle Fergie Burn, Stratha'an, Highland. Possibly the only tree species to survive the last glacial period in the Younger Dryas, the juniper has been in long decline due to over-grazing by sheep and deer and over-planting by commercial forestry, but is now being encouraged in ecosystem restoration projects.

able location for preserving an area of recent wood coppice to produce barrel hoops (mainly for locally manufactured herring barrels) from the regrowth. In 1790, proposals were made to construct boundary dykes from shore level to the rocky escarpments that fringe Loch Aline and the Sound of Mull from Aoineadh Achadh Rainich in the north to Aoineadh Mòr in the south-east. The intention was to enclose the oakwood that the Argyll Furnace Company had recently felled for charcoal manufacture, and to clear the commercially valueless brushwood species from the woods.[47] Such management ensured the survival of the Atlantic oakwood along this coast – it is still there – but the estate record indicates that there was heavy intervention to remove the brushwood species that thrived among the recently coppiced woods. Again, the character of the native woodland was being changed through estate interventions to maximise the yield from the most commercially sought-after species. The wood needs of estate tenants were secondary in this activity.

Such thin tree-cover as was described in Moidart around 1790 is suggestive of a post-felling native woodland landscape where regeneration had either just begun or where other pressures, perhaps from grazing animals, was preventing recovery. This seems to have been the situation in some of the central Highland pinewoods that had experienced heavy

---

47 Ibid., pp. 158, 165. The duke's response to the proposal to clear the brushwood was to ask if it could be cut that year, due to the general shortage of fuel (i.e. peat) that had been caused by the wet weather (pp. 166, 168).

Loch Aline from the south-west, Highland. The woodland flanking the sea-loch is still a mix of Atlantic oakwood with birch and other small tree species, representing the regenerated successors of the wood felled by the Argyll estate at the end of the eighteenth century and hagged on a rolling basis for the next century.

episodes of felling in the middle of the eighteenth century, especially in Abernethy and Rothiemurchus.[48] Analysis of map evidence for woodland in Abernethy suggests recurrent pulses of expansion and contraction through cycles of felling, with sharp decline of around 50 per cent of the wooded area from 1750 to 1812 and halving almost again between 1812 and 1830, before recovering to almost its full 1750 extent by 1858.[49] As trees in excess of 200 years old are still growing in some areas shown as treeless in 1812 and 1830, however, we should be cautious of accepting at face value what the surveyors mapped; it is likely that some of the supposedly cleared, formerly afforested areas by 1830 were either carrying light woodland similar to that reported in contemporary Moidart or by Charles Cordiner in Glen Quoich and Glen Lui in the late 1770s (discussed above, Chapter 7, pp. 212–214), or were areas where regeneration was just commencing. It is probable that this kind of extended cycle of clearance and regrowth had been happening in these woods since the Middle Ages, but that view needs to be qualified by recognition of the likely greatly increased extent of the felling that occurred into the second quarter of the nineteenth century and the possibility that regeneration was also encouraged through estate planting that introduced non-native species alongside the indigenous varieties.[50] Such introductions, again

---

48 Smout et al., *History of the Native Woodlands*, pp. 214–216.
49 Ibid., p. 215 map 8.2.
50 The scale of operations in the Abernethy, Glenmore and Rothiemurchus woods is explored in detail in Smout et al., *History of the Native Woodlands*, chapters 8, 9 and 10. See also Hampson and Smout, 'Trying to understand woods', pp. 89–90.

mainly pedunculate species in oakwoods, has given rise to hybridity, but there is nowadays a suspicion that the evident failure of many western Highland oak woodlands to regenerate in recent decades might relate to genetic issues arising from such planting of apparently superior species that in the long term have proved unsuitable for the environmental conditions of the Atlantic coasts. Felling, however, even on the intensified scale experienced in the late eighteenth and earlier nineteenth centuries, was evidently not the existential threat to these forests that we might think and, despite the inroads made down to the 1830s, they survived with much the same footprint into the last quarter of the nineteenth century. Where regeneration was not prevented by over-grazing or conversion to arable, neither of which were major problems in this part of Strathspey and Abernethy in the first half of the nineteenth century, regrowth probably started immediately, though perhaps hindered by the poor growing conditions in the many cold and wet years down to 1850.

A material of multiple functions

So far, discussion has focused principally on planting new trees, on the transformative impact of that process on the Scottish landscape, and on the juxtaposition of ornament and investment in the thousands of hectares and millions of trees planted from the 1790s to the 1840s. This emphasis itself reflects a general thrust of much secondary literature, which has concentrated on the visual transformation of the land and the creation of the wooded districts with which we are still familiar today. It is all very 'green' but tinged with ecological critique of the impact of monoculture planting across swathes of land that in many cases had not been treed for millennia. It tends also to be preoccupied with the socio-economic impact of commercialised forestry in districts where the land had been otherwise exploited by generations of farmers, and, more recently, with disdainful dismissal of the exotic or alien species introduced and condemnation of the almost systematic undermining of the native species they replaced, which are now promoted as solutions to many of Scotland's ecological ills. In most discussions of woodland, however, it is simply the planting, nurturing and felling cycle that is considered. What the wood was used for and how it helped to further the transformation of Scotland's environment for better or for worse has made its way only infrequently until recently, with notable exceptions, into the wider conversation.[51]

Charcoal and tanbark, as discussed in Chapter 7, continued to be the primary ways in which woodland products other than timber extracted from Scotland's woodland were used. As well as being a specialist fuel used to generate high temperatures for iron-smelting in the era before 1828 and the introduction of James Neilson's coal-fired hot-blast furnace, charcoal was a major component in gunpowder. Already rising demand for gunpowder in the second half of the eighteenth century had climbed steeply after 1792 and the start of the French wars. Although the significance of its manufacture in some districts where a broad range of 'inferior' tree species were available has been noted,[52] the late eighteenth- and earlier nineteenth-century gunpowder manufactory in Scotland is an under-researched topic in respect of its relationship with woodland exploitation.[53] The archaeological and cultural heritage dimension of such sites, however, has begun to be recognised and researched, as in the case of the rise and fall of Scotland's earliest large-scale producer, the Stobsmill works at Gorebridge in Midlothian, which was in operation from 1794.[54] There, the owners entered agreements with the neighbouring Arniston and Vogrie estates for areas of land for new plantations for trees for charcoal, receiving access to the scrub woodland of the Crichton Moss until such time as the plantings were sufficiently developed to start a rolling cycle of coppice. Stobsmill was followed by further mills at Roslin, Marfield near Carlops and at Fauldhouse, all of which were in operation by 1812.[55] For all of these Lothian-based operations, long-exploited woodland was drawn upon while new plantations matured, but elsewhere it was old-managed woodland that provided most of the smallwood used in producing charcoal for the powder mills. Several were located in southern Argyll, where the oak- and birchwoods in particular had been managed since the mid 1700s to deliver high volumes of charcoal, principally for iron-smelting. Gunpowder production was a secondary activity at Furnace, running in the Lochfyne Powder Works alongside iron production until the closure of the site in the 1880s, with other mills at Clachaig in Glen Lean near Dunoon, Melfort near Kilmelford, and Millhouse near Tighnabruaich.[56] Long after charcoal production

---

51 Especially Stewart, 'Using the woods, 1600–1850 (1)' and 'Using the woods, 1650–1850 (2)'; Smout et al., *History of the Native Woodlands*, chapter 4.
52 For example, Stewart, 'Using the woods (2)', p. 124, references the Vale of Leven between Dumbarton and Loch Lomond, which was a 'hot spot for textiles and for gunpowder production, which used alder in particular, but also birch, hazel, willow and rowan'.
53 For a detailed examination of the industrial history of the Argyll operations and Melfort in particular, see J. Robertson, 'The powder mills of Argyll', *Industrial Archaeology Review*, 12:2 (1990), pp. 205–215.
54 Gorebridge Community Development Trust, Stobsmill Gunpowder Works: An Introduction (no date): https://gorebridge.org.uk/heritage/stobsmill-gunpowder-works-an-introduction/.
55 G. Crocker, 'Scotland', in Gunpowder Mills Gazetteer; Black Powder Manufacturing Sites in the British Isles (1996 and 2008), https://catalogue.millsarchive.org/uploads/r/null/8/5/8/858f75adab794e799564ab57ddaf71aee15efaaf073a6c852046f95468cf0a9c/_home_artefactual_digi_objects_Rest_25550.pdf.
56 Robertson, 'Powder mills of Argyll'; Crocker, 'Scotland'; K. McConnell, 'The people of the powder mill', https://www.secretscotland.org.uk/index.php/Secrets/ArgyllGunpowderIndustry; Anon., 'A chemical legacy: a banging beautiful trip round the west coast of Scotland', https://chemicallegacy.wordpress.com/2019/11/01/a-banging-beautiful-trip-round-the-west-coast-of-scotland/.

Bonawe Ironworks, Argyll and Bute. Perhaps the best known of the consumers of the wood products from the Argyll oakwoods, the iron furnaces were only one element in an early industrial complex which included gunpowder manufacture and the production of the pyroligneous chemicals that were in demand by Scottish textile manufacturers.

for the Argyll-based iron industry had effectively ceased as new coal-fired technologies took over, woodland in the west – and in Lothian – continued to be coppice-managed to meet the demand for the production of 'black powder' explosives, with the last works closing only in the 1900s.

Wood-derived chemicals obtained from the processing of coppice-wood and wood trimmings were increasingly important from the later eighteenth century, far exceeding the limited turpentine production that had been tried in some of the Argyll pinewoods in the second half of the eighteenth century (see above, p. 212). There was a high degree of symbiosis between intensification of coppice management in some districts, for example in southern Dunbartonshire, and the rapid expansion of textile production in nearby centres like Bonhill, as coppice smallwood was in demand for everything from cogs in mill machinery to thread and yarn bobbins and wicker packing-cases.[57] The heaviest demand, however, was for raw materials for wood distillation to produce the acetic acid that was a basic

57 Stewart, 'Using the woods (2)', pp. 123–124.

component in dye mordants for coloured and printed textile manufacture. Distilling by-products were charcoal, tar and vinegar, which were also consumed largely by cloth manufacturing, but with charcoal also supplied to foundries and tar sold for agricultural use in sheep-smearing.[58] Pyroligneous or pyrolignous acid processing works were established close to dependable large-scale sources of wood from Perthshire to Ayrshire. Perhaps the best known of the woodland works is the plant at Balmaha at the south-east side of Loch Lomond, where the well-established oakwoods on the Buchanan estate of the dukes of Montrose in particular supplied the raw materials into the 1920s,[59] with the woods of the Colquhouns of Luss and Macfarlanes of Glen Falloch similarly exploited but on a smaller scale. But the ability to transport the raw materials from the woods to new processing works closer to the sources of demand in the early 1800s saw the establishment of new, urban producers who consumed raw materials that were transported to the cities. Turnbull and Ramsay's 'Pyroligneous Acid Works' in Camlachie Street off Great Eastern Road in Glasgow, possibly the earliest of these, was founded c.1806–13.[60] Through 'destructive distillation' of wood (mainly oak), it produced pyroligneous acid, wood spirit and charcoal, which was then processed for the manufacture of acetic acid, methylated wood spirits, cloth-printing dyes and mordants, and ground charcoal for use in iron foundries and the making of gunpowder. The Turnbull company, however, also established pyroligneous works closer to sources of wood for processing, principally at Kilkerran in southern Ayrshire, in operation by 1845, where it took out a twenty-one-year lease of woodland on the Fergusson of Kilkerran estate.[61] The locating of a plant at Kilkerran might seem counterintuitive in arguments for greater centralisation, but the completion in 1840 of the Glasgow to Ayr railway and the presence of a substantial woollens industry around Galston and Newmilns allowed the production from the plant to be transported to the consumption centres in and around Glasgow as well as also opening a new, local market. Until the development of newer, chemical dyes in the second half of the nineteenth century and the growth in the industrial processing of coal for chemical by-products including tar, such operations proliferated in most districts where there was both a supply of wood and proximity to a market. By then, the spread of another voracious consumer of wood had completely changed the dynamics of production and supply, hastening the demise of many of the smaller plants that lay remote from the main communications networks.

Although the railway system that spread through Scotland from the later 1830s and which provided the logic for the siting of the Kilkerran works might be viewed as one of the greatest symbols of the new Industrial Age, it was founded almost literally on Scottish wood. Demand from the railway companies opened a new market for material that the great planters of the 1790s and earlier 1800s had once thought was destined for shipbuilding. Larch found new proponents as an element in the revolutionary new transport system of the steam-powered railways, with its durable and rot-resistant wood used for the sleepers that carried the rails. In the 1830s, the Mackenzie estate at Brahan in Easter Ross was providing larch timbers for the London to Birmingham railway, and in the 1840s railway sleepers were

---

58 V. Georg, 'On smearing sheep', *The Farmer's Magazine*, 7:26 (Edinburgh, May 1806), pp. 172–175.
59 Smout et al., *History of the Native Woodlands*, p. 263.
60 J.R. Hume, *The Industrial Archaeology of Glasgow* (Glasgow, 1974).
61 D.C. McClure, 'Kilkerran Pyroligneous Acid Works 1845 to 1945', *Ayrshire History* http://www.ayrshirehistory.org.uk/AcidWorks/acidworks.htm.

one of the biggest uses of the Breadalbane larches.[62] It was pine, however, that provided most of the sleepers for the rail-lines that transected the Highlands, supplied from the plantations and semi-natural woodland through which they ran.[63] But much of that consumption of pine lay in the 1860s and later, with the more southerly larch plantations being used to source the wood needed for the first burst of railway construction between c.1835 and 1850. When the volume of timber needed to construct the fleets of mainly wooden rolling-stock is added to the equation, much of it to construct the wagons that were transporting the growing tide of coal from Scotland's mines, we begin to catch a better sense of the indispensable role that Scottish woodland still played in the new era of coal, iron and steel.

• • •

This chapter started with the awe and wonderment expressed by early nineteenth-century travellers who first experienced what to them seemed to be vast swathes of luxuriant woodland that stretched across parts of the Scottish landscape. Most were visitors from less wooded areas of the British Isles, so their sense of scale and extent was perhaps somewhat skewed, but they are our primary witnesses to the transformative impact that tree-planting was having across Highland and Lowland zones in the first quarter of the century. While for them, these were sublime, romantic landscapes – quite a change from half a century earlier when Sir William Burrell had sniffed at anything that looked disorderedly unaesthetic – for their planters they were investments banked for future benefit and were rigidly managed to that end. The spread of these plantations marked a final transition away from the loose management of the semi-natural native woodland of diverse species, tree age, shape and size which had evolved since prehistory into the intense management regimes of the nineteenth and twentieth centuries, where only some valued species were planted and grown and the unvalued ones were rooted out as 'weeds' and 'vermin'. With the imposition of this 'Improved' planting policy came an intensification of management of some of the last areas of old native woodland, the diminution of the species range they contained, and the introduction of a major question mark against their 'semi-natural' character. By 1850, perhaps only around half the hectarage of native woodland that had existed in 1750 still survived and, of that, much had been transformed beyond recovery into the ecologically stunted, species-impoverished woods that have struggled into the twenty-first century. While we decry that loss, to our ancestors such transformation was the very essence of Improvement, a reflection of that central belief in humanity's ability to enhance nature and better harness it to the needs of human society. There is no question that much was planted with an eye to beauty and effect, as the seemingly artless but wholly contrived design of the woodland around Blairadam underscores, and there was much emulation of the freer, 'natural' disorder that Adam and his friend and admirer Sir Walter Scott advocated. Yet, apart from the groves and vistas planted to provide framing for views, drawing the eyes to distant historical landmarks or natural features, even those woods were ultimately considered to be cash crops in continuation on a grand scale of the previous two centuries of estate policy. Planting oak for future coppicing was, as Sir Robert Monteath advocated, a major

---

62   NRS GD46/1/393, GD46/1/394; GD112/16/11/5 (51), GD112/53/1 (20). The enquiry from Laurence Oliphant in 1841 about the sale of larch wood for the Scottish Central Railway noted that 'shipbuilding is at present very dull', as iron-hulled vessels were overtaking timber-built ships.

63   Smout et al., *History of the Native Woodlands*, p. 272.

component of many landowners' property portfolios and, while Scott decried the monoculture that such investment involved, preferring mixed woodland for its seasonal colour and texture variation and differences in shape and height, even Scott's many admirers chose more often to ignore his advice and follow Monteath for all planting other than in the immediate vicinity of their houses. As Scott discovered through his own financial embarrassment, planting for beauty and effect alone was ruinously costly and utterly unprofitable. A sea of Sitka still lay far in the future but already by the early 1800s the forestry theorists like Monteath had identified that single-species plantations were the rational model for the best financial returns.

Planting trees with an eye for future profit was not new; it had, indeed, driven kings and parliaments from the 1420s in setting out the legislative frameworks that encouraged woodland plantation. What was new, however, was the scale on which it was being undertaken and the speed with which the new plantations were being made. The new forests being created for the dukes of Atholl, earls of Breadalbane, and Grants of Freuchie dwarfed anything that John Hay had achieved on the Yester and Tweeddale estate. What was also new was that much was single-species plantation, initially of natives like Scots pine or introduced varieties of native species like oak, but chosen for its commercial value alone rather than any recognition of the multifunctional roles performed by the native, semi-natural woods that already existed on their land. A significant amount of the surviving Scots pine woodland outside the central and north-west Highland redoubts of that species are survivors of and successors to trees planted in this era. That early commercial decision-making marked the first step in the trend towards species monoculture that has dominated large-scale forestry to the present and also marked the start of a bias against other native species – like birch, geanies (wild cherry), alder etc. – which lacked the market value of the larger timber trees. It was not only with the arrival of the Sitka at the end of the nineteenth century that the persecution of 'valueless' scrub species began. But with this new, hard-nosed commercialism also came more rigorous control of the woods and an extension of policies of exclusion that had started in the eighteenth century. Improvement thinking did not see woodland as the right place for cattle pasture nor, indeed, for unwelcome tenants who still sought to exercise traditional rights to wood to satisfy their domestic timber needs. Like the peat mosses in the previous century, woodland was increasingly excluded from commons and the increasingly few surviving commonties, and managed as reserved components of an estate portfolio.

Large-scale planting was, thus, changing not just the physical character of the landscape and the ecological balances within large tracts of country, but also changing cultural practice and behaviour. Landowners and their estate managers looked to protect their investments, fencing off plantations and patrolling them to protect them from animal and human depredations. This trend reinvented compartmentalisation of the land that had existed in the Middle Ages – between reserved hunting land and areas designated as 'waste' or under intensive cultivation or grazing – but added physical, often impassable, barriers to divisions that had previously been notional. And, in this purposeful creation of blocks of woodland, much of it planted on land deemed otherwise of low economic value, we see Improvement extend far beyond the settled zones of old agriculture and pasture and, like the draining and clearance of the Lowland wetlands and mosses, change the perception of what could and should be done with upland zones. It was a century yet, as will be explored in the volume 'Scotland 1850 to COP 26', before commercial forestry reached its climax of spread over Scotland's landscape, but the first half of the nineteenth century had opened the eyes of estate owners and, ultimately, government to the potential of what the future could look like.

Scots pine woodland, Glenmore, Highland. Estate planting as well as natural regeneration sustained the pinewoods along the north-western edge of the Cairngorms, where the Gordons and Grants had been exploiting the trees commercially since the middle of the eighteenth century.

Conclusion

As with the climate transitions of the mid thirteenth and early fifteenth centuries, there was little immediately evident to the people of Scotland in their daily experience of weather and its impacts on their lives that the grim years of the mid and later 1840s had marked a turning point that represented the first step on the road to the conditions we are witnessing today. Successive generations had endured many previous peaks and troughs across the 450-year 'little ice age' and past experience suggested that there was no guarantee of a positive change that would be more than transitory. Based on just the prevailing weather patterns across the four decades down to 1850, there was no reason to suppose that improving times for the majority who still lived and worked on the land or the growing number of town-dwellers who were dependent on others to produce their food lay just ahead. It is only through the rapidly growing volume of instrumentally recorded meteorological data gathered from the second quarter of the nineteenth century onwards, and from modern analyses of Greenland ice-cores and other climate proxy records, that the scientists of the later twentieth century could see in retrospect that the late 1840s had marked the final extreme episode of what we now label the 'little ice age'. At the time of transition, regardless of whether you had profited or struggled to subsist through that last bitter sting in the tail, there was little discernible difference in weather conditions as Scotland entered the new decade; winters remained severe and summers unpredictable for many years yet. Given that the majority of Scotland's people were still rural-dwellers and in many parts of the country still primarily employed in working on the land, the climb out of the depths of the prolonged cold and high precipitation conditions of previous decades was protracted and, for some, exceptionally difficult. Even with the continuing expansion of improved farming methods, rapid growth in the availability and use of lime and organic fertilisers, further development of field drainage and plough technology, and experimentation with seed types, harvests in both Highland and Lowland zones continued to be variable and the precarity of the rural economy remained a concern for landowners and their tenants alike. The chancy nature of the rural regime was made yet more unstable by the rapid expansion of Britain's global imperial trade network and its sources of cheaper and more abundant sources of food and raw materials. Landowners who had pursued profits in wool or who relied on rents from sheep-farmers or grain-growers looked to pivot rapidly into new modes of working their assets that relied less on livestock and crops and more on the growing demand for access to game shooting and fishing among the moneyed classes. Scotland was on the threshold of entering a new phase in the continuing anthropogenic refashioning of major components of its land and the ecological balances upon it.

From our modern vantage point and unencumbered by the same pressures as drove landowners and land-users down paths that we now know had no long-term sustainable future, it is too easy to fall into that teleological mindset that sees an almost predetermined trajectory in human actions that lead inevitably to our present environmental position. But such a way of reading the historical record fails to acknowledge the complex interplay of human choices and responses in the face of continuing climatic and wider environmental change that unfolded across the years after 1850. It also fails to recognise that alternative paths were considered but, for a variety of reasons, were not chosen. Despite the supposed switch towards greater negativity that came from the impact of Malthusian theory and the failed experiments of Improvement that littered Highland – and Lowland – Scotland, it can be seen that far from all Scottish landowners and their tenants had become risk-averse; the experience of the eighteenth and earlier nineteenth centuries was written in terms of experimental risk-taking and, for many, that continued in the late nineteenth century, which saw a progressive amelioration in annual mean temperatures as global climate systems moved into a new configuration.

The environmental history of that era of rapid and ever-accelerating change is explored in the volume 'Scotland 1850 to COP 26', where the narrative follows the impacts of that new episode, which lasted into the second half of the twentieth century, and its legacies of degradation, over-exploitation and missed opportunities. In 1850 Prince Albert and his advisers were still striving to bring into reality the dream of the Great Exhibition, in which the triumphs of modern science and engineering – chiefly British – would be showcased to the world. Meanwhile most contemporaries would have given little thought to any associated environmental costs in a project that celebrated human science, reason and what might have been considered as the end to the millennia-old struggle to achieve mastery of nature. If any recognised that they had just experienced one of the great transitions in British and wider European history, they would not have thought in terms of climate and weather. Most educated Victorians saw instead a shift in entirely human terms, in the social order, in political structures or economic systems, providing a human societal exemplar of what from 1859 onwards would be understood as Darwinian evolutionary theory.

In that human societal shift, pre-existing social tensions and failures of government had been aggravated by the environmental stresses and climatic extremities of the 1840s that had reached their nadir in the 1848 'year of revolutions'.[1] A distressing number of our modern leaders still think in similar terms, despite our heightened awareness of the interplay of anthropogenic and non-anthropogenic factors in shaping our wider environment and daily, lived experiences. But the mindset of the ruling classes, captains of industry and scientific community of the early Victorian Age was not predisposed to reflect on the environmental context of the changes they were witnessing, let alone the environmental consequences of the new order they envisaged. They had the power and the will to change nature, to remodel it as both servant and plaything best organised to meet human needs. To them, scientific endeavour had finally freed humanity from the age-old struggle to subsist in the face of the forces of capricious nature. With hindsight, we can understand their confidence but also wonder at their folly, for we are now experiencing the extent to which nature, and

---

1 For the European experience, analysed from a social, economic and political perspective, see M. Rapport, *1848: Year of Revolution* (London, 2008). The impact of environmental factors in stimulating the wider social crisis is mentioned only on pp. 36–37.

especially climate and how we respond to its opportunities and constraints, has made – and continues to make – human history.

Throughout this book and the volume 'Scotland AD 400–1400', it has been the interplay of those factors in the constant reconfiguration of the world around us and our experience of it that has been stressed. That stance has also been qualified by recognition throughout that climate was not and is not the sole determinant affecting either human society or the environment in which it functions. As we have seen, however, it was a major factor and probably the single most important influencing agent; we have seen how changes in mean temperature, precipitation and storminess affected availability of fuel, restricted or encouraged plant growth, influenced patterns of fish migration, and enhanced or diminished the viability of flocks and herds and numbers of wild animals and birds, both indirectly and directly. Centuries of response to changing conditions are evident in the fact that agriculture never entirely failed across even the worst crisis years of the seventeenth century, although in some of the most environmentally marginal and economically vulnerable communities famine, displacement and death were widespread. Techniques to conserve soil fertility and prevent erosion, even if through labour-intensive and in some senses environmentally destructive methods, had developed in the later Middle Ages and sustained Scotland's agricultural base through the worst episodes, down to the point where new methods adopted by Improving tenants removed dependence on those successful but constraining past technologies and practices. As the 1810s, 1820s and 1840s had shown, however, the balance of traditional and Improvement practice had barely proved sufficient to prevent Scotland from sliding into widespread supply crisis. In essence, in an age when human populations were still largely, if not wholly, dependent on solar energy and the resources it delivered on an annual basis, extreme climate change as experienced in that last phase of the LIA was the primary variable affecting human welfare.

Yet it was how human populations chose to respond to those shifting conditions that mattered most. As cultures as diverse as the Greenland Norse and the Hohokam of Arizona and northern Mexico exemplify, dependency traps, cultural rigidity, poor decision-making and failure to adapt sufficiently fast or radically enough to changing conditions led only to what we might euphemistically style 'unfavourable outcomes' in times of climate stress.[2] Even during the worst phases in the fifteenth, seventeenth and nineteenth centuries, however, Scotland never suffered the cold/wet or hot/arid extremes that affected Greenland or the south-western region of North America, and consequently avoided the existential threats that ended cultures whose fundamental structures – from communal behaviours to dietary staples – had been laid down in better times. From the vantage point of 1850, our ancestors looked back upon the factors that had given a more favourable outcome to Scotland. They ignored the recurrent subsistence crises that wracked the nation from the sixteenth century to their own times, or explained them away in terms of the moral failures of an inferior culture, and celebrated the triumphs of Reformation, Revolution, Enlightenment and Union. Looking across that same span of the 'little ice age', however, we can better understand that Scotland benefited from a range of human and natural factors that ameliorated the impact of those successive subsistence crises. Our ancestors might not have appreciated their relative good fortune, especially

---

2 The 'collapse' thesis was expounded by Jared Diamond through his concept of 'ecocide' in *Collapse: How Societies Choose to Fail or Survive* (New York, 2005) and has been critiqued negatively for its overly deterministic interpretation of the role of climate in social and cultural failure. Most studies of such phenomena since have placed greater emphasis on the role of human agency and choice in framing outcomes.

during the dearth- and disease-ravaged 1690s, but geography, topography and social sophistication enabled a range of responses to the novel conditions of the 'little ice age' that allowed many of them to break free from old dependencies.

Looking back from 1850 on this record of crises averted, few observers could have failed to notice the scale of the change that had been wrought in Scotland's environment. Industrialisation might have been what caught most attention, with the smoke of central Scotland's expanding manufacturing towns and their sprawl of workers' housing signalling the presence of something radically new and unfamiliar. That air of unfamiliarity, however, extended far beyond the zones of urbanisation, for the spreading towns were set in the midst of a rural pattern that was itself settling into a new prescription of large, enclosed fields and orderly single-tenant farms. The old patterns of unenclosed fields clustered on the limited areas of easily cultivable land in which our ancestors had invested centuries of energy had given way in many places to an intrusion of agricultural endeavour that penetrated all but the most intractable reaches of the land. Even those few areas beyond the cultivator's ability were subjected to new and more rigorous forms of exploitation as rough pasture or sporting playground. Lime-, phosphate- and potassium-enriched fields spread into wetlands and uplands, replacing moss, carr and wet meadow with drained and enclosed arable cultivation. New woodland containing many alien species, imported from around Britain's expanding overseas empire for their symbolism of global domination as much as their exotic colour and shape, or inherent timber value, signalled another dramatic adjustment, and continued and accelerated a change that had started in the late seventeenth century. Colour and texture would also have struck observers, sometimes jarringly, as draining, liming and soil-enrichment adjusted soil chemistry and nutrient availability, leading to richer, lusher and greener grassland, more abundant crops and the capability of the soil to support new and exotic plant life. And, as that new agricultural regime spread, so the old landscape and its familiar tones and shapes diminished.

But there was no bland uniformity, for enlightenment and reason presented different solutions for different conditions. In the north-east, fattening cattle studded the park enclosures around the new farms and planned villages, while in the northern Highlands sheep displaced cattle and people, with a scattering of shepherds' houses standing in the depopulated townships of the rejected mixed-farming tradition. That might or might not have been 'Enlightenment's Frontier', for even the failure of the Improvement effort in much of the Highlands invited alternative responses that were founded on those fundamental Enlightenment principles of reason and usefulness to deliver what were believed to be optimal outcomes for all parties involved, albeit applied with inhuman rigour and inhumane cost. And already by 1850 that very recent transition was itself yielding to a newer, utterly pragmatic prescription, where great tracts of land were emptied of both the remaining people and the recently introduced sheep, to be reserved as deer forest in which a privileged few could hunt for a few weeks each year. Improvement in a Lowland sense might have reached the end of the road in the Highlands by the 1820s, but practicality, pragmatism and rationality still had decades of course to run.

Returning to the opening observations of this Conclusion, what was all the more remarkable is the extent to which this transformation had occurred against a backdrop of dynamic climate change during which the Scottish weather had produced some of the most extreme ranges of conditions, summer and winter, that challenged Improving landowners, farmers and fishermen. And this is what serves to make the degree of change all the more remarkable, for it happened against the countervailing force of unfavourable weather, where flood could still strip bare the soil from even well-drained 'improved' fields, drought parch newly planted hardwood trees, and cold suppress growth on pastures sown with nutri-

Glen Callater, Aberdeenshire: the mosaic of muirburn on the heather-clad slopes above the loch-side grassland symbolises successive transitions across the centuries, from seasonal grazing for cattle by the peasant communities of lower Glen Clunie and Braemar district to sporting estate for grouse-shooting, leased to wealthy southern tenants.

tionally richer grass and clover mixes. Adding further to the challenges were the political and economic stresses: price inflation during the era of the Corn Laws, which encouraged cultivation of even marginally productive land; price collapse affecting the cattle economy of large parts of the western Highlands; boom-to-bust episodes caused by international war and diplomatic tensions, as with the rise and fall of kelping; and the bounties and dearths of the sea-fishery. Much has been made of this peak period of 'Improvement', especially the oft-repeated mantras of the triumphs of enlightenment and reason supposedly manifest in everything from the brutal 'rationality' of the Clearances to the simple practicality and pragmatism of the changes in the exploitation regimes implemented by small, Improving tenants scattered from Berwickshire to Caithness. And, certainly, there was a strong element of conviction and determination based on new learning, on repeated and observed experiment, on exchange of ideas, and on trial and error. It was absolute conviction in the superiority of larch that drove the duke of Atholl to plant the north Perthshire uplands with (according to different estimates) between 14 and 27 million seedlings of his favoured species. No less was it absolute conviction that led the estate managers for the countess of Sutherland to advise the rationalisation of land-use and introduction of sheep that led to the mass eviction of populations from across the interior of her sprawling estate. But it was also the willingness of the experimenters to take risks and load up their debt-burden to fund the improvements. And the lenders' liquidity provided that funding, much of it derived from the tainted profits of slave-worked plantations in the West Indies, the siphoning off to Britain of the riches of an expanding global empire in India, Africa and Australia, and the profits of war-plunder. All these gave impetus to the wider reconfigurations that were occurring within Scotland. As explored in the volume 'Scotland 1850 to COP 26', the trends of this era of sustained 'Improvement' were first consolidated in the post-1850 era and, as the twentieth century dawned, re-evaluated, often regretted, and progressively reviled.

Bibliography

Unpublished primary sources

Caithness Archives Centre

CC7 Wick Parochial Board minutes
  CC7/10/1 https://www.highlifehighland.com/nucleus-nuclear-caithness-archives/wick-grain-riots-1847/

Glasgow, Mitchell Library

MS.246130 Extracts from the Burgh Council minutes, Elgin

National Library of Scotland [NLS]

Aberdeen, William, *A Chart of the Orkney Islands* (1769). https://maps.nls.uk/coasts/chart/829
Anonymous, 'Map of Blairdrummond Moss showing progress of clearing in 1813'. https://maps.nls.uk/estates/rec/7861
Bruce, Alexander and Cooper, Richard, 'A plan of Loch Sunart . . . become famous by the greatest national improvement this age has produc'd / survey'd &c. by Alexr. Bruce; R. Cooper sculp.' (Edinburgh, 1733). https://maps.nls.uk/view/00000545
Farquharson, John, 'A Map of the forrest of Mar survey'd &c. / Done by John Farquharson of Invercald, in Anno 1703'. https://maps.nls.uk/counties/rec/198
Gordon, Robert, of Straloch, 'Brae of Angus, [and] The height of Anguss, M.T.P. Height of Anguss'. https://maps.nls.uk/view/00000664
Gordon, Robert, of Straloch, 'Glen Yla, Glen Ardle, Glen Shye, out of Mr. T. Pont's papers yey ar very imperfyt'. https://maps.nls.uk/view/00000665
Gordon, Robert, of Straloch, 'Map of River Avon' https://maps.nls.uk/view/00000355
Johnston, W. & A.K. Ltd, 'Johnston's Map of the Orkney Islands' (Edinburgh: W. & A.K. Johnston, [1850?]). https://maps.nls.uk/counties/rec/7256
Langlands, George, 'This map of Argyllshire taken from actual survey is . . . dedicated to . . . John Duke of Argyll . . . / by . . . George Langlands & Son . . . (Engraved by S.J. Neele)' (Campbeltown, 1801). https://maps.nls.uk/joins/7382.html
Map of Blairdrummond Moss showing progress of clearing in 1813. https://maps.nls.uk/estates/rec/7861
Pont, Timothy, [Ben Lawers; Glen Tanar; Strath Avon] – Pont 7. https://maps.nls.uk/view/00002296
Pont, Timothy, [Elgin and North-East-Moray]. https://maps.nls.uk/rec/302
Pont, Timothy, [Forest of Atholl] – Pont 19. https://maps.nls.uk/rec/279
Pont, Timothy, [North Esk; South Esk] (front) – Pont 30. https://maps.nls.uk/view/00002327
Pont, Timothy, Pont Maps of Scotland, ca. 1583–1614 – Pont texts, 'Lochew and Letyr-ew', pp. 119v–120r. https://maps.nls.uk/pont/texts/transcripts/ponttext119v-120r.html
Pont, Timothy, [South Uist; Inverkeithing] – Pont 36 (front). https://maps.nls.uk/view/00002338
Roy, William, General Roy Military Survey of Scotland 1747–1755. https://maps.nls.uk/geo/explore/#zoom=13&lat=56.28581&lon=-3.92479&layers=3&b=1
'Scottish Woodlands': https://maps.nls.uk/pont/subjects/woodlands.html
Thomson, John, *Atlas of Scotland*, 'Orkney Islands' (Edinburgh: J. Thomson & Co., 1822). https://maps.nls.uk/atlas/thomson/476.html
Thomson, John and Johnson, William, 'Northern Part of Argyll Shire. Southern Part' (Edinburgh, 1832). https://maps.nls.uk/atlas/thomson/506.html

National Records of Scotland [NLS]

CH1 – Records of the General Assembly of the Church of Scotland
  CH1/2/62
CH7 – Papal Bulls
  CH7/39
CS226 – Court of Session: Carmichael & Elliot processes, Mackenzie office, unknown extractors
  CS226/362
  CS226/6414
E41 – Exchequer Records (Orkney Rentals and Accounts)
  E41/24
E704 – Exchequer Records: Forfeited Estates 1745: General Management: Barons of Exchequer: Treasury Warrants and Correspondence
  E704/8

E713 – Exchequer Records: Forfeited Estates 1745: General Management: Barons of Exchequer: Annexed Estate Business
   E713/3
E727 – Exchequer Records: Forfeited Estates 1745: General Management: Commissioners for the Annexed Estates: Letters
   E727/10
E730 – Exchequer Records: Forfeited Estates 1745: General Management: Commissioners for the Annexed Estates: Papers relating to Improvements
   E730/30
E732 – Exchequer Records: Forfeited Estates 1745: General Management: Commissioners for the Annexed Estates: Accounts
   E732/24
E746 – Exchequer Records: Forfeited Estates 1745: Particular Management: George Earl of Cromarty: Cromarty Estate, Ross and Cromarty Counties
   E746/98
   E746/126
   E746/127
   E746/176
E777 – Exchequer Records: Forfeited Estates 1745: Particular Management: Perth Estate, Perth County
   E777/139
   E777/281
   E777/313
E783 – Exchequer Records: Forfeited Estates 1745: Particular Management: Alexander Robertson of Struan, Struan Estate, Perth County
   E783/93
   E783/98
GD1/58/2 – Miscellaneous Small Collections
   GD1/58/2/23
GD1/168 – Miscellaneous Small Collections: Invergarry Iron Works
   GD1/168/11
GD1/369 – Documents Relating to the Family of Menzies of Weem
   GD1/369/137
GD1/582 – Miscellaneous Small Collections: Edinburgh and Linlithgow Turnpike Roads
   GD1/582/15
GD1/221 – Miscellaneous Small Collections: Jacktoun Writs
   GD1/221/52
GD1/938 – Miscellaneous Small Collections: Smith Family (Roxburghshire) Papers
   GD1/938/4
GD1/1003 – Miscellaneous Small Collections: MacLaine of Lochbuie Papers
   GD1/1003/5
GD3 – Papers of the Montgomerie Family, Earls of Eglinton.
   GD3/2/115
   GD3/10/4/1
GD5 – Papers of the Bertram family of Nisbet, Lanarkshire
   GD5/267
   GD5/510
   GD5/511
   GD5/513
GD6 – Papers of the Brooke family of Biel, East Lothian
   GD6/826

GD9 – Records of the British Fisheries Society
   GD9/166
   GD9/196
   GD9/270
   GD9/275
GD12 – Title deeds of the Swinton family of Swinton, Berwickshire.
   GD12/38
GD14 – Papers of the Campbell family of Stonefield
   GD14/17
GD16 – Airlie Muniments
   GD16/26/1/206
   GD16/36/3
   GD16/41/9
   GD16/41/681
GD21 – Papers of the Cunninghame Family of Thorntoun
   GD21/452
GD22 – Papers of the Cunninghame Graham family of Ardoch, Dunbartonshire 1400–1892
   GD22/1/455
GD23 – Warrand of Bught
   GD23/4/270
   GD23/6/438
   GD23/6/564
GD24 – Papers of the family of Stirling Home Drummond Moray of Abercairny, Perthshire
   GD24/1/553
   GD24/1/782 (21)
   GD24/1/802
   GD24/3/237
GD25 – Papers of the Kennedy Family, Earls of Cassillis (Ailsa Muniments)
   GD25/1/26
   GD25/8/215
   GD25/8/862
   GD25/9/18
GD26 – Papers of the Leslie Family, Earls of Leven and Melville
   GD26/7/439
GD30 – Papers of the Shairp family, of Houston, West Lothian
   GD30/1879
GD40 – Papers of the Kerr Family, Marquises of Lothian
   GD40/2/14/10
   GD40/9/302
GD44 – Papers of the Gordon family, Duke of Gordon (Gordon Castle Muniments)
   GD44/39/30
   GD44/41/65
   GD44/42/4/37/4 (12–17, 19)
   GD44/43/16 (59)
   GD44/43/50 (15–16)
   GD44/43/101 (20)
   GD44/43/232 (10 and 68)
   GD44/43/241 (34–36)
   GD44/43/270
   GD44/43/334 (12)
   GD44/43/420
   GD44/44/23
   GD44/51/169/17
   GD44/51/358/5
   GD44/51/361/3

GD44/51/465/2 (5)
GD44/51/498/16/12
GD44/51/509/2 (14–16)
GD44/51/568/1
GD44/51/739 (50–52)
GD45 – Papers of the Maule family, Earls of Dalhousie
GD45/16/2194
GD46 – Papers of the Mackenzie family, Earls of Seaforth
GD46/1/194 (3)
GD46/1/307
GD46/1/366
GD46/1/391
GD46/1/393
GD46/1/394
GD46/1/418
GD46/1/476
GD46/1/539
GD46/13/129–135
GD46/15/41
GD46/15/90 (4)
GD46/15/237 (10, 11)
GD46/17/50
GD46/17/76
GD51 – Papers of the Dundas Family of Melville, Viscounts Melville (Melville Castle Papers)
GD51/5/198/1–3
GD51/5/210
GD68 – Murray of Lintrose
GD68/1/8
GD75 – Papers of the Dundas family of Dundas, West Lothian
GD75/326
GD77 – Papers of the Fergusson family of Craigdarroch, Ayrshire 1473 1924
GD77/221
GD79 – Records of the King James Hospital, Perth (Blackfriars)
GD79/1/15
GD79/1/24
GD87 – Papers of the Mackay Family of Bighouse
GD87/2/22
GD110 – Papers of the Hamilton-Dalrymple family of North Berwick
GD110/892
GD110/968
GD112 – Breadalbane Muniments
GD112/1/786
GD112/1/787
GD112/1/788
GD112/1/799
GD112/7/16 (5)
GD112/10/1/3 (56)
GD112/10/1/4 (2–3)
GD112/11/1/4/96
GD112/11/2/5/38
GD112/11/2/5/59
GD112/11/2/5/84
GD112/11/3/2/61
GD112/11/3/2/64
GD112/11/4/2/47
GD112/11/5/2/117
GD112/11/6/3/20
GD112/11/6/4/95
GD112/11/7/5/53
GD112/11/7/7/10
GD112/11/8/3/21
GD112/11/9/2/43
GD112/11/10/14/33
GD112/14/6/7 (25)
GD112/14/13/7 (4)
GD112/15/22 (56)
GD112/15/161 (12)
GD112/15/230 (36 and 38)
GD112/15/242 (14 and 21)
GD112/15/255 (6 and 56)
GD112/15/265
GD112/15/266 (41)
GD112/15/275/22
GD112/15/284/16
GD112/15/301/27
GD112/15/353 (72)
GD112/16/2/1 (3)
GD112/16/4/4 (5)
GD112/16/4/6 (49, 50, 53)
GD112/16/10/6
GD112/16/11/1
GD112/16/11/2 (12, 14–16, 32–33, 35–40, 43)
GD112/16/11/3 (18–33)
GD112/16/11/5 (51)
GD112/16/13/7
GD112/27/46 (20)
GD112/29/66
GD112/39/19/9
GD112/39/32/9
GD112/39/32/14
GD112/39/33/5
GD112/39/33/11
GD112/39/33/27
GD112/39/182/22
GD112/39/192/4
GD112/39/200/15
GD112/39/222/25
GD112/39/231/15
GD112/39/234/13
GD112/39/250/2
GD112/39/273/32
GD112/39/275/4
GD112/39/288/25
GD112/39/289/22
GD112/52/43 (21)
GD112/53/1 (20)
GD112/59/20
GD112/74/19 (4)
GD112/74/70 (6)
GD112/74/72 (15)
GD112/74/145 (4, 37)
GD112/74/232 (7)
GD112/74/370
GD112/74/401
GD112/74/418 (21)
GD112/74/423 (16–23)
GD112/74/486 (9)
GD112/75/163

GD113 – Papers of the Innes family of Stow, Peeblesshire
  GD113/3/265 (8)
  GD113/3/269 (10)
  GD113/5/161e (20 and 21)
  GD113/4/167 (191, 213, 216, 263
    and 337)
  GD113/5/448 (129)
  GD113/5/457 (126)
  GD113/5/483 (3)
GD122 – Papers of the Gilmour family of Craigmillar and Liberton, Midlothian
  GD122/3/13
GD124 – Papers of the Erskine Family, Earls of Mar and Kellie
  GD124/1/527
  GD124/10/243
  GD124/10/250
  GD124/10/255
  GD124/10/260
  GD124/10/292
  GD124/10/301
  GD124/10/321
  GD124/10/324
  GD124/15/942
  GD124/15/1735 (50)
  GD124/17/203
GD128 – Fraser-Mackintosh Collection
  GD128/47/5
GD132 – Paper of the Robertson family of Lude, Perthshire
  GD132/674 (3)
  GD132/871
GD136 – Papers of the Sinclair family of Freswick, Caithness
  GD136/855
GD138 – Papers of the Stewart family, Earls of Galloway (Galloway Charters)
  GD138/3/21
GD150 – Papers of the Earls of Morton
  GD150/3381
  GD150/3482/58
GD160 – Papers of the Drummond family, earls of Perth (Drummond Castle papers)
  GD160/186 (28)
  GD160/229
GD164 – Papers of the Sinclair Family, Earls of Rosslyn
  GD164/621
  GD164/628
GD170 – Papers of the Campbell family of Barcaldine
  GD170/431
  GD170/438
  GD170/629
  GD170/772
GD174 – Maclaine of Lochbuie Papers
  GD174/1411
GD176 – Papers of the Mackintosh of Mackintosh family
  GD176/1288
GD178 – Papers of the family of Horsburgh of Horsburgh, Peeblesshire
  GD178/9/1 (15)
GD190 – Papers of the Smythe family of Methven, Perthshire
  GD190/3/52
  GD190/3/285

GD199 – Papers of the Ross family of Pitcalnie
  GD199/88
GD201 – Papers of the MacDonald family of Clanranald (Clanranald Papers)
  GD201/1/54
  GD201/1/281
  GD201/2/15
  GD201/2/16
  GD201/2/25
  GD201/2/36
  GD201/2/38
  GD201/2/41
  GD201/2/42
  GD201/2/45
  GD201/2/49
  GD201/2/50
  GD201/2/51
  GD201/2/55
  GD201/2/56
  GD201/2/60
  GD201/5/1232
GD220 – Papers of the Graham Family, Dukes of Montrose (Graham Muniments)
  GD220/1/H/2/1/2
  GD220/1/H/2/1/3
  GD220/1/H/3/1/1
  GD220/1/H/5/1/1
  GD220/1/H/6/3/1
  GD220/1/H/6/3/5
  GD220/5/806 (4)
  GD220/5/882 (14, 21–22)
  GD220/5/884 (20)
  GD220/5/1018 (4)
  GD220/5/1051 (3)
  GD220/5/1221 (2)
  GD220/5/1342
  GD220/5/1584 (20)
  GD220/6/579
  GD220/6/585 (1 and 17)
  GD220/6/837
  GD220/6/1015 (27)
  GD220/6/1025 (5)
  GD220/6/1624 (12)
GD224 – Papers of the Montague-Douglas-Scott Family, Dukes of Buccleuch
  GD224/388/5
  GD224/393/6
  GD224/394/6
  GD224/395/6
  GD224/522/3 (66–74, 85, 86)
  GD224/588/7 (17)
  GD224/628/5 (53)
GD237 – Records of Messrs Tods Murray and Jamieson WS, lawyers, Edinburgh
  GD237/20/11 (4)
GD248 – Papers of the Ogilvy family, earls of Seafield
  GD248/47/2 (65)
  GD248/47/4 (Grant of Grant Correspondence)
  GD248/61/1 (51)
  GD248/81/1 (36–37)
  GD248/97/4 (42)

BIBLIOGRAPHY

GD248/104/2 (4)
GD248/104/8 (7)
GD248/105/3 (1 and 11)
GD248/160/12 (14)
GD248/168/3 (7 and 25)
GD248/179/1 (77)
GD248/238/5 (14)
GD248/456/5
GD248/530/1 (8–10)
GD248/563/68 (41)
GD248/563/69 (18)
GD248/564/72 (5)
GD248/672/6 (16)
GD248/680/2
GD248/707/6/1

GD254 – Papers of the Lindsay Family of Dowhill
GD254/31
GD347 – Papers of the Sutherland family of Rearquhar
GD347/91
GD406 – Papers of the Douglas Hamilton family, Dukes of Hamilton and Brandon
GD406/1/4128
GD406/1/4279
GD406/1/9080
GD406/1/10855
GD427 – Papers of the Gillanders family of Highfield
GD427/66
GD427/87
GD427/105–132
GD427/136–138
GD427/183
GD427/197
GD427/233
GD427/252
GD430 – Papers of the Napier Family
GD430/43
GD430/185
GD455 – Hathorn Family of Castle Wigg, Wigtownshire
GD455/27 (5 and 10)
GD504 – Papers of the Sprot family of Haystoun, Peeblesshire
GD504/3/78
GD504/3/150
RH4 – Microfilms
RH4/189 – Papers of James, Andrew and George Meikle, millwrights
RH9/14 – Edinburgh and Leith Papers
RH9/14/83
RH15/27 – Business Correspondence of James Elder, Merchant in Perth and Aberdeen
RH15/27/87
RH15/31 – Miscellaneous Paper, Smollett of Bonhill
RH15/31/88
RH15/31/91
RH15/32 – Arthur Clephane, seed merchant in Edinburgh
RH15/32/77
RH15/32/96
RH15/32/127
RH15/32/131
RH15/32/136A

RH15/44 – Papers of David Ross, Accountant to the GPO, Edinburgh
RH15/44/209
RH15/54 – Miscellaneous Papers, Edward Burd, merchant in Leith
RH15/54/21
RHP427 – Plan of the loch of Spynie and adjacent grounds, Moray
RHP1665/10 – Plan of Dalbog, the property of the Earl of Panmure
RHP1666/4 – Plan of Clochie, the property of the Earl of Panmure
RHP1666/6 – Plan of Tillydovie, the property of the Earl of Panmure
RHP1666/7 – Plan of the separate farms of Drumfurris, Newbigging, Townhead and Mill of Lethnot, the property of the Earl of Panmure
RHP1743 – Plan of the farm of Kirkland, Dalton, Dumfriesshire
RHP2191 – Plan and section of proposed road from Allt Calder to Dalveg
RHP2379 – Volume containing architectural plans and elevations of Gordon Castle, Morayshire
RHP2379/6
RHP3371 – Photostat copy of plan of the Park and Forest of Darnaway and Woods Lands &c. of Dounduff
RHP8982 – Plan of the farms of Auchnahannet (Achnahannet), Auchnagallin (Achnagaul), Ballinlagg (Bellenluig) and Knockannakeist (1802)
RHP31465 – Photocopy of volume of plans (6) as undernoted of lands of Kintrae, Leggat, Ardgye (Ardgay), etc.
RHP31465/4
RHP94440 – Plan of march between lands of Ardgye (Ardquoy), Kintrae and Leggat and lands of Mosstowie, Newton and Quarry Wood, Moray
RHP98239 – Sketch plans of Easter and Wester Delliefure (Dellyfures), improvements of Mein and Sheinmore [not located] and farms of Knockanniekiest, Knockannacardich (Knockanakardich) and Ballinlagg (Bellenluig) (July 1767)

Orkney Library and Archive, Kirkwall

GD150 – Papers of the Earls of Morton
GD150/2520
SC11/5 – Kirkwall Sheriff Court, Processes
SC11/5/1800/4
SC11/5/1802/93
SC11/5/1804/115
SC11/5/1805/5
SC11/5/1805/55
SC11/5/1806/119
SC11/5/1806/124
SC11/5/1806/130
SC11/5/1806/163

Perth and Kinross Archives, A.K. Bell Library

B59 – Records of Perth Burgh
B59/24/15/13
B59/25/3/39

St Andrews University Library Department of Special Collections

B13 – Records of Cupar Burgh
B13/22/57

Printed and digital primary sources and editions

Adam, W., *Remarks on the Blair-Adam Estate* (private circulation, 1834).

Annals of Connacht. http://www.ucc.ie/celt/published/T100011/index.html

Annals of Loch Cé. http://www.ucc.ie/celt/published/T100010A/index.html

Annals of Ulster. https://celt.ucc.ie/published/T100001A/

Anderson, James [Agricola], *Miscellaneous Observations on Planting and Training Timber-Trees; Particularly Calculated for the Climate of Scotland. In a Series of Letters*, by Agricola (Edinburgh: Charles Elliot, 1777).

Anonymous (ed.), *Reports on the State of Certain Parishes in Scotland, 1627* (Edinburgh: Maitland Club, 1835).

Anonymous (ed.), *The Black Book of Taymouth* (Edinburgh: T. Constable, 1855).

Anonymous (ed.), *Correspondence of Sir Robert Kerr, First Earl of Ancram, and his son William, Third Earl of Lothian*, 2 volumes (Edinburgh: Bannatyne Club, 1875).

The Auchinleck Chronicle, printed in C. McGladdery, *James II*, 2nd edition (Edinburgh: John Donald, 2015).

Bain, J. and Rogers, C. (eds), *Liber Protocollorum M. Cuthberti Simonis AD 1499–1513 and Rental Book of the Diocese of Glasgow*, 2 volumes (London: Grampian Club, 1875).

Bald, Adam, 'Journal of Travels and Commonplace Book, 1790–99', in A.J. Durie (ed.), *Travels in Scotland 1788–1881. A Selection from Contemporary Tourist Journals*, Scottish History Society (Woodbridge: The Boydell Press, 2012), pp. 41–99.

Bald, R., *A General View of the Coal Trade of Scotland, Chiefly that of the River Forth and Mid Lothian* (Edinburgh, 1808).

Barrow, G.W.S. (ed.), *The Charters of David I: The Written Acts of David I King of Scots, 1124–53, and of His Son Henry, Earl of Northumberland, 1139–52* (Woodbridge: The Boydell Press, 1999).

Boutcher, W., *A Treatise on Forest-Trees: Containing Not Only the Best Methods of Their Culture Hitherto Practised, But a Variety of New and Useful Discoveries, the Result of Many Repeated Experiments: as Also, Plain Directions for Removing Most of the Valuable Kinds of Forest-trees etc.* (Edinburgh: R. Fleming, 1775).

Bower, Walter, *Scotichronicon*, ed. D.E.R. Watt and others, 9 volumes (Aberdeen: Aberdeen University Press, 1987–8).

Brodie, Alexander and Brodie, James, *The Diary of Alexander Brodie of Brodie, MDCLII–MDCLXXX., and of His Son, James Brodie of Brodie, MDCLXXX–MDCLXXXV: Consisting of Extracts from the Existing Manuscripts, and a Republication of the Volume Printed at Edinburgh in the Year 1740* (Aberdeen: Spalding Club, 1873).

Burt, Edmund, *Burt's Letters from the North of Scotland*, ed. A. Simmons, with 'Introduction' by C.W.J. Withers (Edinburgh: Birlinn, 2012).

Calderwood, A.B. (ed.), *Acts of the Lords of Council*, volume 3 (Edinburgh: HMSO, 1993).

Calendar of Entries in the Papal Registers Relating to Great Britain and Ireland: Papal Letters, xv, 1484–1492, ed. M.J. Haren (Dublin, 1978).

Chambers, W. (ed.), *Charters and Documents Relating to the Burgh of Peebles, with Extracts from the Records of the Burgh. AD 1165–1710* (Edinburgh: Scottish Burgh Records Society, 1872).

Charlton, E., *Travels in Shetland 1832–52*, ed. W. Charlton (Lerwick: The Shetland Times, 2007).

Chronicles of the Frasers: The Wardlaw MS., ed. W. Mackay (Edinburgh: Scottish History Society, 1905).

Colville, J. (ed.), *Letters of John Cockburn of Ormistoun to his Gardener, 1727–1744* (Edinburgh: Scottish History Society, 1904).

Cordiner, C., *The Antiquities and Scenery of the North of Scotland in a Series of Letters to Thomas Pennant* (London, 1780).

Cramond, W. (ed.), *The Records of Elgin*, 2 volumes (Aberdeen: New Spalding Club, 1903–8).

Cregeen, E.R. (ed.), *Argyll Estate Instructions – Mull, Morvern, Tiree – 1771–1805* (Edinburgh: Scottish History Society, 1964).

Defoe, Daniel, *A Tour thro' the whole island of Great Britain*, 3 volumes (London, 1724–7).

Dekker, T., *The Great Frost. Cold Doings in London, Except it be at the Lotterie. With Newes Out of the Country. A Familiar Talke betwene a Country-man and a Citizen Touching this Terrible Frost and the Great Lotterie, and the Effects of Them. The Description of the Thames Frozen Over* (London: Henry Gosson, 1608).

Dekker, T., *The Cold Yeare 1614: A Deepe Snow: In Which Men and Cattell have Perished . . . Or of Strange Accidents in this Great Snow* (London: W.W. for Thomas Langley, 1615).

Dickinson, W.C. (ed.), *The Sheriff Court Book of Fife 1515–1522* (Edinburgh: Scottish History Society, 1928).

Dickinson, W.C. (ed.), *The Court Book of the Barony of Carnwath, 1523–1542* (Edinburgh: Scottish History Society, 1937).

Donaldson, G. (ed.), *Protocol Book of James Young, 1485–1550* (Edinburgh: Scottish Record Society, 1952).

Donaldson, J., *Husbandry Anatomized* (Edinburgh: John Reid, 1697).

Dunbar, J.G. (ed.), *Sir William Burrell's Northern Tour, 1758* (Edinburgh: European Ethnological Research Centre, 1997).

Dundas, J. and Orr, D.G. (eds), *'Quite Happy'. The Diary of James Fyffe, Cattle Dealer, 1836–1840* (Dundee: Abertay Historical Society, 2016).

Dunlop, E. and Kamm, A. (eds), *The Scottish Collection of Verse to 1800* (Glasgow: Richard Drew Publishing, 1985).

Durie, A.J. (ed.), *Travels in Scotland 1788–1881. A Selection from Contemporary Tourist Journals*, Scottish History Society (Woodbridge: The Boydell Press, 2012).

Eagles, J. (ed.), *The Chronicle of Perth* (Llanerch: Llanerch Publishers, 1996).

Evelyn, John, *Sylva; or, A Discourse of Forest Trees and the Propagation of Timber in His Majesty's Dominions* (London: John Martyn, 1664).

Fea, J., *The Present State of the Orkney Islands considered, with An Account of their advantageous Situation, and Conveniences for TRADE; the Improvements they are capable of &c* (1st edition Edinburgh, 1775; reprinted as *The Present State of the Orkney Islands Considered and An Account of the New Method of Fishing on the Coast of Shetland* (Edinburgh: William Brown, 1884)).

Ferrerius, *Ferrerii Historia Abbatum de Kynlos*, ed. W.D. Wilson (Edinburgh: published for the Bannatyne Club, 1839).

Fletcher, Andrew, *Two Discourses Concerning the Affairs of Scotland; written in 1698* (Edinburgh, 1698).

Fraser, W. (ed.), *The Red Book of Menteith*, 2 volumes (Edinburgh: David Douglas, 1880).

Gairdner, J. (ed.), *Historical Collections of a Citizen of London in the Fifteenth Century (William Gregory's Chronicle)* (London: Camden Society, 1876).

Georg, V., 'On smearing sheep', *The Farmer's Magazine*, 7:26 (Edinburgh, May 1806), pp. 172–175.

Gilpin, W., *Observations, relative chiefly to picturesque beauty, made in the year 1776, on several parts of Great Britain; particularly the High-lands of Scotland* (London: R. Blamire, 1789).

Gilpin, W., *Remarks on forest scenery, and other woodland views (relative chiefly to picturesque beauty), illustrated by the scenes of New Forest in Hampshire*, 3 volumes (London: R. Blamire, 1791).

Gilpin, W., *Three essays: on picturesque beauty; on picturesque travel; and on sketching landscape: to which is added a poem, On landscape painting* (London: R. Blamire, 1792).

Gough Map of Great Britain. http://www.goughmap.org/map/

Gray, J.M. (ed.), *Memoirs of the Life of Sir John Clerk of Penicuik, 1676–1755* (Edinburgh: Scottish History Society, 1892).

Guttridge, G.H. (ed.), *The Correspondence of Edmund Burke*, volume 3 (Cambridge: Cambridge University Press, 1961).
Hamilton, H. (ed.), *Selections from the Monymusk Papers (1713–1755)* (Edinburgh: Scottish History Society, 1945).
Hannay, R.K. (ed.), *Rentale Dunkeldense. Being Accounts of the Bishopric (A.D. 1505–1517) with Mylns 'Lives of the Bishops' (A.D. 1483–1517)* (Edinburgh: Scottish History Society, 1915).
Hett, F.P. (ed.), *The Memoirs of Sir Robert Sibbald (1641–1722)* (London: Oxford University Press, 1932).
Historic Manuscripts Commission, *14th Report*, Appendix, volume 3, *The Manuscripts of the Duke of Roxburghe* (London: HMSO, 1894).
Home, Henry, Lord Kames, *Historical Law Tracts*, 1st edition (Edinburgh: printed for A. Millar, London; and A. Kincaid and J. Bell, Edinburgh, 1758).
Hume Brown, P. (ed.), *Early Travellers in Scotland* (Edinburgh: David Douglas, 1891).
Imrie, J., Rae, T.I. and Ritchie, W.D. (eds), *The Burgh Court Book of Selkirk 1503–45*, 2 volumes (Edinburgh: Scottish Record Society, 1960–9).
Johnson, Samuel, 'A Journey to the Western Islands of Scotland', in Samuel Johnson and James Boswell, *A Journey to the Western Islands of Scotland* and *The Journal of a Tour to the Hebrides*, ed. P. Levi (London: Penguin Books, 1984).
Lamont, L. (ed.), *The Diary of Mr John Lamont of Newton 1649–1671* (Edinburgh: Maitland Club, 1830).
Lindsay, Robert, of Pitscottie, *The Historie and Cronicles of Scotland from the Slaughter of King James the First to the Ane thousand five hundreith thrie scoir fyftein zeir*, ed. Æ.J.G. Mackay, 3 volumes (Edinburgh: Scottish Text Society, William Blackwood and Sons, 1899–1911).
Loch, D., *Essays on the Trade, Commerce, Manufacture and Fisheries of Scotland* (Edinburgh: Walter and Thomas Ruddiman, 1778).
Loudon, J.C., *Arboretum et fruticetum Britannicum; or, the trees and shrubs of Britain, native and foreign, hardy and half-hardy, pictorially and botanically delineated, and scientifically and popularly described; with their propagation, culture, management, and uses in the arts, in useful and ornamental plantations, and in landscape gardening; preceded by a historical and geographical outline of the trees and shrubs of temperate climates*, 8 volumes (London: printed privately, 1838).
Mac Carthaigh's Book. http://www.ucc.ie/celt/published/T100015/index.html
Macdonald, Sir James, *General View of the Agriculture of the Hebrides or Western Isles of Scotland* (Edinburgh: Silvester Doig and Andrew Stirling, 1811).
MacKay, W. (ed.), *The Letter-Book of Bailie John Steuart of Inverness 1715–1752*, Scottish History Society 2nd series (Edinburgh: Scottish History Society, 1915).
Mackay, W. and Boyd, H.C. (eds), *Records of Inverness*, 2 volumes (Aberdeen: New Spalding Club, 1911–14).
Macpherson, J., *Fingal, an ancient epic poem, in six books: together with several other poems, composed by Ossian the son of Fingal. Tr. from the Galic language by James Macpherson* (London: T. Becket and P.A. De Hondt, 1762).
Macpherson, J. (ed.), *The Works of Ossian the Son of Fingal*, 2 volumes, 3rd edition (London: T. Becket and P.A. De Hondt, 1765).
Martin Martin, *A Description of the Western Islands of Scotland Circa 1695, by Martin Martin, Gent. Including a Voyage to St Kilda by the same author and A Description of the Western Isles of Scotland by Sir Donald Monro*, ed. D.J. Macleod (Stirling: Eneas Mackay, 1934).
Marwick, J.D. (ed.), *Extracts from the Records of the Burgh of Edinburgh*, volume 1, *AD 1403–1528* (Edinburgh: Scottish Burgh Record Society, 1869).

Masterton, Francis, 'Masterton Papers, 1660–1719', ed. V.A. Noël Paton, *Miscellany of the Scottish History Society*, volume 1 (Edinburgh: Scottish History Society, 1893), pp. 449–493.
Maxwell, Robert, of Arkland, *Select Transactions of the Honourable the Society of Improvers in the Knowledge of Agriculture in Scotland, 1743*, reprint (Edinburgh: The Grimsay Press, 2003).
Menzies, J., *Answers for James Menzies of Culdares, and Angus Macdonald of Kenknock, to the petition of John Earl of Breadalbane* (Edinburgh: private circulation, 1738).
Michie, J.G. (ed.), *Records of Invercauld* (Aberdeen: New Spalding Club, 1901).
Millar, A.H. (ed.), *A Selection of Scottish Forfeited Estate Papers 1715; 1745* (Edinburgh: Scottish History Society, 1909).
Mitchell, A. (ed.), *Geographical Collections Relating to Scotland Made by Walter Macfarlane*, 3 volumes (Edinburgh: Scottish History Society, 1906–7).
Monteath, R., *Miscellaneous Reports of Woods and Plantations* (Dundee: J. Chalmers, 1827).
Mountgomery, Alexander, *The Cherry and the Slae, with other Poems* (Glasgow: Robert and Andrew Foulis, 1751).
Myln, A., *Vitae Dunkeldensis Ecclesiae Episcoporum* (Edinburgh: Bannatyne Club, 1831).
The New Statistical Account of Scotland (Edinburgh: Church of Scotland/Committee for the Society for the Sons and Daughters of the Clergy, 1834–45).
Nicoll, John, *A diary of Public Transactions and other Occurrences, Chiefly in Scotland from January 1650 to June 1667*, ed. D. Laing (Edinburgh: Bannatyne Club, 1836).
Ochterlony, John, of Guynd, 'Account of the Shire of Forfar, c.1682', *The Spottiswoode Miscellany: A collection of Papers and Tracts Chiefly Illustrative of the Civil, and Ecclesiastical History of Scotland*, volume 1 (Edinburgh: The Spottiswoode Society, 1844), pp. 311–355.
Pearson, M. (ed.), *More Frost and Snow: The Diary of Janet Burnet 1758–1795*. Sources in Local History 2 (Edinburgh: Canongate Academic, 1994).
Peterkin, A. (ed.), *Rentals of the Ancient Earldom and Bishoprick of Orkney* (Edinburgh: John Moir, 1820).
Preest, D. (ed.), *The Chronica Maiora of Thomas Walsingham (1376–1422)* (Woodbridge: The Boydell Press, 2005).
Prevost, W.A.J. (ed.), 'A Journie to Galloway in 1721 by Sir John Clerk of Penicuik', *Transactions of the Dumfriesshire and Galloway Natural History and Antiquarian Society*, 3rd series, 41 (1962–3), pp. 186–200.
Regesta Regum Scottorum, volume 5, *The Acts of Robert I 1306–1329*, ed. A.A.M. Duncan (Edinburgh: Edinburgh University Press, 1988).
Register of the Privy Council of Scotland, volume 5, *1592–1599*, ed. D. Masson (Edinburgh: H.M. General Register House, 1882).
Register of the Privy Council of Scotland, volume 8, *1607–1610*, ed. D. Masson (Edinburgh: H.M. General Register House, 1887).
Register of the Privy Council of Scotland, volume 23, *1622–1625*, ed. D. Masson (Edinburgh: H.M. General Register House, 1896).
Register of the Privy Council of Scotland, 2nd series, volume 1, *1625–1627*, ed. D. Masson (Edinburgh, 1899).
Register of the Privy Council of Scotland, 2nd series, volume 2, *1627–1628*, ed. P. Hume Brown (Edinburgh: H.M. General Register House, 1909).
Reid, A.G. (ed.), *The Diary of Alexander Hay of Craignethan 1659–1660* (Edinburgh: Scottish History Society, 1901).
Reid, John, *The Scots gard'ner in two parts, the first of contriving and planting gardens, orchards, avenues, groves, with new and profitable wayes of levelling, and how to measure and divide land: the second of the propagation & improvement of forrest, and fruit-trees, kitchen*

hearbes, roots and fruits, with some physick hearbs, shrubs and flowers: appendix shewing how to use the fruits of the garden: whereunto is annexed The gard'ners kalendar / published for the climate of Scotland by John Reid (Edinburgh: David Lindsay and his partners, 1683).

Rogers, C. (ed.), *Rental Book of the Cistercian Abbey of Cupar-Angus*, 2 volumes (London: Grampian Club, 1879).

Romanes, C.S. (ed.), *Selections from the Records of the Regality of Melrose*, 3 volumes (Edinburgh: Scottish History Society, 1914–17).

Ross, A., 'Two 1585 × 1612 surveys of vernacular buildings and tree usage in the Lordship of Strathavon, Banffshire', *Miscellany of the Scottish History Society*, 14 (2012), pp. 1–52.

RPS: *The Records of the Parliaments of Scotland to 1707*, ed. K.M. Brown et al. (St Andrews, 2007–23).

Scott, W., 'On Planting Waste Lands', *Quarterly Review* (October 1827).

Sibbald, Sir R., *Provision for the Poor in Time of Dearth*, 1st edition (Edinburgh, 1699).

Sinclair, J. (ed.), *The Statistical Account of Scotland*, 20 volumes (Edinburgh: William Creech, 1791–9).

Smith, G.G. (ed.), *The Poems of Robert Henryson*, volume 2 (Edinburgh: Scottish Text Society, 1906).

Stevenson, D. (ed.), *The Diary of a Canny Man 1818–1828. Adam Mackie, Farmer, Merchant and Innkeeper in Fyvie*, compiled by William Mackie (Aberdeen: Aberdeen University Press, 1991).

Stuart, J. (ed.), *Extracts from the Council Register of the Burgh of Aberdeen, 1398–1570* (Aberdeen: Spalding Club, 1844).

Symson, A., 'A large description of Galloway by the parishes in it', in A. Mitchell (ed.), *Geographical Collections Relating to Scotland Made by Walter Macfarlane*, volume 2 (Edinburgh: Scottish History Society, 1907), pp. 51–128.

The Aberdeen Journal, online at British Library Newspapers, 1800–1950. https://www.newspapers.com/paper/aberdeen-journal-and-general-advertiser-for/7832/

The Scots Magazine and Edinburgh Literary Miscellany (February 1814).

Thomson, T. (ed.), *The Acts of the Parliaments of Scotland*, volume 1, A.D. 1124–1424 (Edinburgh: H.M. General Register House, 1844).

Turnbull, George, 'The diary of the Rev. George Turnbull, Minister of Alloa and Tyninghame, 1657–1704', ed. R. Paul, in *Miscellany of the Scottish History Society* (Edinburgh: Scottish History Society, 1893), pp. 295–445.

Walker, C.K. (ed.), *Breaking Away: Coleridge in Scotland* (Newhaven and London: Yale University Press, 2002).

Whyte, A. and Macfarlan, D., *General View of the agriculture of the County of Dumbarton* (Glasgow: James Hedderwick & Co., 1811).

Wordsworth, D., *Recollections of a Tour Made in Scotland*, ed. C.K. Walker (London and Newhaven: Yale University Press, 1997).

Secondary sources

Adams, I.H. (ed.), *Directory of Former Scottish Commonties* (Edinburgh: Scottish Record Society, 1971).

Aitchison, P. and Cassell, A., *The Lowland Clearances: Scotland's Silent Revolution 1760–1830* (East Linton: Tuckwell Press, 2003).

Albritton Jonsson, F., *Enlightenment's Frontier: The Scottish Highlands and the Origins of Environmentalism* (Newhaven: Yale University Press, 2013).

Anonymous, 'A chemical legacy: a banging beautiful trip round the west coast of Scotland'. https://chemicallegacy.wordpress.com/2019/11/01/a-banging-beautiful-trip-round-the-west-coast-of-scotland/

Atkinson, J., 'Ben Lawers: an archaeological landscape in time. Results from the Ben Lawers Historic Landscape Project, 1996–2005', *Scottish Archaeological Internet Reports*, 62 (2016). https://archaeologydataservice.ac.uk/library/browse/issue.xhtml?recordId=1137495&recordType=Monograph Series [Accessed 16 January 2020].

Baker, A., Hellstrom, J.C., Kelly, B.F.G., Mariethoz, G. and Trouet, V., 'A composite annual-resolution stalagmite record of North Atlantic climate over the last three millennia', *Scientific Reports*, 5 (2015). DOI: 10.1038/srep10307.

Bangor-Jones, M., 'From clanship to crofting: landownership, economy and the Church in the province of Strathnaver', in J.R. Baldwin (ed.), *The Province of Strathnaver* (Edinburgh, 2000), pp. 35–99.

Bangor-Jones, M., 'Plantations on the Sutherland estates', *Native Woodland Discussion Group, Scottish Woodland History Conference Notes XVIII, Plantations in Scotland* (2016), pp. 34–39.

Barrett, P. and O'Regan, H., 'Environmental archaeology in the wild: making space for the past in the new conservation movement', *The Archaeologist*, 119 (2023), pp. 3–5.

Bennett, K.D., Fossit, J.A., Sharp, M.J. and Switsur, V.R., 'Holocene vegetational and environmental history at Loch Lang, South Uist, Western Isles', *New Phytologist*, 114 (1990), pp. 281–298.

Beveridge, E., *North Uist: Its Archaeology and Topography*, facsimile edition (Edinburgh: Birlinn Origin, 2018).

Boardman, S.I., '"Pillars of the community": Clan Campbell and architectural patronage in the fifteenth century', in R.D. Oram and G.P. Stell (eds), *Lordship and Architecture in Medieval and Renaissance Scotland*. (Edinburgh: Birlinn, 2005), pp. 123–60.

Boardman, S.I., *The Campbells 1250–1513* (Edinburgh: John Donald, 2006).

Bowler, D., Coleman, R., Perry, D. and Robertson, N., *Perth: The Archaeology and Development of a Scottish Burgh*. Tayside and Fife Archaeological Committee Monograph 3 (Perth: Tayside and Fife Archaeological Committee, 2004).

Briffa, K.R., Jones, P.D., Schweingruber, F.H. and Osborn, T.J., 'Influences of volcanic eruptions on northern hemisphere summer temperatures over the past 600 years', *Nature*, 393 (1998), pp. 450–455.

Briffa, K.R., Jones, P.D., Vogel, R.B., Schweingruber, F.H., Baillie, M.G.L., Shiyatov, S.G. and Voganov, E.A., 'European tree rings and climate change in the sixteenth century', *Climate Change*, 43 (1999), pp. 151–68.

Briggs, A., *The Age of Improvement 1783–1867* (London: Longmans, Green & Co., 1959).

Brochard, T., 'Plantation: its process in relation to Scotland's Atlantic communities, 1590s–1630s', *Journal of the North Atlantic*, Special Volume no. 12 (2019), pp. 73–94.

Brogden, W.A., 'Switsur, Stephen', *Oxford Dictionary of National Biography*. https://doi.org/10.1093/ref:odnb/26855 [accessed 3 April 2022].

Cage, A.G. and Austin, W.E.N., 'Marine climate variability during the last millennium: the Loch Sunart record, Scotland, UK', *Quaternary Science Reviews*, 29 (2010), pp. 1633–1647.

Campbell, B.M.S., *The Great Transition: Climate, Disease and Society in the Late-Medieval World* (Cambridge: Cambridge University Press, 2016).

Carracedo, J.C., Rodriguez Badiola, E. and Soler, V., 'The 1730–1736 eruption of Lanzarote, Canary Islands: a long, high-magnitude basaltic fissure eruption', *Journal of Volcanology and Geothermal Research*, 53 (1992), pp. 239–250.

Carson, R., *Silent Spring* (Boston: Houghton Mifflin, 1962).

Catt, J.A., 'Long-term consequences of using artificial and organic fertilisers: the Rothamsted experiments', in S. Foster and T.C. Smout (eds), *The History of Soils and Field Systems* (Aberdeen: Scottish Cultural Press, 1994), pp. 119–134.

Cheape, H., ' "Every timber in the forest for MacRae's house": creel houses in the Highlands', *Vernacular Building*, 37 (2013–14), pp. 31–50.
Chronology of British Hydrological Events. https://cbhe.hydrology.org.uk/
Clarke, E., 'Switser, Stephen', *Dictionary of National Biography*, volume 55 (Oxford: Oxford University Press, 1898), pp. 241–242.
Comrie, J.D., *History of Scottish Medicine*, 2 volumes (London: The Wellcome Historical Medical Museum, 1932).
Coull, J.R., 'The herring fishery', in J.R. Coull, A. Fenton and K. Veitch (eds), *A Compendium of Scottish Ethnology*, volume 4: *Boats, Fishing and the Sea* (Edinburgh: John Donald, 2008), pp. 208–235.
Coull, J.R., 'White fishing', in J.R. Coull, A. Fenton and K. Veitch (eds), *A Compendium of Scottish Ethnology*, volume 4: *Boats, Fishing and the Sea* (Edinburgh: John Donald, 2008), pp. 253–276.
Cowan, I.B., *The Parishes of Medieval Scotland* (Edinburgh: Scottish Record Society, 1967).
Crawford, I.A., *The West Highlands and Islands: A View of 50 Centuries. The Udal (North Uist) Evidence* (Cambridge: Great Auk Press, 1986).
Crocker, G., 'Scotland', in Gunpowder Mills Gazetteer; black powder manufacturing sites in the British Isles (1996 and 2008). https://catalogue.millsarchive.org/uploads/r/null/8/5/8/858f75adab794e799564ab57ddaf71aee15efaaf073a6c852046f95468cf0a9c/_home_artefactual_digi_objects_Rest_25550.pdf
Crone, A. and Fawcett, R., 'Dendrochronology, documents and the timber trade: new evidence for the building history of Stirling Castle, Scotland', *Medieval Archaeology*, 42:1 (1998), pp. 68–87.
Crone, A. and Watson, F., 'Sufficiency to scarcity, 500–1600', in T.C. Smout (ed.), *People and Woods in Scotland* (Edinburgh: Edinburgh University Press, 2003), pp. 60–81.
Crone, A., Grieve, N., Moore, K. and Perry, D.R., 'Investigations into an early timber-frame roof in Brechin, Angus', *Tayside and Fife Archaeological Journal*, 10 (2004), pp. 152–165.
Cullen, K.J., *Famine in Scotland: The 'Ill Years' of the 1690s*, Scottish Historical Review Monograph series 16 (Edinburgh: Edinburgh University Press, 2010).
Cunningham, I.C. (ed.), *The Nation Survey'd: Timothy Pont's Maps of Scotland* (East Linton: Tuckwell Press, 2001).
Cushman, G.T., *Guano and the Opening of the Pacific World: A Global Ecological History* (Cambridge: Cambridge University Press, 2013).
D'Arrigo, R., Klinger, P., Newfield, T., Rydval, M. and Wilson, R., 'Complexity in crisis: the volcanic cold pulse of the 1690s and the consequences of Scotland's failure to cope', *Journal of Volcanology and Geothermal Research*, 389 (2020), 106746.
Davidson, D.A., 'Soils as cultural resources', in G. Fellows-Jensen (ed.), *Denmark and Scotland: The Cultural and Environmental Resources of Small Nations* (Edinburgh and Copenhagen: Royal Society of Edinburgh and Royal Danish Academy, 2001), pp. 171–180.
Davidson, D.A. and Simpson, I.A., 'Soils and landscape history: case studies from the Northern Isles of Scotland', in T.C. Smout and S. Foster (eds), *The History of Soils and Field Systems* (Aberdeen: Scottish Cultural Press, 1994), pp. 66–74.
Davidson, D.A., Dercon, G., Stewart, M. and Watson, F., 'The legacy of past urban waste disposal on local soils', *Journal of Archaeological Science*, 33 (2006), pp. 778–783.
Davies, A.L., 'Upland agriculture and environmental risk: a new model of upland land-use based on high spatial-resolution palynological data from West Affric, NW Scotland', *Journal of Archaeological Science*, 34 (2007), pp. 2053–2063.
Davies, A.L. and Watson, F., 'Understanding the changing value of natural resources: an integrated palaeoecological–historical investigation into grazing–woodland interactions by Loch Awe, Western Highlands of Scotland', *Journal of Biogeography*, 34 (2007), pp. 1777–1791.
Dawson, A.G., *So Foul and Fair a Day: A History of Scotland's Climate and Weather* (Edinburgh: Birlinn, 2009).
Dawson, A.G., Kirkbride, M.P. and Cole, H., 'Atmospheric effects in Scotland of the AD 1783–84 Laki eruption in Iceland', *The Holocene*, 31:5 (2021), pp. 830–843.
Dawson, S., Dawson, A.G. and Jordan, A.T., 'North Atlantic climate change and Late Holocene windstorm activity in the Outer Hebrides, Scotland', in D. Griffiths and P. Ashmore (eds), 'Aeolian archaeology: the archaeology of sand landscapes in Scotland', *Scottish Archaeological Internet Reports*, 48 (2004). https://doi.org/10.9750/issn.2056-7421.2011.48.
de Vries, J., *European Urbanisation, 1500–1850*, 1st edition (London: Methuen and Co., 1984).
Deary, H., 'Restoring wildness to the Scottish Highlands: a landscape of legacies', in M. Hourdequin and D.G. Havlick (eds), *Restoring Layered Landscapes: History, Ecology, and Culture* (Oxford: Oxford University Press, 2016), pp. 95–111.
Dennison, E.P., 'Burghs and burgesses: a time of consolidation?', in R.D. Oram (ed.), *The Reign of Alexander II 1214–49* (Leiden: Brill, 2005), pp. 253–284.
Dennison, E.P., Ditchburn, D. and Lynch, M. (eds), *Aberdeen Before 1800: A New History* (East Linton: Tuckwell Press, 2002).
Devine, T.M., *The Great Highland Famine: Hunger, Emigration and the Scottish Highlands in the Nineteenth Century* (Edinburgh: John Donald, 1988 and subsequent editions).
Devine, T.M., *The Transformation of Rural Scotland: Social Change and the Agrarian Economy 1660–1815* (Edinburgh: John Donald, 1994).
Devine, T.M., *The Scottish Clearances: A History of the Dispossessed 1660–1900* (London: Allen Lane, 2018).
Diamond, J., *Collapse: How Societies Choose to Fail or Succeed*, 1st edition (New York: Viking, 2005).
Dick Lauder, T., *An Account of the Great Floods of August 1829, in the Province of Moray and Adjoining Districts*, 3rd edition (Elgin: R. Stewart, 1873).
Ditchburn, D., 'Cargoes and commodities: Aberdeen's trade with Scandinavia and the Baltic, c.1302–c.1542', *Northern Studies*, 27 (1990), pp. 12–22.
Dixon, P., 'Field systems, rigs and other cultivation remains in Scotland: the field evidence', in S. Foster and T.C. Smout (eds), *The History of Soils and Field Systems* (Aberdeen: Scottish Cultural Press, 1994), pp. 26–52.
Dixon, P., *Southdean, Borders: An Archaeological Survey* (Edinburgh: RCAHMS, 1994).
Dodgshon, R.A., *Land and Society in Early Scotland* (Oxford: Oxford University Press, 1981).
Dodgshon, R.A., *From Chiefs to Landlords: Social and Economic Change in the Western Highlands and Islands, c.1493–1820* (Edinburgh: Edinburgh University Press, 1998).
Dodgshon, R.A., 'Livestock production in the Scottish Highlands before and after the Clearances', *Rural History*, 9 (1998), pp. 19–42.
Donnelly, T., 'Arthur Clephane, Edinburgh merchant and seedsman, 1706–1730', *Agricultural History Review*, 18:2 (1970), pp. 151–160.
Dougall, M. and Dickson, J., 'Old managed oaks in the Glasgow area', in T.C. Smout (ed.), *Scottish Woodland History* (Edinburgh: Scottish Cultural Press, 1997), pp. 76–85.
Duckham, B.F., *A History of the Scottish Coal Industry*, volume 1, 1700–1815 (Newton Abbot: David & Charles, 1970).
Duncan, A.A.M., *Scotland: The Making of the Kingdom*. The Edinburgh History of Scotland, volume 1 (Edinburgh: Oliver & Boyd, 1975; paperback edition 1978).

Dunlop, J., *The British Fisheries Society, 1786–1893* (Edinburgh: John Donald, 1978).
Durie, A.J., 'The markets for Scottish linen, 1730–1775', *Scottish Historical Review*, 52:1 (1973), pp. 30–49.
Eddy, J.A., 'The Maunder Minimum', *Science*, 18:192 (June 1976), pp. 1189–1202.
Emerson, R.L., 'Sir Robert Sibbald Kt, the Royal Society of Scotland and the origins of the Scottish Enlightenment', *Annals of Science*, 45 (1988), pp. 41–72.
Emerson, R.L., 'Scottish cultural change 1660–1710 and the Union of 1707', in J. Robertson (ed.), *A Union for Empire: Political Thought and the British Union of 1707* (Cambridge: Cambridge University Press, 1995), pp. 121–144.
Fabricius, J., *Syphilis in Shakespeare's England* (London: Jessica Kingsley Publishers, 1994).
Fagan, B., *The Little Ice Age: How Climate Made History 1300–1850*, paperback edition (New York: Basic Books, 2002).
Fairlie, S., 'The nineteenth-century Corn Law reconsidered', *Economic History Review*, 18:3 (1965), pp. 562–575.
Fell, A., *The Early Iron Industry of Furness and District* (Ulverston: H. Kitchin, 1908).
Fenton, A., 'Seaweed as fertiliser', in J.R. Coull, A. Fenton and K. Veitch (eds), *A Compendium of Scottish Ethnology*, volume 4: *Boats, Fishing and the Sea* (Edinburgh: John Donald, 2008), pp. 135–150.
Ferguson, W., *Scotland: 1689 to the Present* (Edinburgh: Oliver & Boyd, 1968).
Fielding, P., *Scotland and the Fictions of Geography: North Britain 1760–1830* (Cambridge: Cambridge University Press, 2008).
Flinn, M.W. (ed.), *Scottish Population History from the Seventeenth Century to the 1930s* (Cambridge: Cambridge University Press, 1977).
Fowler, J., *Landscapes and Lives: The Scottish Forest Through the Ages* (Edinburgh: Canongate, 2002).
Franklin, J., Troy, C., Britton, K., Wilson, D. and Lawson, J.A., *Past Lives of Leith: Archaeological Work for Edinburgh Trams* (Edinburgh: City of Edinburgh Council, 2019).
Fraser, G., *Lowland Lore; or the Wigtownshire of long ago* (Wigtown: Gordon Fraser, 1880).
Fraser, I., 'Three Perthshire water meadows: Strathallan, Glendevon and Bertha', *Tayside and Fife Archaeological Journal*, 7 (2001), pp. 133–144.
Fraser Darling, F., 'History of the Scottish forests', *Scottish Geographical Magazine*, 65:3 (1949), pp. 132–137.
Fraser Darling, F., *West Highland Survey: An Essay in Human Ecology* (Oxford: Oxford University Press, 1955).
Fraser Darling, F. and Boyd, J.M., *The Highlands and Islands* (London: Collins, 1964).
Geikie, A., *Textbook of Geology*, 3rd edition (London: Macmillan and Co., 1893).
Gemmill, E. and Mayhew, N., *Changing Values in Medieval Scotland: A Study of Prices, Money, Weights and Measures* (Cambridge: Cambridge University Press, 1995).
Gibson, A.J. and Smout, T.C., *Prices, Food and Wages in Scotland 1550–1780* (Cambridge: Cambridge University Press, 1995).
Gilbert, J.M., 'Falkland Park to 1603', *Tayside and Fife Archaeological Journal*, 19/20 (2014), pp. 78–102.
Gilbert, J.M., *Hunting and Hunting Reserves in Medieval Scotland* (Edinburgh: John Donald, 1979).
Gilbertson, D.D., Schwenninger, J.-L., Kemp, R.A. and Rhodes, E.J., 'Sand-drift and soil formation along an exposed Atlantic coastline: 14,000 years of diverse geomorphological climatic and human impacts', *Journal of Archaeological Science*, 26 (1999), pp. 439–469.

Gilliam, Terry and Jones, Terry (Directors), *Monty Python and the Holy Grail* (Motion picture). EMI Films (1975).
Gorebridge Community Development Trust, 'Stobsmill Gunpowder Works: an introduction' (no date). https://gorebridge.org.uk/heritage/stobsmill-gunpowder-works-an-introduction/
Gow, S.M., *James Moodie Younger of Melsetter: An Orkney Laird at the Time of the Union* (Kirkwall: Kirkwall Grammar School, 1977).
Grattan, J. and Brayshay, M., 'An amazing and portentous summer: environmental and social responses in Britain to the 1783 eruption of an Iceland volcano', *Geographical Journal*, 161 (1995), pp. 125–134.
Gray, M., *The Fishing Industries of Scotland 1790–1914: A Study in Regional Adaptation*. Aberdeen University Studies series 105 (Oxford: Oxford University Press, 1978).
Haldane, A.R.B., *The Drove Roads of Scotland*, new edition (Edinburgh: Birlinn Origin, 2021).
Hampson, A. and Smout, C., 'Trying to understand woods', in T.C. Smout (ed.), *Understanding the Historical Landscape in its Environmental Setting* (Dalkeith: Scottish Cultural Press, 2002), pp. 87–95.
Harris, R., 'Scotland's herring fisheries and the prosperity of the nation, c.1660–1760', *Scottish Historical Review*, 79:1 (2007), pp. 39–60.
Harrison, J., *A Historical Background of Flanders Moss*. Scottish Natural Heritage Commissioned Report No. 2 (Edinburgh: Scottish Natural Heritage, 2003).
Hindmarch, E., Oram, R., Haggarty, G., Hall, D. and Robertson, J., 'Eldbotle: the archaeology and environmental history of a medieval rural settlement in East Lothian', *Proceedings of the Society of Antiquaries of Scotland*, 142 (2012), pp. 245–300.
Hodgson, G.W.I., Smith, C. et al., *Perth High Street Archaeological Excavation 1975–1977, Fascicule 4: Living and Working in a Scottish Medieval Burgh. Environmental Remains and Miscellaneous Finds* (Perth: Tayside and Fife Archaeological Committee, 2011).
Hodgson, J., *A History of Northumberland*, part 3 volume 2 (Newcastle-upon-Tyne: Edward Walker, 1828).
Hoffmann, R.C. and Ross, A., 'This belongs to us! Competition between the Royal Burgh of Stirling and the Augustinian Abbey of Cambuskenneth over salmon fishing rights on the River Forth', in E. Bhreathnach, M. Krasnodębska D'Aughton and K. Smith (eds), *Monastic Europe AD1100–1700: Communities, Landscape and Settlement* (Turnhout: Brepols, 2020), pp. 451–476.
Holmes, G., *Europe: Hierarchy and Revolt 1320–1450* (London: Fontana Books, 1975).
Hope, D., Picozzi, N., Catt, D.C. and Moss, R., 'Effects of reducing sheep grazing in the Scottish Highlands', *Journal of Range Management*, 49:4 (1996), pp. 301–310.
Hopkins, D.R., *The Greatest Killer: Smallpox in History*, revised edition (Chicago: University of Chicago Press, 2002).
Hopkins, P., *Glencoe and the End of the Highland Wars* (Edinburgh: John Donald, 1998).
House, S. and Dingwall, C., '"A nation of planters": introducing the new trees, 1650–1900', in T.C. Smout (ed.), *People and Woods in Scotland* (Edinburgh: Edinburgh University Press, 2003), pp. 128–157.
Huizinga, J., *The Waning of the Middle Ages* (London, 1924).
Hume, J.R., *The Industrial Archaeology of Glasgow* (Glasgow: Blackie & Co., 1974).
Hunter, J., *On the Other Side of Sorrow: Nature and People in the Scottish Highlands* (Edinburgh: Mainstream Publishing, 1995).
Hunter, J., *Set Adrift Upon the World: The Sutherland Clearances* (Edinburgh: Birlinn, 2015).

Hunter, J., *Insurrection: Scotland's Famine Winter* (Edinburgh: Birlinn, 2019).
Hutchison, I.G.C., *Industry, Reform and Empire: Scotland 1790–1880* (Edinburgh: Edinburgh University Press, 2020).
Hyslop, J. and Hyslop, R., *Langholm As It Was: A History of Langholm and Eskdale from the Earliest Times* (Sunderland: Hills & Co., 1912).
Jervise, A., *Memorials of Angus and the Mearns* (Edinburgh, 1861).
Jiang, Y. and Xu, Z., 'On the Spörer Minimum', *Astrophysics and Space Science*, 118 pts 1–2 (January 1986), pp. 159–162.
Jones, A.K.G., 'The fish bone', in D. Perry et al. (eds), *Perth High Street Archaeological Excavation 1975–1977*, fascicule 4: *Living and Working in a Medieval Scottish Burgh. Environmental Remains and Miscellaneous Finds* (Perth: Tayside and Fife Archaeological Committee, 2011), pp. 53–67.
Kirk, W., 'Prehistoric sites at the sands of Forvie, Aberdeenshire: a preliminary examination', *Aberdeen University Review*, 35 (1953), pp. 150–71.
Kirk, W., 'Sands of Forvie', *Discovery and Excavation in Scotland, 1957* (Dundee: Council for British Archaeology, Scottish Regional Group, 1957), p. 4.
Kirk, W., 'Sands of Forvie', *Discovery and Excavation in Scotland, 1958* (Dundee: Council for British Archaeology, Scottish Regional Group, 1958), p. 2.
Kirk, W., 'Forvie', *Discovery and Excavation in Scotland, 1960* (Dundee: Council for British Archaeology, Scottish Regional Group, 1960), p. 2.
Klein, L.E., 'Politeness and the interpretation of the British eighteenth century', *The Historical Journal*, 45:4 (December 2002), pp. 869–898.
Kurlansky, M., *Cod: A Biography of the Fish that Changed the World* (London: Random House, 1999).
Lamb, H., *Climate, History and the Modern World*, 2nd edition (London: Methuen, 1995).
Lamb, H., *Historic Storms of the North Sea, British Isles and Northwest Europe* (Cambridge: Cambridge University Press, paperback edition 2005).
Lamb, H., *Weather, Climate and Human Affairs* (London: Taylor and Francis, 1988; reprinted London: Routledge, 2012).
Lapointe, F., Bradley, R.S., Francus, P., Balascio, N.L., Abbott, M.B., Stoner, J.S., St-Onge, G., De Coninck, A. and Labarre, T., 'Annually resolved Atlantic sea surface temperature variability over the past 2,900 yrs', *Proceedings of the National Academy of Sciences*, 117:44 (2020), pp. 27171–27178.
Lenman, B.P., *Integration, Enlightenment and Industrialization: Scotland 1746–1832* (London: Edward Arnold, 1981); reprinted as *Integration and Enlightenment* (Edinburgh: Edinburgh University Press, 1992); reissued as Lenman, B.P., *Enlightenment and Change: Scotland 1746–1832*, 2nd edition (Edinburgh: Edinburgh University Press, 2009).
Leopold, J.W., 'The Levellers Revolt in Galloway in 1724', *Journal of the Scottish Labour History Society*, 14 (1980), pp. 4–29.
Lewis, J.H., 'The charcoal-fired blast furnaces of Scotland: a review', *Proceedings of the Society of Antiquaries of Scotland*, 114 (1984), pp. 445–463.
Lindsay, J.M., 'Charcoal iron smelting and its fuel supply: the example of the Lorn Furnace, Argyllshire, 1753–1876', *Journal of Historical Geography*, 1 (1975), pp. 283–298.
Lindsay, J.M., 'The iron industry in the Highlands: charcoal blast furnaces', *Scottish Historical Review*, 56 (1977), pp. 49–63.
Lindsay, J.M., 'The commercial use of woodland and coppice management', in M.L. Parry and T.R. Slater (eds), *The Making of the Scottish Countryside* (London: Croom Helm, 1980), pp. 271–289.
Livingston, A., 'The Galloway Levellers: A Study of Their Origins, Events and Consequences of Their Actions', unpublished MPhil (Research) thesis, University of Glasgow (2009).
Longnon, J. and Cazelles, R., *The Très Riches Heures of Jean, Duke of Berry* (New York: George Braziller, 2008).
Lynch, M., *Scotland: A New History* (London: Century, 1991).
McClure, D.C., 'Kilkerran Pyroligneous Acid Works 1845 to 1945', *Ayrshire History*. http://www.ayrshirehistory.org.uk/AcidWorks/acidworks.htm
McConnell, K., 'The people of the powder mill'. https://www.secretscotland.org.uk/index.php/Secrets/ArgyllGunpowderIndustry
McCormick, F., 'Calf slaughter as a response to marginality', in C. Mills and G. Coles (eds), *Life on the Edge: Human Settlement and Marginality* (Oxford: Oxbow Books, 2006), pp. 49–51.
Macdougall, N., *James IV* (East Linton: Tuckwell Press, 1997).
McGladdery, C.A., *James II*, 2nd edition (Edinburgh: John Donald, 2015).
Mann, A.J., *James VII, Duke and King of Scots, 1633–1701* (Edinburgh: John Donald, 2014).
Marrs, R.H., Lee, H., Blackbird, S., Connor, L., Girdwood, S.E., O'Connor, M., Smart, S.M., Rose, R.J., O'Reilly, J. and Chiverrell, R.C., 'Release from sheep-grazing appears to put some heart back into upland vegetation: a comparison of nutritional properties of plant species in long-term grazing experiments', *Annals of Applied Biology*, 77:1 (2020), pp. 152–162.
Matthes, F.E., 'Report of Committee on Glaciers, April 1939', *Transactions, American Geophysical Union*, 20:4 (1939), pp. 518–523.
Mayewski, P.A., Rohling, E.E., Stager, J.C., Karlén, W., Maasch, K.A., Meeker, L.D., Meyerson, E.A., Gass, F., van Kreveld, S., Holmgren, K., Lee-Thorp, J., Rosqvist, G., Rack, F., Staubwasser, M., Schneider, R.R. and Steig, E.J., 'Holocene climate variability', *Quaternary Research*, 62 (2004), pp. 243–255.
Meeker, L.D. and Mayewski, P.A., 'A 1,400-year high resolution record of atmospheric circulation over the North Atlantic and Asia', *The Holocene*, 12 (2002), pp. 27–66.
Miller, H., *Scenes and Legends of the North of Scotland or the Traditional History of Cromarty* (London: Adam and Charles Black, 1835).
Mitchison, R., *Lordship to Patronage: Scotland 1603–1745* (London: Edward Arnold, 1983); reprinted (Edinburgh: Edinburgh University Press, 1990).
Miyahara, H., Masuda, K., Muraki, Y., Kitagawa, H. and Nakamura, T., 'Variation of solar cyclicity during the Spörer Minimum', *Journal of Geophysical Research*, 111:A10 (2006), A03103.
Miyahara, H., Tokanai, F., Moriya, T., Takeyama, M., Sakurai, H., Horiuchi, K. and Hotta, H., 'Gradual onset of the Maunder Minimum revealed by high-precision carbon-14 analyses', *Scientific Reports*, 11:5482 (2021).
Monbiot, G., 'Sheepwrecked', *The Spectator*, 30 May 2013.
Monbiot, G., 'English countryside "shagged" by the "white plague" of sheep', *The Week*, 20 October 2015.
Morrison, I.A., 'Galloway: locality and landscape evolution', in R.D. Oram and G.P. Stell (eds), *Galloway: Land and Lordship* (Edinburgh: Scottish Society for Northern Studies, 1991), pp. 11–16.
Morrison, J., Oram, R. and Oliver, F., 'Ancient Eldbottle unearthed: archaeological and historical evidence for a long-lost early medieval East Lothian village', *Transactions of the East Lothian Antiquarian and Field Naturalists' Society*, 27 (2008), pp. 21–45.
Morrison, J., Oram, R. and Ross, A., 'Gogar: archaeological and historical evidence for a lost medieval parish near Edinburgh', *Proceedings of the Society of Antiquaries of Scotland*, 139 (2009), pp. 229–256.

Morton, A.S., 'The Levellers of Galloway', *Transactions of the Dumfriesshire and Galloway Natural History and Antiquarian Society*, 3rd series, 19 (1933–5), pp. 245–254.

Nairne, D., *Memorable Floods in the Highlands During the Nineteenth Century with some Accounts of The Great Frost of 1895* (Inverness: The Northern Counties Printing & Publishing Co. Ltd., 1895).

National Museums of Scotland, *The Story of the Plough*. https://www.nms.ac.uk/explore-our-collections/stories/science-and-technology/ploughs

Newland, K., 'The Acquisition and Use of Norwegian Timber in Seventeenth Century Scotland, with reference to the Principal Building Works of James Baine, His Majesty's Master Wright', unpublished PhD thesis, University of Dundee (2010).

Nicholson, R., *Scotland: The Later Middle Ages*, paperback edition (Edinburgh: Oliver and Boyd, 1978).

Ogilvie, J.F., 'Sir Walter Scott and forestry', *Scottish Forestry*, 39:1 (1985), pp. 13–22.

Ogilvie, J.F., 'Sir Walter Scott and Scottish forestry', *Native Woodland Discussion Group, Scottish Woodland History Conference Notes XVIII, Plantations in Scotland* (2016), pp. 40–46.

Ogurtsov, M.G., 'The Spörer Minimum was deep', *Advances in Space Research*, 64 (2019), pp. 1112–1116.

Oliver, S., *Walter Scott and the Greening of Scotland: Emergent Ecologies of a Nation* (Cambridge: Cambridge University Press, 2021).

Oram, R.D., *Torhousemuir Historical Account: Report on Programme of Historical Research undertaken by Dr Richard Oram/Retrospect Historical Services on behalf of Scottish Natural Heritage* (February 1995).

Oram, R.D., ' "It cannot be decernit quha are clean and quha are foulle." Responses to epidemic disease in sixteenth- and seventeenth-century Scotland', *Renaissance and Reformation*, 30:4 (2007), pp. 13–40.

Oram, R.D., 'A fit and ample endowment? The Balmerino estate 1228–1606', *Comentarii Cistercienses* (2008), t.59 fasc. 1–2, pp. 61–80.

Oram, R.D., 'Abondance inépuisable? Crise de l'approvisionnement en combustible et réactions en Ecosse du Nord entre environ 1500 et environ 1800', in J.-M. Derex and F. Grégoire (eds), *Histoire économique et sociale de la tourbe et des tourbières* (Cordemais: Æstuaria, 2009), pp. 31–44.

Oram, R.D., 'Disease, death and the hereafter', in E.J. Cowan and L. Henderson (eds), *Handbook of Scottish Medieval History* (Edinburgh: Edinburgh University Press, 2011), pp. 196–225.

Oram, R.D., 'Perth in the Middle Ages: an environmental history', in D. Strachan (ed.), *Perth 800* (Perth: Perth and Kinross Heritage Trust, 2011), pp. 19–28.

Oram, R.D., 'Social inequality in the supply and use of fuel in Scottish towns c.1750–c.1850', in G. Massard-Gilbaud and R. Rodger (eds), *Environmental and Social Inequalities in the City, 1750–2000* (Banbury: White Horse Press, 2011), pp. 211–231.

Oram, R.D., 'Waste management and peri-urban agriculture in the early modern Scottish burgh', *Agricultural History Review*, 59 part 1 (2011), pp. 1–17.

Oram, R.D., 'The salt industry in medieval Scotland', *Studies in Medieval and Renaissance History*, 3rd series, 9 (2012), pp. 209–232.

Oram, R.D., 'Arrested development? Energy crises, fuel supplies, and the slow march to modernity in Scotland, 1450–1850', in R.W. Unger (ed.), *Energy Transitions in History: Global Cases of Continuity and Change*. Rachel Carson Center Perspectives 2 (Munich: Rachel Carson Center, 2013), pp. 17–24.

Oram, R.D., 'Estuarine environments and resource exploitation in eastern Scotland c.1125–c.1400: a comparative study of the Forth and Tay estuaries', in E. Thoen, G.J. Borger, A.M.J. de Kraker, T. Soens, D. Tys, L. Vervaet and H.J.T. Weerts (eds), *Landscapes or Seascapes? The History of the Coastal Environment in the North Sea Area Reconsidered*. Comparative Rural History Network, volume 13 (Turnhout: Brepols, 2013), pp. 353–377.

Oram, R.D., 'Between a rock and a hard place: climate, weather and the rise of the Lordship of the Isles', in R.D. Oram (ed.), *The Lordship of the Isles* (Leiden: Brill, 2014), pp. 40–61.

Oram, R.D., 'From "Golden Age" to Depression: land, lordship and environmental change in the medieval Earldom of Orkney', in H.C. Gulløv (ed.), *Northern Worlds* (Copenhagen: National Museum of Denmark, 2014), pp. 203–214.

Oram, R.D., 'Environmental change, resource conflicts and social change in Scotland c.1250–1500', in C.-F. Mathis and G. Massard-Gilbaud (eds), *Sous le Soleil: Systèmes et transitions énergétiques du Moyen Âge à nos jours* (Paris: Editions de la Sorbonne, 2019), pp. 211–224.

Oram, R. and Adderley, P., 'Lordship and environmental change in central Highland Scotland c.1300 to c.1450', *Journal of the North Atlantic*, 1 (2008), pp. 74–84.

Oram, R.D. and Adderley, P., 'Lordship, land and environmental change in West Highland and Hebridean Scotland c.1300–c.1450', in S. Cavaciocchi (ed.), *Economic and Biological interactions in Pre-Industrial Europe from the 13th to the 18th Centuries* (Florence: University of Florence Press, 2010), pp. 257–268.

Oram, R.D. and Adderley, P., 'Re Innse Gall: A Norse colony in the Irish Sea and Hebrides?', in S. Imsen (ed.), *The Norwegian Domination and the Norse World c.1100–c.1400* (Trondheim: Rostra, 2010), pp. 125–148.

Parry, M.L., 'Changes in the extent of improved farmland', in M.L. Parry and T.S. Slater (eds), *The Making of the Scottish Countryside* (London: Croom Helm, 1980), pp. 177–199.

Pearson, M.G., 'The winter of 1739–40 in Scotland', *Weather*, 28 (1973), pp. 20–24.

Pearson, M.G., 'Snowstorms in Scotland: 1831–1861', *Weather*, 33 (1978), pp. 392–399.

Phillipson, N., *Hume* (London: Weidenfeld & Nicolson, 1989).

Pribyl, K., *Farming, Famine and Plague: The Impact of Climate in Later Medieval England* (Cham (Switzerland): Springer International Publishing, 2017).

Proctor, C.J., Baker, A., Barnes, W.L. and Gilmour, M.A., 'A thousand year speleothem proxy record of North Atlantic climate from Scotland', *Climate Dynamics*, 16 (2000), pp. 815–820.

Rackham, O., *Trees and Woodland in the British Landscape: The Complete History of Britain's Trees, Woods and Hedgerows*, revised paperback edition (London: Phoenix Press, 2001).

Ramsay, S., Miller, J.J. and Housley, R.A., 'Palaeoenvironmental investigations of Rispain Mire, Whithorn', *Transactions of the Dumfriesshire and Galloway Natural History and Antiquarian Society*, 3rd series, 81 (2007), pp. 35–55.

Rapport, M., *1848: Year of Revolution* (London: Little, Brown, 2008).

Richards, E., *The Highland Clearances*, new edition (Edinburgh: Birlinn, 2018).

Robertson, J., 'The powder mills of Argyll', *Industrial Archaeology Review*, 12:2 (1990), pp. 205–215.

Rorke, M., 'Scottish Overseas Trade, 1275/86–1597', unpublished PhD thesis, University of Edinburgh (2001).

Rorke, M., 'The Scottish herring trade, 1470–1600', *Scottish Historical Review*, 84 (2005), pp. 149–65.

Ross, A., 'Scottish environmental history and the (mis)use of soums', *Agricultural History Review*, 54 (2006), pp. 213–228.

Ross, A., 'Improvement on the Grant estates in Strathspey in the eighteenth century: theory, practice and failure', in R.W. Hoyle (ed.), *Custom, Improvement and the Landscape in Early Modern Britain* (London: Routledge, 2011), pp. 289–311.

Ross, A., *Land Assessment and Lordship in Medieval Northern Scotland* (Turnhout: Brepols, 2015).

Ross, S., *The Culbin Sands: Fact and Fiction* (Aberdeen: Centre for Scottish Studies, 1992).

Ross, S., 'The Culbin Sands: a mystery unravelled', in W.D.H. Sellar (ed.), *Moray: Province and People* (Edinburgh: Scottish Society for Northern Studies, 1993), pp. 87–204.

Rössner, P.R., 'The 1738–41 harvest crisis in Scotland', *Scottish Historical Review*, 90 part 1 (2011), pp. 27–63.

The Royal Commission on the Ancient and Historical Monuments of Scotland. Argyll: An Inventory of the Ancient Monuments: volume 2: *Lorn* (Edinburgh: HMSO, 1975).

The Royal Commission on the Ancient and Historical Monuments of Scotland. Argyll: An Inventory of the Monuments. Volume 7: *Mid-Argyll and Cowal: Medieval and Later Monuments* (Edinburgh: HMSO, 1992).

The Royal Commission on the Ancient and Historical Monuments of Scotland. Mar Lodge Estate, Grampian: An Archaeological Survey (Edinburgh: HMSO, 1995).

The Royal Commission on the Ancient and Historical Monuments of Scotland, *'Well sheltred and watered': Menstrie Glen, A Farming Landscape near Stirling* (Edinburgh: RCAHMS, 2001).

Ruddock, T., 'Repair of two important early Scottish roof structures', *Proceedings of the Institute of Civil Engineers*, 110 (1995), pp. 296–307.

Ryder, M.L., 'Sheep and the Clearances in the Scottish Highlands: a biologist's view', *Agricultural History Review*, 16:2 (1968), pp. 155–158.

Rydval, M., Loader, N.J., Gunnarson, B.E., Druckenbrod, D.L., Linderholm, H.W., Moreton, S.G., Wood, C.V. and Wilson, R., 'Reconstructing 800 years of summer temperatures in Scotland from tree rings', *Climate Dynamics*, 49:9–10 (2017), pp. 2951–2974.

Sansum, P., 'Argyll oakwoods: use and ecological change, 1000–2000 AD – a palynological-historical investigation', *Botanical Journal of Scotland*, 57 (2005), pp. 83–97.

Saville, A.C., Martin, M.D. and Ristaino, J.B., 'Historic late blight outbreaks caused by a widespread dominant lineage of *Phytophthora infestans* (Mont.) de Bary', *PLoS One*, 11:2 (2016). doi:10.1371/journal.pone.0168381

Scottish Government, *Defining Rewilding for Scotland's Public Sector: Research Findings* (published 4 July 2023). https://www.gov.scot/publications/defining-rewilding-scotlands-public-sector/

Serjeantson, D., *Farming and Fishing in the Outer Hebrides AD 600 to 1700: The Udal, North Uist*, Southampton Monographs in Archaeology, New Series 2 (Chandlers Ford: The Highfield Press Southampton, 2013).

Sharples, N. (ed.), *A Norse Farmstead in the Outer Hebrides: Excavations at Mound 3, Bornais, South Uist* (Oxford: Oxbow Books, 2005).

Shaw, J., 'Manuring and fertilising the Lowlands 1650–1850', in T.C. Smout and S. Foster (eds), *The History of Soils and Field Systems* (Aberdeen: Scottish Cultural Press, 1994), pp. 111–118.

Shepherd, C., *The Late Medieval Landscape of North-East Scotland: Renaissance, Reformation and Revolution* (Oxford: Windgather Press, 2021).

Shiel, R., 'Science and practice: the ecology of manure in historical retrospect', in R. Jones (ed.), *Manure Matters: Historical, Archaeological and Ethnographic Perspectives* (Farnham: Ashgate, 2012), pp. 13–24.

Shrewsbury, J.F.D., *A History of Bubonic Plague in the British Isles* (Cambridge: Cambridge University Press, 1971).

Simpson, G., 'Seeing the wood for the trees: Poland and the Baltic timber trade, c.1250–1650', in A. Roznowska-Sadraei (ed.), *Medieval Art, Architecture and Archaeology in Cracow and Lesser Poland* (Leeds: Maney, 2014), pp. 235–254.

Simpson, I.A., 'Relict properties of anthropogenic deep top soils as indicators of infield management in Marwick, West Mainland, Orkney', *Journal of Archaeological Science*, 24 (1997), pp. 365–380.

Sjölund, M.J., González-Díaz, P., Moreno-Vellena, J.J. and Jump, A.S., 'Understanding the legacy of widespread population translocations on the post-glacial genetic structure of the European beech, *Fagus sylvatica* L.', *Journal of Biogeography*, 44:11 (2017), pp. 2475–2487.

Slack, P., *The Invention of Improvement: Information and Material Progress in Seventeenth-Century England* (Oxford: Oxford University Press, 2014).

Slonosky, V.C., Jones, P.D. and Davies, T.D., 'Atmospheric circulation and surface temperature in Europe from the 18th century to 1995', *International Journal of Climatology*, 21 (2001), pp. 63–75.

Smith, B., 'Shetland Archives and sources of Shetland history', *History Workshop*, 4 (1977), pp. 203–214.

Smith, P.J., 'The foul burns of Edinburgh: public health attitudes and environmental change', *Scottish Geographical Journal*, 91 (1975), pp. 25–37.

Smout, T.C., *Scottish Trade on the Eve of Union* (Edinburgh: Edinburgh University Press, 1963).

Smout, T.C., 'The Highlands and the roots of Green consciousness, 1750–1990', *Scottish Natural Heritage Occasional Paper* No. 1 (1990).

Smout, T.C., 'Highland land-use before 1800: misconceptions, evidence and realities', in T.C. Smout (ed.), *Scottish Woodland History* (Edinburgh: Scottish Cultural Press, 1997), pp. 5–23.

Smout, T.C., 'Cutting into the pine: Loch Arkaig and Rothiemurchus in the eighteenth century', in T.C. Smout (ed.), *Scottish Woodland History* (Edinburgh: Scottish Cultural Press, 1997), pp. 115–125.

Smout, T.C. (ed.), *People and Woods in Scotland: A History* (Edinburgh: Edinburgh University Press, 2003).

Smout, T.C., 'Woodland in the maps of Pont', in I.C. Cunningham (ed.), *The Nation Survey'd: Timothy Pont's Maps of Scotland* (Edinburgh: John Donald, 2006), pp. 77–92.

Smout, T.C., 'Managing the woodlands of East Lothian, 1585–1765', *Transactions of the East Lothian Antiquarian and Field Naturalists' Society*, 26 (2006), pp. 41–53.

Smout, T.C., 'Bogs and people in Scotland since 1600', in T.C. Smout (ed.), *Exploring Environmental History: Selected Essays* (Edinburgh, 2009), pp. 99–113.

Smout, T.C., 'Energy rich, energy poor: Scotland, Ireland and Iceland, 1600–1800', in T.C. Smout (ed.), *Exploring Environmental History: Selected Essays* (Edinburgh, 2009), pp. 113–133.

Smout, T.C., 'The pinewoods and human use, 1600–1900', in T.C. Smout (ed.), *Exploring Environmental History: Selected Essays* (Edinburgh, 2009), pp. 71–85.

Smout, T.C. (ed.), *Exploring Environmental History: Selected Essays* (Edinburgh: Edinburgh University Press, 2009).

Smout, T.C., 'A new look at the Scottish Improvers', *Scottish Historical Review*, 91 (no. 231 part 1) (April 2012), pp. 125–149.

Smout, T.C. and Stewart, M., *The Firth of Forth: An Environmental History* (Edinburgh: Birlinn, 2012).

Smout, T.C. and Watson, F., 'Exploiting semi-natural woods, 1600–1800', in T.C. Smout (ed.), *Scottish Woodland History* (Edinburgh: Scottish Cultural Press, 1997), pp. 86–100.

Smout, T.C., MacDonald, A.R. and Watson, F., *A History of the Native Woodlands of Scotland, 1500–1920* (Edinburgh: Edinburgh University Press, 2005).

Smylie, M., *Herring: A History of the Silver Darlings* (Stroud: The History Press, 2004).

South Dakota State University, 'Undocumented volcano contributed to extremely cold decade from 1810–1819', *ScienceDaily*, 7 December 2009. www.sciencedaily.com/releases/2009/12/091205105844.htm

Stell, G.P. and Hay, D.H., *Bonawe Iron Furnace* (Edinburgh: HMSO, 1995).

Steven, H.M. and Carlisle, A., *The Native Pinewoods of Scotland* (Edinburgh: Oliver and Boyd, 1959).

Stewart, M., 'Using the woods, 1600–1850 (1): The community resource', in T.C. Smout (ed.), *People and Woods in Scotland* (Edinburgh: Edinburgh University Press, 2003), pp. 82–104.

Stewart, M., 'Using the woods, 1600–1850 (2): Managing for profit', in T.C. Smout (ed.), *People and Woods in Scotland* (Edinburgh: Edinburgh University Press, 2003), pp. 105–127.

Stone, J., 'Timothy Pont: three centuries of research, speculation and plagiarism', in I.C. Cunningham (ed.), *The Nation Survey'd: Timothy Pont's Maps of Scotland* (East Linton: Tuckwell Press, 2001), pp. 1–26.

Tait, A.A., *The Landscape Garden in Scotland 1735–1835* (Edinburgh: Edinburgh University Press, 1980).

Thick, M., 'Garden seeds in England before the late eighteenth century. II: The trade in seeds to 1760', *Agricultural History Review*, 38 (1990), pp. 58–71.

Thomson, W.P.L., *History of Orkney* (Edinburgh: The Mercat Press, 1987).

Thomson, W.P.L., *The New History of Orkney*, 3rd edition (Edinburgh: Birlinn, 2008).

Thorarinsson, S., *Hekla: A Notorious Volcano*, trans. Jóhann Hannesson and Pétur Karlsson (Reykjavík: Almenna bókafélagið, 1970).

Thordarson, T. and Self, S., 'Atmospheric and environmental effects of the 1783–1784 Laki eruption: a review and reassessment', *Journal of Geophysical Research: Atmospheres*, 108 (2003), pp. 1–29.

Tipping, R., 'The form and fate of Scotland's woodlands', *Proceedings of the Society of Antiquaries of Scotland*, 124 (1994), pp. 1–54.

Tipping, R., 'Cereal cultivation on the Anglo-Scottish Border during the "Little Ice Age"', in C.M. Mills and G. Coles (eds), *Life on the Edge: Human Settlement and Marginality*. Oxbow Monograph 100 (Oxford: Oxbow Books, 1998), pp. 1–11.

Tipping, R., 'Palaeoecology and political history: evaluating driving forces in historic landscape change in southern Scotland', in I.D. Whyte and A.J.L. Winchester (eds), *Society, Landscape and Environment in Upland Britain*. Society for Landscape Studies supplementary series 2 (2004), pp. 11–20.

Tipping, R., *Bowmont: An Environmental History of the Bowmont Valley and the Northern Cheviot Hills, 10,000 BC – AD 2000* (Edinburgh: Society of Antiquaries of Scotland, 2010).

Tittensor, R.M., 'History of the Loch Lomond oakwoods. 1: Ecological history', *Scottish Forestry*, 24 (1970), pp. 100–118.

U.S. Geological Survey Fact Sheet 2005–3045 (2005), U.S. Geological Survey and the U.S. Forest Service – Our Volcanic Public Lands: Pre-1980 Eruptive History of Mount St. Helens, Washington. https://pubs.usgs.gov/fs/2005/3045/#:~:text=Kalama%20Eruptive%20Period%20(A.D.%201479,form%20during%20the%20Kalama%20Period.

Veale, L., Endfield, G. and Bowen, J., 'The "Great Snow" of winter 1614/1615 in England', *Weather*, 73:1 (2018), pp. 3–9.

Wagner, S. and Zorita, E., 'The influence of volcanic, solar and CO_2 forcing on the temperatures in the Dalton Minimum (1790–1830): a model study', *Climate Dynamics*, 25 (2005), pp. 205–218.

Walker, B. and Ritchie, G., *Exploring Scotland's Heritage: Fife, Perthshire and Angus*, 2nd edition (Edinburgh: HMSO, 1995).

Watson, A. and Allan, E., 'Depopulation by clearances and non-enforced emigration in the North East Highlands', *Northern Scotland*, 10:1 (1990), pp. 31–46.

Watson, F. and Dixon, P., *A History of Scotland's Landscapes* (Edinburgh: Historic Environment Scotland, 2018).

Watson, R., *The Literature of Scotland* (Basingstoke and London: Macmillan and Co., 1984).

Watts, S.H., 'High mountain trees: altitudinal records recently broken for eleven different tree species in Britain', *British and Irish Botany*, 5:2 (2023), pp. 167–179.

Whitehouse, N., Jenkins, E. and Barratt, P., 'Rewilding: the historic environment and moving beyond "wilderness"', *Archaeology Scotland*, 47 (Summer 2023), pp. 14–17.

Whittington, G. and Edwards, K.J., 'Climate change', in K.J. Edwards and I.B.M. Ralston (eds), *Scotland After the Ice Age: Environment, Archaeology and History, 8000 BC – AD 1000*, paperback edition (Edinburgh: Edinburgh University Press, 2003), pp. 11–22.

Whyte, I.D., *Agriculture and Society in Seventeenth Century Scotland* (Edinburgh: John Donald, 1979).

Whyte, I.D., 'The emergence of the new estate structure', in M.L. Parry and T.S. Slater (eds), *The Making of the Scottish Countryside* (London: Croom Helm, 1980), pp. 117–135.

Whyte, I.D., 'George Dundas of Dundas: the context of an early eighteenth-century Scottish improving landowner', *Scottish Historical Review*, 60 (1981), pp. 1–13.

Whyte, I.D., *Scotland Before the Industrial Revolution: An Economic and Social History c.1050–c.1750* (Harlow: Longman Group, 1995).

Williamson, J.G., 'The impact of the Corn Laws just prior to repeal', *Explorations in Economic History*, 27:2 (1990), pp. 123–156.

Wilson, R., Loader, N.J., Rydval, M., Patton, H., Frith, A., Mills, C.M., Crone, A., Edwards, C., Larsson, L. and Gunnarson, B.E., 'Reconstructing Holocene climate from tree rings: the potential for a long chronology from the Scottish Highlands', *The Holocene*, 22:1 (2011), pp. 3–11.

Winchester, A.J.L., *The Harvest of the Hills: Rural Life in Northern England and the Scottish Borders, 1400–1700* (Edinburgh: Edinburgh University Press, 2000).

Wiseman, A.E.M., '"A noxious pack": historical, literary and folklore traditions of the wolf (*Canis lupus*) in the Scottish Highlands', *Scottish Gaelic Studies*, 25 (2009), pp. 95–142.

Wood, G., *Tambora: The Eruption that Changed the World* (Princeton: Princeton University Press, 2014).

Woodward, D., 'A comparative study of the Irish and Scottish livestock trades in the seventeenth century', in L.M. Cullen and T.C. Smout (eds), *Comparative Aspects of Scottish and Irish Economic and Social History 1600–1900* (Edinburgh: John Donald, 1976), pp. 147–164.

Woodward, D., '"Gooding the earth": manuring practices in Britain 1500–1800', in T.C. Smout and S. Foster (eds), *The History of Soils and Field Systems* (Aberdeen: Scottish Cultural Press, 1994), pp. 101–110.

Wormald, J., *Court, Kirk and Community: Scotland 1470–1625* (London: Edward Arnold, 1981); reprinted (Edinburgh: Edinburgh University Press, 1991).

Worrell, R. and Mackenzie, N., 'The ecological impact of using the woods', in T.C. Smout (ed.), *People and Woods in Scotland* (Edinburgh: Edinburgh University Press, 2003), pp. 195–213.

Index

Abbotsford (Borders), 371, 372, 374, 375
Aberdeen, 35, 36–37, 48, 75–76, 80, 133, 150, 186–187, 265, 293, 339, 345, 349, 351
Aberdeen Intelligencer, 279
Aberdeenshire, 28, 39, 102, 103, 148, 186–187, 191, 205, 215, 253, 265, 268, 293, 298, 300, 301, 302, 308, 327, 329, 341, 359
Aberfeldy (Perth and Kinross), 196, 366
Abernethy (Highland), forest of, 220, 384, 385
Acharn, Falls of (Perth and Kinross), 170, 224, 368
acorns: *see* oak trees
Adam, William, 371, 389
Aden House (Aberdeenshire), 327
Africa (*see also* enslaved people; guano), 165, 230, 270, 321, 322, 396
Age of Improvement: *see* Improvement
Agnew, William, of Wigg, 239
agriculture and agricultural land, 27, 36–42, 100, 157, 163, 186, 189, 269, 288, 390, 394
　　arable, 6, 9, 27, 29, 39–40, 70, 77, 82, 84, 94–95, 97, 99–108, 122, 123, 157, 160, 164, 167, 173, 180, 238–239, 240, 249, 252, 254, 255, 258, 263, 266, 268, 270, 275–276, 280, 289, 301, 306, 311, 312, 316, 320, 326, 328, 335, 340, 351, 365, 368, 377, 385, 395
　　expansion of, 39, 102, 240
　　illicit, 39, 77
　　Improvement, 103, 161, 163, 251, 252, 254–256, 268, 272, 275, 309, 310–311, 313–326, 327, 351, 368
　　loss of, 96, 185
　　subsistence, 182, 191, 234, 269, 276, 339
　　traditional, 12, 170, 309, 322
agricultural improvement: *see under* Improvement
Ailsa Craig (Ayr), 115
Aird, the (Inverness-shire), 21
Airthrey Castle (Stirling), 370
Albert, Prince, 393
alder trees, 69, 136, 140, 142, 196, 222, 226, 313, 315, 372, 381, 382, 390

Alexander III, king of Scots, 12
Alloa (Clackmannanshire), 48, 67, 181, 345
Alps and Alpine, 5, 8
Anderson, Alexander, of Strichen, 265
Anderson, James (Agricola), 217, 218
Angus (district), 33, 67, 74, 102, 116, 118, 125, 136, 138, 148, 179, 180, 202, 241, 267, 272, 305, 319, 329, 359
　　fuel use in, 148–149, 240, 272, 319, 352
　　glens, 67, 118, 138, 307; *see also* Esk
　　woodland in, 136–138, 202, 226
Annan (Dumfries and Galloway), 185, 196
Annexed Estates, 170, 268
anthrosols, 104, 105, 107, 272, 273, 317
apple (trees), 140, 186–187, 205
arable: *see under* agriculture and agricultural land
arboretums, 368, 370, 372
arboriculture, 157, 199, 230
Arbroath (Angus), 71, 148, 241
Arcan, Bog and Mains of (Highland), 315
Ardeonaig (Perth and Kinross), 319
Ardkinglass (Argyll and Bute), 370
Ardonald (Aberdeenshire), limestone quarry, 319
Ardtalnaig (Perth and Kinross), 319
Ardtornish (Highland), 382,
Argyll, earls and dukes of (*see also under* Campbell family), 224, 227, 246, 295, 326, 349
　　Colin Campbell, 1st earl of Argyll, 73
Arnbathie (Perth and Kinross), peat moss, 51
Aros (Argyll and Bute), 327
Arran, island of, 138, 139, 180
　　deer forest, 138
ash trees, 63, 103, 127, 132, 136, 140, 186, 187, 196, 218, 229, 381, 382
aspen trees, 127, 210, 367, 381
Assynt (Highland), 119, 334
Atholl
　　earls, marquises and dukes of: *see under* Murray family
　　estate, 122, 216, 220, 226, 262–263, 368, 377–380
　　Forest of, 136
Atlantic coast, 15, 18, 27–28, 29, 49, 56, 91, 95, 108, 139, 185, 280, 295, 307, 385

Atlantic oakwood, 221, 227, 382, 383
'Auchinleck Chronicle', 24, 25, 33, 37
Auldhame (East Lothian), 115–116
Ayr, river (Ayr), 74

Baden, Lewis (mine engineer), 242
Baillie of Dochfour family, 315
Bald, Robert, 345
Baldoon (Dumfries and Galloway), 123–124, 257
Baleshare (North Uist), 28, 185, 188
Balfour, Andrew, 200
Baltic (region), 36, 84, 125, 132, 281
Banff (Aberdeenshire), 148, 150, 212, 273
Banffshire, 176, 220
barilla (opposite-leaved saltwort), 354, 356, 357
bark, 61, 111, 125, 218, 221, 223, 224, 226, 229, 354, 366, 368, 377
　　peeling, 125
barley, 12, 36, 88, 177, 306, 309
barony courts, 63, 119, 147
Barra, island of, 18, 28, 139
barrier islands, 28, 95
Barrisdale (Highland), 279, 356
basketry, 138, 226
bass (fish), 283
Bass Rock (East Lothian), 115
beans, 36, 103, 187, 255
Beauly (Inverness-shire), 21, 315
　　Firth, 315
beech trees, 7, 9, 196, 202, 204–206, 208, 209, 218, 230, 372, 377
　　mast, 204, 205, 206, 208, 209
　　seedlings, 208
beef, 123, 125, 257, 261, 264, 330
beehives, 10, 110, 111, 116–117
bees, 10, 110, 116–117, 130
beeswax, 10, 111, 116
Benbecula, island of, 27, 28
Benshie, wood of (Ayr), 228
bere, 278, 279
Berry, Duc de, 46, 47
Berwickshire, 59, 180, 321, 396
Bilbao, 281, 282
birch trees, 127, 136, 138, 139, 140, 142, 186, 187, 196, 206, 210, 218, 229, 313, 367, 372, 381, 382, 390

411

birk/Birkin: *see* birch trees
Black Douglas family, 38, 67, 76
Blaeu brothers (Willem Janz and Jan), 135, 144, 146
Blair Atholl (Perth and Kinross), 136, 196
Blair Drummond (Stirling), 160, 313, 345, 365
Blairadam (Fife), 370, 371, 389
Board of Trustees for Fisheries, Manufactures and Improvements in Scotland, 280–281
Boddin (Angus), 267, 319
bog, 139, 164, 167, 275, 276, 315
    raised, 51, 365
Bonawe furnace, 223, 224, 246–248
Bonhill (Dunbartonshire), 260, 387
Bordeaux (France), 34
Borders, Scottish (district), 30, 31, 39, 49, 68, 123, 132, 161, 194, 237, 265, 304, 339, 372
    agriculture in, 30–31, 39, 123, 265–266
    fuel supplies in, 49, 237
    woodland in, 68, 132, 161, 372
Bornais (South Uist), 27, 49, 71
Borrowmoss (Dumfries and Galloway), 99, 105
Borve (Barra), 18
Boutcher, William (seedsman), 160, 208
Bowden Moor (Borders), 140
Bower, Walter, 11, 12, 13, 17, 20, 21, 22, 23, 24, 33, 34, 37, 47, 66, 67, 77, 85, 138
    Scotichronicon, 13, 33
Bowhill House (Scottish Borders), 377
Bowmont Valley (Scottish Borders), 30–31, 41, 51, 52, 56, 102, 315
Braan, river (Perth and Kinross), 169
Braemar (Aberdeenshire), 212, 214
Brahan (Highland), 315, 368
Breadalbane (district and estate), 30, 119, 122, 173, 179, 206, 208, 211, 223–224, 227, 229, 262, 263, 268, 279, 292, 294, 301, 319, 325, 326, 351, 368, 379, 389
    earls and marquesses of, 122, 162, 169–170, 173, 205, 206, 208, 220, 221, 223–224, 247, 249, 261–262, 366, 377, 379, 390; *see also* Campbell family, Campbell of Glenorchy
Brechin (Angus), 67, 148
Brims (Highland), 260
British Fisheries Association/Society, 281, 282, 358, 362
British Isles, 2, 5, 15, 19, 290, 311, 364, 389
    harvest failures in, 94
    weather patterns in, 15, 16, 19, 20, 87, 181, 185, 293, 307
Brodie, Alexander, of Brodie (diarist), 91
Brora (Highland), coal mine, 244
broom (*brume*), 43, 61, 62, 63, 65–66, 102, 129, 151, 196, 221
    fuel, 50, 65
    sowing of, 43, 61, 129
Buccleuch, earls and dukes of, 123, 132
    estate, 123, 316, 377, 380
Buchan (Aberdeenshire), 28, 76, 290, 326, 327, 353, 361, 365
Buchan, earl of: *see* Erskine, David

Buchanan (Stirling), 204, 208, 218, 388
Burke, Edmund, 165, 168
Burnet, George, of Kemnay, 186
Burnet, Janet, of Kemnay and Disblair, 186–187, 293, 341
Burntisland (Fife), 360
Burray (Orkney), 258
Burrell, Sir William, 104, 194, 196–197, 202, 206, 208, 209, 221, 252, 257, 266, 273, 276, 278, 389
Burt, Edmund, 164, 166
busses (fishing-boats), 72, 281, 359

Cadzow Park, 69
Cairngorm Mountains, 16, 85, 90, 91, 139, 142, 144, 177, 178, 184, 189, 211, 301, 304, 305, 361
    pine tree-ring sequences from, 16, 24, 91, 121, 172, 184, 212
    temperatures in, 24, 139, 188, 189
Caithness, 57, 71, 118, 176, 190, 256, 260, 296, 308, 313, 355, 358, 360–361, 396
    fishery, 358, 360–362
Caledon, Great Wood of, 135, 210
calf-slaughter, 55–56, 77, 125
Calluna heath, 51, 55
Cambuskenneth Abbey, 74
Campbell family, 72, 73, 212, 246, 278, 326, 349
    Archibald, earl of Ilay, 275, 276
    Argyll, earls and dukes of: *see main entry*
    Campbell of Airds, 276
    Campbell of Barcaldine, 247
    Campbell of Glenorchy, 122, 223, 250, 326; *see also* Breadalbane, earls and marquesses of
    Campbell of Knockbuy, 326
    Campbell of Lochnell, 223, 246, 250
    Campbell, Colin, 1st earl of Argyll, 73
    Campbell, Duncan, of Aros, 328
    Campbell, Duncan, of Glenure, 226
    Campbell, John, 3rd earl of Breadalbane, 169, 245–246
Campsie (Perth and Kinross), 63
Canada, 217, 370
Cantick Head (Orkney), 258, 260
Cara, Park of (Orkney), 258–260
Cardney (Perth and Kinross), 58
Caribbean plantations (*see also* enslaved people), 10, 116, 281, 364
    sugar, 116, 364
Carlingwark Loch, 270, 272
carr, 313, 372, 395
Carrick (district), 136, 196
Carron, river (Highland), 74
Carron, river (Stirling), 74
carse and carseland (*see also* Gowrie, Carse of), 38, 49, 58, 60, 87, 98, 103, 315
Castle Grant (Highland), 160, 177, 205, 209, 210, 265
Castle Lachlan (Argyll and Bute), 72–73
cattle (*see also* dairying, livestock), 21, 22, 40, 48, 52, 55, 56, 63, 65, 84, 103, 121, 122, 123–125, 164, 178, 181, 211, 221, 227, 234, 257–263, 264, 279, 292, 301, 303, 305, 306, 309, 320, 326–331, 332, 333, 334, 353, 390, 395, 396

enclosures or parks: *see* parks
cereals (*see also* barley, bere, oats, wheat), 40, 44, 49, 53, 56, 74, 84, 89, 103, 276, 278, 280, 301, 311, 312, 313, 316, 335
charcoal and charcoal manufacture, 8, 48, 126–128, 193, 212, 221, 223–224, 227, 229, 245–249, 250, 341, 377, 380, 382, 383, 386–388
Charles II, king of Scots, 102, 123, 133, 140, 157, 201
Charlestown, limekilns (Fife), 318–319
chestnut trees (horse), 202, 205
Cheviot Hills, 30, 40, 102
chroniclers and chronicles: *see* 'Auchinleck Chronicle'; Bower, *Scotichronicon*; Fordun, John of; *Fortingall, Chronicle of*; *Frasers, Chronicle of the*; Irish chroniclers, chronicles and annals; *Perth, Chronicle of*; Pitscottie; Walsingham
Cille Pheadair (South Uist), 27
Clackmannan, 173
    Forest of, 48, 67
Clackmannanshire, 48, 98, 162, 323
Clann Domhnuill, 76
Clanranald: *see* MacDonald of Clanranald
Clearances, 252, 396
    Highland and Island, 7, 324, 331–335, 338, 396
    Lowland, 252–254
clearance
    peat and wetland, 167, 277, 313, 343, 345, 351, 353, 365, 390
    stone, 326
    woodland, 7, 29–30, 70, 77, 136, 150, 193, 229, 331, 334, 368, 384
Clephane, Arthur (seedsman), 202, 205, 208, 260, 263, 264, 277
Clerk, John, of Penicuik, 180, 181, 257–258, 268
climate proxies (*see also* speleothems), 16, 18, 22, 24, 76, 85, 90, 172, 212, 392
clover (purple, red, yellow and white), 255, 260, 263–266, 276, 396
Clunie, laird of, 149
Clunie, Loch of (Perth and Kinross), 58
Clyde, river, 74, 136, 260, 275, 298, 300, 351
    Firth of, 73, 114, 115, 273, 281, 362, 363
    fishery, 281, 282, 363
    islands, 136, 139, 249
    valley, 136, 249
Clydesdale, 25, 242, 249
Cnip (Lewis), 18
coal and coal-mining, 46, 48, 49, 51, 147, 150, 154, 160, 162, 233, 234, 235, 237, 238, 241, 242–245, 268, 272, 289, 293, 306, 318–319, 340, 341–343, 344–353, 365, 377, 386, 387, 388–389
    coked, 249
    duty, 241, 243–244, 272, 319, 344–345, 347, 348
coastal erosion: *see under* erosion
Cockburn, John, of Ormiston, 182, 263
cod, 71–72, 93, 178, 281–283, 338, 358, 361
Coigach (Highland), 279, 356
Coleridge, Samuel Taylor, 381
Coll, island of, 227, 326, 349

INDEX

colonies and colonialism, 158, 165–166, 193, 252, 255, 281, 286, 287, 309, 321
Colquhoun family, of Luss, 160, 221, 388
common land, 102, 104, 105
   grazing: *see main entry*
   muir, 148, 238, 240, 256
commonties, 102, 233, 239, 252, 254, 255, 256–258, 260, 261, 337, 340, 390
   division of, 233, 254, 255–257, 340
Conon, river (Highland), 74, 315
coppice and coppicing (*see also* haggs), 9, 62, 66, 69–70, 77, 128, 129, 136, 138, 142, 198, 221, 223–224, 226, 227, 229, 247, 250, 375, 381, 382, 383, 386–387, 389
Cordiner, Charles, 212–214, 384
Corn Laws, 308, 311, 312, 316, 396
Corpach (Highland), 246
Cotyards (Perth and Kinross), 63
Coupar Angus Abbey (Perth and Kinross), 21, 38, 58, 63, 66, 118, 122, 123, 130, 140
Coxton (Moray), 242
Craigentinny Midlothian), 320
Crail (Fife), 71, 72, 359
Craleckan (Argyll and Bute), iron furnace, 246–247, 248
Crieff (Perth and Kinross), 196, 261, 349, 379
cruives, 54, 74
Culbin (Moray), 28, 97, 104, 106, 174
Cullen (Aberdeenshire), 220; *see also* Grant family, estates; Seafield estate
Cumberland, 244, 247
Cumming family, of Altyre, 209, 220, 304
Cupar (Fife), 181, 255
Cunninghame (district), 136

dairying, 55–56, 77, 88, 123, 125, 262–263, 299
Dalguise (Perth and Kinross), 58
Dalkeith Park (Midlothian), 69–70, 190
Dalrymple, Sir John, 345, 353
Dalvorar (Aberdeenshire), 144
Darby, Abraham, 249
Darnaway (Moray), castle and forest, 67, 218, 272
Darien Scheme, 175
David, duke of Rothesay, 33
dearth, 23, 24, 25, 36, 37, 38, 45, 76, 80, 82, 85, 88, 89, 93, 99, 102, 113, 114, 138, 175, 176, 186, 289, 291, 306, 307, 360, 395
Dee, river (Aberdeenshire), 74, 213, 214, 270, 301, 303
Dee, river (Dumfries and Galloway), 272
deer, 8, 21, 68, 110, 112, 113, 122, 133, 136, 144, 189, 213, 381
deer forest, 133, 138, 330, 331, 335, 338, 381, 395
Defoe, Daniel, 201
Dempster, George, of Dunnichen, 272
dendrochronology, 69, 178
Denmark, wood from, 133
depopulation, 254, 324, 331, 335
Description of the Western Isles of Scotland (Martin), 158
Description of the Western Isles of Scotland (Monro), 138
Deveron, river (Aberdeenshire), 74, 282, 303, 319
Dickson, Robert, of Hassendeanburn (seedsman and nurseryman), 218, 226

Dirleton (East Lothian), 97
Disblair House (Aberdeenshire), 186–187
disease (human), 2, 3, 11, 12, 31–35, 36, 42, 79, 93, 138, 320, 407
   'boch', 34
   cholera, 33, 34
   Covid-19, 31, 33
   dysentery, 33, 34
   epidemic, 2, 31–35, 42, 93, 108, 312
   'Great Mortality': *see* plague (bubonic)
   legislation: *see main entry*
   pestilence, 15, 33, 34
   plague (bubonic): *see main entry*
   potato: *see* potato late blight fungus
   'qwhew', 33–34
   smallpox, 34
   syphilis, 34–35
   Treponema pallidum, 34–35
   Variola, 34
   'wame ill', 33
disease (tree), 201
ditches and ditching, 63, 68, 276, 337
domestic livestock: *see* cattle; sheep
Don, river (Aberdeenshire), 74, 177, 303, 314
doocots, 117
Doon, river (Ayr), 74
Dornoch Firth, 181
Douglas, Archibald, of Douglas, 218, 226
Douglas, David, 368–370
Douglas Castle (Lanarkshire), 218, 226
Douglas fir, 369, 370
Doune (Stirling), 58, 140
   castle and park, 140–142
Doune of Rothiemurchus (Highland), 210
Drainie (Moray), 240
drains and draining, 30, 100–101, 148, 154, 163, 167, 194, 239, 240, 252, 255, 276–277, 309, 313–315, 316, 317, 323, 327, 337, 343, 351, 353, 365, 374, 390, 392, 395
Drave / Lammas Drave (herring fishery), 283, 359, 360, 361, 362
Dronner's Dyke: *see under* Montrose
drought: *see under* weather
drove-roads, 260, 262, 328, 330
drovers and droving (cattle), 124, 258, 260, 261–264, 305–306, 309, 326–330
droving stances, 262, 326, 328, 330apter
Drumlanrig Castle (Dumfries and Galloway), 196, 375, 377
Drummond Castle (Perth and Kinross), 196, 201
Drummond family, of Blair Drummond, 160, 313, 345
   Drummond, George, 345, 365
Dryburgh Abbey, 65
Drymen (Stirling), 204
Duff, William, Lord Braco, 242
Dumbarton (Dunbartonshire), 28, 73
Dunbar, Sir David, of Baldoon, 123
Dunbar, Sir George, of Mochrum, 275–276
Dunbar, Sir William, 256
Dunbartonshire, 125, 223, 254, 263, 328, 351, 387
Dundas family, 160
   Dundas, George of Dundas, 126
   Dundas, Henry, 379
Dundee, 29, 33, 48, 74, 76, 136, 150, 181, 267, 339, 345

dunes: *see* sand dunes
dung (fuel), 48, 49, 51, 234, 341
Dunkeld (Perth and Kinross), 58, 169, 206, 216, 224, 366, 378, 379
   bishops and bishopric, 57–58, 135
Dunninald (Angus), 267–268, 272, 319
Dunrobin Glen (Highland), 372
dykes, 63, 98, 99, 140, 151, 162, 227, 263, 305, 325, 326, 330, 337, 363
Dysart (Fife), 76

Eachkamish (Baleshare), 188
Eagle, Archibald (seedsman), 160
Earn, river, 29, 90, 181
East Lothian, 13, 28, 33, 96, 107, 173, 174, 199, 201, 206, 263, 355, 356, 359
Easter Ross, 67, 85, 188, 189, 261, 308, 313, 315, 316, 339, 355, 388
Edinburgh, 32, 34, 35, 48, 49, 58, 72, 80, 88, 90, 93, 103, 156, 158, 160, 179, 181, 182, 202, 208, 226, 260, 261, 265, 278, 293, 301, 320, 326, 339, 345, 356
   booksellers in, 160
   fuel use in, 48–49, 150, 235, 345, 351
   Physic Garden, 200, 201, 216
   Royal Society of, 158
   seedsmen in: *see* Boutcher, William; Clephane, Arthur; Eagle, Archibald
Eglinton (Ayr), 93, 196
Eildon Hills (Borders), 140
El Niño/La Niña, 25
Eldbotle (East Lothian), 28, 97
Elder, James, 265
Elgin (Moray), 32, 147, 150–152, 235, 240, 241–242, 275, 276, 319
   fuel use in, 147, 150–154, 235, 237, 240, 241–242, 352
Elgin, earls of, 318
   5th earl, 318
Elie (Fife), 97
elm trees, 103, 127, 186, 204, 205, 208, 209, 218, 228, 229, 230, 305, 372, 381, 382
   Dutch, 204, 205, 208, 230
   English, 208, 230
   wych, 204, 208
embankment, 290, 305, 313, 314–315, 316, 317, 326
emigration and emigrants, 288, 294–295, 309
enclosure, 63, 65, 66, 102, 109, 112, 113, 117, 125, 162, 163, 194, 218, 254, 255–256, 257, 258–261, 315, 325, 326, 328, 330, 337, 395
   1661 Act, 102, 123, 233, 255, 257
   cattle: *see under* parks
   deer: *see under* parks
   woodland (*see also* haining), 68–69, 112, 129, 130, 133, 136, 139–142, 197, 204, 221, 227–228, 229
England, 82, 157, 175, 193
   agricultural advances in, 157, 263
   cattle trade with, 124, 257, 261, 326, 328
   coal from: *see* coal and coal mining
   coke use in, 249
   epidemics in, 33–34

fish trade and consumption in, 71, 73, 74, 281, 359
   garden design in, 160
   ironmasters and tanners from, 124, 221, 229, 245
   meadows in, 58, 316
   merchants from, 73, 82, 215, 216
   mining and mining technology in, 242, 244
   trees and woodland in, 198–199, 202, 204–205, 208, 224, 230, 372, 374
   weather in, 15, 181, 183, 293, 296
Enlightenment, Age of, 155–156, 165, 337, 394, 395, 396
enslaved people, 10, 252, 286, 287, 322
Eoligarry (Barra), 28
erosion
   coastal, 26–28, 91, 94
   gulley, 30, 52
   soil, 28, 29–30, 52, 315, 332, 394
   water, 26, 28, 29, 94–99, 104, 106, 121, 154, 316
   weather, 26, 27, 28, 76, 151, 351, 352
   wind, 27, 28, 94–95, 106, 313, 351
Erroll (Perth and Kinross), 87
Erskine family, 48, 100
Erskine, David, 11th earl of Buchan, 155, 280
Esk, rivers (Angus), 74
   North, 74, 303
   South, 74, 100, 101, 303
Ettrick (Borders), 123, 377
Evelyn, John, 199, 201
Ewesdale (Dumfries and Galloway), 123, 377

Falkirk (Stirling), 261
Falkland (Fife), 33, 58, 133
   Wood of, 68–69, 132
famine, 2, 3, 11, 22, 25, 33, 36–42, 44, 76, 79–80, 84, 88, 91, 108, 174–175, 176, 182, 189, 191, 286, 287, 288, 307
   Highland, 287–288, 307, 309, 394
   potato: see potatoes
Farquharson, John, of Invercauld, 144, 211, 212, 220
Farrar, river, 74
fermtouns, multiple-tenancy, 122, 151, 252, 253, 257, 324, 337
fertiliser and fertiliser use (*see also* guano; lime and liming; manure; middens and midden waste; sea-ware and seaweed; turf; waste (land)), 103, 107, 270, 272–274, 275, 276, 317, 320, 321, 327, 354, 392
Fife, 33, 39, 49, 60, 68, 89, 91, 97, 102, 107, 116, 122, 132, 181, 202, 226, 256, 266, 267, 273, 278, 281, 283, 304, 345, 351, 359, 362
   earls and dukes of, 189, 211, 212, 220
Findhorn Bay (Moray), 97
Findhorn, river, 29, 74, 174, 272, 303, 304, 305, 315
Findhorn village (Moray), 174
Findrassie (Moray), 240
Finlarig Castle (Stirling), 208, 224

fir or 'fur' trees (*see also* Douglas fir), 9, 66, 138, 196, 202, 208, 209, 218, 223, 327, 366, 370, 378
fish: *see* bass; cod; haddock; herring; ling; salmon; *see also* fishing and fisheries
fishing and fisheries (*see also under* Forth; Iceland; Moray Firth), 5, 6, 7, 9, 14, 43, 71–76, 80, 90, 93, 101, 115, 158, 178, 183, 275, 280–284, 290, 291, 295–296, 307, 358–364, 392, 396
   Basque, 72
   boats, 73, 178, 281–283, 296, 359, 361–362; *see also* busses
   Breton, 72
   Dutch, 72, 73, 280, 281, 359
   English, 73, 359
   haaf (Orkney), 283
   lobster, 282, 283
   salmon: see main entry
   Scottish legislation, 71, 74
   Shetland, 282–283
   west-coast, 72–74
   Western Isles, 280–284
fish-traps, 74
Flanders, 74
Flanders Moss (Stirling), 167, 252, 266, 365
Fletcher, Andrew, of Saltoun, 93
floods and flooding (*see also under* Tay, river; *see also* Findhorn, river; Forth, river; Spey, river), 13, 15, 25, 80, 86, 87, 89, 90, 98, 103, 174, 178, 180, 290, 293, 297, 302, 303–305, 314, 315, 353, 395
   sea, 27, 91, 94, 97, 173, 174
   spate, 19, 29, 30, 80, 86, 90, 98, 173, 174, 177, 180, 185, 290, 298, 301, 302–304, 305
Fochabers (Moray), 190, 208
fodder, 12, 55–57, 59, 65, 68, 77, 93, 106, 112, 121, 255, 263, 266, 277–278, 279, 288, 290, 292, 296, 300, 305, 306, 309, 313, 316, 320, 330, 337, 340, 353, 367
foggage, 56, 57, 58, 59, 121, 123, 306
Foljambe, Joseph, 322
food security, 4, 103, 110, 161
food shortages, 12, 14, 23, 25, 36, 45, 80, 85, 88, 93, 103, 112, 113, 114, 175, 181, 182, 188, 288
Fordun, John of (chronicler), 138
Fordyce, Thomas (glass-maker), 355
forest (hunting and deer) (*see also under* Arran; Atholl; Clackmannan; *see also* Darnaway; Mamlorn; Mar), 48, 67, 112, 122, 136, 138, 144–145, 211, 214, 261, 331, 334, 335, 338, 381, 395
   free forest jurisdiction, 62, 112, 261
Forester's Guide and Profitable Planter: *see* Monteath, Sir Robert
Forfeited Estates, 186, 189, 245, 268, 355
Forres (Moray), 209, 272, 305
Forth
   carse, 58, 98, 103, 315
   estuary, 49, 98, 115, 233, 235, 242, 243, 275, 297
   Firth of, 6, 11, 13, 35, 89, 90, 107, 114, 147, 150, 181, 267, 281, 282, 283, 304, 315, 358, 359

fishermen and fishery, 115, 178, 281–282, 283, 358–362
   ports, 72, 178, 235, 275
   river, 74, 80, 98, 173, 181, 238, 245, 277, 344, 345
   salt and salt-making, 49
   Valley, 136, 243, 319
Forth and Clyde canal, 297, 315, 360
Fortingall, Chronicle of, 85, 86
Forvie (Aberdeenshire), 28
foxes, 21, 118
France, 15, 34, 243, 281, 352, 356, 377
   fish trade to, 281
Fraser Darling, Frank, 8, 193, 246, 331–334, 337
Frasers, Chronicle of the, 88
Freuchie (Highland), 139, 142, 210, 390; *see also* Castle Grant
frost: *see* weather
'frost fairs', 12, 87
fuel (*see also* charcoal and charcoal manufacture; coal and coal mining; dung; peats (fuel); seaweed (fuel); smallwood; wood (fuel); *see also under* broom; heather; turf), 44–51, 65, 80, 99, 105, 106, 108, 111, 127, 147–154, 181, 221, 233–250, 255, 268, 272, 276, 287, 289, 291, 297, 300, 306, 308, 309, 313, 315, 318, 319, 325, 327, 337, 338, 339, 340–353, 354, 356, 365, 380, 381, 386, 394
   poverty, 147, 149, 233, 235, 351
   supply crisis, 44, 51, 146, 233, 235, 289, 341, 351, 365
furnaces, 127, 223, 224, 243, 246–248, 249, 383, 386
   Argyll Furnace Company, 383
   Bonawe, 223, 224, 248
   Craleckan, 247, 248
   Furnace, Argyll, 386
   Glen Kinglass, 223
   Invergarry, 246
   Loch Maree, 127
   Netherhall, 247
Fyffe, James, 305–306
Fyvie (Aberdeenshire), 298, 300

Gaelic language and literature, 163, 167, 262
Galbraith: *see* Murphy and Galbraith
gales: *see* weather, wind and gales
Galloway, 76, 123–124, 147, 194, 196, 206, 233, 235, 237, 240, 243, 244, 245, 253, 256, 257–258, 260, 272, 273, 290, 301, 316, 324, 330, 339, 352, 356, 365
   earl of, 258–259
game (hunting), 14, 110, 111–114, 180, 189, 290, 367, 392
gannets (solan geese), 113, 114–116
Garlies (Dumfries and Galloway), 259
Garmouth (Moray), 184
Garry (Highland), river and glen, 246
Garry (Perth and Kinross), river and glen, 29, 136
gean/geanies (wild cherry), 186, 210, 381, 390
geese, 113, 114
Gilpin, Rev. William, 374

INDEX

Girvan Valley (Ayr), 136, 196
glaciers, 1, 5, 8
Glamis, castle and estate (Angus), 226, 267
Glasgow, 181, 221, 251, 278, 298, 339, 345, 364, 388
    flooding, 298
    processing and industrialisation in, 116, 223, 364, 388
Glen Affric (Highland), 335
Glen Almond (Perth and Kinross), 122
Glen App (Ayrshire), 196
Glen Brighty (Angus), 122, 123
Glen Derry (Aberdeenshire), 211, 214
Glen Dochart (Stirling), 292, 351
Glen Falloch (Dunbartonshire), 388
Glen Garry (Highland), 246
Glen Garry (Perth and Kinross), 136
Glen Goulandie (Perth and Kinross), 162
Glen Kinglass (Argyll and Bute), 223, 246
    furnace, 223–224, 246
Glen Lochay (Stirling), 293
Glen Lui (Aberdeenshire), 211, 214, 384
Glen Lyon (Perth and Kinross), 135, 301
Glen Noe (Argyll and Bute), 221
Glen Quaich (Perth and Kinross), 294
Glen Quoich (Aberdeenshire), 213, 214, 384
Glen Shee (Perth and Kinross), 122
Glencoe (Highland), 91
    Massacre, 91–93
Glenisla (Angus) (*see also* Isla, river), 21, 58, 118, 122, 123, 138, 149
Glenluce (Dumfries and Galloway), 243
Glenmore (Highland), 146, 189, 200
Glenorchy, lord (*see also* Breadalbane, earls and marquesses of; Campbell family, Campbell of Glenorchy), 122, 139, 223, 261
    estate, 212, 221, 250, 326
Glenorchy (district), 326
    oakwoods, 212, 223–224
    pinewoods, 212
Glenorchy Firwood Company, 212
Gometra, island of, 227
'Gonial Blast', 293
Gordon, Alexander, of Culvennan, 272
Gordon, James, of Rothiemay, 135
Gordon, Robert, of Gordonstoun, 152
Gordon, Robert, of Straloch, 129, 135, 138
Gordon, Thomas, 313–314
Gordon estate, 244, 319
Gordon family, earls of Aberdeen, 327
Gordon family, earls and marquises of Huntly and dukes of Gordon, 142, 146, 160, 189, 205, 220–221, 304, 312, 377
    George, 1st duke of Gordon, 208, 265
Gordonstoun (Moray), 152, 240
Gough Map, 21
Gowrie, Carse of, 38, 60, 87
Graham estate, 202–204
Graham family (*see also* Montrose, dukes of), 202
    Graham, Hugh, of Arngomery, 167, 252
    Graham, James, 1st duke of Montrose, 202
    Graham, Mungo, of Gorthie, 202
    Graham of Orchill, 160
Grangemouth (Stirling), 98, 318

Grant family, of Freuchie and Castle Grant, earls of Seafield, 139, 142, 144, 146, 160, 209, 210, 220–221, 265, 368, 390
    estates, 184, 205, 209, 220–221, 381
    Grant, Sir Francis, 220–221
    Grant, Sir James, 244
Grant family, of Monymusk, 160, 215, 220
    Grant, Sir Archibald, 162, 177, 205, 215, 218, 268, 274, 278–279
Grant family, of Rothiemurchus, 215, 377
grass and grassland, 40, 43, 51, 52, 53, 55–57, 79, 82, 85, 86, 89, 93, 121, 125, 142, 151, 163, 174, 182, 183, 186–187, 229, 240, 255, 257, 258, 261, 263–266, 276, 301, 306, 308, 320, 322, 328, 371, 396
    salt, 98
grazing, 8, 14, 30, 39, 51, 52, 55, 56, 60, 63, 77, 84, 95, 99, 102, 105, 106, 111, 121–122, 123, 124, 125, 132, 140, 154, 181, 204, 227, 254, 257, 258–260, 261, 262, 292, 316, 326, 328, 330, 331–335, 381, 383, 390
    coastal or links, 95, 98, 103, 125
    common, 55, 77, 121–122, 255, 261, 337
    over-grazing and grazing pressure, 77, 95, 121, 136, 211, 263, 292, 331, 340, 385
    summer, 56, 95, 121, 125, 133, 261, 263, 325
    upland, 40, 52–53, 77, 82, 84, 125, 261, 263, 325
    winter, 15, 300, 306
    woodland, 63, 102, 122, 129, 211, 213, 214, 221, 227, 230; *see also under* pasture and grazing
Great Exhibition, 393
Great Michael (warship), 132
'Great Mortality': *see* plague (bubonic)
'Great Spate' or 'Great Flood', 302–304, 305
Greenland, 16, 392, 394
Great Yarmouth (Norfolk), 33
Greenock (Renfrewshire), 181, 275, 362
grouse, 113, 330
guano, 270, 321–322, 327
'guld': *see* marigold
gunpowder manufacture, 377, 386, 388
    mills, 386–387

HMS *Athole*, 379
Haddington (East Lothian), 13, 34
Haddo House (Aberdeenshire), 327
haddock, 283, 284, 338, 358–359
haggs (*see also* coppice and coppicing), 69, 223, 224, 226, 228–229, 250, 381
hail, 178, 183
haining, 63, 129, 132, 139, 140, 226, 228
Hamburg, 282
'Hamburg' barrels, 75–76
Hamilton family, 69, 114, 159
Hamilton (Lanarkshire), 298
    Palace, 196, 201
Hamilton-Dalrymple of North Berwick, 160, 208
hares, 112, 303
Harris, island of, 138, 139, 309
Harrows, 104, 136, 326

harvests, 5, 12, 15, 56, 57, 74, 86, 229, 305, 306, 309, 320
    abundant/successful, 19, 26, 36, 89, 91, 182, 191, 299, 300, 301, 306, 308
    failure, 5, 12, 13, 20, 24–25, 26, 36, 37, 38, 44–45, 79, 81–82, 89, 91, 93–94, 97, 172, 175–176, 279, 293, 306, 309
    poor, 20, 25, 85, 86, 87, 89, 174, 175–176, 180–181, 183, 184, 186, 188, 296, 298
    timing of, 12, 38, 54, 55, 85, 89, 91, 123, 172, 173, 187, 296, 299, 301, 306
    yields, 12, 28, 36, 38–39, 40, 62, 77, 81, 84, 90, 99, 102, 104, 108, 109, 113, 161, 252, 254, 264, 266, 294, 296, 299, 301, 316, 322, 377
Hassendeanburn (Peebles-shire), nursery, 218, 226
hawks, 114
hawthorn, trees and hedges, 63, 186, 187, 204, 361, 382
hay, 57–59, 60, 93, 163, 186, 296, 301
Hay family, earls and marquises of Tweeddale
    4th marquis, 208
    Hay, John, 2nd earl, 140, 199, 201, 206, 375
Hay, Sir George, 127–128, 129, 193
heather (*hathir*), 151, 214, 275
    fuel, 49, 50, 51
Hebrides (Inner and Outer), 48, 56, 71, 76, 114, 138, 139, 158, 257, 264, 279, 288, 289, 308, 323, 324, 328, 330, 349, 356; *see also named islands*
hedges, 43, 61, 62, 63, 129, 130, 140, 206, 296, 330, 351
Heisgeir, island of, 28
Hekla (volcano, Iceland), 93
herring, 71–74, 77, 93, 114, 178, 281–284, 290, 307, 358–364, 366, 363
    Drave: *see main entry*
    'winter herrin', 359, 360, 361
hoarding (food), 5, 12, 25, 82
holly trees, 127, 208, 367
Holyroodhouse, Palace of, 133, 200
Henry, Home: *see* Kames, Lord
honey, 10, 110, 111, 116–117
Honourable Society of Improvers in the Knowledge of Agriculture in Scotland, 159, 161, 162, 242, 260, 267–268, 273, 275
    Transactions, 268, 273, 275
hornbeam trees, 205, 208, 230
horse-stud, 21
horses, 57, 118, 150, 152, 171, 226, 300, 313, 320, 322, 323
hounds (wolf), 21, 118
Houston and Killallan (Renfrewshire), 351
Huizinga, Johan, 2, 3
Hume, David, 156
hunting, 113, 114, 118, 138, 189
    deer, 213
    hunting rights, 14, 112, 114, 134
    illicit and poaching, 14, 110, 112, 113, 114
    land, 112, 113, 138, 390; *see also* forest laws, 112
    over-hunting, 113, 114
    wolf, 21, 118, 120

415

Huntly, earl of: *see* Gordon family, earls and marquises of Huntly

ice: *see under* weather
Iceland, 34, 71, 93, 178, 190, 292, 293, 359
    epidemics in, 34
    fishery, 178, 359
    volcanoes in: *see* Hekla; Laki eruption
Improvement, 3, 10, 57, 66, 80, 103, 111, 125, 154, 155–170, 171, 186, 191, 192, 197, 206, 215, 218, 226, 228, 230, 233, 234, 240, 242, 243, 244, 251, 252–254, 258, 260, 263–264, 266, 268, 269, 270, 272, 274–278, 280, 284–285, 287, 290, 292, 304, 309, 310–311, 313, 314–317, 324, 326, 327, 329–330, 331, 337, 338, 339–340, 344, 347, 349, 356, 364, 365, 368, 371, 373–374, 389, 390, 393, 394, 395, 396
    agricultural, 103, 157, 160, 234, 240, 252–254, 269, 272, 275, 280, 324, 337, 368
    literature, 234, 243, 317
Inchaffray (Perth and Kinross), 103
Inchcolm Abbey (Fife), 11, 13
Inchkeith, island, 35
infield, 103, 266
Ingliston, cattle park (Midlothian), 260, 263, 326
Innes, George, of Stow, 218, 226
Innes, Gilbert, of Stow, 266
Inveraray (Argyll and Bute), 73, 212
Invergarry (Highland), 210, 246
Inverness, 21, 67, 118, 148, 194, 238, 241, 257, 282, 381
Inverness-shire, 181, 302, 315, 351
Irish chroniclers, chronicles and annals, 15
    Annals of Connacht, 15
    Annals of Loch Cé, 15
    Annals of Ulster, 17, 21, 23
    Mac Carthaigh's Book, 15
iron manufacture and ironmasters (*see also* furnaces; Hay, Sir George), 8, 127, 193, 221, 223–224, 229, 231, 243, 245–247, 248, 249, 332, 353, 382, 386–388
Irvine (North Ayrshire), 32
Isidore of Seville, 161, 162, 164
Isla, river (*see also* Glenisla), 29
Islay, island of, 123, 138, 176, 307

Jacobites, Jacobitism and Jacobite risings, 2, 155, 156, 234, 268, 280, 355
James, duke of Albany, 157, 200
James I, king of Scots, 24, 36, 55, 66–67, 76, 126
    legislation, 36–37, 55, 61, 125–126
James II, king of Scots, 67, 68, 76
James III, king of Scots, 59, 68, 76
James IV, king of Scots, 80, 126, 129, 132
James V, king of Scots, 80, 82, 84, 117, 129, 130, 132, 140, 368
James VI, king of Scots, 80, 113, 140, 170, 245, 358
James VII, king of Scots: *see* James, duke of Albany
Johnson, Samuel, 166, 170, 194, 196–198
juniper, 381–382

kailyards, 104, 278, 279
Kames, Lord (Henry Home), 252, 277–278, 345, 365
kelp and kelping, 295, 309, 338, 339, 350, 353–358, 362, 396
Kelso Abbey (Scottish Borders), 50
Kelton (Dumfries and Galloway), 270
Kennedy family, lords of Dunure and earls of Cassillis, 115
Ker, Walter, of Caverton, 50
Kerr, William, 3rd earl of Lothian, 98
Kildonan (Highland), 334
Kilkerran (Ayrshire), 388
Killin (Stirling), 179, 351
kilns, 82
    corn-drying, 42, 44, 82, 174
    lime, 162, 242, 268, 272, 275, 318–319, 327
Kincardine, Woods of (Perth and Kinross), 226
Kincardineshire, 178, 181, 283, 329
King's Lynn (Norfolk), 33
Kinlochmoidart (Highland), woodland, 246, 382
Kinnoull (Perth and Kinross), 58
Kinross-shire, 122, 371
Kintyre (Argyll and Bute), 248
Kirk Yetholm (Borders), 256
Kirkcaldy (Fife), 90
Kirkcudbright (Dumfries and Galloway), 258, 270
Kirkintilloch (Dunbartonshire), 150
Kirkmaiden (Dumfries and Galloway), 243
Kirkwall (Orkney), 282
Kirriemuir (Angus), 138, 320

Laich/Laigh of Moray: *see under* Moray
Lairg (Highland), 334
Laki eruption (Iceland), 189–191, 288–289
Lammas Drave: *see* Drave
Lamont, John, of Newton (diarist), 88–90
landscape aesthetics and philosophies, 163–164, 166, 169–170, 194, 209, 370–377, 378
Lanarkshire, 39, 85, 125, 201, 218, 254, 256, 316
'larch autarky', 217
larch trees, 186, 187, 197, 200, 209, 216–218, 220, 226, 230, 366, 368, 372, 377–380, 388–389, 396
    seed and seedlings, 209, 220, 372, 378, 379
Largo Bay (Fife), 97
Latheron (Highland), 361
Lauder (Scottish Borders), 105
Lauder, Sir Robert, of the Bass, 115–116
leather and leather-processing, 193, 212, 263, 354
Leblanc, Nicolas, 357
legislation, 14, 21, 24, 25, 26, 31–32, 36–39, 41, 42, 43, 44, 53–55, 61–63, 66, 68, 72, 74, 82, 102, 104, 106, 108, 111, 112, 113–116, 117, 118, 125–126, 129–130, 133, 140, 154, 198, 230, 255, 257, 260, 311
    agriculture and crop-sowing, 14, 36, 157, 255, 257, 260
    fishing, 72, 74
    food-stuffs, 25, 26, 36, 82, 102

'guld law', 38
hunting, 113–116
muirburn control, 41, 53–55
Passenger Vessels Act 1803, 294–295
plague control, 1456 Act, 31–32, 35
planting and enclosing land, 1661 Act, 102–103, 123, 133, 140
tree-planting, 14, 24, 43, 61, 63, 111, 125–126, 157, 198, 230: 1428 Act, 126; 1458 Act, 24, 61–63, 66, 68, 126, 129–130; 1535 Act, 82, 129–130, 140; 1607 Act, 130, 140; 1661 Act, 133
turf-stripping, 104, 106
wolves: 1428 Act, 21, 24, 118; 1458 Act, 21, 118
legumes, 109, 254, 263, 265, 277
Leith (Midlothian), 66, 90, 176
Levellers and Levellers' Revolt (Galloway), 258
Leven, river (Dumbartonshire), 74
Lewis, island of, 18, 139, 157, 242, 261, 281, 350, 363
    fishery, 281, 363
lime and liming, 103, 108, 109, 162, 242, 252, 266–269, 275, 317, 318–320, 327, 337, 392, 395
    building, 242, 266, 268, 318, 319
lime trees, 202, 205, 208
limekilns, 160, 162, 268, 272, 318
limestone, 162, 172, 233, 266–269, 272, 317, 318, 319
    quarries, 268, 272, 318
Lindsay, Sir Robert, 132
ling, 283, 338
Linlithgow (West Lothian), 66, 67, 80, 133, 229
livestock (*see also* cattle, sheep), 43, 51–60, 77, 82, 84, 93, 118, 121–125, 132, 133, 154, 174, 252, 254, 255, 258, 262, 265, 270, 278, 290, 291, 300, 305, 309, 320, 326, 327, 330, 331, 337, 367, 377, 392
    mortality, 42, 86, 118, 120, 296, 300
lobster, 282, 283
Loch A'an, 144, 146
Loch Aline, 383
Loch Alsh, 127
Loch Arkaig, 215
Loch Awe, 179, 221
Loch Builg, 144, 146
Loch Carron, 127
Loch Duich, 127
Loch Eil, 246
Loch Erisort, 139
Loch Etive, 223, 224, 246
Loch Fyne, 72, 73, 246, 363–364
Loch Goil, 363
Loch Lang (South Uist), 95
Loch Leven Castle (Perth and Kinross), 372
Loch Lomond, 167, 202, 221, 264, 388
Loch Long, 363
Loch Maree, 127, 128, 129, 193
Loch Moy, 315
Loch na Sealga, 212
Loch Ness, 21
Loch Ordie, 379
Loch Rannoch, 135
Loch Stornoway, 139

INDEX

Loch Sunart, 16, 24, 227, 382
Loch Tay, 170, 208, 224, 294, 319
Loch, David, 280
Loch, James, 372
Lochaber (district), 210, 215, 221
Lochar Moss (Dumfries and Galloway), 276–277
Lochbay, Skye (Highland), 358, 362
Lochbroom (Highland), kelping, 356
Lochnell (Argyll and Bute), 206, 223, 250
London, 34, 87, 160, 282, 359, 379, 388
    epidemics in, 34
Lords Auditors, court of, 39, 59
Lorn (Argyll and Bute), 273
Lossie, river, 147, 303
Loth (Highland), 361
Lothian, 20, 39, 44, 48, 68, 89, 90, 91, 93, 103, 116, 124, 172, 180, 199, 201, 202, 254, 263, 264, 266, 281, 283, 293, 345, 351, 362, 387
    earl of, 98
    East: see main entry
    West: see main entry
Loudon, John, 368
Lovat, Lord, 238
lucerne, 255, 263, 265
Lunan Water, 100, 319
Lyon, Margaret, Lady Lovat, 21, 148

machair, 18, 27, 28, 56, 95–96, 98, 185
Machars (Dumfries and Galloway), 123, 240, 244, 257, 324, 352
MacDonald of Clanranald, 246, 247
MacDonald of Kenknock, 261
MacDonald of Sleat, 158
MacGeorge, William, 351
McHaffie, James, 324
Macintyre, Duncan Ban, 262
Mackenzie family, 261, 350
    earl of Seaforth, 127, 261
    Mackenzie, George, 93
    Mackenzie, Sir George, of Rosehaugh, 201
    Mackenzie, John, 350
Mackenzie/Seaforth estate, 127–128, 312–313, 315, 350, 388
Mackie, Adam, 298–301, 305
Mackintosh of Mackintosh family, 315, 381
Maclaine of Lochbuie family, 247
Maclean estate (Coll), 326
MacLeod family, of Dunvegan, 158
Macpherson, James, 167, 169
Mamlorn, Forest of, 261
Man, Isle of, 244
Mansfield, earl of, 370
manure (see also fertiliser and fertiliser use), 104, 108, 274, 275, 305, 317, 320, 321, 338
    sheep, 320
manuring practice, 105, 163, 194, 255, 274, 278, 319, 320, 337
Mar, Forest of, 144–145, 211, 214, 261
marigold, corn or field (*Glebionis segetum* or 'guld'), 38
marl and marling, 266, 269–270, 272, 319
marram grass, 98, 151
Martin, Martin, 139, 158, 159, 160, 162, 163, 170, 227, 272, 280
    'Brief Account of the Advantages the Isles Afford by Sea and Land and Particularly for a Fishing Trade', 158, 162
    Description of the Western Isles of Scotland, 158
Marwick (Orkney), 104, 105
Mary, Queen of Scots, 80, 114, 133
Maxwell, Robert, of Arkland, 160, 275, 320
meadows, 57, 58–59, 68, 106, 275, 276, 316, 320
Mearns, 319, 359
Melrose, 132
    abbey, 65, 123, 140
Melville estate (Fife), 226
Menstrie Glen (Clackmannanshire), 162, 323
Menzies family, 261
Menzies, Sir Robert, of Menzies, 216
Mercer family, of Aldie, 227
Mhic Lachlainn family, 72, 73
Mid Calder (West Lothian), 353
middens and midden waste, 104, 272, 274, 275, 320
Middle Lix (Stirling), 351
milk (see also dairying), 55, 56, 77, 264, 300
mills and milling, 13, 23, 88, 181, 268, 287, 303, 305, 387
    (gun)powder mills, 386
Minch, the, 281, 363
Moidart (Highland), 248, 382, 383, 384
Monadhliath, mountains, 303, 305
Monro, Donald, archdeacon of the Isles, 138, 139
Monteath, Sir Robert, 373, 374, 389, 390
Montgomerie family, of Eglinton, 93, 196
Montreathmont, Muir of (Angus), 148, 240, 241
Montrose (Angus), 67, 100–101, 241, 315
    Basin, 100–101, 315
    Dronner's Dyke, 100–101
Montrose, dukes of (see also Graham family), 202, 208, 221, 226, 388
    estate, 180, 204, 218, 226, 388
Monymusk (Aberdeenshire), 187, 205, 215–216, 220, 268; see also Grant family, of Monymusk
Moodie, James, 258–260
Moorfoot Hills, 296, 345
Moray, 28, 67, 76, 91, 97, 98, 104, 106, 119, 147, 151, 174, 178, 184, 190, 208, 220, 235, 240, 253, 281, 282, 296, 302, 303, 304–305, 308, 312, 313, 315, 329, 361
    'Great Spate': see main entry
    Laich/Laigh of, 98, 240, 313, 315
Moray Firth, 67, 76, 178, 281, 296, 303, 305, 361
    fishery, 281–282, 361
Morayshire, 80
Morvern (Highland), 273
moss and mossland (see also peat; peat mosses and muirs), 47, 48–49, 51, 108, 111, 147, 148–154, 167, 214, 233, 234, 235, 237–240, 242, 245, 249, 250, 252, 266, 270, 275–277, 290, 301, 313, 327, 340, 343–345, 347, 349–353, 365, 373, 386, 390, 395
    lowland, 327
    upland, 275
Mount St Helens (volcano), 26
Mount Westdahl (volcano), 293
Moy (Highland), 315
Moy (Moray), 305
muir and muirland, 14, 31, 54, 102, 147, 148, 149, 237, 238, 240, 241, 256, 278, 345
muirburn, 53–55, 334
Mull, island of, 138, 227, 247, 273, 328, 350, 355, 358, 382, 383
Mull, Ross of, 350
'mulones' (cod?), 71
Murphy, Roger, and Galbraith, Arthur, 223–224, 229, 246
Murray family
    earls, marquises and dukes of Atholl, 122, 159, 215–217, 220, 238, 377, 380, 390
    Murray, James, 2nd duke of Atholl, 215, 216
    Murray, John, 3rd duke of Atholl, 169–170, 215
    Murray, John, 4th duke of Atholl, 215, 216, 218, 366, 368, 377–379, 390, 396
    Murray, Sir Patrick, 122
Murthly Castle, 370

NAO (North Atlantic Oscillation), 16, 19, 20, 24, 25, 29, 81, 85, 90, 91, 93, 172, 176, 177, 178, 180, 181, 183, 188, 289, 290, 307
Nairn (Highland), 105
Nairne family and estate, 206
Napoleonic Wars, 356, 377
Ness, river, 74
Netherhall furnace (Cumbria), 247
Nethy, river, 305
New Deer (Aberdeenshire), 327, 353, 365
Newburgh (Fife), 74, 297
Newhaven (Midlothian), 275
Newmilns (Ayrshire), 388
Newton Stewart (Dumfries and Galloway), 259
Nicoll, John (diarist), 88–91
Niddry Castle (West Lothian), 201
Nithsdale (Dumfries and Galloway), 136, 195, 377
nitrogen, 273, 321, 322
nitrogen-fixing, 103, 109, 254, 263, 266, 273
North America, 165, 200, 294, 370, 379, 394
North Berwick (East Lothian), 71, 116, 160, 208
North Sea coast, 15, 16, 28, 108
Northern Isles (see also Orkney Islands, Shetland Islands), 15, 57, 304
    hay production in, 57
    fishery, 71, 282, 283, 305
    peat use in, 48
Northumberland, 181, 183
    coalfield, 48, 243, 345
Norway, 93
    fishery, 72, 359
    spruce, 231

oak trees, 8, 67, 69, 103, 127–128, 132–133, 136, 196, 205, 208, 212, 218, 221–224, 226, 228–229, 230, 250, 354, 367, 372, 373, 374, 377, 379, 381, 382, 385, 386, 388, 389, 390

417

acorns, 204, 205, 209, 372
Atlantic oakwoods: see main entry
bark, 111, 125, 218, 221, 223–224, 226, 229, 354, 366, 368, 377
English, 204–205, 224, 374
European, 224
Irish, 372, 374
pedunculate, 205, 224, 230, 374, 377, 385
sessile, 205, 221, 224, 374, 381
see also timber
oats, 12, 36, 42, 44, 57, 86, 88, 186–187, 278, 279, 309, 312, 313
black, 276
Ochil Hills, 162, 175, 323
Ogilvy, James, of Clunie, 149
Ogilvy family, of Airlie, 148
'On Planting Waste Lands' (Walter Scott), 373
orchards, 63, 117, 129, 139, 140, 163, 205
Orkney Islands (see also Northern Isles), 26, 27, 104, 258, 264, 269, 279, 296, 321, 327, 355, 356
coastal and wind erosion in, 26–27, 28, 80, 94, 104
commercial grazing in, 258–260
fishing, 281, 282, 283
kelp manufacture in, 355, 356
sea-ware use in: see sea-ware and seaweed
turf-stripping in, 104, 106
Orrin, river, 315
osier (tree), 63, 130
Ossian, 167, 169
Ossian's Hall and the Hermitage (Perth and Kinross), 169
Ouchterlony, John, of Guynd, 136, 139
Outer Hebrides: see Western Isles
outfield, 40, 41, 65, 77, 103, 255, 265, 266, 321

Pacific Ocean, 25, 26
Paisley (Renfrewshire), 278, 351
Papa Westray (Orkney), 95
Panmure (Angus)
castle, 202
earls of, 202
parks (enclosures), 63, 117, 130, 132, 136, 186, 187, 395
cattle, 103, 113, 123–124, 125, 257–260, 261, 263, 326, 328
deer, 68, 133, 140, 142
woodland, 68–69, 132, 136–137, 150, 221; see also main entry
partridge, 113, 114
pasture and grazing, 6, 14, 40, 51, 52, 68, 84, 103, 104, 106, 121, 123, 162, 164, 182, 211, 221, 239, 254, 255, 256, 258, 261, 262, 263, 270, 276, 290, 311, 314, 315, 326, 328, 330, 340, 365, 390, 395
common, 122, 263
extent, 103, 104, 121, 123
illicit, 14, 63
summer, 121, 122, 123, 263
winter, 121, 123, 301, 306
wood, 211, 213, 215, 220, 229, 230, 261, 331
peas, 36, 103, 163, 186–187, 255

peat, 8, 33, 47, 48, 51, 66, 106, 111, 139, 147, 149, 151, 167, 214, 233, 237, 242–243, 244, 255, 276, 333, 335, 351, 352, 365
peats (fuel), 14, 44, 47, 48–51, 99, 111, 146–154, 162, 234, 235–243, 244, 249, 276, 289, 300, 306, 308, 315, 335, 340, 341–345, 347–350, 352–353, 364, 365, 381
common sources, 147, 148–151, 154, 235, 237–238, 240, 249, 328, 341, 344, 347–348, 353, 365
depletion, 48–50, 148, 233, 244, 341, 344, 348–349, 351, 352
supply conflicts, 148–149, 151–152, 238–242
thermal efficiency, 48, 243, 341
peat mosses and muirs, 14, 33, 48, 49, 51, 99, 108, 139, 147, 148–152, 154, 235, 238, 240–242, 250, 276, 313, 315, 328, 343–345, 347–353, 364–365, 390
scalped, stripped or flayed, 51, 105, 235, 238–240, 244–245, 249, 276–277, 327, 341, 343, 348, 351, 365
waterlogged, 139, 149, 154, 214, 218, 249, 314, 352
Peebles, 32, 375
Peeblesshire, 296, 316
Pennant, Thomas, 167, 197
Perth, 24, 38, 66, 67, 71, 82, 135, 148, 206, 221, 226, 256, 275, 370
bridge, 86, 206
Dominican friary, 51, 60
epidemics in, 34
fish trade in, 71, 76
fuel use in, 48, 49, 51, 148, 150, 235, 244
port, 29–30, 221, 226
Tay freezing at, 87, 88, 181, 297
Perth, Chronicle of, 86, 98
Perthshire, 80, 85, 118, 135, 136, 138, 148, 167, 173, 197, 200, 201, 202, 206, 215, 226, 261, 268, 276, 293, 301, 328, 351, 366, 388, 396
pestilence, 15, 33, 34
phosphates, 108, 273, 321, 322, 395
Physic Garden: see under Edinburgh
Phytophthora infestans: see potato late blight fungus
pinetums, 370
pinewoods: see under woodland
Pitgaveny (Moray), 240
Pitscottie (Fife), Sir Robert Lindsay of, 132
Pittenweem (Fife), 71, 105
plague (bubonic), 11, 31–32, 33–34, 35, 42, 70, 81
plane trees, 103, 186, 187, 229
plantation: see under woodland
ploughs, 136, 313, 322
'Rotherham', 322
'Scots', 322
Pluscarden (Moray), 242
poaching: see under hunting
'policy' and 'politeness', 66, 80, 125, 129, 130, 132, 157, 164, 234, 289, 368, 379
Pont, Timothy, 126–128, 129, 135–136, 138, 144, 146, 212
poplar trees, 136, 218, 226, 377
Port Seton (East Lothian), glass-works, 355
Portsoy (Aberdeenshire), 176

potassium, 108, 273, 321, 322, 395
potato late blight fungus (*Phytophthora infestans*), 9, 280, 287, 306–309, 322, 362
potatoes, 279–280, 284, 306–309, 362
prices, 36–37, 38, 42, 82, 85, 89, 113, 306, 311, 349, 377, 396
fuel, 48, 233, 238, 242, 306, 353
grain, 36–37, 81, 308, 309, 311
Privy Council, 130, 157
Pultneytown (Highland), 358, 362
pyroligneous acid, 388

Queensberry, duke of, 277, 375
Queensferry Narrows, 181, 297

Raasay, island of, 138
rabbits, 96, 113, 114, 303, 334
railways, 327, 353, 388–389
rain and rainfall: see under weather
Rannoch (Perth and Kinross), 135, 268
Wood of, 135
Rannoch Moor, 167, 255, 268
Rattray (Aberdeenshire), 28
Rattray (Perthshire), 63
Rawlinsons of Graithwaite, iron-founders, 246
Reay, Lord, 276
Red Head (Angus), 241, 244, 319
reeds, 100, 103, 151, 365
Reform Act (1832), 4
Reformation, 2, 80, 84, 115, 394
Reid, John, 200–202, 205, 226
Renfrewshire, 124, 254, 351
Repton, Humphrey, 374
Restenneth (Angus), 272
Retford, Thomas (nurseryman), 218
'rewilding', 7, 9, 10, 284
rig-and-furrow, 28, 69, 256, 322, 323
Rispain Mire (Dumfries and Galloway), 235, 237
Robert III, king of Scots, 11
Robertson of Struan, forfeited estate, 167, 268
Rollo, Lord, 276
Romantics and Romanticism, 5, 125, 161, 164, 168, 169, 332, 368, 371, 373–374, 375, 378, 389
Rona, island of, 138
Ross: see Easter Ross; Wester Ross
Rossie (Angus), 101, 267
Rosslyn, earls of, 267
Rothiemurchus (Highland), forest of, 210, 215, 220, 384
Rotterdam, 202, 257
Roy, General William, 7, 194, 196, 206, 212, 223, 254, 255, 323
Royal Navy, timber needs, 210–211, 216, 220, 379–380
Rum, island of, 139
Russia, 15

sainfoin, 265
St Alban's (Hertfordshire), 15
St Andrews (Fife)
archdiocese, 107
cathedral-priory, 13, 17
town, 105, 197
St Margaret's Hope (Orkney), 260
salmon fishing and fishery, 74–76, 77, 283

INDEX

salt and salt-making, 49, 72, 100, 101, 243, 357, 362
saltpans, 49, 90, 242, 244
Samalas eruption (Gunung Rinjani, Lombok), 26, 178
sand-blow (*see also* Baleshare; Culbin; Eldbotle; Udal), 18, 26–27, 28, 31, 97, 104
sand dunes, 27, 28, 95, 96, 97, 98, 104, 106, 185, 188
Sandray, island of, 28
sawmills, 184, 213–214
Scalpay, island of, 138
Scandinavia, 20, 125, 181, 185, 290
    timber imports from, 125
Scone (Perth and Kinross), 256
    abbot of, 38, 42, 51
    Palace, 370
Scones Lethendy (Perth and Kinross), 245
Scotichronicon, 13, 33
Scots, kings and queens of: *see* Alexander III, Charles I, Charles II, James I, James II, James III, James IV, James V, James VI, James VII, Mary, Robert III, William
Scots gard'ner: *see* Reid, John
Scots pine trees, 8, 206, 208, 209, 214, 218, 220, 231, 368, 372, 379, 381, 390
    seeds and seedlings, trade in, 206, 218, 220, 230
Scott family
    3rd duke of Buccleuch, 375, 377
    earls and dukes of Buccleuch, 123, 132
    Scott, Sir Patrick, of Rossie, 267
    Scott, Sir Robert, of Dunninald, 267–268, 272
    Scott, Sir Walter, 322, 370–375, 378, 389, 390
    Scott of Balwearie, 59
    Scott of Dunninald, 267
    Scott of Rossie, 267
Seafield estate, 220, 226, 377
sea-floods: *see* floods and flooding
sea-ice, 93, 178, 181
sea-ware and seaweed (fertiliser), 104, 107–108, 125, 163, 272–274, 321, 354, 355
seaweed (fuel), 48, 234, 341
seedsmen, 160, 205, 208, 216, 230, 260, 263, 265, 277, 370
Select Society of Edinburgh, 156, 158, 160–161
Selkirk (Scottish Borders), 34, 377
'Seven Ill Years', 91, 125, 154, 155, 172, 175, 180, 191, 290, 341
Sheader (Sandray), 28
sheep (*see also* livestock), 8, 21, 22, 38, 40, 48, 52, 55, 104, 114, 118, 121, 122–123, 135, 181, 227, 258, 262, 290, 292, 293, 303, 310, 320, 326, 330, 331–335, 338, 353, 377, 388, 392, 395, 396
    dung, 48, 320
    flocks, 12, 19, 21, 23, 55, 58, 76, 77, 80, 104, 108, 110, 118, 122–123, 182, 263, 290, 292, 320, 326, 331–337, 394
Shetland Islands (*see also* Northern Isles), 93, 176, 293, 295, 305, 365
    fishery, 93, 178, 283, 290, 292, 295–296, 307, 361

harvest failures, 93
    peat use in, 237, 352
    sea-ware use in, 107; *see also* sea-ware and seaweed
shielings and shieling grounds, 52, 77, 121, 125, 133, 155, 263
Shin, river (Highland), 74
Sibbald, Sir Robert, 93, 158, 159, 200
Siberia, 16, 290
Sidlaw Hills, 51
Sierra Nevada (California), 1
Sinclair, George, of Ulbster, 256
Sinclair, Sir John, of Ulbster, 155, 281, 310, 327, 331, 359, 381
Sinclair family, of Freswick, 313
Sitka spruce, 231, 379, 390
Skye, island of, 138, 139, 158, 358, 363
slaves: *see* enslaved people
Slezer, John, 101
smallpox, 34
smallwood, 48, 66, 221, 341, 386, 387
Smith, Adam, 156
Smollett, George, of Ingliston, 260
Smollett, Sir James, of Bonhill, 260, 261, 263
Snelsetter (Orkney), 258, 260
snow and snowfall: *see under* weather
Sobieski-Stuart brothers, 8
Society for the Importation of Foreign Seeds, 202
Society of Free British Fishery, 282
soda ash (*see also* kelp and kelping), 354–357;
sodium carbonate (*see also* kelp and kelping), 354, 357
soil
    acidification, 51, 53
    erosion: *see under* erosion
    waterlogging, 15, 52, 53, 66, 121, 139, 149, 154, 214
solan geese, 114–116
solar minimum, 25, 191
    Dalton, 288–289, 293, 296
    Maunder, 56, 80–81, 82, 85, 88, 99, 118, 121, 125, 288
    Spörer, 25–26, 79, 80, 81, 82, 85, 108, 118, 121, 341
    Wolf, 12
Solway, Firth, 74, 98, 185, 293, 298
South Ronaldsay (Orkney), 258
South Walls (Orkney), 258
Southern Uplands, 25, 48, 52, 76, 111, 121, 129, 147, 209, 293, 298, 330, 334, 377, 380
speleothems (*see also* Traligill; Uamh an Tartair), 16, 85, 90, 172, 176, 177, 178, 183
Spey, river, 29, 74, 178, 184, 303, 304, 319
Speyside, 177, 184, 188, 205, 220, 265
Spynie (Moray), 240, 314
    Loch of, 153, 240, 313, 315
'stadial' history, 164, 168, 170
Statistical Account of Scotland ('Old' and New), 155, 270, 281, 288, 310, 327, 344, 351, 353, 359, 381
Steuart, John, 257, 281, 282
Stewart, Sir James, of Burray, 258–260
Stirling (Stirling), 33, 58, 67, 74, 140, 150, 275, 277, 345
    castle, 66, 80, 133
    fuel use in, 150

King's Park, 140
Stirling, John, of Herbertshire, 229
Stobsmill (Midlothian), 386
storms and storminess: *see under* weather
Stotfield (Moray), 296
Stratha'an Aberdeenshire and Moray), 142–144
Strathardle (Perth and Kinross), 122, 162
Strathbraan (Perth and Kinross), 169
Strathconon (Highland), 315
Strathcononish (Argyll and Bute), 211
Strathearn (Perth and Kinross), 103, 276
Strathmore (Angus *and* Perth and Kinross), 63, 123
Strathnaver (Highland), 118, 334
Strichen (Aberdeenshire), 265
subsistence crises, 1, 12, 37, 91, 94, 123, 155, 188, 191, 286, 288–289, 307, 308, 356, 394
sugar (cane), 10, 116, 364
Sutherland, 16, 85, 90, 139, 183, 206, 244, 276, 281, 301, 334, 372
Sweden, 34, 133
Swindon Hill, Bowmont Valley (Borders), 315
Swinton family, 59
Switzer or Sweetzer, Stephen (garden designer), 160
sycamore trees, 208, 218
Sylva Abbotsfordiensis (Walter Scott), 372, 374
Symson, Andrew, 124, 233, 243, 352
syphilis, 34–35

Tambora (volcano), 289, 296, 298
tanbark (*see also* bark), 193, 212, 227, 231, 360, 382, 386
tanbarkers, 245, 332
Tarradale (Highland), 315
tathing, 40, 41
Tay, river, 21, 29, 58, 63, 74, 86–88, 90, 135, 181, 221, 282, 297, 304, 315, 378
    carse, 315
    floods and spates, 29, 86–88, 90, 297, 304
    salmon fishery, 74
Taymouth Castle (Perth and Kinross), 196, 205, 208, 224, 227, 301, 319, 377
Tayside, 89
teak, 217, 379
Teviotdale (Borders), 123, 136, 308
thermal energy needs and sources, 44, 45, 48, 51, 111, 148, 149–150, 153, 233, 234, 235, 243, 249, 251, 287, 340, 341, 344, 347, 348, 352, 365
Tilt, river, 29
Timanfaya (volcano), 178
timber (*see also* oak trees; woodland, pine), 60–61, 65–69, 90, 101, 103, 109, 111, 126, 127–128, 132–133, 135, 138, 140, 142, 151, 192–193, 197, 199, 205–206, 210–212, 214–215, 217, 218, 221, 223–224, 226, 227, 229, 230, 240, 250, 366, 367, 368, 370, 372, 377, 379–381, 386, 389, 390, 395
    Baltic and Scandinavian, 132, 133
Tiree, island of, 227, 278, 349, 356, 382
Tobermory (Argyll and Bute), 358, 362
Tomphubil (Perth and Kinross), 162, 319

419

Torhousemuir (Dumfries and Galloway), 324
Traligill (Sutherland), 16, 24, 85, 90, 91, 172, 176, 178, 180
Tranent (East Lothian), 49
tree-lines, 8, 15
trees and shrubs: *see individual species names*
   barren, 140, 215
Treponema pallidum, 34–35
Très Riches Heures du Duc de Berry, 16
Tugnet (Moray), 319,
Tummel, river, 29, 378
turf, 28, 96, 97, 99, 104, 133, 142, 147, 275, 326
   fuel, 48, 50, 148, 151
   as soil deepener, 104–106, 151, 240
Turnbull and Ramsay's 'Pyroligneous Acid Works', 388
turpentine, 212, 387
Turner, John (seedsman), 205
turnips (swede, yellow), 251, 277, 278–279, 305, 320, 330
Tweed, river, 74, 197, 372, 375
Tweeddale, earl of: *see under* Hay family estate, 390

Uamh an Tartair (Sutherland), 16
Udal (North Uist), 18, 96
Uists, islands, 27, 28, 95, 121, 139, 185, 355
   machair: *see main entry*
   North Uist, 18, 28, 56, 95–96, 185
   South Uist, 49, 71, 95
Ullapool (Highland), 358, 362
Ulster Plantation, 123, 157
Ulva, island of, 227
Union of the Parliaments, 2, 155, 156, 170, 175, 176, 193, 197, 202, 326, 394
Ural Mountains, 16, 20
Usan (Angus), 267

violence and social disturbance, 14, 39, 76, 148–149, 261
volcanic eruptions (*see also* Hekla; Laki eruption; Mount St Helens; Mount Westdahl; Tambora; Timanfaya), 26, 93, 178, 189–191, 288, 289, 293, 296, 298
volcanic forcing, 25, 178, 293, 296

Walsingham, Thomas, 15
Wardrope, Thomas, of Gothens, 51
Wars of Independence, 12
waste (land), 252, 258, 338, 365, 373, 390
waste (refuse), 49, 104, 272, 274, 275, 320
water erosion: *see under* erosion
water-meadows: *see* meadows
waterlogging, 15, 52, 53, 66, 90, 121, 139, 149, 154, 214, 218, 249, 314, 315, 352, 372
wattles and wattling, 133, 138, 228, 275, 313
weather, 1, 3, 4, 5, 6, 9, 11, 12, 13, 14–27, 28, 31, 33, 36, 41–42, 43, 44–45, 47, 49, 51, 52, 56, 66, 71, 72, 76–77, 79–91, 93–98, 99, 100–101, 106, 108–109, 110–111, 112, 113, 114, 116, 118, 119, 121, 139, 149, 151, 153, 171–191, 228, 243, 249, 280, 283, 287, 288–307, 309, 311, 327, 340, 341, 349, 350, 351, 356, 359, 392, 393, 395

bloody rain, 89
cold, 13, 15, 16, 20–24, 27, 33, 47, 56, 62, 80–81, 85, 87, 88, 93, 108, 116, 125, 139, 149, 153, 172–173, 175, 178–184, 186, 187, 188, 191, 243, 249, 283, 287, 288, 290, 291, 292–293, 296, 298, 300–301, 304, 306, 307–308, 309, 311, 315, 341, 350–351, 359, 385, 392, 394, 395
drought, 5, 15, 19, 55, 76, 90, 121, 164, 174, 177, 290, 298, 300, 301
frost, 12, 23, 32, 80, 86, 87, 90, 172, 176, 177, 179, 180, 181, 186, 187, 188, 290, 293, 300, 306, 308, 379
hail and hailstorms, 178, 183
ice, 4, 5, 12, 22, 87, 93, 178, 179, 181, 290, 297, 300, 392
rain and rainfall, 4, 16, 25, 29, 30, 42, 44, 56, 82, 85, 86, 87, 89–90, 98, 115, 154, 171, 173, 175, 177, 178, 183, 184, 185–187, 289, 290, 293, 294, 296, 298, 301, 302, 305–309, 314, 352
snow and snowfall, 4, 5, 8, 12, 15, 19, 80, 87, 90, 98, 112, 171, 172, 173, 175, 176, 178–181, 183–184, 185, 186–187, 189, 191, 286, 290, 291–296, 298, 299–302, 306, 308
storms and storminess, 8, 12, 14, 15, 17, 18, 27, 56, 66, 76, 79, 80, 81, 90, 94–95, 101, 108, 125, 139, 172, 176, 178, 180, 183, 189, 191, 214, 283, 290, 292–293, 296, 300, 302–304, 307, 308, 351, 394
wind and gales, 8, 9, 15, 17, 18, 20, 22, 27, 33, 47, 55–56, 80, 86, 90, 91, 94–96, 98, 108, 171, 172, 173–174, 178, 180, 181, 183, 185, 187, 191, 286, 290, 292–293, 296, 298, 299, 300, 302, 303, 305, 306, 309, 310, 313, 327, 351
weather erosion: *see under* erosion
Wemyss (Fife), 90
West Indies, 396
West Lothian, 201, 263, 345
Wester Ross, 127, 211, 212, 281, 282, 358
Western Isles *or* Outer Hebrides (*see also* Barra; Benbecula; Heisgeir; Lewis; Sandray; Uists), 15, 18, 27, 55, 71, 80, 94, 95, 107, 125, 127, 138, 160, 185, 227, 234, 258, 272, 279, 282, 294, 308, 321, 352, 353
Description of the Western Isles of Scotland, 158
fishery, 71, 158, 281–282, 363
kelp manufacture in: *see* kelp and kelping
machair districts in: *see* machair
sand-blow: *see main entry*
soil fertilisation in: *see* sea-ware and seaweed
Westfield (Moray), 240
Westray (Orkney), 94–95
wheat, 12, 36, 37, 38, 44, 103, 163, 251, 278, 305, 316
Whitefield (Perth and Kinross), 162
Whithorn (Dumfries and Galloway), 147, 235, 237, 239, 240, 352
Wick (Highland), 256–257, 358, 361–362

Wigg, laird of: *see* Agnew, William
Wigg, Moss of (Dumfries and Galloway), 239, 240
Wigtown (Dumfries and Galloway), 99, 105, 123
   Bay, 99
wildfowl, 23, 100, 110, 113, 114, 164, 180, 313, 365
   hunting, 113, 114
'wildwood', 7, 128, 135
willow trees, 63, 103, 130, 140,142, 196, 210, 218, 226, 313, 315, 372, 381
wind: *see under* weather
wind erosion: *see under* erosion
'winter herrin', 359, 360, 361
wolf and wolves, 9, 21, 24, 76, 111, 118–120, 121
wood (fuel), 69, 215, 250
wood pasture: *see under* pasture and grazing
wood-cutting (illicit), 14, 61, 63, 139, 154, 227–228, 238, 351, 377
woodland
   ancient, 7, 198, 213
   birch, 136, 139, 313
   coppice: *see* coppice and coppicing; haggs
   pine, 8, 209–215, 223, 383, 387
   plantation, 63, 66, 77, 109, 111, 130, 140, 154, 157, 192–221, 224–231, 233, 234, 238, 240, 251, 252, 254, 255, 302, 303, 305, 325, 327, 351, 365, 368, 370, 371–372, 377–378, 380, 381, 382, 386, 389–390, 396
   enclosure: *see main entry*
   regeneration, 7, 8, 15, 61, 62, 66, 67, 108, 111, 128, 129, 139, 180, 193, 212–214, 220, 223, 224, 229, 230, 250, 331, 332, 367, 375, 381, 383–385
wool and woollens, 76, 84, 123, 263, 268, 335, 355, 388, 392
Wordsworth, William and Dorothy, 374, 378
Wrey, Sir Bourchier, 366

Yairs, 74
Yarrow Valley (Scottish Borders), 377
Yester (East Lothian), 140, 199, 201, 202, 206, 208, 390
yew trees, 202, 205
yields: *see under* harvests
Ythan, river, 28, 74